D1329671

Rice in West Africa

Rice in West Africa

POLICY AND ECONOMICS

Scott R. Pearson · J. Dirck Stryker
Charles P. Humphreys

Patricia L. Rader · Eric A. Monke · Dunstan S. C. Spencer
Kathryn Craven · A. Hasan Tuluy
John McIntire · John M. Page, Jr.

Universitas
BIBLIOTHECA
ttaviensis

STANFORD UNIVERSITY PRESS
Stanford, California 1981

HD
9066
·A462
P4
1981

Stanford University Press
Stanford, California

©1981 by the Board of Trustees of the
Leland Stanford Junior University

Printed in the United States of America
ISBN 0-8047-1095-3
LC 80-50906

Preface

Sub-Saharan Africa was the only region in the world to have experienced a decline in food production and consumption per capita between 1960 and 1975. Projections of future African food supplies and needs foretell a continuation or even a worsening of this pattern unless prompt and effective remedial actions are taken. This book attempts to address several significant aspects of this crucial problem, and results from a research project, "The Political Economy of Rice in West Africa," carried out jointly by the Food Research Institute, Stanford University, and the West Africa Rice Development Association (WARDA) between June 1976 and September 1979. The project was funded by the U.S. Agency for International Development (AID) under contract number AID/afr-C-1235.

The chapters that follow present the results of an innovative approach to agricultural policy analysis that could have wide applicability throughout Africa—or, for that matter, in any low income, predominantly agricultural country. The approach embodies three essential characteristics. Initially, microeconomic analyses are carried out to determine the economic efficiency of alternative techniques and locations of production in integrated systems encompassing farming, processing, and distribution. Next, the incentive effects of government trade, price, tax/subsidy, and investment policies are analyzed by measuring transfers of income to or from farmers, millers, and traders occasioned by each type of policy. Finally, comparative analyses are undertaken of the productive efficiency of various techniques and locations and of the effectiveness of policies in advancing objectives, first within individual countries, and then among a number of countries within a region. The approach thus incorporates the full spectrum of issues in agricultural policy analysis, starting with the efficiency of farm production and concluding with the comparative effectiveness of alternative policies.

The research results offer something of interest to commodity specialists, Africanists, and policy makers in West Africa and foreign aid agencies. For five West African countries, the study investigates major economic and political influences on the expansion of rice production and the efficiency of existing and proposed methods of growing, milling, and marketing rice. The book explores the objectives of governments concerned with expanding rice output, the constraints limiting achievement of these objectives, and the effectiveness of recent and prospective policies used to advance the objectives in light of these constraints. In this three-year research effort, the contributing authors have compiled an unusually detailed set of economic information and comparative analyses. The authors hope that the results presented in this book will make a contribution toward the development of more coherent policies to help resolve Africa's growing food problem.

While this book was in preparation we accumulated a long string of intellectual debts. An untold number of farmers, traders, government officials, foreign experts, and all-purpose raconteurs in West Africa indulged our fantasies and ignorance as we attempted to learn how and why rice was produced, traded, and consumed. Their names are far too numerous to mention, even if we had recorded them all, and we suspect some are undoubtedly pleased to remain anonymous and not be implicated in this effort. Others might find our results interesting, important, or perhaps even exciting and thus be disappointed that their help is not being specifically acknowledged. We are enormously grateful to those in both groups who provided us with insights and sometimes saved us from mistakes.

We thank individuals from both WARDA and AID for their collaboration and support, which were essential for the undertaking of the study, and we are especially grateful to Djibril Aw, Sidi Coulibaly, Jacques Diouf, and Jean-Claude Levasseur of WARDA and to Stephen Klein, Stanley Krause, W. Haven North, and Norman Ulsaker of AID. As is usual in applied economic analysis, none of these individuals nor their organizations necessarily agrees with all of our conclusions, and it is possible that some officials in both agencies agree with little or none of what we recommend. Only time will tell as we observe what kinds of future rice development projects these groups choose to support in West Africa.

Our research was immeasurably aided by a number of other colleagues. Three participants in the study, Walter P. Falcon, William O. Jones, and Robert Netting, were instrumental in shaping the research design, guiding the field work, and assisting interpretation of the results. With only a slight shift in the division of responsibility, any of them would have become a co-author of this book. Several other colleagues at Stanford and elsewhere offered useful comments at various stages of the project. Notable in this group are Carl H. Gotsch, Bruce F. Johnston, Leon A. Mears, W. H. M. Morris, Anne E. Peck, John Tarrant, and T. Kelley White. In addition, unusually high quality research assistance was provided by Bhaskon Khanna, Gerald C. Nelson, David Trechter, and Laurian J. Unnevehr, and excellent translations from English to French were done by Douglas Bouloch and Myrna Rochester. Transferring research results from the hand-written draft to the typed manuscript and then to the published page requires additional skills, and we have been fortunate to work with exceptional people doing these tasks. The secretarial staff at the Institute and at WARDA executed their duties very well, often under great time pressure, and Stanford University Press chipped in with their renowned efficiency and professionalism. Carmen Bermodes and Wilma McCord of the Institute and our editor Elizabeth Spurr deserve special mention in this regard.

There is one final person who requires a category unto herself. Linda W. Perry did no field work. But she was involved in just about all other facets of the project and throughout performed extraordinarily well. Administration, typing, research assistance, editing, and final preparation of the manuscript for the press all fell within her bailiwick. We are extremely grateful to her and to all of the superbly competent people who assisted our work.

S.R.P.

J.D.S.

C.P.H.

Contents

PART FIVE   Mali

PART SIX   Comparative Analysis

Tables

Maps and Figures

Rice in West Africa

Introduction

Scott R. Pearson, J. Dirck Stryker,
and Charles P. Humphreys

In 1973–74 rice policy in West Africa was at a crossroads. Countries in the Sahelian zone were in the last of six years of severe drought, which caused, among many unfortunate consequences, a sharp increase in food imports.[1] Concurrently, international prices of wheat and rice, the principal traded foodgrains, had temporarily soared to unexpected and previously unknown heights, four times their levels in the late 1960's.[2] Many West African governments were faced with the unhappy prospect of needing substantial foodgrain imports at a time when prices were high. Moreover, these governments and their supplies of foreign exchange were hit doubly hard, since the high bills for imported foodgrains were accompanied, coincidentally, by a quadrupling in the price of petroleum.

The importance of rice policy in West Africa grew substantially following the unanticipated confluence of events in 1973–74. Because very little wheat is grown in the region, rice is the most important foodgrain whose imports could be substituted for directly by increases in local production and marketings. Rice has traditionally been the principal staple food in a contiguous group of western coastal countries, running in an arc south and east from the Gambia through western Ivory Coast, and along the banks of the Niger River in Mali and Niger.[3] After World War II, rice cultivation and consumption expanded throughout the region.

[1] The Sahel, which takes its name from the Arabic word for shore—in this case the shore of the Sahara—is a semiarid expanse of land covering some 5.25 million square kilometers and stretching south from the Sahara Desert to the woodlands from the African savannah. Of the five countries considered in this book, two—Mali and Senegal—are in the Sahelian zone.

[2] See Falcon and Monke (1) for an analysis of principal influences affecting the world market for rice and of the relationships between the international markets for rice and wheat. The main causes of the dramatic price increases for both commodities were unusually bad weather in many import foodgrain-producing areas, leading to a reduction in supplies and the very low stock positions in wheat and rice resulting from large drawdowns in the early 1970's.

[3] One type of rice, *Oryza glaberrima*, is indigenous to West Africa, probably originating in the interior delta of the Niger River in what is now Mali. Traditional methods of cultivating *O. glaberrima* have been practiced in parts of the region for perhaps 15 centuries or longer. Most rice currently grown in West Africa is *Oryza sativa*, which is indigenous to Asia. *O. sativa* was probably introduced into West Africa by Arab traders five or six hundred years ago, but the spread of its cultivation was limited until it was encouraged by the French and British colonial rulers beginning in the mid-nineteenth century. In this book, the adjective "traditional" is used to describe systems of cultivating both types of rice that have been practiced in West Africa for at least several decades.

By 1974, all governments in West Africa began to reassess policies affecting their nations' production, consumption, and trade of rice. Rice imports had suddenly become very expensive and their future price seemed uncertain. Governments responded by raising prices of rice to their consumers and producers, though not always by the full extent of the rise in corresponding import prices. In spite of record high prices, however, several countries continued to import large quantities of rice.

The picture in West Africa shifted again in 1975. Consumers reacted to the higher prices by switching partially from rice to other foodstuffs. Producers who received higher prices for their paddy responded with increases in output and, especially, marketings. Importing agencies, which in some countries had overbought in 1973 or 1974, were forced to carry over stocks into 1975. All of a sudden, there seemed to be too much rice for sale, rather than too little, especially in the eyes of governments which wished to mitigate the large downward fluctuation in prices that would have cleared the markets.

This concern with recent shortages and apparent gluts of rice was evidenced in 1975 at the annual meeting of the West Africa Rice Development Association (WARDA),[4] when its Governing Council called for a study of the prospects for future intraregional trade in rice and of its contribution to the objective of ultimate regional self-sufficiency. Since a number of WARDA countries had achieved or regained self-sufficiency in rice in 1975 and some even had small exportable supplies, it seemed desirable to examine prospects for trade within the region. The Food Research Institute agreed to carry out the study.

Researchers at both WARDA and the Institute initially thought that the most significant results from the project would involve intraregional trade in rice, and the research was designed accordingly. Lessons from the Institute's earlier study of rice in Asia influenced the design. Microeconomic study of alternative techniques for rice production, milling, and marketing received major attention, as did analysis of rice policy. Both of these investigations were integrated directly into the trade analysis.

Midway through the project, researchers at WARDA and the Institute prepared a study of prospects for intraregional trade of rice in West Africa (7). The results of this study confirmed that, even based on reasonably optimistic assumptions for growth of production, steadily increasing consumption of rice meant that the WARDA region was likely to require increasing amounts of rice imports during the period ending in 1990. This conclusion, coupled with large levels of rice imports into the region in

---

[4] WARDA is an intergovernmental organization of which all 15 countries in the West Africa region are members. When WARDA was established in 1971, a primary goal of the new organization was to provide rice research and development services to member countries and thereby to assist the region in achieving self-sufficiency in rice.

1977 and 1978, reemphasized the importance of analyzing government policies aimed at increasing the production, milling, and marketing of rice and at influencing rice consumption.

From the beginning, primary emphasis was placed on a group of WARDA-member countries in which rice was an important staple food and which during 1966–75 had been relatively large importers of rice—Ivory Coast, Liberia, and Senegal—or had good prospects of becoming rice exporters within the region—Mali and Sierra Leone. Project researchers carried out detailed fieldwork in the first four of these countries and updated earlier work in Sierra Leone done by Njala University College (University of Sierra Leone) and Michigan State University. This book is the result of that fieldwork.

The remainder of this introductory chapter contains five parts. The first two are concerned with analyzing rice policy in West Africa. In one, a framework for policy analysis is presented, and in the other, the organization of the country studies of rice policy is outlined. Attention in the succeeding two parts of the chapter is on the other principal focus of the study, a comparative analysis of the economics of alternative methods of rice farming and marketing. One part contains a brief summary of the analytical method employed to measure economic efficiency and the impact of policy, and the other sets out the guideposts used in organizing the country studies of the economics of rice. The fifth and last part of this chapter is a reader's guide to the entire volume, which describes the rationale underlying the book's organization and the location of various kinds of information and analysis.

Framework for Analyzing Objectives, Constraints, and Policies

Readers of essays growing out of the Institute's earlier study of the political economy of rice in Asia are already familiar with the methodological framework for policy analysis that emphasizes interactions among a nation's objectives, constraints, and policies.[5] This framework helps to structure use of economic logic to examine government policies, but it is not a model that can be used to predict government actions.[6] Governments are viewed as having several objectives that they try to achieve within a framework of constrained optimization. Constraints are limits on the availability or deployment of resources and on the flexibility of consumer preferences that prevent the full attainment of all objectives. Poli-

---

[5] This approach is summarized in Timmer (5, pp. 191–96).

[6] In his essay that concludes the set of policy studies of the Asia rice project, Timmer concludes (6, p. 420) that hope that the "framework of the policy-making process might actually be implemented . . . has vanished."

cies are the instruments used by governments to achieve objectives by
influencing the allocation of resources and patterns of consumption. Constraints on resources thus limit the extent to which policies succeed and
hence the degree to which objectives are attained. The method of implementing policies can also affect their success or failure. Policy analysis
consists of identifying the relevant government objectives, specifying the
nature of resource or consumer constraints, delineating the policy options, and tracing the interactions.

Problems arise in applying the framework because it is not always clear
in practice how to distinguish between a policy and a constraint and because objectives held by governments sometimes act as constraints. For
this reason some of the authors of the country studies in this volume have
followed the approach less closely than others. Nevertheless, the framework is used in Chapter 11, which concludes the set of policy studies, because of its clear advantages in organizing complex comparative material.

Objectives

A government at any given time might have several objectives which
often involve trade-offs. It is useful to categorize these objectives into two
groups—fundamental and proximate. Three fundamental economic
objectives—efficient generation of income, more equal distribution of
income, and security (the probability of obtaining income)—are distinguished. All other objectives are considered to be proximate. The purpose
is not to claim supremacy for the fundamental objectives, since other goals,
such as avoiding excessive inflation, might be more important to
governments. Nevertheless, these proximate objectives are desired chiefly
because of their contribution to one or more fundamental goals rather than
as ends in themselves.

The validity of this categorization is well illustrated by considering the
objective of self-sufficiency in rice, which is a stated goal of almost all
governments in West Africa. Each of the three fundamental objectives is
influenced by self-sufficiency, the reduction of imports to zero in most
years by increasing domestic production. Substitution of domestic production for rice imports can generate income if the country has a comparative advantage in rice because it can produce it efficiently; alternatively,
if domestic rice is more costly than imports, potential national income
declines. Likewise, promotion of rice development to achieve self-sufficiency might change the distribution of income—among regions, between the public and private sectors, and between producers and consumers—in ways that either serve or contravene advancement of this
government objective. Finally, greater reliance on local supplies of rice
could lead to greater security of income to producers if new techniques

reduce the variation of domestic production, and of food supplies to consumers if expanded local supplies are cheaper on average than imports. But it could have the opposite effect if local production varies more than the price of imports and if there are other, more secure opportunities to use domestic resources to generate foreign exchange.

Self-sufficiency, therefore, can contribute positively or negatively to all three fundamental objectives. Although governments might place a high priority on this proximate objective, policy analysis is better served by measuring costs and benefits of increased self-sufficiency in terms of its effects on the three fundamental objectives. A similar argument could be made with respect to several other proximate objectives, such as the generation of government revenue, price stability, and regional development. As with self-sufficiency, it is preferable to examine the influence of each of these intermediate objectives on income, distribution, and security rather than to consider it as an end in itself.

Constraints

The analysis of rice policy in West Africa is based on an interpretation of constraints which follows closely that used in the Asia project. In that study, constraints on policy implementation were considered to encompass a vast range of influences, including demand and supply parameters—especially resource endowments and technical relationships—and social, legal, and political factors (2, 3). In principle, of course, any of this panoply of influences, from high capital costs to consumer pressure in the cities, can limit the ability of a government to carry out a policy. It is useful, however, to distinguish between fixed constraints and those that are more elastic.

Fixed resources create an absolute barrier to successful implementation of a policy in the absence of measures to augment or substitute for the limited resource. For example, highly variable and low rainfall in an area precludes greater yields until investments in irrigation are undertaken. In contrast, constraints imposed by resources in elastic supply can be overcome at the margin by paying more—e.g. to attract labor from other employment. The producer or consumer in question simply bids up the price in order to obtain the input or final product.

Given enough investment or the passage of sufficient time, most constraints can be overcome. Unfortunately, policy making is done in the short run, and governments are thus faced with a wide array of constraints in selecting policies.

Whereas the number of applicable constraints is almost limitless, the range of options available to governments in West Africa for implementing rice policy is quite narrow. This phenomenon is not restricted to rice or to West Africa. There happen to be only a few methods, each with several variations, by which a government can intervene to influence production or consumption of a commodity.[7] For convenient analysis, policies are categorized here in three areas—trade and price policies affecting the commodity, taxes or subsidies on current inputs, and public investments to augment resources.

Policies

Foreign trade policies include restrictions (duties or quantitative controls) or subsidies on imports and taxes or subsidies on exports that cause the domestic price of a commodity to differ from its international price. In West Africa, most governments intervene to limit imports of rice and thus raise the domestic price of rice above the international level. Domestic price policies are interventions that cause producer prices of rice to diverge from consumer prices by more (or less) than normal marketing margins. For example, price controls or rationing by a government store tend to alter the market price of rice—which may have already been modified by a trade policy—and permit consumers with access an opportunity to purchase at less than that price. Another example of price policy is the use of milling or marketing subsidies, which lowers prices to consumers beneath domestic market levels that would otherwise prevail.

Another type of policy either taxes or subsidizes current inputs used in producing rice. West African governments have established a number of input subsidies in rice production that reduce the prices of the subsidized inputs and thus cause returns to farmers to exceed what they would be if farmers paid full costs. Furthermore, most imported inputs are neither restricted nor heavily taxed, and relatively few other taxes on inputs for rice farming exist in the region.

Investment policies are inherently longer-run and involve subsidized capital creation in production, milling, and marketing, as well as in supporting infrastructure. These policies have effects similar to those of input subsidies, but unlike subsidies to current inputs, they are multiperiod and resource-augmenting. With the assistance of foreign aid donors, West African governments have made significant investments in rice farming and milling, especially during the past decade.

[7] In the Asia study, Timmer listed seven possible forms of intervention (5, p. 194): (1) consumer programs (including subsidies for rice or substitutes); (2) farm production programs (either intensification or diversification); (3) domestic marketing investments; (4) concessional foreign trade (including exchange rate biases); (5) direct taxation or other forms of fiscal transfers; (6) price controls by legal fiat and/or market operations, including (a) floor price, (b) ceiling price, (c) buffer stock as stabilizer, usually in conjunction with (a) and (b); and (7) physical controls, including rationing and nonprice collections and disbursements.

Organization of the Country Studies of Rice Policy

Although the organization of the five country studies of rice policy that follow is similar, authors naturally depart from the common approach according to individual circumstances. The country studies open with a brief introduction that summarizes their principal themes and organization. The second section contains a discussion of the technical and economic setting, including descriptions of economic geography (population, land, and climate) and rice farming systems, marketing and milling techniques, and consumption patterns.

Policy analysis begins in the third section of each paper, which investigates the history of rice policy by dividing the colonial and independence eras into several periods, each characterized by easily identifiable directions of rice policy. For two countries the selection of periods is based on changes in government, whereas for the three others, which have had continuous leadership since independence, the choice turns on other criteria. Within each period, authors identify the main objectives, constraints, and policies influencing the rice sector. They also examine how conditions in earlier periods affect the choices of policies and objectives and the importance of constraints in subsequent periods.

This look at the history of rice policy sets the stage for an evaluation of recent policies, which is the heart of the analysis in each paper. Drawing on the historical section, authors discuss the policy options available to the government and the main constraints limiting their implementation. Efforts are made to assess the effectiveness of these policies in achieving national objectives. Economic welfare effects of principal policies are analyzed with respect to gains, transfers, and losses of income.

The analysis is aided by attempts to examine the political significance of the effects of rice policies on the distribution of income—among regions of the country, between the private sector and the government treasury, between consumers and producers, among different kinds of producers, and between urban and rural inhabitants. These distributional effects, whether intended or not, often have a more immediate political influence than does progress toward achieving the two other fundamental goals, efficient income generation and security. Farmers, for example, might not know or care whether an input subsidy policy contributes to national income or food security, but they can be expected to know whether the policy benefits them. Similarly, consumers are typically sensitive to increases in prices of staple foods, including rice, as recently evidenced by the riots in Monrovia, Liberia, that followed announcement of a higher official wholesale price for rice. Consequently, governments often give greater attention to the distributional impact of their policies than to other effects. This political reality helps to explain why certain policies are chosen when several alternatives appear to be available.

Each of the country studies concludes with a prognosis of future rice policies based on the analysis of past and current experience. The focus, however, is more on suggesting recommendations for feasible policy changes than on attempting to predict likely choices of rice policy by existing or future governments. Some authors are more adventuresome than others in offering advice and assessing its impact. The hope, of course, is that bold conclusions based on careful analysis will challenge those intimately involved in formulating rice policy in West Africa to reexamine the effectiveness of existing policies in furthering national objectives.

Framework for Analyzing Economic Efficiency and Comparative Advantage

As with the framework for policy evaluation, the detailed microeconomic analysis of alternative techniques and locations of rice production, processing, and distribution in each of the five West African countries builds on the methodology of the Institute's study of rice in Asia (4). The principal purpose of this analysis is to discover how and where rice output can be expanded efficiently. Authors of the country studies measure economic efficiency with a common methodology, which is discussed in Appendix A, "A Methodology for Estimating Comparative Costs and Incentives." The approach, illustrated below, is based on the social valuation of rice produced and of the costs of all inputs used in production. Actual values employed in the country studies are discussed in detail in Appendix B, "Shadow Price Estimation."

The measurement of economic efficiency has important applications both as a means of evaluating the efficacy of past policy initiatives and as a guide to future investment opportunities. One of the striking features of the rice economies of West Africa is that government policies and investments have created a wide variety of farming and marketing techniques within individual countries. A comparison of the relative efficiency of these techniques is therefore of interest in evaluating their success in meeting basic policy objectives. It is also important in understanding which, if any, of the West African countries possess a comparative advantage in rice production, thus serving as a guide to future investment and as a basis for evaluating gains from intraregional rice trade.

The approach compares estimates of private profitability—the difference between returns and costs in actual prices facing farmers, millers, or traders—with estimates of social profitability—the residual remaining when costs and returns are evaluated in social prices. Two kinds of adjustments are made to convert private costs and returns into social prices. First, outputs and tradable inputs are valued in comparable world prices to eliminate the transfers caused by government policies. Rice output, for example, is measured in terms of what the country must pay for its imports (or can receive for its exports) of rice instead of the actual market price that prevails domestically. Similarly, an input, such as fertilizer, that can be purchased from abroad is valued at its true import cost in place of a subsidized (or taxed) market price.

Second, labor, capital, and land are valued with respect to their social opportunity costs, which represent the value in world prices of the output forgone from not using the resources in their best alternative employment. For example, farmers might receive government credit at an interest rate of 6 percent when the government could otherwise have used the capital in a development project yielding a 15 percent social rate of return. In this instance the social price of capital would be 15 percent instead of the 6 percent actually paid by the farmers for credit.

Having made these two kinds of adjustments, one can compare social benefits with social costs and determine whether social profitability is positive or negative and thus whether or not it is efficient to produce rice. If the social benefits are calculated with reference to a competitive international price in an export market, furthermore, positive social profitability implies an international comparative advantage because the country can produce rice efficiently both for its own use and for export.

Organization of the Microeconomic Country Studies

Following a brief introduction, each of the country studies of the microeconomics of rice production begins with a description of production and post-harvest techniques, including area, output, types of inputs, scale of operation, and credit and input delivery systems. Tables summarize the key characteristics of each technique, notably input-output coefficients and unit costs. Discussion encompasses farm production, collection of paddy from farmer to mill, milling, and distribution of rice from mill to wholesaler. A number of alternatives to be analyzed are then defined by combining various production and post-harvest techniques.

Following a description of incentives provided by policies and a review of the methodology and social prices of domestic factors and rice, the principal analytic results are presented. Private benefits, costs, and profitability of alternative ways of producing, milling, and distributing rice are compared with evaluations in social terms. Divergences between private and social profitability measure the incentive effects of specific policies. The reliability of these results is tested by carrying out sensitivity analysis on yields, shadow prices of domestic factors, milling outturns, and the world price of rice.

The remainder of each country study is concerned with interpreting the efficiency analysis in a broad policy context. Emphasis is on the effectiveness of each rice production activity in contributing to various national objectives. Expansion of techniques that use resources inefficiently, for example, might nevertheless contribute positively to enhancing food security or to improving income distribution even though they decrease national income. The microeconomic analysis of rice in each country thus provides support for and feeds back into the policy analysis.

A Reader's Guide to the Book

The interdependence of the two chapters—one on policy and the other on microeconomics—for each country should be clear from the foregoing discussion. Each country chapter draws on its counterpart for information and interpretation to such an extent that there is no inherently logical order for reading one before the other. The policy analysis provides both background and broad interpretation, whereas the microeconomic study contains detailed results that both depend on and form the basis for conclusions in the policy discussion. Readers interested principally in a particular country should read the two chapters for that country and then turn to the comparative analysis in the two final chapters.

In spite of the interrelatedness of the pairs of country studies, the chapters in the book are arranged so that readers concerned only with policy analysis—or, alternatively, only with microeconomic analysis—can select the relevant set of country studies and then move on to the corresponding comparative chapter. These specialist readers are encouraged to read both comparative chapters, however, to understand the relationships between the microeconomics and government policies. In addition, those who are mainly interested in the microeconomic analysis should begin with Appendixes A and B on methodology and social prices.

Finally, a recommended strategy for readers in a hurry is to skip directly from this introductory chapter to the two comparative chapters. As interest and time permit, these readers can review individual country chapters for detailed description and analysis of policies and production techniques.

Citations

1 Walter P. Falcon and Eric A. Monke, "The Political Economy of International Trade in Rice." Stanford/WARDA Study of the Political Economy of Rice in West Africa, Food Research Institute, Stanford University, Stanford, July 1979.

2 *Food Research Institute Studies*, 14, No. 3 (1975).

3 *Food Research Institute Studies*, 14, No. 4 (1975).

4 Scott R. Pearson, Narongchai Akrasanee, and Gerald C. Nelson, "Comparative Advantage in Rice Production: A Methodological Introduction," *Food Research Institute Studies*, 15, No. 2 (1976).

5 C. Peter Timmer, "The Political Economy of Rice in Asia: A Methodological Introduction," *Food Research Institute Studies*, 14, No. 3 (1975).

6 ———, "The Political Economy of Rice in Asia: Lessons and Implications," *Food Research Institute Studies*, 14, No. 4 (1975).

7 West Africa Rice Development Association (WARDA) and Food Research Institute, "Prospect of Intraregional Trade of Rice in West Africa." WARDA/77/STC7/9, Monrovia, Liberia, September 1977.

PART ONE

The Ivory Coast

1. Rice Policy in the Ivory Coast

Charles P. Humphreys and Patricia L. Rader

Ivorian economic growth springs from its agriculture. Since colonial times, agriculture has been carefully promoted by planning, research, and investments, aided by significant inflows of foreign labor and capital and by high world prices for export crops. A balance-of-payments constraint never posed serious problems and balanced budgets have generally helped to avoid inflation, despite high government expenditures.

Rice, however, is an exception to the Ivorian success story. This lack of success scarcely shows, for production has greatly increased. But increased production has been extracted at a high cost to the economy and to the government budget. In the face of constraints caused by factor endowments and geography, government policies have failed to introduce or develop the basic technological changes necessary to make the growth of rice production efficient.

The anatomy of these policies is the subject of this paper. It begins with a summary description of the country and its rice sector. A critical review of the historical evolution of government objectives and policies sets the stage for the analysis and evaluation of past and present interventions in the rice sector. A discussion of lessons and alternative policies concludes the paper.

Setting

Population

In 1975 the Ivory Coast had a total population of nearly 7 million, of which one-third lived in urban areas and two-thirds in rural areas. Of this total, approximately 5 million lived in the more prosperous forest zone while the remaining 2 million lived in the northern savannah zone, giving densities of 30 and 12 inhabitants per square kilometer, respectively. In the major rice-producing areas, population densities are among the highest.[1] Thirty-five percent of the rural population is involved in rice production, two-thirds of which is in the forest zone (23).

[1] See 22. Population densities for the rice-producing areas are: Man, 40 inhabitants per square kilometer; Gagnoa, 38; Daloa, 24; and Korhogo, 22.

TABLE 1.1. *Population, Rice Consumption, and Rice Production in the Ivory Coast, 1960–77*

Year	Current estimated population (000)	Total rice consumption[a] (000 mt)	Actual per capita rice consumption (kg)	Early population predictions[b] (000)	Projected consumption[c] (000 mt)	Actual domestic production of rice[d] (000 mt)	Percent shortfall (−) or surplus (+) of production
1960	3,865	109.7	33.0	—	—	—	—
1961	3,984	116.6	29.3	—	—	—	—
1962	4,107	124.4	30.3	—	—	—	—
1963	4,234	145.4	34.3	—	—	—	—
1964	4,865	173.4	39.7	—	—	—	—
1965	4,500	207.5	46.1	3,688	170	129	−24%
1966	4,684	213.2	45.5	3,781	172	131	−24
1967	4,876	167.6	34.4	3,875	137	144	+5
1968	5,076	227.9	44.9	3,972	178	181	+2
1969	5,284	247.9	46.9	4,071	191	192	+1
1970	5,500	237.8	43.2	4,172	180	159	−13
1971	5,725	259.4	45.2	4,277	193	162	−19
1972	5,959	276.4	46.4	4,384	203	205	+1
1973	6,202	285.1	46.0	4,494	207	168	−23
1974	6,456	182.0	28.2	4,606	130	175	+26
1975	6,720	201.8	30.0	4,721	142	222	+36
1976	6,950	309.6	44.5	4,839	210	242	+13
1977	7,300	342.5	46.9	4,960	230	224	−3

SOURCES: Data for current estimated population, for total rice consumption, and acutal domestic production of rice are taken from 7. The early population predictions are based on data in 20, p. 16.

[a] Equivalent to total net rice availability.

[b] Estimated from the 1960 figure using the predicted 2.5 percent growth rate. Figures are consistent with those used in 27, p. 139, but are slightly lower than later revisions made in 26.

[c] Calculated from the early population predictions using actual per capita consumption rates.

[d] The previous year's production net of seeds and losses.

Although the overall population growth rate is high—4.1 percent in 1975—over a third is contributed by significant in-migration from neighboring countries. Non-Ivorian African migrants make up between one-quarter and one-third of the country's total population as well as most of the hired unskilled labor force in both urban and rural areas (23).

There are also two major internal migration patterns—from savannah to forest, and from rural to urban areas. Growth rates for the two ecological zones—5 percent per annum in the forest and only 2 percent in the savannah—are quite disparate. Census results reveal negative growth rates for the adult male population in the savannah zone (23). Urban growth rates average 8.5 percent per annum, but Abidjan, the capital city, is growing at an estimated 10 percent (23).

Consumption

The analysis of Ivorian rice consumption presents a paradox. Whereas aggregate rice consumption has increased with population growth, there has been no pattern of growth in per capita consumption figures since 1965, in spite of large increases in real per capita income and a high rate of urbanization.[2] Per capita rice consumption has averaged slightly more than 40 kilograms (kg) during this period (Table 1.1) (7, p. 19). Unfortunately, little is known about how consumption is distributed between rural and urban areas. Since 1965, the percentage of total calories that each of the several main starchy staples represents has not changed significantly, as shown in Table 1.2. Rice comprises about one-quarter. The large overall increase in available calories is accounted for by the high population growth and by upward revisions in production estimates after 1974. These data suggest that growth of income and urbanization may be less important than relative prices in determining the demand for rice.

Rice and bread are the only two starchy staples for which the government sets official consumer prices, seemingly well controlled through an extensive system of small retail outlets.[3] In the face of rapidly increasing prices for yams, cassava, and plantains since 1974, government policy has

[2] Data on income and urbanization are taken from Humphreys and Rader (7, pp. 2–3):

	Thousands of constant CFA francs per capita	Urban population as percent of total
1965	78	21%
1975	93	32
Percent change	19	52

[3] There are approximately 600–650 outlets in the distribution system under Ministry of Commerce supervision (20).

TABLE 1.2. *Annual Consumption of Starchy Staples, Selected Years*
(Calories in billions)

Starchy staple	1949–51 Calories	1949–51 Percent	1960–61 Calories	1960–61 Percent	1964–66 Calories	1964–66 Percent	1975–76 Calories	1975–76 Percent
Rice	211	12%	414	20%	725	24%	1,184	23%
Wheat	32	2	20	1	226	8	252	5
Corn	111	6	264	13	409	14	668	13
Millet/sorghum	125	7	105	5	112	4	181	4
Fonio	9[a]	0.5	13	0.5	20	1	6	—
Yams	514	28	597	28	671	23	1,100	22
Banana plantains	173	9	186	9	225	7	411	8
Cassava	598	33	427	20	471	16	1,101	22
Taro (coco yams)	30	2	69	3	86	3	128	3
Sweet potatoes	9[a]	0.5	11	0.5	13	—	8	—
TOTAL	1,812	100%	1,909	100%	2,612	100%	4,474	100%

SOURCE: Data for 1949–51 are based on production and import data from Ivory Coast, Government of, Ministère du Plan, Service de la Statistique, *Inventaire Economique de la Côte d'Ivoire, 1947–56* (Abidjan, 1958). Data for other years are from 7.

NOTE: Edible quantities are calculated by deducting seeds, distribution losses, milling wastes, and other preparation losses (e.g. peeling) using coefficients based on assumptions in Food and Agriculture Organization of the United Nations, *Food Balance Sheets, 1964–66 Average (Ivory Coast)* (Rome, 1971), pp. 291–92, and Ivory Coast, Government of, Ministère du Plan, Departement des Etudes de Developpement, *Les Produits Vivriers de Base dans l'Alimentation en Côte d'Ivoire—Modes de Préparation, Coefficients de Transformation*, by J. P. Chateau (Abidjan, 1973). Calories per kg are based on the U.S. Department of Health, Education, and Welfare, and the Food and Agriculture Organization of the United Nations, *Food Composition Table for Use in Africa*, by Woot-Tsuen Wu Leung (Bethesda, Md., 1968).

[a] Estimated at 0.5 percent of total staple food calories, using the 1960–61 average.

made rice, along with bread, the cheapest starchy staple in terms of calories per CFA franc.[4]

Geography

The Ivory Coast covers an area of 322,500 square kilometers, about one-half of which is arable (25). It is divided into two main ecological zones, which correspond roughly with the forest and savannah zones depicted in Map 1.1. The forest zone consists of two parts extending over approximately the southern half of the country. There is a narrow coastal belt with high rainfall and acidic soils which produces industrial crops such as oil palm, pineapples, and rubber. The forest north of this coastal belt benefits from the most fertile soils in the country and produces the major cash crops—cocoa and coffee—and most of the food crops, particularly rice. This region receives 1,500 to 1,600 millimeters (mm) of rainfall per annum beginning in March or April and ending in October or November, split by a short dry season in July and August (34).

The northern savannah zone also comprises two parts.[5] The center, around Bouaké, is a transition zone of highly variable rainfall over one or two seasons. It produces cotton, some coffee and cocoa, and food crops, especially yams. The second area in the savannah covers the northern third of the country and is the driest and least fertile region, characterized by latosols, which, owing to lack of protective vegetation and intense rainfall, are susceptible to erosion and leaching. Crops grown here include cotton, the main cash crop, and cereals, which are mainly for home consumption. In the savannah zone, the rainy season extends from May or June through September or October, providing 1,100 to 1,500 mm per year (34).

Production

Most agricultural production in the Ivory Coast comes from small family farms. It is generally land-extensive, based on shifting cultivation with bush fallows. In more densely settled areas, which are also centers of rice production, these fallow periods are only two or three years for each

[4]The supporting data are as follows (CFA francs per kg):

	Food price index	Rice	Bread	Yams	Plantains	Cassava	Corn
1964–66	123	51	42	26	17	21	38
1976–77	213	117	120	91	66	98	106
Percent increase	75	130	230	250	288	367	179

[5]The department of Bouaké is included in the savannah zone, even though it dips into the forest. Its vegetation and rainfall patterns make it similar to the rest of the savannah zone, which includes the departments of Touba, Séguéla, Katiola, Dabakaha, Bouna, Korhogo, Ferkessédougou, Odienné, Boundiali, and Bondoukou.

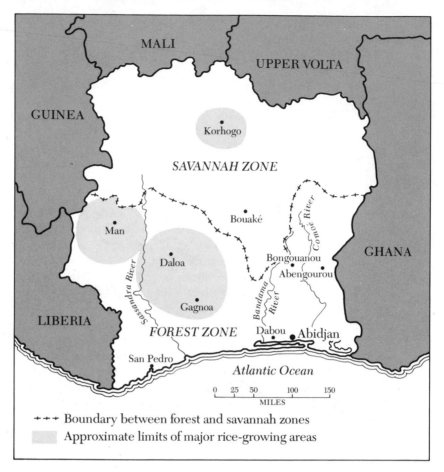

MAP 1.1. Major Rice-Producing Areas of the Ivory Coast. Based on Ivory Coast, Government of, Ministère du Plan, *La Côte d'Ivoire en Chiffres, Edition 1977–1978* (Abidjan, Société Africaine d'Edition, 1977).

one to two years of production. These shorter fallows do not reflect an overall shortage of land but indicate localized scarcities around population centers. Investments in equipment are quite small, and lack of working capital may constrain expansion.

Farms vary greatly but typical characteristics can be noted, especially when comparing the forest and savannah zones. Farms, especially in the south, are multiple-crop enterprises, producing both food and industrial crops and hiring outside labor. Although families are smaller in the forest zone, farm sizes are larger, reflecting in part the extensive cultivation of coffee and cocoa and in part the use of temporary hired labor. Some aggregate and average indicators are given in Tables 1.3 and 1.4.

TABLE 1.3. *Summary Characteristics of Ivorian Agriculture*

Characteristic	Entire country	Forest zone Quantity	Pct.	Savannah zone Quantity	Pct.
General					
Total farm pop. *(000)*	3,683	2,198	60%	1,485	40%
Ave. family size	6.7	6.6	—	6.9	—
No. of family farms *(000)*	550	335	61	214	39
Ave. farm size *(ha)*	5.01	6.08	—	3.33	—
Est. farm revenue *(000 CFA francs)*	219	248	—	92	—
Rice-related					
Farm pop. growing rice *(000)*	1,601	1,051	66	550	34
Farms growing rice *(000)*	237	162	68	75	32
Pct. farms growing rice	43	48	—	35	—
Rice land per farm growing rice *(ha)*	1.29	1.35	—	1.15	—

SOURCE: 7.

Agricultural land use is divided between industrial and food crops, the latter composing about 45 percent or 4.5 million hectares (ha), a share that has changed little since 1965 (Table 1.5). Rice represents a relatively small part of this amount, comprising roughly one-fifth of food-crop land in both forest and savannah zones. Total land in agricultural production, including fallow, may be as high as 10 million ha, or nearly two-thirds of estimated arable land.[6]

Government intervention in agriculture has largely been oriented toward expanding the production of industrial crops through marketing programs and increasing their yields by supplying modern inputs. Input distribution for important cash crops is assured by government-owned development companies. Purchases of these crops are at floor prices fixed at the top level of government and effectively maintained. Rice is the only staple food crop for which there has been a state development company, SODERIZ, and fixed producer prices.[7]

Although official minimum agricultural wages have been in effect since 1955, they seem to have little impact on the cost of labor. Private wages are often as much as 100–200 CFA francs higher than the legal rate.[8]

Agricultural credit is furnished mainly by the government through the

[6]The Ministry of Planning (28, pp. 212–13) estimated that total land in cultivation cycles was 7.6 million ha in 1965 and predicted an increase to 9.6 by 1975. The Ministry of Agriculture (18) indicated that growth of cultivated land has been even more rapid. Another planning document (31) has given lower estimates of land in cultivation cycles (less than 50 percent of the arable land).

[7]SODERIZ, the acronym for Société pour le Développement de la Riziculture, was the state company responsible for increasing rice production from 1970 to 1977.

[8]Legal wages, at 250 CFA francs/day in 1976 for agricultural labor in coffee, cocoa, and rice, are published in a government publication (11). Local currency is the CFA (Communauté Financière Africaine) franc, which is tied to the French franc. The exchange rate used in this paper is 250 CFA francs per U.S. dollar.

TABLE 1.4. *Rice Production in the Ivory Coast, by Technique, 1976*

Production technique	Area[a] (000 ha)	Yield (mt paddy/ha)	Production (000 mt paddy)	Percent paddy
Traditional rainfed, forest	222	1.1	249	59%
Traditional rainfed, savannah	107	0.9	95	22
Fertilized rainfed, forest	5	2.0	9	2
Fertilized rainfed, savannah[b]	8	1.7	13	3
Total rainfed	342		366	86
Irrigated, no fertilizer, forest	5	2.5	12	3
Irrigated, no fertilizer, savannah	5	2.4	11	3
Fertilized irrigated, forest	5	3.3	16	4
Fertilized irrigated, savannah	3	3.7	11	3
Total irrigated	18		50	12
Flooded[c]	5	1.7	9	2
TOTAL[d]	364	1.2	425	100%

SOURCES: Totals are from *19*. The breakdowns by technique are based on information in *15* and Ivory Coast, Government of, Ministère de l'Agriculture, Compagnie Ivoirienne pour le Développement des Textiles, *Rapport Annuel d'Activité, Compagnie 1976–77* (Bouaké, August 1977).
[a] In pure-stand equivalents. Area harvested.
[b] Includes manual (17 percent), ox (30 percent), and tractor (53 percent) cultivation.
[c] Estimates, which includes fertilizer area (64 percent).
[d] Totals may vary slightly due to rounding.

Banque Nationale de la Développement Agricole (BNDA), usually at preferential interest rates.[9] Half of the agricultural credit is used to finance government buying operations, and a fourth is loaned to farmers.

Virtually all traditional rice is rainfed, primarily upland. Of this rice, probably 90 percent is *Oryza sativa* and the rest *Oryza glaberrima*. The small amount of flooded rice in the northwest is relatively unimportant, and there are no indigenous water control systems. As shown in Table 1.4, four-fifths of paddy produced in 1976 came from this rainfed technique, although the proportion was higher in 1975. Most past expansion has usually been associated with demographic growth.

Historically, rainfed paddy production has been localized in the northern savannah around Korhogo and extending to the west toward Guinea—now accounting for about one-fourth of traditional paddy production—and in the southwest near Man and the Liberian border. Climatically, these two areas are most suited to rainfed rice production (*4*). Since 1900, when the French introduced rice to feed the forced labor on coffee and cocoa plantations, rice has spread eastward through the forest zone. Natural conditions favor rice more in the forest zone than in the savannah: rainfall is greater and more regular; the growing season is gen-

[9] Interest rates, including insurance fees, range from 8.5 to 10.5 percent per annum. Hungry-season loans, which comprise about one-third of the lending, carry an annualized rate of 15 percent (*9*).

TABLE 1.5. *Crop Production in Thousand Hectares, 1965 and 1974*

Crop	1965 Production[a]	Pct.	1974 Production[a]	Pct.
Coffee	680	32.4%	1,121	30.6%
Cocoa	424	20.2	745	20.3
Corn	186	8.9	167	4.6
Banana plantains	120	5.7	480	13.1
Rice	209	10.0	258	7.0
Yams	142	6.8	149	4.1
Cassava	56	2.7	169	4.6
Taro (coco yam)	42	2.0	306	8.4
Millet, sorghum, and fonio	63	3.0	94	2.6
Oil palm	19	0.9	22	0.6
Cotton (traditional)[b]	41	1.0	—	—
Cotton (improved)	21	2.0	34	0.9
Peanuts	37	1.8	34	0.9
Coconut	11	0.5	26	0.7
Rubber	10	0.5	16	0.4
Sugar	0	0.0	9	0.2
Others[c]	36	1.7	34	0.9
TOTAL	2,097	100.1%	3,664	99.9%

SOURCES: 1965 data are from *28*, pp. 212–13, except for improved cotton, which is from Ivory Coast, Government of, Ministère de l'Agriculture, Compagnie Ivoirienne pour le Développement des Textiles, *Rapport Annuel 1975–76* (Bouaké, Aug. 1977); 1974 data are from *18*, 1, pp. 116–19, except for rubber, which is from *19*, p. 103.

NOTE: To account for mixed stands, areas have been reduced by 40 percent for banana plantains; cassava; taro; millet, sorghum, and fonio; and traditional cotton (also for ground peas and sweet potatoes under "Others"). To account for multiple-cropping, areas have been reduced by 20 percent for corn, rainfed rice, yams, improved cotton, and peanuts.

[a] All figures are for area harvested with the exception of rubber in 1974, which represents area planted.
[b] Not indicated for 1974.
[c] The 1965 figure is for ground peas, tobacco, bananas, sweet potatoes, and pineapples only, in order of importance; specific crops are not indicated for 1974.

erally longer; and traditional yields are 25 percent higher (1.1–1.3 metric tons [mt] paddy/ha).

The manual cultivation techniques and crop calendar used for improved rainfed rice differ little from those used in traditional production, although yields increase by 70 percent owing to modern inputs, and sickle harvesting saves labor. Government subsidies on modern inputs, including extension, for rainfed rice average about 40 percent of actual costs (*6*). Use of fertilizers and selected seeds in upland production has greatly expanded in recent years.[10]

The government has attempted to introduce large-scale mechanization into rainfed rice production through partial subsidization (18 percent) of tractor services (*6*). Mechanized land clearing, which had exceeded

[10] The improved rainfed seeds are: one-half Moroberekan, an improved local variety with a cycle of 145 days used since 1960; one-third Iguapé Cateto, a Brazilian variety with a cycle of 135 days; and one-sixth Dourado, another Brazilian vareity with a cycle of 105 days (*35* and *16*). The seeding rate is 60 kg/ha, except for the short-season rice, which uses 80. Fertilizers used are 150 kg/ha of 10-18-18 and 75 kg/ha of urea.

10,000 ha by 1976, is fully subsidized. Mechanization represents an effort to stabilize upland farming by using rotations, primarily with cotton and fodder. Ox cultivation is expanding rapidly, although its use in the forest zone is largely precluded by the higher incidence of trypanosomiasis and denser tree cover.

Despite the overwhelming predominance of rainfed rice, government efforts to introduce modern rice cultivation have been focused on irrigated rice.[11] The most widespread irrigation systems divert streams onto surrounding lowland areas of 10 to 15 ha. Investment costs are moderate ($1,500/ha), three-fourths being subsidized. The partial water control assures only one crop in most cases, although forest farmers average about 1.3 crops per year. By 1976, nearly 20,000 ha had been developed. Since the early 1970's, the government has also made considerable investments in dams, sometimes augmented by large pumps, to assure double-cropping—especially in the savannah. Such projects usually cover 100 to 200 ha, and by 1976 about 4,500 physical [12] ha had been developed at a cost of approximately $5,000/ha, virtually fully subsidized (7, p. 8). Recent estimates indicate that about 60,000 hectares could still be developed for irrigation at fairly constant costs,[13] although public financing will be required to realize such expansion.

Production techniques for irrigated rice differ mainly with respect to the degree of mechanization, although cultivation is usually manual on holdings of one-quarter to one-half hectare. Improved inputs appear to be used on only about half the irrigable area (7). There is a limited use of power-tillers, and there have been experiments with tractors and combine-harvesters on the areas irrigated by pumps. Accordingly, labor input is high.

Marketing and Milling

Two parallel marketing systems—the government sector with industrial-scale mills, and private transporters, merchants, and small-scale

[11] Selected seeds for irrigated rice include the following: two-thirds IR5, an IRRI rice with a 140-day cycle; one-tenth Jaya, a rice from India with a 120-day cycle; and one-fifth CS6, an Ivorian cross being phased out. Transplanting is at the rate of 40–50 kg/ha. Recommended fertilizer dosage rates are the same as for rainfed rice. Furadan is the insecticide used against stem borers, applied at the rate of 28 kg/ha. Herbicides are rarely used, but the most common are propanil and 2-4-D, applied at the rates of 7–8 liters and 1–1.5 liters/ha, respectively.

[12] Physical hectares are area measurements that do not take account of number of crop seasons per year.

[13] Roughly one-third of the 60,000 ha could have complete water control, enabling double-cropping. The remainder are in small-scale diversion irrigation schemes (13, 16). In a governmental study (14), 40,000 square kilometers were surveyed in 1966 around Kohogo, Odienné (northwest corner), and Man, locating roughly 15,000 ha of irrigable swampland. Because these zones are relatively favorable for rice, it is difficult to generalize these results to estimate total irrigable swampland.

millers—exist in the Ivory Coast.[14] Before the 1973–74 price increases, government buying programs failed to compete with home consumption as well as private merchants and millers who handled virtually all marketed rice, over both short and long distances.

The domestic price policy established in 1974–75, which fully subsidizes SODERIZ for its collection, milling, and distribution costs, has given the government control of the paddy market and permitted it to purchase a quarter of national production and perhaps half to three-quarters of all paddy sold. Although government policy has forced much of the traditional milling sector into dormancy, private paddy buyers and transporters remain important, as do private rice wholesalers.[15]

Since the mid-1960's, the government has increased its industrial-scale milling capacity two and one-half times to over 150,000 mt paddy in 1976 (7, p. 11). Outturns of rice have also improved from an estimated 45–50 percent to an estimated 66 percent in 1975/76. In addition, some 1,700 to 3,000 private small steel-cylinder mills have been installed by private operators, mostly in the forest zone.[16] Although total capacity probably exceeds 500,000 mt, current utilization for the sector may be as low as 10–15 percent.[17] Outturns are slightly lower than those of government mills, but the rice produced is often fresher and sells at a premium of as much as 10 percent above the official price.

Before 1974, the government market channel handled mostly imported rice, purchased and distributed by a cartel of import houses working with the government. As much as 50 percent of the imports are estimated to have been consumed outside Abidjan (21). Any domestic rice that did flow to Abidjan was channeled through the traditional private sector. Abidjan and the government mills served as basing points for official consumer prices. After 1975, when imports were temporarily halted, the distribution system had to be reversed. Government mills in the interior supplied Abidjan with rice, and domestic rice replaced imports. Because official consumer prices were equalized throughout the country, the gov-

---

[14] Industrial-scale mills are large, integrated units, including cleaners, hullers, whitening cones, and sorters. Capacity is usually 2-4 mt paddy per hour.

[15] The government pays 75 CFA francs/kg of paddy delivered mill (65 CFA francs official farmgate price) and sells white rice for 87 CFA francs/kg, ex-mill. Total production costs, including paddy and its conversion to rice, are established at 139 CFA francs/kg of rice. The subsidy is thus 52 CFA francs (139-87), which more than covers costs of collection and milling. In addition, transport costs from mills to Abidjan are fully subsidized.

[16] Estimates of traditional mills vary considerably. The low figure is based on information supplied by the local manufacturer and takes account of imports before 1968, local production of units since then, and exports. A government publication (18) estimated 2,900, which is probably biased upward. On the other hand, coffee hullers, of which there are some 13,000, can also be adapted to process paddy.

[17] Total small-scale milling can be estimated at about 50,000 mt of paddy per year, based on 1,500 operating units with a 0.2 mt rice-per-hour capacity, 135 days of operation annually, 10-hour days, and a 63 percent milling outturn.

ernment rationed shipments from its mills to urban centers, and the Caisse de Péréquation subsidized transport costs to Abidjan.[18] A market in privately milled rice continued to exist but operated over shorter distances, serving areas near producing centers. The resumption of imports in 1977 had no immediate effect on the price structure, although it can be expected to ease pressure to ship rice to Abidjan from government mills.

Neither storage nor transport is a constraint. Transport of both paddy and rice is assured by a large private sector.[19] Storage is decentralized in the private channels, and capacity and quality are unknown. Public storage was a constraint when the government began to dominate paddy marketing after 1974, but the government reacted quickly by increasing capacity from about 15,000 mt to a projected 200,000 mt by 1979 (7, p.11; 14).

History

The early evolution of Ivorian rice policy was gradual from the colonial period until 1974. Despite ambitious planning statements, self-sufficiency in rice was not viewed as an urgent goal before 1974. Government policy makers planned to increase rice production through a gradual process of technological change, since world prices were consistently low and Ivorian production was not competitive. But in 1973–74, the quadrupling of world rice prices and a doubling of imports catapulted the government into a new policy of price support which radically increased the scope and level of its intervention (Table 1.6).

Three distinct periods of Ivorian rice policy emerge: the colonial period; the period between 1960 and the 1973–74 price increases; and that after 1974. This section describes these three periods, highlights government objectives and policies, and offers explanations for particular government strategies.

The Colonial Period

The colonial period was a time of investment in both physical infrastructure and basic agronomic research. By 1960 an extensive internal road system had been completed, and the construction of the Vridi Canal in 1951 had reduced the cost of agricultural exports and inaugurated the rapid growth of Abidjan as a major West African port. Research institutes

---

[18] The Caisse Général de Péréquation des Prix des Produits et Marchandises de Grande Consommation (Caisse de Péréquation), which was consolidated in 1971 from several earlier price equalization boards, is charged with assuring constant prices of several commodities that are considered critical consumer goods, such as rice, sugar, flour, cement, gasoline, and iron reinforcing rods. In the cases of rice, sugar, bread, and gasoline, prices are the same throughout the country. Transport differentials are paid by the Caisse de Péréquation.

[19] Although transport is probably efficient, real costs of collecting paddy may be higher than the official margin of 10 CFA francs/kg paddy.

were established to improve the export tree crops—primarily coffee and cocoa—but also coconuts, oil palm, and rubber. During this period coffee, cocoa, and timber were established as the foundations of the Ivorian economy. These three commodities accounted for most of the 9 percent average annual real growth rate of exports between 1950 and 1974, with coffee the most important (37, 38).

Initial colonial objectives with respect to rice were not clear-cut, but the commitment to increase production had emerged before independence. Early documents indicated a large variation in output from year to year. Unlike the production of other foodstuffs, however, rice production was not growing.[20] At the same time consumption habits were changing. Because rice was used increasingly as a wage good for laborers on coffee and cocoa plantations, imports became necessary to meet domestic consumption. In the decade before independence, imports steadily increased from zero to 35,000 mt (24). Toward the end of the era, the desire to replace growing imports with local production became a more definite long-term goal, and development of irrigated rice was chosen as the main policy instrument to achieve this objective. The Third Territorial Plan, dating from 1958, focused on rice, allocating it nearly a quarter of the agricultural investment budget and virtually all of the funds devoted to foodcrops.[21]

The main legacies of the colonial period were not only its policy orientation of gradually increasing rice production but also its institutions and instruments for implementing this increase. The Second Territorial Plan, dating from 1953, organized the first systematic action for rice.[22] Research stations were established in the savannah zone and the western forest zone, the centers of indigenous rice production. Studies were carried out in the savannah zone to locate areas that could be developed into small-scale diversion irrigation projects. SATMACI,[23] the state company for agricultural modernization, was created, and it introduced nitrogen fertilizer into savannah rice production in 1953 and distributed selected seeds there starting in 1954 (20). Numerous private farmers were attracted to mechanized rice production in the northwest during the 1950's, only to abandon the technique as too costly, given the uncertain climate. Several cooperative mills were installed in northern towns, and in 1955

---

[20] During 1947–58, paddy production varied from 71,000 to 147,000 mt with no clear trend. It averaged 105,000 mt, with a coefficient of variation of 0.23 (24, 20).

[21] See planning document (30), which reports that 1,482 million CFA francs went to rice, out of an agricultural budget of 6,466 million.

[22] See 24. "Etudes de Riz" appeared for the first time, and they were allocated 5 percent of the agricultural investment budget.

[23] SATMACI is the acronym for Société d'Assistance Technique pour la Modernisation Agricole de la Côte d'Ivoire. It is the state company responsible for the coffee and cocoa sector of the economy and, since 1977, all crops grown in the middle forest zone.

TABLE 1.6. *Selected Historical Series for the Ivorian Rice Economy*

Year	Gross domestic product (GDP) per capita[a] (CFA francs)	Agriculture as pct. of GDP	Paddy as pct. of GDP	Rice[b] Area[c] (000 ha)	Rice[b] Yield (mt paddy/ha)	Rice[b] Production (000 mt paddy)	Net rice imports[d] Total (000 mt)	Net rice imports[d] Value (000,000 CFA francs)	Consumer price index Total	Consumer price index Food
1960	58,458	40.0%	n.a.	194.1	0.82	160.0	35	868	102.5	105.4
1961	62,185	35.4	n.a.	n.a.	n.a.	156.0	33	1,008	114.1	125.6
1962	63,221	32.1	n.a.	n.a.	n.a.	230.0	43	1,418	112.4	118.7
1963	71,406	30.8	n.a.	n.a.	n.a.	220.0	25	786	112.4	118.3
1964	81,973	31.7	n.a.	247.6	1.00	247.9	58	1,949	113.9	118.9
1965	77,818	30.6	1.9%	261.3	0.96	250.0	78	2,216	117.0	122.4
1966	80,768	29.5	1.9	258.0	1.07	274.9	82	3,094	121.9	127.6
1967	79,653	26.3	2.3	300.9	1.15	344.6	24	869	124.6	126.6
1968	88,177	25.6	2.2	299.0	1.22	365.4	47	1,872	131.4	134.8
1969	88,731	23.3	1.7	288.3	1.05	303.0	56	1,878	137.1	141.6
1970	92,165	22.5	1.5	289.1	1.09	315.5	79	2,031	148.9	163.0
1971	95,326	21.9	1.8	282.0	1.37	385.1	97	2,200	147.7	160.1
1972	98,781	21.0	1.7	282.0	1.13	320.0	77	2,192	148.2	158.3
1973	97,309	20.4	1.7	290.0	1.16	334.9	145	8,496	164.1	186.3
1974	95,850	n.a.	3.7	310.7	1.36	422.1	65[e]	7,781[e]	193.1	220.0
1975	93,154	n.a.	3.9	360.6	1.28	460.9	2	210	215.0	242.8
1976	100,577	n.a.	2.5	364.0	1.17	425.5	−30[f]	201	241.4	260.5
1977	n.a.	n.a.	n.a.	n.a.	n.a.	400.0	159[g]	10,191[h]	307.4	364.7

SOURCE: 7.

[a] At constant 1973 prices.
[b] Calculated at official prices.
[c] In pure-stand equivalents.
[d] Considerable variation exists in reports of total imports. E.g. 20, p. 5, reports 51,000 mt for 1961; 33, p. 4, gives 34,000 mt for 1963, 59,000 mt for 1965, and 43,000 mt for 1967; and Ivory Coast, Government of, Ministère de Commerce, Caisse Générale de Péréquation des Prix des Produits et Marchandises de Grande Consommation, "Riz d'Importation—Evolution des Prix CAF, la TM" (Abidjan, 1977), gives 100,000 mt for 1970, 106,000 mt for 1971, 88,000 mt for 1972, and 80,000 mt for 1974. The authors believe the Customs' figures reported here to be the most accurate and complete.
[e] Includes 9,000 mt of paddy imported at a cost of 1,145,000,000 CFA francs, converted to rice equivalent using a coefficient of 0.63.
[f] "Value" column is positive despite net exports because imports were valued much more highly than exports.
[g] Gross imports (export figures unavailable) include paddy imported from Mali converted to rice using the actual milling outturn of 0.56.
[h] Preliminary estimate.

the government set an official price for hulled rice in the savannah zone. Finally, a price equalization board was formed in 1955 to compensate for price fluctuations in rice imports, and a rice commission composed of private traders was established at the same time to manage the importation of rice (2, p. 91).

Independence to 1974

Objectives. Policies instituted in the late colonial period were largely continued during the first decade after independence. Although the new government placed a greater emphasis on high rates of growth of revenue per capita, the external orientation of the economy was preserved. Foreign trade and financing remained key instruments in obtaining this fundamental growth objective. Both budgetary equilibrium (to help control inflation) and a positive trade balance (to permit unrestrained capital flows) were considered essential in preserving an environment attractive to foreign investors. This outward-oriented growth policy was focused around the diversification of agricultural exports to reduce excessive reliance on coffee, cocoa, and wood. There was also a policy to establish an industrial sector to process agricultural production and replace manufac-

TABLE 1.7. *Investment in Agriculture and in Rice, 1960–77*
(Millions of CFA francs)

Year	Total investment budget	Agricultural investment	Rice investment	Rice as pct. of total	Rice as pct. of agriculture
1960	8,000	n.a.	110[b]	1%	n.a.
1961	8,000	n.a.	110[b]	1	n.a.
1962	5,000	n.a.	60[b]	1	n.a.
1963	5,000	500	60[b]	1	12%
1964	12,000	2,200	710	6	32
1965	13,000	500	210	2	42
1966	15,000	1,800	840	6	47
1967	18,000	3,100	860	5	28
1968	20,000	6,200	990	5	16
1969	27,000	4,200	1,740	6	41
1970	44,000	11,400	2,070	5	18
1971	42,000	6,200	400	1	7
1972	35,000	4,500	450	1	10
1973	37,000	6,900	2,390	7	35
1974	44,000	8,000	2,030	5	25
1975	54,000	13,000	2,530	5	20
1976	60,000	11,600	1,780	3	15
1977[a]	123,000	17,200	1,960	2	11

SOURCES: 1960–63 data are from *11*; 1964–77 agricultural investment data are from Ivory Coast, Government of, Ministère de l'Economie, des Finances, et du Plan, *Budget Spécial d'Investissement et d'Equipement* (Abidjan, yearly).

NOTE: All figures are based on "planned expenditures," not appropriations. Neither Caisse de Stabilisation et de Soutien des Prix des Produits Agricoles (CSSPPA) investments before 1974 nor investments made directly by development companies are included.

[a] Includes CSSPPA funds invested during 1974, 1975, 1976, and 1977.
[b] Foreign funds only.

TABLE 1.8. Achievement of Rice-Planning Projections in the Ivory Coast

Rice planning projections	1965		1970		1975		1980	1985
	Projected	Actual	Projected	Actual	Projected	Actual	Projected	Projected
Third Four-Year Plan, 1958–62:								
Expenditures (000,000 CFA francs)	1,482	410[a]	—	—	—	—	—	—
Irrigated rice (000 ha)	30	8	—	—	—	—	—	—
Ten-Year Perspective, 1960–70:								
Expenditures (000,000 CFA francs)	1,100[b]	1,150[c]	3,286[b,a]	6,500	5,441[b,e]	7,800	—	—
Irrigated rice (000 ha)	11	8	31	10[f]	51	21[f]	—	—
Paddy from irrigation (000 mt)	28	24[g]	78	30[g]	128	63[g]	—	—
Total paddy production (000 mt)	220	258	295	335	395	434	—	—
Rice imports (000 mt)	10	73	0	77	0	12	—	—
New milling capacity (mt/hour)[i]	12	14	22	16	—	—	—	—
Five-Year Plan, 1971–75:								
Expenditures (000,000 CFA francs)	—	—	—	—	8,724[b]	7,800	4,910[b]	—
Irrigated rice (000 ha)	—	—	11	10[f]	27	21[f]	37	—
Paddy from irrigation (000 mt)	—	—	40	30[g]	186	63[g]	276[b]	—
Total rice production (000 ha)	—	—	303	286	297	345	303	—
Total paddy production (000 mt)[h]	—	—	359	335	524	434	680	—
Rice imports (000 mt)[h]	—	—	40	77	17	12	0	—
Draft Five-Year Plan, 1976–80:								
Expenditures (000,000 CFA francs)	—	—	—	—	—	—	17,000	—
Irrigated rice (000 ha)	—	—	—	—	29[j]	—	72	80
Paddy from irrigation (000 mt)	—	—	—	—	85[j]	—	250	285
Total paddy production (000 mt)[h]	—	—	—	—	450	—	695	1,030
Rice imports (000 mt)[h]	—	—	—	—	—	—	66[k]	0

SOURCES: The planning projections are from 30, 27, 28, and 29. Data for the actual area in total production and total paddy produced, for actual area in irrigated rice, for mill installations, and for rice imports are taken from 7. Actual expenditures under the Third Four-Year Plan, 1958–62, are from 11. The actual expenditures through 1965 indicated under the ten-year perspectives, 1960–70, also include the "planned expenditures" of Ivory Coast, Government of, Ministère de l'Economie, des Finances, et du Plan, Budget Spécial d'Investissement et d'Equipement (BSIE) (Abidjan, 1964 and 1965). The actual expenditures for 1970 and 1975 are the "planned expenditures" of BSIE for the years 1966–70 and 1971–75, respectively.

[a] Grant assistance, primarily for rice, 1959–63.
[b] Projections for the five-year period preceding and including the year indicated.
[c] Based on foreign aid only during 1961–63.
[d] Includes planned rice mill investment of 186 million CFA francs about 1965.
[e] Includes planned rice mill investment of 341 million CFA francs about 1970.
[f] Cropped hectares based on physical area by assuming 20 percent has been abandoned, 90 percent of what remains is cultivated, and there are an average of 1.2 crop seasons per year.
[g] Production based on area in irrigated rice and assumed average yields of 3.0 mt/ha.
[h] This number is calculated from a three-year average, centered on the year indicated, from 7.
[i] Theoretical capacity rating.
[j] Based on developments that were to be completed by the end of 1975.
[k] Calculated as a residual based on projected total paddy production and consumption and converted to rice at 63 percent.

tured imports. Except for possible investments in milling, the rice sector was not viewed as being affected by these two policies.

The new government adopted secondary objectives which did involve the rice sector. The first was to increase the incomes of farmers, particularly those in the northern savannah zone. Rice was identified, along with sugar, cotton, and tobacco, as a vehicle for achieving this goal of regional income distribution. The second was to help maintain a positive trade balance by reducing food imports. Eventual self-sufficiency was a continuation of one colonial objective, and rice, as the only domestically produced staple food that was imported, seemed a natural focus for this objective (11, 26, 27, 37). These objectives were retained during the 1971–75 Plan, which further stressed import substitution, largely because of concern over the foreign exchange cost of growing rice imports. Rice, with sugar, meat, and fish, was earmarked as a commodity for which growing demand could be met by increasing domestic production (28).

As Table 1.7 shows, the government budgeted sizable funds for rice compared with its share of either agriculture or GDP. These expenditures also tended to exceed targeted funds, as shown in Table 1.8, suggesting a continuing commitment to promote rice production. Yet expanded rice production was not critical to achieve major objectives, nor was it even a necessary policy to achieve secondary objectives. Moreover, cheap rice imports made Ivorian rice uncompetitive and should have defused concern over growing consumption. Why did the government desire to expand rice production?

First, the Ivory Coast is traditionally a rice-producing and -consuming country (Tables 1.1, 1.2, and 1.3). Planners apparently felt that natural conditions favored rice development and that rice production should be protected as an infant industry.

Second, research begun by colonial institutions and continued after independence provided the foundation for assembling a technical package of selected seeds, fertilizers, and irrigation systems.[24] Green Revolution technology was also imported from Asia during the late 1960's (37, p. 144).

Finally, planning documents toward the end of the period express a growing fear that excessive reliance on large imports of rice could be financially destabilizing. The ratio of the value of rice imports to the net trade surplus averaged over 12 percent, rising in some years to 20 percent. On the other hand, rice import prices—even in nominal terms—actually declined during 1960–72, and the cost of rice imports was never a

---

[24]The share of funding devoted to research was significant. During 1961–63, 38 percent of French grant aid for rice went to studies. Of the 577 million CFA francs allocated to rice in the 1964–65 French-guaranteed loan for agricultural development, 45 percent was concentrated on research (11).

large share of export earnings, averaging only 2.4 percent for the period 1960–76 and never being larger than 4.4 percent (7; 25, p. 58).

Policies. Four types of policies—research, institutional, investment, and pricing—were emphasized during the 1960–74 period. Although the last two are the more important quantitatively, they would not have been effective without research and the creation of supporting institutions.

Research policies sought ways to shift downward and outward the supply curve for rice through technological improvements. This shift was considered essential in order to increase rice production, compete efficiently with imports, and leave adequate supplies of labor and land available for use in export crop production. The most important result was the identification of investment projects. Taiwanese technical experts helped develop viable small-scale irrigation schemes and extension programs in the north starting in 1963 and in the forest after 1967. Improved seeds from Brazil, India, and the Philippines were introduced beginning in 1966 as part of a constant search for higher yields, shorter growing seasons, and greater disease resistance. Experimentation with fertilizers was pursued in the north, west, and center. Five regional studies were contracted about 1970 to identify profitable production systems and methods of development. Out of this experience and research came four investment projects in 1971–73.

Institutional policies were designed to assure input delivery and marketing of output. Management of the rice sector was consolidated under SATMACI in 1963 and was consigned in 1970 to SODERIZ, a move that had been suggested in planning and budget documents during the previous ten years.

SODERIZ instituted a contract system for supplying subsidized modern inputs to rice production. The development company provided an input package which was paid for by the farmer at harvest in either cash or paddy. Farmers were not obligated to pay the participation fee if they failed to obtain specific minimum yields.[25] Partially subsidized inputs included selected and treated seeds, fertilizers, insecticides, sales of capital equipment, land development, extension services, maintenance of irrigation works, and mechanized cultivation services.[26] All land under con-

[25] The guaranteed yields were 2 mt paddy/ha in rainfed rice and 4 mt paddy/ha in irrigated rice. Although there was considerable variation in practice, the guarantee seemed to be applied only in cases of complete crop failure. Where average yields for a farmer were positive but below standard, the guarantee was calculated by assuming that part of the field failed totally and the other part produced the guaranteed level, giving the lower average. The fee was waived only on that share of land assumed to have had no harvest.

[26] Fertilizer and seeding rates are given above. Extension density is a function of the contracts accepted by farmers, but ideally is one agent per 50–80 cropped ha; recently it has been higher. Irrigation water is not measured, so payment is a function of land area, not volume of water delivered.

tract was supposed to be uniformly fertilized regardless of its place in rotation or natural soil fertility. But because the system was based on family farms, producers retained almost full control over crop calendars, actual dosage rates, and sales. Partially as a result of SODERIZ initiatives and also owing to higher paddy prices, use of improved seeds and fertilizers grew consistently during this period.

In an effort to improve paddy marketing, often viewed by government officials as a critical bottleneck, an official paddy price of 18 CFA francs/kg was introduced in 1966. After 1967, to promote sales of domestic rice, import licenses were granted only after importers had first purchased available local rice from government mills. Quality buying standards for paddy were introduced in 1968. These institutional changes facilitated future investment and the implementation of domestic price policy.

Investment policy was not important initially. Until 1971–72, major investments were limited to irrigation projects in the savannah zone and to eight industrial-scale mills located throughout the country. The early savannah zone project, financed by a 1963 German loan for 1.6 billion CFA francs, was a direct extension of previous colonial efforts. The two phases of mill investments, which began in roughly 1965 and 1970, represented easily defined projects financed by supplier credits. Both the irrigation investments and the mills seemed to exhaust the readily available investment possibilities.

For most of the period, investment in production was not a major policy instrument. As Table 1.8 shows, actual development of irrigated land and improved production fell far below targets, often attaining only half the area planned. The government was able to obtain significant foreign financing—on fairly soft terms—only after the state rice development agency was created to centralize planning and the rice studies of the late 1960's completed project designs based on data from the previous decade (Table 1.9).[27] Of the six billion CFA francs received, less than one-sixth was allocated to rainfed rice, reflecting the government's emphasis on irrigated rice development. Irrigation, especially by the end of the period, was favored because it allowed more secure water control, provided higher yields, and made extension work easier.

Although both trade and domestic price policy were used to encourage

[27] Foreign financing came from the following sources (13, 16):

Organization	Amount (billion CFA francs)	Year committed	Terms (percent)
International Coffee Organization	1.468	1971	0 over 30 years
European Development Fund	2.928	1971	grant
Caisse Centrale (CCCE, France)	0.860	1972	3.5 over 10 years with 3 years grace
KFW (German)	1.013	1973	2.25 over 20 years with 8 years grace

TABLE 1.9. *Financing of Foreign Investments in Rice Development in the Ivory Coast, 1960–77*

Year	Foreign funds allocated to the rice sector (000,000 CFA francs)	Foreign funds as a pct. of total rice investment	Foreign funds as a pct. of total investment budget
1960	90[a]	n.a.	n.a.
1961	120	n.a.	n.a.
1962	150	n.a.	n.a.
1963	10	n.a.	n.a.
1964	560	79%	50%
1965	70	33	46
1966	380	45	27
1967	310	36	39
1968	350	35	40
1969	1,310	75	56
1970	1,390	67	34
1971	150	37	48
1972	30	7	51
1973	1,250	52	51
1974	1,480	73	59
1975	1,710	68	48
1976	850	48	37
1977	490	25	40

SOURCE: 1960–63 data are from *11;* 1964–77 data are from Ivory Coast, Government of, Ministère de l'Economie, des Finances, et du Plan, *Budget Spécial d'Investissement et d'Equipement* (Abidjan, yearly).
    NOTE: Before 1970 foreign funds refer only to loans.
    [a]This figure may include some non-rice monies.

additional production, trade protection afforded to domestic rice producers was more important. Official paddy prices were set beneath market prices, providing no incentive effect. On the other hand, during the first 13 years of independence the nominal protection coefficient averaged 1.3. Nevertheless, planning documents throughout the 1960's argued that additional protection was needed (20, 27).

Domestic price policy was consolidated in 1971 by the reorganization of the Caisse de Péréquation. This organization fixed retail rice prices, which it defended through imports, compensating for any fluctuations in import prices. The official paddy support price was linked to the retail rice price, which required a modest subsidy paid to government rice mills. The subsidy was financed by the Caisse de Péréquation out of revenue earned from the difference between import and higher retail prices (Table 1.10).[28] Despite the subsidy, competition from small-scale private millers maintained the market price of paddy above the support

[28]Theoretically, during 1960–73 rice imports generated about 10 billion CFA francs in tariff revenue (see Table 1.10). Imports were controlled by a cartel, which may have appropriated most of the implicit tariff revenue before the Caisse de Péréquation began to function.

TABLE 1.10. *Government Revenue from Rice Imports and Nominal Protection on Rice, 1960–77*

Year	C.i.f. rice prices[a] (CFA francs/kg)	Estimated Abidjan wholesale buying price[b] (CFA francs/kg)	Abidjan retail selling price[c] (CFA francs/kg)	Implicit tariff[d] (CFA francs/kg)	Estimated total tariff revenue[e] (million CFA francs)	Nominal protection coefficients[f]
1960	25	41	45	15	435	1.6
1961	32	49	54	16	336	1.5
1962	32	46	51	13	455	1.4
1963	31	49	54	17	425	1.5
1964	33	41	46	7	301	1.2
1965	30	46	51	15	900	1.5
1966	38	50	56	11	572	1.3
1967	36	55	61	18	216	1.5
1968	41	52	58	10	430	1.2
1969	33	55	61	21	1,134	1.6
1970	25	67	74	41	2,870	1.6
1971	21	45	50	23	1,587	2.1
1972	28	45	50	16	1,152	1.6
1973	59	57	63	−4	−552	0.9
1974	112	107	116	−8	−496	0.9
1975	50	97	108	46	9	1.9
1976	30[g]	87	100	56	6	2.9
1977	69	87	100	16	1,472	1.2

SOURCE: 7.

[a] Prices for 25–30 percent brokens.
[b] Based on official prices for 1973 and later years; before 1973, estimated at 90 percent of the retail price.
[c] Prices after 1970 appear to be official prices.
[d] The difference between the Abidjan wholesale buying price and the c.i.f. price plus landing costs (evaluated in 1975 at 2.7 percent of the c.i.f. price).
[e] Based on imports of 25–30 percent brokens only. Actual government revenue is probably less, since the series really indicates transfers from the consumer sector to other sectors.
[f] Defined as the sum of the c.i.f. price and implicit tariff divided by the c.i.f. price.
[g] This price may be unreliable because it is based on very low imports.

price,[29] and amounts purchased and milled by the government mills were not very large. As a result, the total budgetary cost of the subsidy was quite low.[30]

Post-1974

Until 1974, the Ivory Coast followed a fairly cautious rice development strategy, which could be summarized as an effort to help its infant industry slowly mature. Domestic price support had been limited to small increases in 1972 and 1973, linked to import price rises. In 1974 several factors converged, shifting the government policy focus from technology transfer, investment, and institution building to high price supports maintained with government subsidies.

First, world rice prices suddenly began to rise in 1973. By 1974 they had quadrupled. The government reacted in 1973 by doubling imports, afraid that prices would go even higher. These large imports confirmed fears of excessive outflows of foreign exchange. In 1974, even though imports returned to previous volumes, prices had doubled from 1973 levels. Outflows of foreign exchange were almost as great as in 1973 (7, p. 12). Substitution of rice imports with domestic production suddenly gained a much higher priority than ever before.

Second, during the previous period imports averaged 65,000 mt per year and were not diminished despite a doubling of domestic production. Even though per capita consumption grew very little, total rice consumption was higher than anticipated for two major reasons: population growth had been underestimated from incomplete census information, and the real price of rice had fallen by 25 percent compared to all other food during 1960–71 (7, p. 18). Irrigated land development and production were both lagging behind projected increases, even though investments were larger than planned. The technical package was not profitable enough to allow significant expansion of the production of improved rice.

The solution for meeting high consumption demands and offsetting insufficient domestic production was to allow consumer prices to rise with the higher import prices, thereby reducing consumption and making domestic rice more competitive without excessive government subsidies. In 1974, the official price was increased by 80 percent to 125 CFA francs/kg,

[29]The domestic price was about 50–75 percent higher than the official paddy price. Budgetary calculations in planning documents 33 and 31 suggest that, at official rice prices, paddy production could be profitable only with very optimistic assumptions (for example, 5 mt paddy/ha per irrigated crop).

[30]For three seasons, 1971–72 to 1973–74, the subsidy varied from 15.7 to 21 CFA francs/ kg of rice. During this same three-year period, less than 30,000 mt of rice were milled in government mills. The total cost of the subsidy was only 522 million CFA francs from 1971 to 1974.

and the official support price of paddy was raised by 130 percent to 65 CFA francs/kg. The government reacted to the increase in world prices by first raising the retail rice price and then deriving the support price of paddy.

These policies elicited expected responses. Per capita consumption fell from nearly 50 kg of rice in 1973 to just under 30 kg in 1974 and 1975, as shown in Table 1.1. The production of paddy rose nearly 25 percent between 1973 and 1975, with marketed supplies probably increasing even more. Imports, partly offset by large carry-overs from the huge 1973 purchases, dropped to zero in 1975 and 1976.[31]

Meanwhile, previously introduced investment and technical policies were continued. Foreign financing worth 2.2 billion CFA francs was committed in 1974 and 1975.[32] Between 1973 and 1974, the use of selected seeds, fertilizers, and insecticides doubled. Investment in irrigation progressed rapidly. Between 1973 and 1976 swamp irrigated land increased by one-third, pump irrigation grew tenfold, and dam irrigation tripled (7, p. 8). The past investment projects were beginning to bear fruit.

Nevertheless, this new domestic price policy, adopted in a brief period of high world prices, was difficult to sustain. Government mills were suddenly confronted by the need to purchase, store, and mill a fourth of Ivorian production—ten times more than a year earlier. Although there were physical constraints in both storing and milling paddy, the past mill investments greatly alleviated them. The major bottleneck was the distribution of milled rice. In the face of high prices, total demand fell dramatically. The government found itself not only with large stocks of imported rice, but also with growing stocks of domestically milled rice. The stock build-up was aggravated by the reluctance of the influential import cartel, accustomed to importing one-third to one-fourth of rice consumption, to handle only domestic rice (1). Imported stocks were expensive to maintain, while unsold domestic stocks took up storage space and tied up funds, impeding further purchases of paddy. Self-sufficiency also put an end to the government revenue from rice imports, which was needed more than ever to pay the growing subsidies to producers.

The immediate solution was to readjust prices. The retail price of rice was lowered from 125 to 100 CFA francs/kg in 1975. Three aspects of this revised policy merit mention. First, there is no evidence that consumer pressure forced the reduction in retail prices; rather, the reduction was an effort to reduce stocks held at government expense. Second, there was no corresponding reduction in paddy prices, perhaps reflecting pressures from influential producer groups as well as from the rice development

---

[31] In these two years, there were minor imports of 100 percent whole-grain rice, which is outside government control.

[32] All this additional foreign financing came from the French Caisse Centrale de Coopération Economique (CCCE), largely as extensions to existing investments.

company, which needed high paddy prices to maintain its new share of the market. Third, rice prices were equalized throughout the country, creating the need to control distribution and ration marketings.

By 1976, the lower consumer price of rice reestablished consumption at previous levels, drawing down surplus stocks. But the stock problem was solved by creating another: the need to support highly subsidized producer prices. The government had inadvertently placed itself in a situation of subsidizing producer prices, which it had previously conscientiously avoided.

In 1978, the government was still attempting to maintain this price relationship, which gave approximately a 100 percent subsidy to domestic rice.[33] But the budgetary costs were high. Producer price subsidies ran 4 to 5 billion CFA francs per year, financed out of coffee and cocoa earnings by the Caisse de Stabilization (CSSPPA).[34] The glut of paddy attracted by the high prices spurred large government investments in new milling and storage capacity, both financed on hard terms.[35] Foreign donors have postponed additional rice investments since 1975, regarding the present price policies as financially unsound. In the wake of financial problems largely spawned by the domestic price policy, the state rice development company was disbanded in 1977. Institutionally, rice had lost its equal status among important cash crops, since management of the sector is no longer consolidated but has been divided among the state companies charged with coffee and cocoa, oil palm, and cotton. It is no longer clear how the extension program will function in the future.

Meanwhile, the initial success of the domestic price policy on production is eroding because of the increased producer prices of other crops. Paddy output has fallen since 1975, despite an expansion of irrigated land under cultivation and greater use of modern inputs. Consumption has also returned to the highest previous levels, in the face of increasingly expensive food substitutes. In 1977, rice imports of 110,000 mt were the largest since 1973 (7, p. 12).

In summary, the history of Ivorian rice policy reveals a long-standing effort to find a strategy to improve productivity and increase output. Given the decision to focus on rice, a noncompetitive sector, the policies

[33] The agreement between SODERIZ and the Caisse de Péréquation established total costs of producing domestic rice by government mills at 139 CFA francs/kg rice, ex-mill. Import prices in 1977 are estimated at about 69 CFA francs, c.i.f. (7, p. 12; 8).

[34] Caisse de Stabilisation et de Soutien des Prix des Produits Agricoles was created to be responsible for exports of major export crops—especially coffee and cocoa—and to assure producer incomes for these crops. Proceeds from exports are used in general development projects.

[35] Theoretical milling capacity was to be increased by 132,000 mt paddy/year, or 63 percent compared with 1976 theoretical capacity (38 mt/hour multiplied by 5,500 hours/year [7, p. 11]). Storage was to be increased by 125,000 mt. Total cost was estimated at 9.5 billion CFA francs (16).

were consistent from colonial times until 1974. The shift to domestic price policies in response to recent short-term world price changes, however, has required numerous ad hoc adjustments and put the future of the entire strategy into question.

Evaluation

Since independence, two important secondary objectives—improving the trade balance and enhancing regional income distribution—have been used to justify government rice policy. Increased domestic rice production has been singled out as a major instrument for attaining these secondary objectives. To assess the success of government policies, it is first necessary to examine how government actions have affected production.

Increased Production

Since 1965, Ivorian paddy production has increased by roughly two-thirds, or 182,000 mt,[36] which is over 5 percent a year. Such growth compares favorably with important industrial crops, such as cocoa and oil palm, although it lags far behind new diversification crops such as cotton. Early planning objectives were achieved, although those established by the end of the 1960's, based on a recalculation of population growth resulting in higher demand, were missed by nearly 20 percent.

The aggregate data are not very helpful in isolating the sources of this growth, but there are informative indications. Perhaps half the growth came simply from increasing inputs of labor and land, while less than half can be attributed to fertilizer, irrigation, and mechanical technologies.

Land area devoted to rice is estimated to have increased by over 100,000 ha, or 40 percent in the period between 1965 and 1976.[37] Until 1974, the growth of land allocated to rice was only about 1.2 percent per annum, much lower than the rural population growth rate. The real impetus for bringing additional land into rice production apparently came from the price increases in 1973 and 1974. These price rises may have augmented area planted by 50,000 to 55,000 ha, or almost half the total expansion during 1965–76.[38] The input of labor also increased with greater land use.

[36] See Humphreys and Rader (7, p. 4). Calculations based on the three-year average centered on 1965 (258,000 mt) and the average of 1975–76 (440,000 mt). Using 1960 as the starting point gives a more dramatic growth of 135 percent and 250,000 mt. Because the accuracy of agricultural statistics was vastly improved after the regional surveys of 1962–64, an analysis based on the 1965–76 period gives a more reliable indication.

[37] See Humphreys and Rader (7, p. 4). The calculation uses a three-year average centered on 1965 and the average of 1975–76. The growth rate is based on 11 years.

[38] The details of this calculation are: (1) 290,000 ha devoted to rice in 1973; (b) 7,000 ha coming into production during 1974 and 1975 because of natural growth, based on the past

The other important factor behind production growth is an increase in yields. The average yield for the 1974–76 period is 1.27 mt/ha, 18 percent higher than for the 1964–70 period.[39] Several things contributed to this increase. Since 1965, irrigated land increased by over 15,000 ha. With water control, yields are on the order of 2.5 mt/ha, giving an additional output of nearly 40,000 mt paddy (see Table 1.4).[40] Fertilizer applications also increased rapidly after independence, reaching levels of about 1,500 mt of nitrogen by 1975–76, one-third of which was used on irrigated rice. Average response rates are difficult to determine, but available agronomic results suggest that the fertilizer may have increased paddy production by about 20,000 mt during this period.[41]

As summarized in the tabulation on p. 41, land, irrigation, and fertilizer account for roughly 150,000 mt, or over 85 percent of the increase in paddy production. The residual could be attributed to extension, improved seeds, pesticides, and better cultivation techniques. What is not measured but important is the amount of additional labor input needed to utilize the increased inputs of other factors.

This increase in production deriving from new technology has occurred more slowly than anticipated and has been expensive. By 1975, the projections from the early 1960's of land under irrigation were less than half realized, and the revised 1970 projections of production from irrigated projects were only one-third achieved. Moreover, expenditures during

---

trend; (c) about 6,500 ha of new riceland with an overall cropping intensity of about 1.3 brought into production in 1974 and 1975 by land development projects, including mechanized clearing (7, p. 8); and (d) 361,000 ha in rice in 1975, giving an increase of 71,000 ha, of which only 15,000 are accounted for by normal growth and government projects. Since prices of paddy increased by 160 percent in 1973 and 1974, the short-run supply elasticity would be 0.12.

[39] The 1964–66 average yield is 1.01 mt paddy/ha (7, p. 4).

[40] Compared to traditional rainfed production, about three-fifths of yield increases under irrigated systems may be attributed to water control along with ancillary labor inputs.

[41] Fertilizer dosage rates are uniformly 50 kg N/ha, two-thirds applied as urea. In SOD-ERIZ (15, p. 237) response rates are reported as 34 kg paddy/kg N for applications up to 35 kg N/ha under irrigated conditions in the savannah zone. IRAT (36, p. 14) reports a response rate of 9–14 paddy per kg N for applications of 200 and 100 kg N/ha, respectively, also under irrigated conditions. For rainfed rice, IRAT also reports 5–10 kg paddy per kg N in the savannah and an average of 14 kg paddy per kg N in the forest zone, both for applications of 50–60 kg N/ha. The figures presented in the text are based on the following data and assumptions:

Type of rice	Percent of fertilizer	Total N (mt)	Assumed kg paddy per kg N	Total increase in paddy (mt)
Forest irrigated	17	255	25	6,375
Forest rainfed	15	225	14	3,150
Savannah irrigated	16	240	25	6,000
Savannah rainfed	52	780	7.5	5,850
TOTAL	100	1,500	14.25	21,375

Input or change in technology	Change in input level	Increase in paddy production (*mt*)	Pct.
Continuation of past trend of land increases	35,000 ha	35,000	20%
1973–74 price-induced land increases	50,000 ha	50,000	29
New land brought under irrigation	15,000 ha		
Production increase due to land		15,000	9
Production increase due to water control		25,000	14
Mechanically cleared land	2,000 ha	2,000	1
Nitrogen fertilizer	1,500 mt	20,000	11
SUBTOTAL	—	147,000	—
Residual (extension, seeds, pesticides, etc.)	—	28,000	16
TOTAL	—	175,000	—

the 1960's exceeded those targeted. During 1971–76, total budgetary support of rice (investments and subsidies) far exceeded funds initially allocated, as shown in Table 1.11.

Three factors help to explain why the increase in rice production has been so expensive. First, contrary to earlier claims (37, p. 143), recent agronomic evidence and experience indicate that the natural conditions in the Ivory Coast are not well suited to rice production. The entire country

TABLE 1.11. *Government Expenditure on Rice, 1971–76*

Expenditure	000,000 CFA francs
Investment budget	9,580
Operating budget	138
Input and paddy marketing support	1,456
Producer price support[a]	14,708
Transport of domestic rice to Abidjan[b]	600
Total expenditure	26,482
Planned expenditure[c]	10,469
Excess expenditure	16,013

SOURCES: Investment budget figure is from Ivory Coast, Government of, Ministère de l'Economie, des Finances, et du Plan, *Budget Spécial d'Investissement et d'Equipement* (Abidjan, yearly). Operating budget figure is from Ivory Coast, Government of, Ministère de l'Economie, des Finances, et du Plan, *Budget du Fonctionnement* (Abidjan, yearly). Figure for input and paddy marketing support is from Ivory Coast, Government of, Ministère du Commerce, Département du Commerce Intérieur, Caisse Générale de Péréquation des Prix des Produits et Marchandises de Grande Consommation, personal communication (Abidjan, 1978) and is for buying operations during 1971-72 to 1976-77. The figure for producer price support is from Ivory Coast, Government of, Banque Nationale de Développement Agricole, personal communication (Abidjan, 1977) and represents buying operations during 1971-72 to 1976-77. The planned expenditure figure is taken from Table 1.8.

[a]This figure covers unpaid principal and interest.

[b]Rice milled by SODERIZ and sold in Abidjan is transported by the Caisse Générale de Péréquation, primarily since the 1975 law equalizing rice prices all over the country. This is estimated, based on the assumption that two-thirds of SODERIZ rice is shipped to Abidjan, over an average distance of 400 km at official tariff rates.

[c]This amount is based on the five-year plan figure, prorated over six years.

TABLE 1.12. *Some Estimated Returns to Labor for Rice and Other Crops*

	Producer price (CFA francs/ kg)	Assumed yield (mt/ha)	Gross revenue (000 CFA francs/ha)	Farmer costs, non-labor (000 francs/ha)	Labor days/ ha	Net CFA francs/ labor day
Improved manual rainfed rice						
Pre-1974 prices	28	1.5–2.2[a]	42–62[a]	29–40[a,b]	95–120[a]	137–183[a]
Post-1974 prices	65	1.5–2.2[a]	98–143[a]	29–40[a,b]	95–120[a]	726–852[a]
Improved manual irrigated rice						
Pre-1974 prices	28	3.5	98	66[b]	240	133
Post-1974 prices	65	3.5	228	66[b]	240	675
Selected coffee	150	0.65	98	43[c]	98	561
Improved cocoa	175	1.0	175	34[d]	100	1,410
Improved cotton						
Manual	70	1.0	70	17[e]	145	366
Animal traction	70	1.2	84	21[e]	93	677

SOURCES: The producer prices and data for rice production are from Chapter 2. Information for coffee and cocoa is contained in Ivory Coast, Government of, Ministère de l'Agriculture, Société d'Assistance Technique pour la Modernisation Agricole de la Côte d'Ivoire, *Manuel de Caféiculture*, Fasc. 4, Gagnoa, 1975, "Prix de Revient du Kilogramme de Cacao," Abidjan, Dec. 9, 1976, and personal communication, Abidjan, 1978. Information for costs, and labor and yields, respectively, in cotton production are based on Ivory Coast, Ministère de l'Agriculture, Compagnie Ivoirienne de Développement des Textiles, "Coût Total de la Culture d'un Hectare de Coton, Campagne 1975–1976," Bouaké, June 16, 1976, and "Temps de Travaux (Région Nord)," Bouaké, Feb. 9, 1977.

[a]The first figure refers to production in the savannah zone and the second to production in the forest zone.

[b]Includes the farmer share of the annuity on land clearing or development, farmer cost of fertilizers, seeds, pesticides, extension, and irrigation maintenance plus tools and working capital.

[c]Includes the annuity on the plantation, cost of tools, and charges for hulling and sorting.

[d]Includes annuity on the plantation, costs of fertilizer, insecticides, equipment, and materials. Extension costs are excluded.

[e]Includes only fertilizer and tools, and annuity on oxen and equipment in the case of animal traction. All other costs are assumed to be borne by the government.

generally suffers from irregular rainfall, high rates of evapotranspiration, and inadequate ground water. No region is naturally suited to two crops of rainfed rice, and only limited areas are suited to one crop per year, making expanded rice production generally risky unless there is expensive investment in water control (4). Adverse climate is compounded by rolling topography and the absence of flood plains larger than several hundred ha. In addition, soils have been found ill-suited to irrigation in many projects (40). As a result, land development has been more expensive than expected, the number of crops per year—even under pump irrigation—has been smaller than studies predicted, and yields have not achieved planned levels.

Second, labor is costly. High demand for labor makes agricultural wage rates among the highest in West Africa, often between $1.50 and $2.00 per day. Labor may have become more expensive in real terms as the Ivorian economy grew and its neighbors to the north developed. Industrial crops, such as coffee, cocoa, and cotton, give high returns to farm labor, requiring that high prices be paid for paddy to make it competitive (Table 1.12).

Third, technological transfers promoted by the government have been insufficient to overcome these high costs because they are relatively labor-using and land-saving, missing the basic resource constraint. Relatively few funds have been devoted to divisible small-scale, labor-saving technologies, such as power tillers, ox cultivation, and herbicides. As could have been predicted from the experiences under the French, heavy mechanization is too expensive to be privately profitable, except with large subsidies. Yield increases have been achieved at high costs as well, and price subsidies are again necessary to make improved production privately profitable. These costs do not appear to be falling: soil acidification and weed infestation may have reduced yields on rainfed rice; irrigation projects have not been properly maintained, reducing their lifetimes; and extension costs have not fallen as predicted in the 1971–75 Plan (28).

Beneath the umbrella of temporarily high world prices and government borrowing, policies were adopted and sustained that encouraged movements along the supply function. The increased production that thus occurred by using more labor and other scarce factors was achieved at increasing costs.

The long-term rice development strategy attempted to shift outward the supply curve, rather than to move along it. More important, because of perceived labor shortages at current wages, the strategy was meant to increase returns to labor by introducing new labor-saving technology. The land use and techniques of 1965 would have given somewhat lower production in ten years than occurred after government programs had introduced new technology. As shown in Table 1.13, traditional rainfed production uses nearly twice the recurrent labor per metric ton of paddy compared with improved techniques. Viewed from this partial perspec-

TABLE 1.13. *Labor Input into Rice*

Production technique	Recurrent labor days per ha[a]	Yield (*mt paddy/ha*)	Labor days/ mt paddy
Traditional rainfed[b]	85, 115	0.9, 1.3	88, 94
Irrigated, unfertilized	210	2.4	88
Rainfed, fertilized[b]	95, 120	1.5, 2.2	59, 63
Irrigated, fertilized[b]	240, 250	3.5, 4.0	63, 69
Mechanized, rainfed, fertilized	30	2.0	15

SOURCE: These figures are based on information in Chapter 2.
[a] Labor times do not include labor equivalents for clearing or land development.
[b] The first figure refers to savannah production, the second to forest production.

tive of expensive labor, government policy to expand production using the new techniques appears appropriate.

However, when account is also taken of land, the other major factor in rice production, the policy followed seems inappropriate. Although output per unit of labor has increased absolutely, it has declined relative to the output per unit of land. In other words, the technological change in Ivorian rice production has been labor-using. Given that land is still relatively abundant and cheap while labor is and may be becoming increasingly expensive, the superior strategy would have been to adopt technologies that increase labor productivity, not only absolutely but also relative to land. Figure 1.1 illustrates these strategies. Traditional rice production occurs along isoquant I at point a, indicating a land-extensive system. Past government policies have introduced technological changes, shifting production to the new isoquant II.[42] Existing factor prices cause improved production to occur at point c. Labor input per ton of paddy has fallen less than land, and the new technique is labor-using in the Hicksian sense (5, pp. 121–22). The more appropriate policy would be the introduction of technology that uses more land relative to labor. In Figure 1.1, this policy is shown by isoquant II'. At the same factor prices, production would occur at point b, which is labor-saving and land-using.

There are at least two reasons why the Ivorian government did not adopt this strategy. First, the labor-saving technology illustrated by isoquant II' primarily requires mechanization of rice, which is often very expensive under tropical conditions. Profitable, easily adopted forms of mechanized technology were not available. On the other hand, land-saving technology in the form of fertilizer and high-yielding varieties was readily available from abroad. Second, Ivorian planners may feel that the supply of land is the fundamental long-run physical constraint.[43]

In conclusion, production has increased rather impressively. However, this has been due more to high prices and subsidies than to introduction of socially profitable techniques. Technological transfers have occurred, but slowly, and of the wrong kind. These transfers have not really lowered production costs or saved labor (6). A fairly high rate of nominal protection permitted rice production to expand during the 1960's, but at a slow pace. Only the high domestic price supports since 1974, including both a higher rate of nominal protection and producer subsidies, have made adoption of the new technology and expanded output privately profitable.

[42] Production along isoquants II and II' equals production along isoquant I.

[43] Planning documents consistently recognize the need to conserve forest land. One such planning document (29, II, pp. 65–67) states that usable forest reserves will disappear by about 1985 if past trends continue, largely owing to pressure from agriculture. A second document (28, pp. 199, 213) argues that food crop production must be intensified and consolidated under modern crop rotation systems.

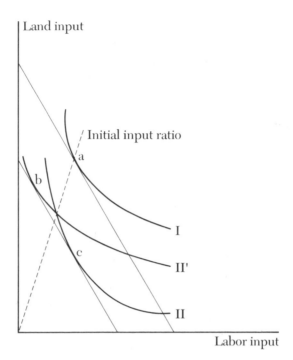

FIGURE 1.1. Factor Bias of Technological Change.

Import Substitution

Except for 1974 and 1975, the Ivory Coast has never consumed less rice than it has produced. The years 1975 and 1976 were the only ones during which there have been no imports owing in part to stock carry-overs from previous years. By 1977 imports were more than 100,000 mt. Why was it that despite large production increases, self-sufficiency was not attained?

On the whole, per capita consumption changed very little between 1967 and 1977. Production increases barely exceeded population growth during the period 1967–76.[44] Hence, it is understandable that imports, on the average, changed very little over this period.[45] Basically, government policies have been oriented to maintain consumption by replacing

[44] For the period 1965 to 1976–77, production of paddy grew at 5 percent per annum, while total population grew at just over 4.0 percent per annum (7, pp. 2, 4).

[45] For the period 1965–70, annual imports of 25–35 percent broken rice averaged 61,000 mt rice. During the next six years, they averaged 59,000 mt, although the annual variability was much greater.

imports with domestic production.[46] When there were shortfalls in production, demand was met by imports rather than curtailed by real price hikes.[47]

Rice import substitution has been seen as a means to make a positive contribution to the balance of payments, thereby helping preserve the liberal exchange policy required to attract foreign capital. In order to increase rice production to reduce imports, domestic supplies of labor, capital, and land must be diverted from other activities. If these other activities are more efficient than rice production at earning or saving foreign exchange, the import substitution policy reduces the capacity of the country to earn foreign exchange. For the Ivory Coast, most of the agricultural export activities—coffee, cocoa, coconuts, cotton, and palm oil—are more efficient earners of foreign exchange than is rice (39). Moreover, it appears that other food crops, such as manioc, corn, and banana plantains, also save foreign exchange more efficiently than can rice production. If balance-of-payments improvement is an objective, increased rice production is a very inefficient means of achieving it (see also 6).

Even the direct savings in foreign exchange—measured as the difference between the value of the rice imports and the foreign exchange costs of imported inputs—have been small. During 1965–76, rice imports averaged 44 CFA francs per kg, c.i.f.,[48] but techniques introduced by the government, such as small-scale irrigation and improved rainfed production, have import costs of over 15 CFA francs per kg of rice.[49] For some of the highly mechanized, irrigated methods of production, the cost of imported inputs exceeds 30 CFA francs per kg rice.

Regional Income Distribution

This distributional objective translates mainly into raising rural incomes in the savannah zone. The major instruments used to encourage paddy production—investment, trade controls, and domestic price support—have increased producer revenue, but only investment favored the savannah relative to the forest zone.

---

[46] Two separate planning documents (27, p. 137–39; 28, pp. 143–45) argued that cereals consumption should be encouraged in order to improve nutrition. Rice production was singled out as the major instrument for reaching this objective.

[47] Despite the large increase in prices in 1974, real rice prices between 1965 and 1977 actually fell. Between 1965 and 1977 consumer prices increased 2.6 times. Nominal retail prices increased from 51 to 134 CFA francs/kg. But by 1978 the government had increased imports sufficiently to lower the retail price of rice to 100 CFA francs/kg, the official price (7, p. 18).

[48] Imports of 25–35 percent brokens only.

[49] These costs include collection, milling by SODERIZ, and delivery to Abidjan wholesalers. See Humphreys (6).

The investment policy, representing a well-developed effort to shift the supply curve, has been clearly focused on the savannah zone. Of the 18 billion CFA francs invested in paddy production projects between 1964 and 1978, 60 percent are localized in the savannah.[50] Government investment per farm family producing rice has been over four times greater in the savannah zone, indicating the government has made—initially at least—strong efforts to realize its income distribution objective.

Trade and price policies—both nominal protection and domestic price supports—have clearly not improved distribution because the forest zone produces about 70 percent of Ivorian rice. Responses to price incentives are probably also higher in the forest, given that rainfall is better, yields are higher, and the marketing infrastructure is better developed than in the north. Higher paddy prices naturally tend to benefit the forest zone relative to the savannah. Of those price subsidies paid out of government funds, southern rice farmers received half again as much as their savannah counterparts, on a per family basis. Roughly 9.5 billion of the 12 billion CFA francs paid out as government price support subsidies during the six buying seasons of 1971–72 to 1976–77 were distributed in the forest zone.[51]

These observations suggest that investment subsidies, rather than domestic price supports and trade control, are a more effective way for the government to use the rice sector as a channel for attaining its regional income distribution objective. Yet rice production using these investments remains relatively unattractive without high, government-supported paddy prices, so that investment policy alone is insufficient. Whether rice production should be used at all to achieve this objective is a different and more important question, considered in the next subsection.

Gain or Loss from Policies

On balance, Ivorian rice policy has conflicted with the country's fundamental objective of stable, long-run economic growth. The pur-

[50] This information has been taken from the listing of projects in SODERIZ (17). The investment breaks down as shown. These figures exclude major investment in milling, which has been more concentrated in the forest zone.

Region	Billion CFA francs	Percent
Forest	6.250	34%
Savannah	10.830	59
General	1.234	7

[51] According to estimates based on the regional distribution of paddy purchases by SODERIZ in 1975, nearly 80 percent originated in the forest zone. This compares to a share in national production of about 70 percent. Many of the beneficiaries were actually migrants from the savannah zone.

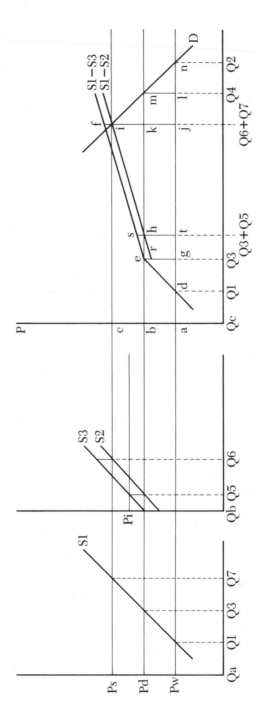

FIGURE 1.2. Effects of Rice Policies: (a) Supply under traditional production and trade protection; (b) Supply from producers receiving subsidized inputs; (c) Aggregate supply.

poses of this section are to trace the transfers of resources into the production of paddy and rice and to examine the efficiency and welfare effects of policies.

Figure 1.2 illustrates—heuristically—the impact of government policies. The supply curve is relatively inelastic at higher production levels, and demand is fairly elastic for staples, both reflecting Ivorian conditions. The world rice price (P_w) is set at its traditionally low level, at which the expected gap between domestic production and consumption would be quite large. As shown in Figure 1.2c, initial conditions, with no government intervention, would have production of Q_1, and imports of Q_2 − Q_1.

The initial condition has never existed because government intervention has established the domestic price of rice at a level higher than the world price, resulting in a nominal protection coefficient greater than unity. At a higher domestic price (P_d), domestic production moves along the elastic part of the supply curve to Q_3, total demand falls to Q_4, and imports are reduced to Q_4 − Q_3. This policy of trade control, which maintains domestic prices above import prices, has four distributional effects: additional domestic resources are pulled into rice production; the government earns revenue from rice imports; producers receive a transfer of welfare from consumers; and the economy loses the consumers' surplus attached to the reduction in rice consumption.

Besides protecting rice, the Ivorian government has also adopted a strategy of subsidizing investments to bring about the technological change necessary to shift the supply curve outward. By subsidizing investment in irrigated land development and the distribution of modern inputs, government policies have effectively shifted the supply curve to S_2 for producers who receive these production subsidies. The true shift, reflecting underlying technological transfer, has been much smaller, so that the real supply function is S_3. Everything beneath this curve represents real resource costs. The vertical difference between the artificial (S_2) and the real (S_3) supply functions represents a transfer per kg of rice from the government to the privileged producers, enabling them to pay the higher costs of production. Under these circumstances, production increases by Q_5 (in Figure 1.2b) implying a real domestic price of P_i (the price that calls forth this output on the new real supply curve S_3 in Figure 1.2b). The per unit government subsidy is thus P_i − P_d.

There are three effects of this input subsidy policy. First, more domestic resources are drawn into production, which now rises to $Q_3 + Q_5$. Traditional production may fall, however, as some traditional farmers switch to techniques using the subsidized inputs. Second, the government budget is affected. Government revenue falls as imports are replaced by the domestic production. More government financing is also needed to pay for the additional, subsidized factors of production. Third, farmers using the subsidized inputs receive a transfer, even though part of the subsidy simply covers additional real costs.[52] Consumer welfare remains unchanged, because consumption and price stay at Q_4 and P_d, respectively.

The 1974–75 domestic price support policy introduced further producer subsidies by maintaining artificially high paddy prices, as illustrated by P_s in Figure 1.2.[53] Paddy production moved along the traditional supply function S_1 to Q_7 (Figure 1.2a) and along the new subsidized supply curve S_2 to Q_6 (Figure 1.2b). Many of the effects mirror previous ones. More domestic resources are bid into rice production. Total costs of government subsidies on inputs increase, since improved production is expanded. Imports fall further, along with government revenue. Consumer welfare remains unchanged, since consumers continue to buy quantity Q_4 of rice at price P_d. In contrast to the input subsidy policy, the output price subsidy benefits a much larger number of farmers, who retain the bulk of the price support as a transfer rather than using it to pay for additional production costs. Assuming an integrated market for paddy, all producers receive the price subsidy $P_s - P_d$ on marketed output.[54] Only in the case of the additional output does the subsidy pay for real costs, that is, the area under the real supply curve.

Figure 1.2c summarizes the overall impact of these policies. Consumers have had a clear loss in welfare (area abmn), although not as much as implied by producer prices because of the government subsidies on rice output.[55] Most of this reduction in consumer welfare has been transferred either to the government as import tariff revenue or to producers in higher prices. Producers have clearly gained, both from the transfer of consumer surplus (area abkj) and from output price supports (representing direct government transfers—area bcik if all paddy were purchased).

---

[52] To the extent that subsidies are in the form of land investment rather than current inputs, farmers are more likely to receive a subsidy sufficient only to cover real costs, the area between S_3 and the domestic price P_d. This is true because land subsidies can be tailored to cover only marginal costs for each producer. In the Ivory Coast, most input subsidies have been channeled through land investment.

[53] The new consumer price was later reduced from 125 to 100 CFA francs/kg, which currently represents a constant real rice price, compared to the cost-of-living index (7, p. 18).

[54] In fact, not all farmers benefit equally from the subsidized paddy prices. Because prices are supported primarily at government mills, more distant farmers receive less. In addition, owing first to inadequate storage and milling capacities and later to insufficient funds, government purchases have been rationed and not all paddy offered has been accepted. Some farmers must therefore accept lower, unsubsidized prices for paddy, which in 1975 averaged about 53 CFA francs/kg in small rural markets (6).

[55] Without the government subsidy per unit of output, $P_s - P_d$, the high producer price (P_s) would have caused a total loss of consumer welfare equal to the area acin.

To the degree that the government was able to tie input subsidies to real costs, these subsidies have probably contributed little to producer welfare, going instead to pay for the higher marginal production costs. The initial government budget gain from the tariff revenue (area geml) dwindled (to area jkml) as domestic production replaced imports. The budget also incurred substantial obligations to subsidize inputs (area efir) and to support producer prices above domestic market prices (area bcik at the maximum). On balance, government expenditures on rice since 1974 have exceeded taxes from the rice sector, requiring other sectors of the economy to finance the interventions in rice.

Deadweight losses to the economy have been large. The consumer surplus attached to the fall in consumption from Q2 to Q4 (triangle lmn) is a net welfare loss, worth perhaps one-half billion CFA francs.[56] This loss may be justified if the government believes that higher rice consumption has negative externalities, but planners' concerns with quality of diet indicate this is not the case. More likely, the government may believe the loss justified because benefits from increases in the income of rice producers and in its own budgetary revenue outweigh the greater losses to rice consumers.

Much more important, there has been a large deadweight loss of productive efficiency (area defj) caused by diverting resources from other uses into domestic rice production. Although rice producers have gained from the transfers, the economy as a whole has lost. It is useful here to measure this loss of efficiency, based on calculations from Humphreys (6). The measurement is based on the concept of net social profitability, defined as the difference between the value added by additional rice production—measured in world prices—and the social opportunity cost (or the value of alternative output forgone) of the domestic factors used in that production. Given the costs prevailing in 1975–76, the average net social profitability was almost a negative 20 CFA francs per kg of rice produced.[57] To calculate the total loss caused by government interventions,

[56] This loss can be roughly estimated by assuming a price elasticity of demand for rice of −0.5 (which is consistent with the fall in consumption from 46 to 28 kg per capita when prices rose from 70 to 125 CFA francs in 1974), a nominal protection coefficient of 1.3 (the average for 1969–73), and a c.i.f. price of 65 CFA francs/kg (a price consistent with the 1975 wholesale price of 87 CFA francs). By increasing the domestic rice price over the import price by 30 percent, or about 20 CFA francs (the vertical distance lm), the trade policy suppressed consumption in the early 1970's by perhaps 50,000 mt per year (the horizonal distance ln). The deadweight loss to consumption (area lmn) was thus about 0.5 billion CFA francs in 1975.

[57] This amount is the weighted average of the net social profitability in all types of rice production except traditional upland that is hand-pounded and consumed on-farm. The weights are the share each contributes to total output. For the range of production changes considered, it is assumed that average production costs equal marginal costs.

it is also necessary to estimate the production increase that has been caused by the policies, which is the distance dj. A reasonable proxy for this increase is the amount produced for sale under traditional techniques plus the production using modern inputs and on land developed through government subsidies. In 1975, this probably amounted to nearly 150,000 mt of rice.[58] Assuming average costs equal marginal costs for each type of production, the estimated annual loss to the economy in productive efficiency (the area beneath this portion of the supply curve) is almost 3.0 billion CFA francs, or one-third of 1 percent of GDP in 1975.[59] It is equivalent to over 25 percent of the additional rice produced, valued at world prices.[60]

We can summarize these welfare and efficiency losses as well as the transfers in terms of the areas of Figure 1.2c. In terms of recipients of transfers from rice consumers, total loss of consumer surplus (abmn) equals gain by producers (abkj) plus government revenue (jkml) plus deadweight consumption welfare loss (lmn). In terms of sources of transfers to rice producers, total gain by producers (acij + efir) equals higher consumer prices (abkj) plus government investment and input subsidies (efir) plus government output subsidy (bcik). In terms of the composition of transfers to rice producers, total gain by producers (acij + efir) equals producer surplus (acihred) plus deadweight efficiency loss (defj). As indicated in previous sections, the budgetary share of these transfers is very important, and most of it comes from sources other than rice consumers. Foreign donors have financed over half the investment and related input subsidy programs. Domestic investment funds come from a variety of sources, but the largest are direct and indirect taxes on domestic trade and production unrelated to rice consumption or production. Government output subsidies were initially funded with revenue from rice imports—that is, by transfers from consumers. But until 1977 the largest share was financed through loans from the BNDA, postponing the ultimate budgetary impact of the policy. These debts have finally been paid by the CSSPPA from profits on coffee and cocoa exports (i.e. taxes on coffee and cocoa producers).

The magnitude of these transfers helps explain why rice in the Ivory

[58] This calculation is based on the 1975 production of 465,000 mt paddy, of which slightly more than 50 percent is estimated to have been produced traditionally, hand-pounded and consumed on-farm. If imported rice were sold at P_w, it could also conceivably replace part of this on-farm consumption if the market price were to fall beneath costs of production.

[59] The aggregate supply curve consists of several discrete steps, each representing a different and increasingly costly production technique. Aggregate net social profitability (NSP) thus equals the product of the NSP and the quantity of rice produced, summed over all the relevant production techniques.

[60] World rice prices are estimated at 75 CFA francs per kg rice, c.i.f. for 25–35 percent brokens. This was the price level used in Humphreys (6).

Coast has become so politicized. Farmers, the former state-owned production company (SODERIZ), paddy assemblers, and millers (especially the large-scale government mills) all developed a strong interest in seeing the trade protection, the investment and input subsidies, and the price supports continue. The subsidized inputs and investments benefited farmers who could acquire them, and the subsidized paddy price further increased their gains. The high paddy price supports attracted so much paddy to government mills that SODERIZ became the tenth largest company in the country by 1975.[61]

Among the losers, consumers stand out as the largest. But the group is diffuse, and incomes have grown sufficiently so there has been little pressure to bring the domestic price in line with the import price.[62] Other sectoral interests, mainly coffee and cocoa, which have been taxed to finance rice production and which have had to compete with rice for production factors and financing, have been more assertive. Their commodity prices have recently risen relative to paddy, and they have now largely appropriated the rice production sector into their own programs after the dissolution of the rice agency. Foreign donors, losers in the sense that they have financed part of the transfers, have reacted by limiting future funding until the price policy is changed to reduce the subsidies. The import cartel, temporarily squeezed out in 1975 and 1976, has managed to regain its historical position as the country returned to a net import position. In all of this, the government budget has been the ultimate residual equilibrating agency. As the financial burden grows, the government has become increasingly reluctant to divert funds from other projects into a sector that has shown few real improvements in efficiency.

Prognosis

Saddled with an expensive legacy of past domestic price supports, the Ivorian government found itself in 1978 faced with the need to revise policies aimed at the rice sector. The budget had become constrained. Past price supports had strong effects. Producers gained enormously from the transfers. The huge price subsidies made it possible for government mills to compete with the private sector and enabled the government to dominate paddy marketing. Control over the bulk of marketed paddy helped the government defend spatial and temporal equality of producer and consumer prices. Finally, large sales to the government were used to jus-

[61] The ranking is based on gross sales in 1975. See the government financial report (22, p. 6).
[62] Real per capita GDP grew by 2.4 percent per year from 1965 to 1977, compared to a constant real price of rice (7, pp. 3, 18).

tify significant investment in government milling and storage capacity, which would probably have been idled without the price subsidies.

The interest groups, which have benefited from past decisions, may make reorientation difficult, probably delaying final decisions. But the budgetary constraint, which is now binding on both recurrent expenditures and investments, greatly impedes efforts to continue past policies designed to expand production. The grace period afforded by loans from the BNDA, by taxes on coffee and cocoa exports, and by foreign aid contracted in the early 1970's is over. A reorientation is inevitable.

In the past, Ivorian rice development was aimed at achieving the secondary objectives of maintaining a positive trade balance and of improving regional income distribution. The strategy was to increase domestic rice production by using a combination of trade protection and investment measures, supported in recent years by producer price subsidies. The basic constraint was the high cost of production, given prevailing factor and commodity prices. It is significant that planners recognized this constraint from the beginning, and much of the Ivorian rice development strategy has been an effort to overcome it.

Although government emphasis on the two objectives on which rice policies have been focused remains unchanged, neither objective has benefited much from these policies despite the increases in domestic rice production. What is worse, the high cost of the increased rice production may have actually depressed overall growth, the fundamental policy objective. There are two lessons here. First, producing rice is an inefficient means of improving the trade balance because the country does not enjoy a comparative advantage in rice production at prevailing world prices. Despite the original strategy to make rice competitive by lowering production costs, the expansion resulted mainly from bringing more factors into production, not from increases in productivity. The resources pulled into additional rice production would have been more efficiently used to earn or save foreign exchange if they had been switched to export diversification or to other import-substituting activities. Second, expanded rice production brought about through high support prices has not been an effective means for improving regional income distribution. Investment and input subsidies could have been more effectively tied to specific geographic locations, thus assuring the income transfer, though high producer prices have been necessary to make investments attractive to farmers. In any event, channeling the transfer through a crop such as rice, which is not competitive with imports, was not efficient.

The failure of the rice sector to contribute to the objectives set for it stems from the inability of government policies to overcome the basic resource constraints. Expensive labor has been and remains a preoccupation of Ivorian farmers as well as planners. Despite large migrations from neighboring countries to the north, the profitability of other agricultural crops and the dynamism of the secondary and tertiary sectors have created a high demand for labor. The technological package—fertilizers, selected seeds, extension services, and irrigation development—was meant to raise labor productivity. The unspoken corollary was that costs would then fall, protection could end, and subsidies would no longer be needed. The strategy failed because costs have not fallen.

The shape of future government policy is difficult to predict, but normative aspects of several alternatives can be examined. Analytically, the most desirable long-run solution would be consistent with the fundamental government objective of economic growth, which would simultaneously contribute to the secondary goals as well. It must also alleviate current pressures on the government budget. A movement toward free trade by reducing trade control and domestic paddy price supports would both contribute to efficient growth of the economy and reduce the burden of rice subsidies on the budget. Imports would increase and consumers would benefit as rice prices fell in line with world prices. Rice producers would tend to lose, although producers of other crops currently taxed to finance rice subsidies would gain.

Over the long run, economic justification for government intervention in rice production rests essentially on government ability to shift the supply function outward. Such intervention will succeed only if new techniques lower per unit production costs, rather than simply shifting costs from labor or land to capital or imported inputs that must be subsidized. The choice of technical innovations depends on the relative prices of land and labor that are expected to prevail during the next few decades.

Because labor is currently expensive relative to land, immediate improvements should focus on reducing the overall share of labor costs. Small-scale mechanization suitable for small farms is one means to facilitate soil preparation, harvesting, and threshing, but tropical conditions tend to raise operating costs and reduce equipment lifetimes. The repair and service infrastructure in the Ivory Coast is still insufficient to maintain small motorized machines in good operating condition. Animal traction may be more suitable to African conditions than mechanized sources of power, but trypanosomiasis impedes its use in the forest zone where the bulk of Ivorian rice is produced. Improved planting and weeding methods and better timing could increase yields without raising labor inputs, but such improvements require an effective extension service, which in the past has been costly and limited mainly to the distribution of improved seeds and fertilizers. Herbicides could save weeding labor but currently appear only marginally cost-effective (6). The range of efficient labor-saving techniques thus appears to be quite restricted at present, giving the government few options to make domestic rice production competitive at current or expected world prices.

Over the longer run land could become a major physical constraint. If the area in perennial plantations and under extensive, shifting cultivation techniques continues to grow at the rate observed between 1965 and 1975, unused arable land may disappear before the year 2000.[63] Costs of land development and use will increase, since land will be cultivated farther from consumption centers and because more intensive cultivation may lead to erosion and weed infestation while lowering fertility. In these circumstances, land-saving technologies such as irrigation investment could become important, if certain conditions prevail. First, rice must be competitive with alternative uses of the scarce land, such as industrial crops. Current studies suggest it is not (6, 9). Second, even if rice production were an efficient use of scarce land, labor-using irrigation makes economic sense only if the value of land increases more rapidly than wages. Given the historical demand of the Ivorian economy for labor, such prospect seems unlikely in the next two to three decades.

Because land scarcity is a future problem, the economic inefficiency and large subsidies of current rice policies can be justified only if a long lead time is necessary for investment and learning in rice production. There is little evidence that such lead times, except in research, are required. Future benefits must compensate for current losses, a condition demanding that rice production become highly competitive, not marginally so. Existing irrigation investments have relied on cheap foreign funds but have been expensive and only partially utilized. Unless capital costs are kept low and irrigation efficiency is improved, savings may be insufficient to make irrigated Ivorian rice competitive with imports even if good rainfed land were scarce.

Although the ideal future reorientation of government rice policy should rely on continued research and limited investment, with imports being used to equate supply and demand at prices close to world levels, other adjustments may be required in the interim. These short-run policies depend more heavily on the price mechanism coupled with trade control to equate the demand and supply of rice. The most obvious immediate change is to raise the consumer price. By the end of 1977, the official retail price of rice had fallen in real terms to only 75 percent of its 1965 level. Relative to the food price index, the real price had fallen even

[63] On the basis of information in Table 1.5, the amount of arable land in a cultivation cycle grew at 2.9 percent per year between 1965 and 1974. Ten million ha were estimated in a cultivation cycle in 1974, based on an estimated 5–7 ha fallow per cultivated ha (28, p. 213). At this growth rate, all arable land could be in a cultivation cycle in only 16 years.

lower.[64] Raising the consumer price to the level of producer price supports would shift the burden of the output subsidy from the government budget to rice consumers. There would be no further decline in economic efficiency although there would be a greater deadweight loss in consumer welfare.[65] Because rice imports would fall without further increases in inefficient rice production, the policy would also save foreign exchange, although these savings may be partially offset by larger wheat imports caused by increased consumption of bread. Finally, investment and input subsidies could still be used to improve regional income distribution until better alternatives are found.

The evaluation of this alternative changes if the world rice price rises. If the world price were as high as the producer support price, there would be no deadweight losses of economic efficiency or consumer welfare from replacing imports with domestic rice production. Government revenue on rice imports would disappear, but the only budgetary obligations would be the input subsidies. Yet even if the world rice price were to be double its 1976–77 levels, it would still be below the level of the current domestic producer price support.[66] Under present technology, import substitution therefore involves a considerable loss to the economy.

The other short-term change is to lower the producer price to the consumer price level, thereby eliminating the very high producer price supports introduced in 1975. Consumption levels and consumer welfare would remain unchanged, and there would be a much smaller loss of production efficiency, since production would decline. In essence, such a reorientation represents a return to the policies of the second historical period, 1960–74, when import substitution was viewed as a long-term goal linked to the growth in the competitiveness of Ivorian rice production. In effect, this alternative is already being implemented. Relative to the producer prices of other export crops, the official producer price of paddy has been greatly eroded. In terms of the consumer price index, the official

[64] In 1965, the retail price of rice was 51 CFA francs per kg, compared to an official price in 1977 of 100 CFA francs. Over the same 12 years the Abidjan African consumer price index rose from 117 to 307. The food price index rose from 122 to 365 (7, pp. 15, 18).

[65] To the extent that there are insufficient government funds to purchase paddy, it is impossible for all farmers to receive the higher support price. An increase in the retail price of rice would shift the burden of the subsidy to consumers and assure that all farmers receive equivalent prices. More farmers would probably receive the higher prices, which would bid additional domestic factors into rice production, causing a greater decline in economic efficiency.

[66] The current official domestic producer price support (P_s) is equivalent to a c.i.f. price of about \$550/mt of rice. The 1976–77 c.i.f. price is estimated at \$250/mt (60–65 CFA francs/kg). The highest c.i.f. price of rice with 25–35 percent brokens since independence was only \$465/mt in 1974 (7, p. 12). The long-run price for rice of this quality is expected to be about \$300 per mt (3).

paddy price has fallen by one-third in real terms.[67] Moreover, the government no longer defends this official price by purchasing all paddy offered for sale.

Unless pressures on the government budget are eased in some other way, either of these alternatives, or some combination, will have to be adopted. In any event, the government rice-milling sector will lose its sheltered position, and the government budget, as well as the coffee and cocoa farmers, will gain. If the world price of rice rises, import substitution becomes not only feasible but desirable, especially if foreign exchange to pay for more expensive rice imports becomes a constraint. As long as the world price is low relative to domestic production costs— which is more likely—it will be difficult to find innovations that can make Ivorian rice competitive. In this circumstance, lower producer prices, increased consumption, and larger imports make the most sense economically. In addition, such a short-run adjustment is consistent with the long-run solution that relies on cheaper imports while a search for more efficient production techniques continues. This strategy recalls that of the 1960's, but with a new focus on labor-saving, rather than land-saving, technical changes.

Citations

1 Chambre de Commerce de Côte d'Ivoire, *Bulletin Mensuel*, Abidjan, monthly.

2 ———, "La Distribution des Marchandises et Produits en Côte d'Ivoire," *Bulletin Mensuel*, 6, June 1973.

3 Walter P. Falcon and Eric A. Monke, "The World Market for Rice in the 1980s," Stanford/WARDA Study of the Political Economy of Rice in West Africa. Food Research Institute, Stanford University, Stanford, July 1979.

4 J. Gigou, "Etude de la Pluviosité en Côte d'Ivoire, Application à la Riziculture Pluviale," *L'Agronomie Tropicale* (Paris), XXVIII, No. 9, Sept. 1973.

5 John R. Hicks, *The Theory of Wages*, 2d ed. Macmillan, London, 1964.

6 Charles P. Humphreys, "Rice Production in the Ivory Coast," Stanford/WARDA Study of the Political Economy of Rice in West Africa. Food Research Institute, Stanford University, Stanford, July 1979; chapter 2.

7 Charles P. Humphreys and Patricia L. Rader, "Background Data on the Ivorian Rice Economy," Stanford/WARDA Study of the Political Economy of Rice in West Africa. Food Research Institute, Stanford University, Stanford, July 1979.

8 Ivory Coast, Government of, Banque Nationale de Développement Agricole, personal communication. Abidjan, Jan. 19, 1978.

[67] Since the official paddy price was last raised in 1974, producer prices for coffee, cocoa, and cotton have increased by 50, 64, and 78 percent, respectively (7, p. 17). Between 1974 and 1977, the Abidjan African consumer price index rose from 193 to 307. In real terms, the 1974 paddy price of 65 CFA francs/kg equaled only 41 CFA francs by the end of 1977 (7, pp. 15, 18). By 1976, increasingly stringent government buying standards for paddy had reduced the actual price received by producers by at least 5 percent (14). Policies toward quality discounts are currently in flux.

9 ———, Banque Nationale de Développement Agricole, personal communication. Gagnoa, Nov. 1977.

10 ———, Banque Nationale de Développement Agricole, *Rapport Annuel 1975–1976*. Abidjan, 1976.

11 ———, Conseil Economique et Social, *Rapport sur l'Evolution Economique et Sociale de la Côte d'Ivoire, 1960–1964*. Abidjan, 1965.

12 ———, *Journal Officiel de la République de Côte d'Ivoire*. Abidjan, Imprimerie Nationale, weekly.

13 ———, Ministère de l'Agriculture, Division d'Etudes Techniques des Projets Agricoles, personal communication. Abidjan, 1978.

14 ———, Ministère de l'Agriculture, *Etude des Bas-Fonds Rizicultivables*, by Bureau pour le Développement de la Production Agricole (BDPA). Paris, 1966.

15 ———, Ministère de l'Agriculture, SODERIZ, *Rapport Annuel, 1976*. Abidjan, 1977.

16 ———, Ministère de l'Agriculture, SODERIZ, Direction de Production, personal communication. Abidjan, 1977.

17 ———, Ministère de l'Agriculture, SODERIZ, *Six Ans Déjà*. Abidjan, 1977.

18 ———, Ministère de l'Agriculture, Direction Générale du Développement Agricole, Direction des Statistiques Rurales, *Recensement Agricole*, 3 vol. Abidjan, Sept. 1976.

19 ———, Ministère de l'Agriculture, Direction des Statistiques Rurales, *Statistiques Agricoles*. Abidjan, 1975 and 1976.

20 ———, Ministère de l'Agriculture et de la Coopération, Bureau d'Etudes et de Réalisations Agricoles, *Plan de Développement de la Riziculture, 1963–1970*. Abidjan, June 1962.

21 ———, Ministère du Commerce Intérieur et de la Distribution, personal communication. Abidjan, Nov. 1977.

22 ———, Ministère de l'Economie et des Finances, Banque des Données Financières, Direction de la Comptabilité Publique et de Trésor, *Centrale de Bilans, 1975*. Abidjan, 1976.

23 ———, Ministère de l'Economie et des Finances, Direction de la Statistique, *Recensement Général de la Population, 1975*. Abidjan, 1975.

24 ———, Ministère des Finances, des Affaires Economiques, et du Plan, Service de la Statistique, *Inventaire Economique et Social de la Côte d'Ivoire, 1958*. Abidjan, 1960.

25 ———, Ministère du Plan, *La Côte d'Ivoire en Chiffres, Edition 1977–1978*. Société Africaine d'Edition, Abidjan, 1977.

26 ———, Ministère du Plan, *Loi Plan de Développement Economique, Social, et Culturel, pour les Années 1967–1968–1969–1970*. Abidjan, 1967.

27 ———, Ministère du Plan, *Perspectives Décannales de Développement Economique, Social, et Culturel, 1960–1970*. Abidjan, 1967.

28 ———, Ministère du Plan, *Plan Quinquennal de Développement Economique, Social, et Culturel, 1971–1975*. Abidjan, 1971.

29 ———, Ministère du Plan, *Projet de Plan Quinquennal de Développement Economique, Social et Culturel, 1976–1980*, 5 vols. Abidjan, May 1976.

30 ———, Ministère du Plan, *3ᵉ Plan Quadriennal de Développement Economique et Social, 1958–1962*. Abidjan, 1958.

31 ———, Ministère du Plan, Département des Etudes du Développement, *Essai d'Actualisation des Perspectives Décannals pour le Riz*. Abidjan, 1968.

32 ————, Ministère du Plan, Département des Etudes de Développement, *Essai d'Analyse des Résultats du Recensement Agricole*. Abidjan, Feb. 1977, preliminary.

33 ————, Ministère du Plan, Département des Etudes de Développement, Sous-Direction de la Plantification Economique, *Travaux Préparatoires au Plan 1971–1975, 2eme Esquisse: Les Objectives de Production Agricole—le Riz*. Abidjan, 1969.

34 ————, Ministère du Plan, Institut de Géographie Tropicale de l'Université d'Abidjan, et Office de la Recherche Scientifique et Technique Outre-Mer, *Atlas de Côte d'Ivoire*. Abidjan, 1971.

35 ————, Ministère de la Recherche Scientifique, Institut de Recherches Agronomiques Tropicales et des Cultures Vivrières, *Fiches des Variétés de Riz*, by B. Leduc. Bouaké, Dec. 1974.

36 ————, Ministère de la Recherche Scientifique, Institut de Recherches Agronomiques Tropicales et des Cultures Vivrières, *Rapport Annuel, 1975*. Bouaké, Oct. 1976.

37 Abdoulaye Sawadogo, *L'Agriculture en Côte d'Ivoire*. Presses Universitaires de France, Paris, 1977.

38 J. Dirck Stryker, "Exports and Growth in the Ivory Coast: Timber, Cocoa, and Coffee," in Scott R. Pearson and John Cownie, eds., *Commodity Exports and African Economic Development*. Lexington Books, D. C. Heath, Lexington, Mass., 1974.

39 J. Dirck Stryker, "Economic Incentives and Costs in Agriculture," West African Regional Project, The World Bank, Washington, D.C., April 1977.

40 West African Rice Development Association (WARDA), Development Department, "Rizières à Perméabilité Elevée dans le Nord de la Côte d'Ivoire." Monrovia, June 1976.

2. Rice Production in the Ivory Coast

Charles P. Humphreys

The Ivory Coast stands as an enviable example of successful agricultural development in Africa. In the decade between 1965 and 1975, production of industrial crops increased by 5 percent per year. Food output grew by 3–4 percent per year during the same period. The growth of food production was nearly double the growth of the rural population, and despite higher incomes and rapid urbanization, domestic food production reduced food imports for urban consumers by nearly one-half (20, p. 44). With an annual increase of more than 5 percent during this decade, the rice sector was one of the leaders in the growth of food output (6, p. 20). By 1975–76, the country had even achieved self-sufficiency in rice, owing in significant measure to increased production.

Government policies have not only promoted this impressive growth but also changed the structure of Ivorian rice production, increasing the share of output supplied by improved seeds, fertilizers, irrigation, and mechanized techniques. These increased tonnages and the adoption of modern methods are often considered testimony to the success of Ivorian rice policies (17, pp. 144–45). An evaluation of Ivorian rice policy, however, depends more on an analysis of the costs of production than on measuring growth rates. Unless rice output can be expanded at costs that are competitive with the imports it replaces and unless the new techniques lower costs of traditional production, policies to bring about these changes cause an inefficient allocation of national resources.

The purpose of this chapter is to examine the efficiency of Ivorian rice production in order to evaluate rice policies of the government. The analysis relies on costs and returns that are fully disaggregated to the microlevel and estimated in both social and private prices. A comparison of private and social profits reveals the impact of government policies on the incentives provided to farmers, merchants, and millers to grow, process, and distribute rice. By using social prices it is also possible to evaluate the relative efficiency of various methods of producing rice domestically. Finally, the gain or loss to the nation from various government programs can be estimated by aggregating social profits of different techniques. These welfare effects and transfers can be used to measure quantitatively the effects of government policies on its objectives. With these results it is possible to draw conclusions that can guide future government programs for the rice sector.

The rest of this paper is divided into seven sections. The next part describes the techniques of rice production, assembly, milling, and marketing. The subsequent section reviews the structure of government incentives affecting rice production. It is followed by a review of the methodology and important empirical assumptions used in the analysis. The empirical results are presented in the succeeding three sections. These results comprise a comparison of private and social costs, an examination of the sensitivity of the results to changes in assumptions, and an evaluation of the effectiveness of rice policies in advancing government objectives. The paper ends with conclusions drawn from the analysis.

Description of Techniques

The techniques described in this section form the basis of the microeconomic analysis of the costs of rice production.[1] In the first part that follows, the production of paddy in the Ivory Coast is divided into several farm techniques, each having different technical coefficients. The subsequent three subsections present information that describes the major methods used to market paddy, mill rice, and distribute rice to consuming centers. The various techniques of these four activities—production, marketing, milling, and distribution—can be combined to describe all the national production of rice in the Ivory Coast. Some of these combinations are discussed in the final subsection.

Production

In 1976, paddy production attained 425,500 metric tons (mt), down about 10 percent from the record harvest of the preceding year. On a national basis, yields averaged 1.2 mt/hectare (ha), although they varied from 0.9 to 4.0 mt/ha, depending on the method of farm production. Table 2.1 gives 1976 output estimates and other descriptive information for 20 farm production techniques. This broad range of techniques reflects the wide diversity of production possibilities in the Ivory Coast. To a large extent, this diversity results from the interplay of four major factors—climate, water control, mechanization, and the use of modern inputs. Table 2.1 shows how these four characteristics vary for the different techniques of farm production.

The Ivory Coast covers two major ecological zones (for details, see 7).

[1] Highly detailed microeconomic budgets have been prepared for each technique and are presented in 3.

The forest zone, extending over the southern half of the country, is favored by high rainfall (1,500–1,600 millimeters [mm] in the major rice areas) distributed over two seasons, the second season having lower and more erratic rainfall. About 65 percent of Ivorian paddy is produced in this region. The other half of the country is covered by savannah, which is characterized by lower and more variable rainfall (1,100–1,500 mm) occurring in only one rainy season. Traditionally, both yields and costs of farm production are lower in the northern, savannah region, as a result of poorer rainfall, lower wages, and easier land clearing.

In the Ivory Coast, there are no indigenous water control systems and paddy production is primarily rainfed, mostly upland.[2] In 1976, upland production covered about 95 percent of the area in rice and contributed about 85 percent of national production. About two-thirds of the upland area is in the forest zone, which provides 70 percent of upland production. Irrigated production has been emphasized since the mid-1960's, and investments depend on the government for initiative and financing. By 1976, over 17,000 ha of rice were harvested in irrigation schemes, adding almost 50,000 mt paddy to the national supply.

Three main types of irrigation have been tried in the Ivory Coast. The first and most prevalent is the lowland irrigation system, which diverts water from small streams onto nearby bottom lands. These developments are small (10–15 ha), are relatively inexpensive to build, rely heavily on farmer participation, and are widely dispersed. The two major disadvantages are that they do not provide complete water control, making double-cropping rare, and they demand regular maintenance, which farmers seldom provide. By 1976, nearly 19,000 ha had been developed under this system of irrigation, some of which has been converted to dam irrigation. About three-fifths is located in the savannah zone (6). Despite some possibilities for double-cropping, only about 90 percent of the available area is actually harvested because of the large number of swamps that have been abandoned.

The second type of irrigation system relies on dams and storage reservoirs. The system is designed for large flood plains, generally exceeding 100 ha. The networks are constructed and managed by government agencies, financed primarily with foreign funds. Farmers contribute to the investment only by participating in the clearing, leveling, and bunding, as they do in the smaller lowland irrigation systems. Although quite expensive to build (1.7 million CFA francs/ha),[3] the systems theoretically assure double-cropping, which is important in the savannah areas, where

---

[2] Although a small amount of flooded rice is grown in the northwest, the quantity is relatively unimportant, and expected yields, given the lack of water control, are similar to upland rice yields.

[3] The local currency is the CFA (Communauté Financière Africaine) franc, which is tied to the French franc. The exchange rate used in this paper is 250 CFA francs per U.S. dollar.

TABLE 2.1. *Key Characteristics of Rice Production Techniques, 1976*

Production technique[a]	Harvested area, 1976 (000 ha)[b]	Paddy production, 1976 (000 mt)	Paddy yield (mt/ha)[c]	Type of water control	Rice crops/ year	Land preparation	Improved seeds	Fertilizer	Pesticides
Forest									
F1 Upland	221.700	248.95	1.12	rainfed	1.00	manual	no	no	no
F2 Upland	4.615	9.23	2.00	rainfed	1.00	manual	yes	yes	no
F3 Lowland	2.787	9.75	3.50	div. irr.[a]	1.30	manual	yes	yes	yes
F4 Lowland	4.809	12.02	2.50	div. irr.	1.30	manual	no	no	no
F5 Lowland	0.603	2.11	3.50	div. irr.	1.30	tiller	yes	yes	yes
F9 Irrigated	1.469	4.04	2.75	d/p irr.[e]	1.85	tractor	yes	yes	yes
Savannah									
F10 Upland	107.300	94.98	0.89	rainfed	1.00	manual	no	no	no
F11 Upland	1.331	2.00	1.50	rainfed	1.00	manual	yes	yes	no
F12 Upland	2.303	4.03	1.75	rainfed	1.00	ox	yes	yes	no
F13 Upland	0.162	0.16	1.00	rainfed	1.00	ox	no	no	no
F14 Upland	4.161	7.28	1.75	rainfed	1.00	tractor	yes	yes	no
F15 Lowland	1.189	4.76	4.00	div. irr.	1.10	manual	yes	yes	yes
F16 Lowland	3.873	9.30	2.40	div. irr.	1.10	manual	no	no	no
F5A Lowland	0.293	1.17	4.00	div. irr.	1.10	tiller	yes	yes	yes
F17 Irrigated	1.328	4.65	3.50	dam irr.	1.85	manual	yes	yes	yes
F18 Irrigated	0.779	1.95	2.50	dam irr.	1.85	manual	no	no	no
F19 Irrigated	0.160	0.44	2.75	pump irr.	1.85	tractor	yes	yes	yes
F10A Flooded	1.824	3.10	1.70	unimp. fl.[f]	1.00	manual	no	no	no
F11A Flooded	2.298	4.60	2.00	imp. fl.[g]	1.00	manual	yes	yes	no
F14A Flooded	0.983	0.98	1.00	imp. fl.	1.00	tractor	yes	yes	no
TOTAL	364.000	425.50							

SOURCES: Total production figures are taken from Ivory Coast, Government of, Ministère de l'Agriculture, Direction des Statistiques Rurales, *Statistiques Agricoles* (Abidjan, 1976). Breakdowns of areas and yields are estimated mainly from information in Ivory Coast, Government of, Ministère de l'Agriculture, Compagnie Ivoirienne pour le Développement des Textiles (CIDT), *Rapport Annuel 1975–1976*; *Rapport Général de Synthèse 1972–1976, Riz-Coton* (Bouaké, May 1976), and Société pour le Développement de la Riziculture (SODERIZ), *Rapport Annuel 1976* (Abidjan, 1976). See also 6.

NOTE: Harvesting is carried out manually in all techniques except F9, F14, and F19, in which combine harvesters are used. Estimates have not been calculated for the techniques whose F designations refer to the numbering system used to identify production techniques in the analytical calculations. Estimates have not been calculated for the techniques whose F designation is followed by the letter A, but the technique indicated by the corresponding number can serve as an approximation.

[a]"F" designations refer to the numbering system used to identify production techniques in the analytical calculations. Estimates have not been calculated for the techniques whose F designation is followed by the letter A, but the technique indicated by the corresponding number can serve as an approximation.

[b]Total cropped area, and therefore includes areas on which second crops are grown.

[c]Yields are estimates for 1976. Slightly different yields, considered more representative, are used in the analysis. See Table 2.2.

[d]"Div. irr." is improved lowland cultivation using diversion weirs. "D/p irr." is irrigation using either dams or large pumps or a combination of the two.

[f]"Unimp. fl." is cultivation relying on natural flooding, mainly in the northwest.

[g]"Imp. fl." is cultivation relying on natural flooding, probably without land improvements, but using modern inputs. One-third may be plowed by oxen.

they are mainly located (6). But because of insufficient water storage and porous soils, they often fail to provide year-round water security (21). By 1976, dam irrigation systems covered about 3,500 physical ha,[4] although data in Table 2.1 indicate that not all was in cultivation.

The third irrigation system uses water pumped from nearby rivers, sometimes in conjunction with dams. Large, diesel-powered pumps (650–700 cubic meters per hour [m³/hr]) are employed to supplement rainfall. To date only three schemes exist, each fairly experimental and small. Together, they amounted to about 1,000 physical ha in 1976. In addition to being expensive to build and operate (the investment is 1.2 million CFA francs/ha), these systems have been plagued by highly variable river flows, porous soils, and extremely inefficient water delivery and use. Though a few farmers have experimented with small portable pumps (15–20 m³/hr), there have been no government efforts to develop a small-scale technology.

The third distinguishing characteristic of rice production techniques is the source of power used for cultivation. Most rice, both rainfed and irrigated, is manually cultivated, using short-handled hoes (*dabas*). Of the several alternatives to manual cultivation, it appears that the most widely used are medium-sized tractors (65 hp), sometimes with combine harvesters. Nearly 7,000 ha are cultivated in this manner, almost exclusively on upland rice in the savannah zone. Tractor cultivation in this region dates from before independence, although it has been greatly expanded in recent years by mechanized land-clearing and tractor-hire services provided at subsidized rates by government agencies. The use of tractors on irrigated lands is limited to a few experimental projects and government seed farms.

The next most widely used alternative to manual cultivation is animal traction. Since the late 1960's in the savannah, the Compagnie Ivoirienne pour le Développement des Textiles (CIDT), the state agency in charge of cotton production, has promoted ox cultivation, which appears to be spreading rapidly for both cotton and cereal production. By 1976, oxen may have been used to cultivate as much as 2,500 ha of rice, both upland and lowland. Constraints to the use of animal traction include trypanosomiasis—which requires that trypano-tolerant taurin oxen be used instead of larger, more powerful Zebu—and the difficulty of destumping fields cleared from heavy forests. These constraints become more severe as the forest cover increases.

The government has recently taken steps to introduce small-scale motorized equipment, primarily power tillers, on irrigated rice fields— mostly in the forest zone. Small, motorized drum-type threshers are also

[4] Physical ha are area measurements only, which do not take into account the number of crops per year.

occasionally used. Efforts to introduce manually operated machines, such as pedal threshers and crank winnowers, have met with little enthusiasm from farmers, largely because they offer no savings of labor time.

The last key characteristic of rice production techniques is the use of modern inputs, such as improved and treated seeds,[5] compound fertilizer, and urea. An insecticide (Furadan) is also used on improved irrigated rice against stem borers. Government programs that deliver these inputs provide an extension service to help assure that recommended practices are followed. The application rates for these inputs are tabulated below:

	Upland	Irrigated
Selected seeds (kg/ha)	60[6]	40
Compound fertilizer (10-18-18) (kg/ha)	150	150
Urea (46-0-0) (kg/ha)	75	75
Insecticide (kg/ha)	—	28
Extension (ha/agent)[7]	100	50

The only differences are between upland and irrigated rice. Recommended fertilizer applications, however, are constant, regardless of climate, water control, or type of mechanization. These modern inputs, provided mainly through government distribution programs, were used on nearly 25,000 ha in 1976, which produced over 50,000 mt paddy.

Of the farm production techniques that make up Ivorian paddy production, only nine will be examined in detail in this paper. These are the techniques that produce the bulk of paddy or that represent major alternatives to traditional production methods. They are summarized below:

F1   Traditional upland, manual, forest

F10 Traditional upland, manual, savannah

[5] The selected seed varieties for rainfed rice include Moroberekan (about 50 percent of the area), Iguapé Cateto (about one-third), and Dourado (about 15 percent). Moroberekan, with a cycle of 145 days, was developed from local varieties of *Oryza sativa* by the Institut de Recherches Agronomiques Tropicales et des Cultures Vivrières (IRAT). It has been in use since 1960. Both Iguapé and Dourado were imported from Brazil, with cycles of 135 and 105 days, respectively. Selected seed varieties for irrigated rice include IR5 (about two-thirds of the area), CS6 (about 20 percent), and Jaya (about 10 percent). IR5 is an IRRI rice with a cycle of about 140 days. CS6, with a cycle of 115 days, was developed at the Centre de Semence at Dabou from IRRI variety IR480-14 and is being phased out. Jaya has a 120-day cycle and was imported from India, where it was developed from Taichung Native 1 (TN-1). Although used in the past, IR8 is no longer recommended because of low yields. Two flooded varieties have also been used—L78 and IM16—with cycles of 150 and 160 days, respectively. The former is being phased out because of low yields, shattering, and lodging (*1, 10, 14*).

[6] The seeding rate for short-season varieties is 80 kg/ha.

[7] These are norms. In reality, an agent is responsible for fewer ha.

F2 Upland, modern inputs, manual, forest
F11 Upland, modern inputs, manual, savannah
F3 Improved lowlands, modern inputs, manual, forest
F17 Dam irrigation, modern inputs, manual, savannah
F12 Upland, modern inputs, oxen, savannah
F14 Upland, modern inputs, tractor, savannah
F5 Improved lowlands, modern inputs, power tiller, forest

Together these techniques account for over 90 percent of paddy production and over 70 percent of the modern inputs used. The important technical and economic coefficients for them are given in Table 2.2.

Traditional production (F1 and F10) is currently the predominant technique in the Ivory Coast. Normal yields in the forest zone are about 1.3 mt/ha reflecting higher and longer rainfall and the fact that rice comes at the beginning of rotations. In the savannah, they are 30 percent less, or 0.9 mt/ha. Upland rice is usually planted only on freshly cleared forest or bush fallow in the forest zone, often in conjunction with the establishment of tree crop plantations. In the savannah, crop rotations average four years, of which rice may be two or more. Threshing times are somewhat higher in the forest as a result of the higher yields. In both regions, cultivation is usually the responsibility solely of women. Because of its land-extensive nature, continued expansion of traditional upland rice depends critically on the availability of forest or fallow land.

Improved upland rice production, while still wholly manual, benefits from modern inputs. The use of fertilizers and improved seeds results in increases in yields of 0.9 mt/ha in the forest and 0.6 mt/ha in the savannah. Cultivation techniques are similar to those used in traditional rice, except that land clearing, seeding, and weeding are more thorough because of improved production systems. Despite more intensive cultivation and higher yields, total labor times are only slightly higher than for traditional production, a result made possible by the use of sickles for harvesting. Probably owing to the use of commercial inputs, men are responsible for the bulk of the labor. Because this type of production is more prevalent and profitable in the forest zone than in the savannah, consideration is focused mainly on its use in the former (F2).

Two techniques are used to illustrate the two primary methods of water control—lowland swamps and dams. Production in both techniques is assumed to be entirely manual and to use modern inputs. Yields, averaging 3.5–4.0 mt/ha, are higher for dams because of better water control. The labor input—about three-fourths male—is 100–150 percent higher per ha than on upland rice. These high labor times are the consequence of yearly land preparation, transplanting, careful weeding, irrigation control, and longer harvesting times needed for the higher yields. Although swamps

TABLE 2.2. *Technical and Economic Coefficients for Farm Production*
(Data given per ha)

Production technique[a]	Yields[b] (mt paddy)	Farm labor[c] (days)	Land development cost[d] (000 CFA francs)	Intermediate inputs[e] (000 CFA francs)	Extension service (000 CFA francs)	Mechanization cost (000 CFA francs)
F1	1.3	120	13.5	3.9	0	0
F2	2.2	121	18	24.7	7.8	0
F3	3.5	240	365	32.7	15.7	0
F4	2.4	209	365	3.3	0	0
F5	4.0	202	420	32.3	15.7	23.6
F9	2.75	34	1,243	46.9	15.7	72.0[f]
F10	0.89	85	7	4.1	0	0
F11	1.5	97	7	25.2	7.8	0
F12	1.8	90	12.7	25.2	7.8	7.7[g]
F13	2.0	30	130	25.2	7.8	43.1[f]
F15	3.5	237	365	33.3	15.7	0
F16	2.4	206	365	3.3	0	0
F17	4.0	247	1,703	33.3	15.7	0
F18	2.7	211	1,703	3.3	0	0

SOURCE: Data for these techniques appear in 3.
   NOTE: All costs are in 1975 CFA francs and represent market delivery costs. Therefore, these costs include indirect taxes and subsidies incurred in the delivery of these inputs but exclude farm level subsidies or charges.
   [a] For explanation of F numbers, see Table 2.1.
   [b] Yields used to convert per hectare subsidies to paddy equivalents are different from those of 1976 shown in Table 2.1, but are considered more representative because 1976 was a drought year.
   [c] Male and female labor days are considered equal in physical terms, but each is valued at a different wage rate.
   [d] Equal to the initial investment.
   [e] Intermediate inputs include seeds, fertilizers, insecticides, and herbicides. Working capital is not included.
   [f] Includes the cost of mechanized harvesting.
   [g] Represents the capital charges, repairs, and maintenance on oxen and equipment.

have been developed in both the forest and the savannah, recent government investments have favored the forest zone, where this type of irrigation (F3) is most effective. The use of dams (F17) is analyzed only for the savannah, where most have been constructed. Production in irrigation systems also occurs without fertilizers and other modern inputs, but these techniques are not viable in the long run.

Three techniques are used to illustrate the three primary mechanized alternatives to manual cultivation—power tillers, oxen, and tractors. All use modern inputs, but the type of water control and the location vary. Power tillers (F5) are studied only in conjunction with lowland irrigation in the forest zone, where they are mainly used. Yields are estimated at 4.0 mt/ha, higher than under manual cultivation largely because better farmers are assumed to use power tillers. Compared to similar manual cultivation, the net labor savings amount to 38 days, all for land preparation. Oxen and tractors are analyzed only for upland rice in the savannah, since their use in the forest zone is negligible. Experience with oxen (F12) indicates that yields are 0.3 mt/ha higher than under manual cultivation, and labor savings—both the land preparation and transport—are estimated at only seven days. Tractors (F14), along with combine harvesters, save two-thirds of the labor (67 days) required for manual cultivation. Every technique except weed control is mechanized, and deeper plowing facilitates that task. In addition, experience shows that yields are even higher than with oxen.

Paddy Assembly

The post-harvest activities include the assembly of paddy, milling, and distribution to consumers, but only about 40 percent of paddy is marketed. Most production is hand-pounded or hulled locally and consumed by the producers. For this share of output, no assembly is necessary.

For the production that is sold, two marketing systems coexist—a government network and a private channel. The state rice agency established some decentralized buying stations to which farmers themselves can deliver their paddy, and these are now maintained by the newly created Office de la Commercialisation des Produits Agricoles (OCPA). The government then assumes responsibility for delivering the paddy to mills owned and operated by the state.

The private marketing system, however, furnishes the bulk of paddy delivered to government mills and all the marketed paddy that is processed in small-scale hullers. Hence, collection techniques in this analysis are based solely on the private channel, which operates through buying agents and a large private transport network. By most accounts, it is highly competitive and efficient (see 20, p. 44). In contrast to the government system that accepts paddy only at buying stations, private merchants purchase directly on the farms and assume responsibility for all aspects of paddy marketing.

The most important features differentiating the private collection techniques are distances between the farm and the merchant's entrepôt and between entrepôt and mill. For paddy delivered to government mills the distance between farms and entrepôts is estimated at 15 kilometers (km), and shipment from the warehouses to mills averages about 100 km. Despite the shorter distance between farm and warehouse, transport costs appear to be 50 percent greater for this stage of collection. For paddy delivered to small-scale hullers, only the initial bulking distance is included because these small mills are widely dispersed. In both marketing techniques, losses are assumed to be 3 percent of the paddy, and average storage periods are estimated at one month. Additional details are given in 3.

Milling

Paddy is converted into rice by three different processes in the Ivory Coast—hand pounding, motorized steel cylinder hullers, and integrated industrial mills. Table 2.3 gives basic information about each milling tech-

TABLE 2.3. *Key Characteristics of Rice-Milling Techniques*

Milling technique	Projected full capacity (mt paddy/year)	Quality of output	Milling ratio	Rice milled (mt milled rice)	Unit cost (CFA francs/mt milled rice)[a]	By-products
M1 Hand pounding, forest	158,000[b]	sometimes par-boiled; sometimes red; often stones	0.69	109,000	32,910	chicken feed
M2 Hand pounding, savannah	63,000[b]	sometimes par-boiled; sometimes red; often stones	0.65	41,000[c]	33,400	chicken feed
M3 Small-scale steel hullers	650,000[d]	sometimes par-boiled; sometimes red; often fresh[e]	0.63	47,000[f]	5,000	usually none; sometimes for cooking fuel
M4 Government in-dustrial mills	157,000[g]	25–35% brokens; white only; not always fresh	0.66	78,803[h]	13,961	brokens sold to breweries; flour sometimes sold for small animal feed; husks and bran are burned

SOURCES: This information is based on data in Chapter 1, Ivory Coast, Government of, Ministère de l'Agriculture, Direction Générale du Développement Agricole, Direction des Statistiques Rurales, *Recensement Agricole* (Abidjan, Sept. 1976), Société d'Assistance Technique pour la Modernisation de l'Agriculture en Côte d'Ivoire (SATMACI), Opération-Riz, "Budget Prévisionnel Rizeries" (Gagnoa, n.d.), and Société pour le Développement de la Riziculture (SODERIZ), "Compte d'Exploitation Prévisionnel 75-76-77" (Abidjan, n.d.), "Détail des Charges d'Exploitation Génerale par Rizerie et par Kilo de Paddy Usine sur la Période du 26-09-76 au 25-03-77" (Gagnoa, Aug. 4, 1977), "Établissement d'un Barème d'Usinage" (Abidjan Jan. 10, 1977), *Projet de Développement de la Riziculture des Bas-fonds en Zone Forestière—Addendum à l'Étude de Factibilité SODERIZ* (Abidjan, n.d.), and personal communication (Direction de l'Industrialisation) (Abidjan and Gagnoa, 1977). See also 3 and 6.

[a] Because milling operations do not have identical cost categories, these unit costs are not strictly comparable. For example, working capital charges for paddy and rice stored between harvest and the time of consumption (an average of six months) are included in hand pounding because there are no assembly and distribution activities associated with this milling technique. For small-scale hullers, which are assumed to do custom milling only, no working capital on paddy and rice is charged, because merchants bear these costs. For the industrial mills, an average period of only three months is used for the calculation.

[b] Projected full capacity is assumed to equal rice milled in 1976. Real capacity is probably sufficient to mill most of the national production.

[c] Calculated as the residual of total production in 1975 (465,000 mt paddy) minus seeds (about 24,000 mt paddy), the quantity assumed to have been milled by small-scale hullers (75,000 mt), and the amount of paddy purchased by government mills (148,000 mt). This residual is allocated to the forest and savannah according to the yields in Table 2.2 and land area in Table 2.1.

[d] Based on estimates of 1,500 operating units with a capacity of 0.2 mt of rice per hour, ten-hour days, 135 days of operation per year, and a milling outturn of 63 percent. This capacity is never fully utilized.

[e] Rice hulled by small-scale mills is usually marketed soon after the harvest. Compared to government rice, which may have had longer storage, this rice smells and tastes fresher.

[f] Based on estimates of 50,000 mt of paddy marketed and 25,000 mt of paddy custom-milled for producers.

[g] Based on 11 months of operation at 500 hours/month, and 75 percent utilization of installed theoretical capacity.

[h] Milled during SODERIZ's fiscal year—September 25, 1975, to September 25, 1976, and includes 565 mt paddy milled at San Pedro. Less then 85 percent of the paddy purchased that year was milled.

nique, including potential capacity, utilization rates, milling outturns, and costs.

Rice consumed by producers and sold in village markets is usually hand-pounded by women with wooden mortars and pestles. Probably over half the rice produced is milled in this manner. The method has a high labor input and is competitive with mechanical milling only because there are no assembly and distribution costs and lower losses.[8] Such rice may sometimes be parboiled, although the practice is not common.

Hand pounding is being replaced by small-scale milling, especially near towns. The mills—manufactured in the Ivory Coast since 1968—employ steel cylinders and screens to separate the hulled rice from husks and bran. The mills are powered by either electric or diesel motors (11 hp), but the latter appear to be preferred. These mills are usually operated on a custom basis, with charges calculated per unit of output. Thus, there are no charges for bags, losses, or working capital. A few mills are, however, vertically integrated with assembly and distribution.

Mills are widely dispersed in rice-growing regions, reducing the distance over which paddy must be transported. As shown in Table 2.3, the private milling sector has an enormous capacity, in part because coffee hullers can be easily adapted to mill paddy. But because of high, subsidized prices paid at government mills since the price increases in 1974, as of 1976–77 these private mills have been relegated to milling for local own-consumption and to supplying local markets when government mills are unable to purchase paddy. As a result, capacity utilization may be less than 15 percent. Milling outturns are relatively low because of mixed varieties and the inadequacy of the adjustment mechanism. But the quality of the rice often exceeds that produced in government mills because the processing and marketing network associated with the private milling is shorter and faster, which enables fresher rice to reach consumers.

The large industrial mills are owned and operated by the government, although the state system still relies primarily on the private marketing and transport sector to deliver paddy and distribute milled rice. Presently, there are ten installations, the largest having an average annual throughput of 15,000–20,000 mt paddy.[9] The mills—usually manufactured by Olmia—consist of dryers, cleaners, hullers, whitening cones, and sorters, all arranged in integrated units and powered by electric motors. Milled rice is packaged in 60 kilograms (kg) plastic fiber bags which are not reused. Losses are estimated at 3 percent, and average storage of

[8] Labor accounts for over half the cost. The other major cost is the capital charge on paddy stored until it is consumed, which is assumed to be for six months and at high, traditional interest rates.

[9] The theoretical rated capacity is 4 mt/hr. In addition, there is a small rubber roller mill at San Pedro, with an operational capacity of 0.5 mt paddy per hour.

paddy and rice may be as long as three months. Only fine and medium brokens are sold as by-products, mostly to breweries.

Construction of these mills was begun in the mid-1960's, but they have been utilized to a significant degree only after high government subsidies enabled them to compete successfully with the private milling sector. These mills acquired an estimated three-fourths of the paddy marketed in 1976. With the improvements installed in the 1970's, milling outturns steadily rose from the low levels of the 1960's and now exceed the milling ratio for steel cylinder hullers. As shown in Table 2.3, however, unit costs remain higher.

Rice Distribution

The major difference among the techniques to distribute rice is the distance between the mill and the consuming center. No distribution is necessary for home consumption. Transportation charges are assumed to be negligible for consumption in villages within the producing areas, and the distribution consists only of handling and storage. For rice consumed in the major urban centers—Abidjan and Bouaké—the distances depend on whether rice is produced in the forest or the savannah zone, as indicated below. (Distances in the forest zone are calculated from Daloa and in the savannah zone from Korhogo.)

Producing region	Consuming center	Estimated average distance (km)
Forest zone	Abidjan	400
Savannah zone	Abidjan	650
Forest zone	Bouaké	250
Savannah zone	Bouaké	300

To both Abidjan and Bouaké, the distance from the forest zone is less than from the savannah zone. Bouaké, however, is closer to the producing regions than is Abidjan.

The distribution of rice from mills to wholesalers relies both on the government and on private merchants and transporters. The government does not own distribution facilities, but arranges for rice to be shipped from state mills by private carriers and delivered to private wholesalers. In this way, the government rations shipments from state mills to assure an adequate supply of rice to Abidjan, reimburses private carriers who actually transport the rice for it, and assumes responsibility for storing rice in Abidjan, usually in private warehouses that it rents. The government also bears the costs of storage, working capital, and losses; the latter are estimated at 2 percent.

Because the government subsidizes the shipment from its own mills, distribution by the private sector is largely limited to the rice—from both government mills and small hullers—that is consumed locally. The private sector techniques differ little from those supported by the government, although losses are assumed to be only 1 percent.

Combination of Techniques

These four activities—production, assembly, milling, and distribution—are integrated in different ways to produce Ivorian rice. Traditional paddy production milled by the government and shipped to Abidjan is the most important combination, accounting for over a third of the rice marketed. Traditional production consumed locally accounts for a larger share of marketings than that shipped to Abidjan. For these local markets, small-scale millers and government mills are equally important. Improved production techniques are estimated to furnish less than one-fourth of paddy marketed, and most is probably milled by the government. Except for improved upland manual production and production from irrigated lowlands in the forest zone, however, no improved technique coupled with government milling provides more than 1 percent of rice marketed.[10]

Government Policies

Although the private returns from rice production are strongly dependent on the underlying technical coefficients summarized in the last section, economic policies of the government intervene to increase or decrease returns from the various techniques. The purpose of this section is to describe the Ivorian policies that are used to promote, or discourage, national rice production.

Government incentives can be classified into four policies—trade control, domestic price support, subsidies (or taxes) on recurrent inputs, and investments financed from public funds. Trade control, by keeping domestic rice prices above import prices, increases returns to farmers and encourages production. Between 1960 and 1973, the nominal protection coefficient (NPC)—which is a measure of the incentive to domestic producers relative to imports—averaged 1.3, although at current domestic prices and expected long-run import prices, the NPC is only 1.13.[11] Trade policy has been maintained by restricting rice imports. Although official import duties on rice are usually suspended, import quotas have been set and allocated since 1955 by the Caisse Général de Péréquation

---

[10] For a more extensive discussion of these combinations and their relative importance in the rice economy, see 3, Tables L1, L2, L3, and L4.

[11] The nominal protection coefficient is defined as the ratio of the c.i.f. import price, plus the implicit tariff on rice imports, to the c.i.f. import price. In the calculation, the long-run import price is 75 CFA francs per kg rice. Data come from 7, Table 1.10, and the concept is explained in greater detail in 15.

des Prix des Produits et Marchandises de Grande Consommation (Caisse de Péréquation) and a cartel of the major private import-export firms. The premium generated by the import quota is now largely received by the government (see 7).

Since 1975, domestic price support for rice output has been much more important than trade control. By heavily subsidizing both milling and distribution costs, the government has been able to raise producer prices for the paddy it purchases without equivalent increases in consumer prices. Producer prices are officially established for the farm level, but they are usually supported at the government mills and sometimes at government buying stations. In 1974, the following official buying prices were established (in CFA francs per kg paddy): farm level, 65; government buying station, 70; and government mill, 75. The price actually received by the farmers, however, depends on their location. For those who are isolated, actual prices received will be less than the official farm level price. Data presented in 3 suggest that at distances of only 35 km between the farm and government buying station (or merchant's entrepôt), the farm price for paddy may be as low as 60 CFA francs/kg. Actual farm prices also depend on the ability of the government to purchase all the paddy delivered to state mills. Owing initially to inadequate storage and milling capacity and subsequently to the failure of the government to provide funds, state mills have been unable to purchase all paddy offered for sale, allowing a parallel market in paddy to operate.[12] Because the private channel for assembly, milling, and marketing does not benefit from government subsidies on milling and distribution, unofficial paddy prices are determined by the consumer price. Since 1975, this price has been effectively maintained by trade policy at the official level of 100 CFA francs per kg rice.[13] Hence, paddy prices in the parallel private market are no more than 50–55 CFA francs per kg, delivered to small rural markets in 1976–77 (9).

By subsidizing state milling and the distribution of rice through government channels, the government has artificially compressed the usual margin between the purchase price of paddy and the selling price of rice, creating a deficit as shown below (in CFA francs per kg rice):

Official cost of paddy	114
Estimated cost of rice, delivered Abidjan	136–38
Official selling price of rice to wholesalers	87
Approximate deficit	49–51

The estimated cost of paddy is based on a purchase price of paddy at state

[12] According to the Société pour le Développement de la Riziculture (SODERIZ), government mills may have refused to purchase as much as 80,000 mt paddy in 1976 (10). Government mills did purchase nearly 150,000 mt paddy in 1975–76 (6, p. 10).

[13] There was a brief period during April–October 1977 when the market price exceeded the official price (11).

mills of 75 CFA francs and a milling outturn of 0.66. The estimated cost of rice is based on state milling costs of 14.5 CFA francs/kg rice and shipping costs of 7.4 and 9.4 from the forest and savannah zones, respectively [see 3].) In order to cover this deficit, the government treasury is obligated to pay to mills operated by state agencies a subsidy of 52 CFA francs per kg rice produced. In addition, it reimburses transport and other charges on rice shipped from these mills to Abidjan and Bouaké. This price policy was originally financed from taxes on rice imports, but more recently funds have come from taxes on exports of coffee and cocoa through the Caisse de Stabilisation et de Soutien des Prix des Produits Agricoles (CSSPPA). By 1976, most of the subsidies had not, in fact, been paid, and the state mills—already deeply in debt to the Banque Nationale du Développement Agricole (BNDA)—were unable to continue to finance their deficit buying operations.

The third type of government policy consists of subsidies on recurrent farm inputs. Since the early 1970's, the government has greatly increased its supply of modern inputs and services to farmers. Inputs like selected seeds, fertilizers, and insecticides—along with extension services and the maintenance of water delivery systems—are delivered as a package. Mechanization, except for animal traction, is delivered as a service. For oxen and implements, the government subsidizes the investment. In return, the farmer is assessed a fixed fee for the input package or service, which varies according to whether production is upland or irrigated. These fees, denominated in both paddy and CFA francs (at 65 per kg), are shown below:

	Kg paddy per ha	CFA francs per ha
Modern inputs for upland rice	350	22,750
Modern inputs for irrigated rice	650	42,250
Power tiller–irrigated rice	262	17,000
Tractor and combine harvester:		
Upland rice	550	35,750
Irrigated rice	785	51,000
Pump irrigation	240	15,600

These fees cover only a portion of total costs, and every type of improved paddy production is subsidized to some extent through government programs to aid farmers.[14]

[14] Beginning in 1977, the government adopted a program to distribute fertilizers for irrigated rice to farmers free of charge. At prices used in the analysis, this new program raises the subsidy rate to almost 60 percent of the value of the modern inputs. In addition to these farm-level subsidies, there is also a government subsidy on compound fertilizer paid at the factory. For 10-18-18, the fertilizer used on rice, this subsidy amounts to roughly 20 percent of production costs. Average subsidy levels on all fertilizers are about 25 percent higher than on rice fertilizer. For bananas, pineapples, sugar, and cotton, factory-level subsidies on fertilizer are above the average.

TABLE 2.4. *Farm-Level Subsidies on Inputs, 1975–76*
(CFA francs, unless otherwise indicated)

Production technique[a]	Total market cost of input package (per ha)[b]	Net farm level subsidy			Subsidy rate (percent of total market cost)
		(per ha)	(per mt paddy)	(per mt rice)[f]	
Modern inputs (F2)[c,d]	37,129	14,379	6,536	9,903	39%
Modern inputs (F11, F12, F14)[c,d]	37,695	14,945	9,963	15,095	40
Modern inputs (F3, F5, F9)[c,e]	63,034	20,784	5,938	8,997	33
Modern inputs (F15, F17, F19)[c,e]	63,119	20,870	5,938	8,997	33
Power tillers (F5)[c]	24,170	6,580	1,645	2,492	28
Animal traction (F12)	9,165	1,885	1,047	1,586	21
Tractors and harvesters (F14)[c]	43,600	7,950	3,975	6,023	18
Tractors and harvesters (F19)[c]	72,929	21,929	7,974	12,082	30

SOURCE: 3.
[a] For explanation of F numbers see Table 2.1.
[b] Includes indirect taxes and subsidies incurred on the manufacture and distribution of these inputs.
[c] Includes a charge for working capital over nine months, plus a 5-10 percent increase to account for exoneration from repayment in case of crop failures except where pump or dam irrigation provides full security. Mechanical services and pump irrigation charges carry an average working capital charge over three months, except for power tillers, where the charge is over six months.
[d] Includes seeds, fertilizer, and extension.
[e] Includes seeds, fertilizer, insecticides, extension, and maintenance of the irrigation canals.
[f] Conversion from paddy to rice based on a 0.66 milling outturn.

In Table 2.4, these input subsidies are presented in terms of various packages delivered by the government. Subsidies are calculated both per ha and per mt of output, and the subsidy rate is also given. Several conclusions can be drawn from this table. First, there is no significant difference in subsidy rates for modern inputs between the forest and savannah, but the level of subsidies per mt of output for upland rice is higher in the savannah. Second, both the level and rate of subsidies on modern inputs for upland rice are higher than for irrigated rice, reflecting the similarity of input packages for the two types of water control but lower yields on upland rice. With respect to tractor services, irrigated rice receives both higher levels and higher rates of subsidies than does upland rice. Finally, absolute subsidies per mt of paddy are much higher for large-scale than for small-scale mechanization.

The fourth type of government policy to provide incentives for paddy production is publicly financed land development, usually made available to rice farmers free of charge. Through this investment policy the government has promoted irrigated production to a level where it accounts for about 10 percent of national production. This program has been extremely costly, and subsidy rates on investment costs have been much higher than those on recurrent inputs and services, as shown in Table 2.5. In all cases, subsidy rates exceed 50 percent. The most heavily subsidized

TABLE 2.5. *Farm-Level Subsidies on Land Development*
(CFA francs, unless otherwise indicated)

Land investment	Initial cost (per ha)	Total market cost per crop[a] (per ha)	Subsidy per crop (per ha)	Subsidy per crop (per mt paddy)[b]	Subsidy rate (percent of total market cost)
Lowland irrigation, forest	365,000	35,800	22,645	6,470	63%
Lowland irrigation, forest, suitable for power tillers	420,000	43,839	22,645	5,661	52
Lowland irrigation, savannah	365,000	42,309	26,762	7,646	63
Pump installation and irrigation network	1,243,000	79,627	79,627	28,955	100
Dams and irrigation network	1,703,000	85,633	77,313	19,328	90
Winch clearing, upland, savannah	12,744	3,785	2,804	1,558	74
Mechanized clearing, upland, savannah	130,000	28,398	23,398	11,699	82

SOURCE: 3.
[a] Includes indirect taxes and subsidies incurred on the inputs used in these land developments.
[b] Subsidies converted to mt by using the yields in Table 2.2.

land developments are the largest and most capital-intensive production systems, both irrigated and upland. Absolute subsidies per mt paddy as well as subsidy rates are highest for systems with full water control.

In addition to these four instruments directed specifically at rice, several other policies affect private incentives to produce rice. These policies consist primarily of tariffs on imported inputs, domestic value-added taxes, and special credit programs. Agricultural equipment and modern inputs receive favored taxation treatment, with very low or zero tariffs and the lowest tax rate on value added. On the other hand, indirect inputs, such as fuel and transportation, are relatively heavily taxed. These indirect taxes reduce direct government subsidies at the farm level by 10–15 percent for improved production.

Low interest credit is provided through the BNDA for farmer working capital and investments and for the purchase of paddy and the distribution of rice by government agencies. Access is largely restricted to approximately 15–20 percent of Ivorian farmers who also receive modern inputs from state agricultural agencies.[15] The majority of farmers finance their investments at traditional interest rates, which are considerably higher than those offered by the BNDA.

The impact of the trade policy and direct subsidies on inputs and investments, as well as the effect of indirect taxes on inputs, is measured by the effective protection coefficient (EPC).[16] For paddy produced by improved techniques, milled by the government, and shipped to Abidjan for consumption, the EPC ranges from 1.17 to 1.29 (3, Table M4). Since the NPC is equal to only 1.13 under the same assumptions, these results show the importance of direct government subsidies on recurrent inputs and on investments to the rice sector.[17] These direct subsidies offset other indirect taxes and augment the incentive provided by trade protection. Moreover, domestic price support policies provide even stronger additional incentives.

Methodology and Major Assumptions

The methodology is designed to measure the economic efficiency and comparative advantage of rice production. Production expenses are divided into categories—the taxes and subsidies stemming from govern-

---

[15] In 1975–76, the BNDA extended hungry-season loans to nearly 90,000 farmers, with an average value per loan of about 15,000 CFA francs. State agencies also used BNDA funds to provide financing for about 90,000 farmers, each receiving on average about 40,000 CFA francs. There is considerable overlap between these two groups of farmers. See 8.

[16] The EPC is the ratio of value added in market prices to value added in world prices. See Page and Stryker (15).

[17] The subsidies included in the calculation of the EPC include those on the tradables used in paddy production and rice distribution. The net subsidy paid to government mills is considered to be part of the domestic price support policy.

ment policies, and the real resource costs—which allow results to be calculated in both social and private prices (15).[18] This measure of social efficiency, labeled net social profitability (NSP), equals the difference between the import value of a kg of rice and the cost of producing it domestically. Resource costs are divided into imported inputs valued at c.i.f. prices and primary factors—labor, capital, and land—valued at social opportunity costs. Because the NSP equals the difference between value added in world prices and opportunity costs to the economy of obtaining this value added, a negative NSP implies a lack of comparative advantage, or a loss of economic efficiency. If efficient economic growth is an objective, activities with a larger NSP should be preferred. Since the magnitude of the NSP depends on the unit used, comparisons with activities other than rice production require a measure that is free of units. One such measure is the resource cost ratio (RCR), which equals the social opportunity cost of domestic primary factors divided by value added in world prices.

The impact of government policies can be assessed by estimating net private profitability (NPP) and comparing it with the NSP. In this methodology, private profitability is defined as the difference between the market value of output and the private cost of all inputs—including capital, where private prices include taxes and subsidies.

This method of analysis requires the critical assumption of fixed input-output coefficients, implying constant costs for each alternative. As a result, average and marginal costs are equal for each technique. Results are interpreted in a partial equilibrium framework. The measures of efficiency refer to marginal changes, and results may differ if changes in output are sufficiently large to alter the aggregate demand for domestic factors. In carrying out the calculations, technical coefficients have been used that are in accordance with normal, long-run conditions (see 3).

In addition to technical production coefficients, prices are needed to estimate social and private profitabilities. The remainder of this section discusses shadow prices for labor, capital, and land, the world price of rice, and private rice prices.

Shadow Prices of Domestic Factors

Estimates of shadow prices—or social opportunity costs—depend on two conditions (19). First, if there is no alternative use for the factor brought into rice production—that is, if it is not fully employed—its shadow price can be assumed equal to zero. This condition applies for

---

[18] Social prices exclude all taxes and subsidies and use shadow prices to value domestic factors. Private prices include the effect of all taxes, subsidies, and other imperfections or distortions in the market. A third category, market prices, includes the effects of indirect taxes or subsidies, but excludes the taxes or subsidies paid directly to the farmer, merchant, or miller.

land in the Ivory Coast. Second, if the factor markets function well and distortions caused by government intervention are minor, market prices offer good approximations to shadow prices. Shadow wage and interest rates in the Ivory Coast rely partly on this assumption.

Shadow wage rates for unskilled agricultural labor are based largely on market wage information for irrigated rice production in the forest zone (19). They are assumed to differ by ecological zone and by sex, as shown below (in CFA francs per day) (3):

	Forest zone	Savannah zone
Men	450	350
Women	350	275

In addition to actual cash wage payments, they include adjustments for meals, search costs, and supervision. Although wages vary by sex and region, there is virtually no evidence of seasonality.[19] These wages are consistent with returns to traditional food crop production for on-farm consumption, which establish a minimum value for wage rates in a land-abundant, agricultural economy. Wages for skilled labor vary according to skill level, and specific rates have not been estimated. However, it is assumed that for skilled labor, shadow wages also equal market wages.

In order to estimate comparative advantage for the future, it is important to evaluate whether these wages are likely to remain constant in years to come. The past growth of rural agricultural wages is difficult to evaluate, owing both to the lack of a reliable time series for wages and to an appropriate income deflator. However, daily agricultural wages appear to have increased faster than the consumer price index (13, 12, 11, 5, 6):

Year	African consumer price index, Abidjan	Daily agricultural wage CFA francs	Daily agricultural wage Index	Growth of real GNP per capita
1960	100	100	100	100
1963	110	125–135	130	122
1976–77	236	300–350	325	172

(The value of the agricultural wage in CFA francs is only the cash portion of the wage. In the last row, the consumer price index in Abidjan and the growth of GNP are for 1976. Also in the last row, the values for the agricultural wage refer mainly to 1977 and apply primarily to the Center West region in the forest zone.) The real growth in wages between 1963 and

[19] The only evidence of seasonality comes from data collected by SODERIZ in its "Terrors-Test" near Touba (the western edge of the Ivory Coast, along the southern fringe of the savannah near Guinea) (10). Wages for weeding may be 30 percent less for the second crop, and wages for harvesting may in some cases be less. Reports from farmers in forest areas indicate no seasonality, probably because coffee and cocoa harvests coincide with the cultivation of the second crop of rice.

1976 amounts to slightly less than half of the growth in real GNP per capita in the same period. If the past growth of the Ivorian economy continues in the future, it is probable that real wages will increase at an average rate exceeding 1 percent per year.

Shadow interest rates are more difficult to estimate because the capital markets are highly segmented into public funds, commercial bank loans, and traditional credit. In addition, selling and buying prices for traditional credit are widely divergent. This segmentation represents market imperfections—rationing of funds by foreign aid donors, limited absorptive capacity, externalities that reduce the cost of lending to farmers in public projects and to borrowers in the commercial sector, and the lack of information and high transaction costs in the traditional sector (19). Because these imperfections are likely to persist in the future, it is preferable to estimate the social value of capital in each segmented market, rather than to estimate an average interest rate for the economy, weighted by the share of capital from the different sources. It is assumed that the market interest rates that currently prevail in each market represent the shadow interest rates when adjusted for expectations of inflation. In the traditional market, lower rates are assumed for self-financed, long-term investments by traditional farmers than for short-term borrowing from moneylenders. The difference reflects the high transaction costs in the traditional market, which are not relevant when capital is self-financed. This segmentation tends to increase the social price of capital in traditional rice production relative to improved production where cheaper government financing is available. These interest rates are the same for both forest and savannah zones and are summarized below in real annual percentage rates (3):

Government funds (loans to SODERIZ)	5
Agricultural development bank funds (BNDA loans to farmers)	8
Investments by farmers using own funds (savings)	15
Short-term financing by farmers using borrowed funds (traditional credit)	25

Although unused land remains available in the Ivory Coast, a private market does exist—especially near urban areas and for developed agricultural land. As data from the Center West in the forest zone show, both land prices and rents can be significant (16):

Type of land and transaction	Private land cost (000 CFA francs/ha)
Sale of forest upland	15–25
Sale of forest improved lowlands	120–280
Rental of forest upland fields for food crops	5
Rental of improved lowlands	25–30

These costs have been used to estimate the following private land rents in rice production (in CFA francs/ha) (3):

Upland, modern inputs, forest	5,000
Improved lowlands, manual	11,845
Improved lowlands, power tillers	28,806
Dam irrigation	16,680

For irrigated land, the share of the investment contributed by the owner is excluded from the rent, since it is a capital charge.

Despite these positive private prices, the shadow price of land has been assumed to be zero because these rents probably reflect site value—the favorable location of the field relative to roads, villages, and markets—rather than scarcity of resources (19). Even including fallow land, probably no more than two-thirds of total arable was in farm production in 1975 (7, p. 5). The rate at which upland forest areas are being brought into production has been increasing, and projections indicate that all land may be in a production cycle by 1990 (7, note 63). Although undeveloped irrigable land is abundant, as much as 60,000 ha (7, p. 7), land for upland rice production may become scarce in the future. At such time, a positive shadow price would be warranted.

World Price of Rice

The world price of rice used in this analysis is assumed to be $300 per mt, c.i.f. Abidjan, or 75 CFA francs per kg rice. The figure is based on the following factors: an estimated long-run price of $350/mt for Thai 5 percent brokens, f.o.b. Bangkok in 1975 dollars (2); Ivorian import of 25–35 percent brokens, priced about 30 percent below the Thai 5 percent brokens;[20] and an estimated transport cost of about $50/mt.[21] In order to compare imported with domestically produced rice, internal prices have been calculated from the $300 base figure. To obtain internal prices, costs are added to the c.i.f. Abidjan price to reflect the tradable inputs used to unload and transport imports to consumption centers (or to consumers for on-farm consumption). These world price equivalents are summarized below in CFA francs/kg rice (3):

[20] During the period 1955–74, the Thai export price for 25–35 percent broken rice averaged 32 percent less than for 5 percent brokens. This discount varied from 16–26 percent in the years after 1974. These calculations are based largely on rice prices released by the Rice Committee Board of Trade in Thailand, and published by the USDA (21).

[21] This transport cost may be too high. During 1960–73, the c.i.f. price, Abidjan, averaged only $25/mt higher than the f.o.b. price, Bangkok. See 6, Table A.8, and 21. On the other hand, it may be that preferential trading arrangements with Italy, which was the largest foreign supplier of rice to the Ivory Coast (17 percent of imports during 1965–76), caused actual import prices to be lower than estimated from Thai exports (6 and 17, p. 155). Since only 4 percent of rice imports came from Thailand in 1965–76, Thai exports affected Ivorian import prices only indirectly (see 6, Table A.8).

Consumption center	Price	Consumption center	Price
Abidjan, c.i.f.	75.0	Korhogo (savannah)	78.3
Abidjan, unloaded	75.1	Forest farms	79.8
Bouaké	77.7	Savannah farms	80.3
Daloa (forest)	77.8		

In order to simplify the analysis, four basing points for consumption have been chosen. The major consumption center is the port city of Abidjan. Consumption in the producing region of the forest is assumed to be centered around Daloa (400 km from Abidjan) and that in the savannah is assumed to be centered around Korhogo (650 km from Abidjan). The secondary consumption center is Bouaké, lying roughly midway between Daloa and Korhogo and 375 km from Abidjan.

Private Rice Prices

Net private profitability depends on the private prices of both inputs and outputs. Since input prices are discussed above, this subsection focuses on rice prices in the domestic market. Several factors determine the choice of private rice prices selected for use in this analysis. First, private margins are assumed to reflect real operating costs because the private commerical sector can be considered competitive, in view of the extensive transport network, the considerably underutilized capacity in small hullers, and the large number of marketing agents. Second, the government usually defends the wholesale rice prices effectively, through its use of trade control and the rationing and subsidization of domestic rice milled by state

TABLE 2.6. *Private Prices, 1976*
(CFA francs/kg)

Marketing stage	On-farm, hand pounding	Small-scale hullers	Far from government mills	Close to government mills
Farm price (paddy)[a]	52	50	60	65
Mill purchase price (paddy)	52[b]	59	75	75
Mill selling price (rice)	103	99	87[c]	87[c]
Equivalent wholesale buying price (rice)[d]	103[b]	100/104[e]	87[f]	87[f]

SOURCES: Available price data are taken from 6 or Charles P. Humphreys, field observations (Abidjan, Bouaké, Gagnoa, and Man, 1976–77). Margins used to estimate prices not observed are based on 3.

[a]The low price refers to isolated farmers (50 or more km from government mills). The higher price refers to farmers close enough to receive the official government price. Buyers paying the higher price are assumed to sell to the government mills. Prices in this range have been reported in 9. For example, during 1976, the average paddy price in small markets was 53 CFA francs/kg.

[b]The collect and distribution activities are nonexistent for hand pounding.

[c]Government milling costs are subsidized.

[d]This is the price at which domestic rice is at the equivalent stage of marketing as imported rice purchased by wholesalers.

[e]The price of 104 implies a retail price of local rice of about 110 CFA francs/kg, which was actually observed in the markets. A price of 100 is used for custom-hulled rice for producers' home consumption, which is a price consistent with retail selling prices.

[f]The distribution cost (mainly transport to Abidjan) is fully subsidized by the Caisse Générale de Péréquation des Prix des Produits et Marchandises de Grande Consommation.

agencies. Third, the inability of the government to purchase all paddy offered for sale, except perhaps in 1975–76, has permitted merchants in the private market to collect and sell paddy to private hullers. The price of paddy milled by the private sector will depend, of course, on the retail price of rice, whereas the government price of paddy will reflect its domestic price policy. These private prices are summarized in Table 2.6.

Private and Social Costs

In this section, results—in both private and social prices—are compared for several techniques of producing and processing rice in different regions. Private costs and profitabilities are discussed first, with particular focus on farm-level production. That subsection is followed by an examination of social costs and profits. The final part compares the two sets of results.

Results in Private Prices

Farm costs. At the farm level, paddy production is uniformly cheaper in the savannah, as shown in Table 2.7. For similar production techniques, private costs are 4–9 CFA francs/kg paddy cheaper in the savannah, or 10–20 percent of total costs. The largest divergence, between improved lowlands in the forest (F3) and upland ox cultivation in the savannah (F12), amounts to 16 CFA francs. These differences occur despite higher yields in the forest zone because the significantly lower wages assumed for the northern area make total costs per kg paddy less than in the forest zone. Labor-saving techniques used on irrigated paddy production in the forest zone do not lower private costs sufficiently to make it competitive with production in the savannah.

In all cases, the private cost of producing paddy with upland techniques is uniformly cheaper than with irrigation. The divergence is greatest in the forest, where upland yields are highest; it costs farmers almost 9 CFA francs more to grow a kg of paddy on improved lowlands (F3), compared with upland cultivation (F2). Irrigation is, of course, more attractive in the drier savannah. Dam irrigation (F17), which allows double-cropping[22] and high yields, is only slightly more expensive than manual, upland alternatives.

The use of modern inputs—mainly seeds and fertilizers—also lowers production costs slightly, depending on the concomitant increase in yields. Because of favorable climatic conditions, yields increase the most

[22] The average number of crops per year is about 1.85 on cultivated land. When account is taken of the 20 percent of land idled for various reasons—like lack of water—the number of crops per year is about 1.5 for all land in projects having dam irrigation.

TABLE 2.7. *Farm Production Costs*
(CFA francs, unless otherwise indicated)

Production technique[a]	Private costs			Social costs		
	Per kg paddy	Per kg rice[b]	Rank	Per kg paddy	Per kg rice[b]	Rank
F1	47.3	71.7	8	47.1	71.4	5
F2	41.7	63.2	6	45.1	68.3	3
F3	50.4	76.4	10	57.6	87.3	9
F5	49.1	74.4	9	52.0	78.8	7
F9	45.5	68.9	7	83.1	125.9	10
F10	38.5	58.3	4	38.4	58.2	1
F11	37.8	57.3	3	46.6	70.6	4
F12	34.2	51.8	1	42.5	64.4	2
F14	35.0	53.0	2	50.4	76.4	6
F17	39.7	60.2	5	56.2	85.2	8

[a] For explanation of F numbers, see Table 2.1.
[b] Paddy is converted to rice at the rate of 0.66.

in the forest. As a result, private costs fall by almost 6 CFA francs/kg paddy, or over 10 percent. In the savannah, where the inputs raise yields less, cost reductions are negligible.

Production techniques that employ labor-saving mechanization are much more attractive in the savannah, despite the lower wages that prevail in that region. The least costly techniques in the savannah use oxen (F12) and tractors (F14), although cost savings are only about 10 percent. The use of power tillers in the forest zone lowers costs only slightly and contributes little to make lowland production, which is labor-intensive, more competitive with upland production.

Post-harvest costs. The private costs of the relevant post-harvest techniques are given in Table 2.8. Since these private costs include government subsidies, they mask the relatively high market cost of government milling and of distribution to Abidjan. Compared to small mills, a large proportion of these high costs is caused by the use of disposable plastic rice bags, higher losses, and longer storage periods. By substantially eliminating seasonal price fluctuations, government price policy has made paddy storage unattractive to private millers and merchants. Storage must now be carried out by state agencies. In addition, the official selling price for rice is 27 CFA francs less than the official purchase price of paddy in rice equivalent. However, mills are subsidized by the government at a rate that more than offsets the elevated operating costs and losses caused by the official price structure. As a result, private costs are negative for government mills.

Distribution costs are basically a function of distance, but there are other differences. Costs in market prices for the private sector are lower

TABLE 2.8. *Post-Harvest Costs*
(CFA francs per kg rice)

Marketing or milling technique	Market costs	Private costs	Social costs
Assembly to small hullers	13.5	13.5	12.3
Assembly to government mills	15.7	15.7	14.1
Milling by hand pounding, forest	30.1	30.1	30.1
Milling by hand pounding, savannah	30.4	30.4	30.4
Milling by small-scale steel cylinder hullers	4.8	4.8	4.0
Milling by government industrial mills	14.5	−38.2	13.2
Village distribution, in forest and savannah	2.1	2.1	2.1
Government shipment to Bouaké, from forest	5.6	0.0	5.6
Government shipment to Bouaké, from savannah	6.3	0.0	6.0
Government shipment to Abidjan, from forest	7.4	0.0	6.9
Government shipment to Abidjan, from savannah	9.4	0.0	8.6

NOTE: Market costs include indirect taxes and subsidies incurred on the distribution and manufacture of inputs used in the techniques. Private costs include market costs plus government transfers made directly to the merchant or miller. Social costs exclude all taxes and subsidies, and value all resources in opportunity costs.

than those in the distribution of rice from publicly owned mills primarily because losses are assumed to be only half as great. However, since the official price of rice ex-mill is the same as the official wholesale buying price, costs of government distribution are fully reimbursed by the Caisse de Péréquation. This policy reduces to zero the total private cost of rice distributed through government channels. To some extent, private merchants overcome the disadvantage caused by subsidies to government mills by producing a higher-quality rice, which can be sold at a premium above the rice milled by the government.

Private profitability. Table 2.9 gives profitability results for nine farm production techniques, combined with alternative milling techniques and marketing destinations. In all cases, net private profitability is positive, although it is very low for improved forest production that is privately hulled for village consumption. The most profitable methods for producing rice rely on oxen and tractors—as well as government milling—whereas the least profitable technique involves traditional upland production in the forest milled by small hullers.

Private costs are also less for rice from the savannah than from the forest. The differences between the two regions tend to be largest for rice from traditional production, amounting to about 15 CFA francs per kg of rice. For rice from improved upland production, however, the difference falls to only 6 CFA francs, indicating that government subsidies for modern inputs decrease the private cost advantage of the savannah zone relative to the forest.

Private profitability for rice processed in government mills is clearly

TABLE 2.9. *Indicators of Private and Social Profitability for Ivorian Rice Production*

(CFA francs/kg milled rice)

Production technique[a]	Private cost	Social cost[b]	NPP	NSP	Effective protection coefficient	Resource cost ratio
Hand pounded, farm consumption:						
F1	82.2	82.0	20.4	−2.1	1.284	1.026
F10	72.1	71.9	30.5	8.5	1.276	0.895
Small hullers, village consumption:						
F1	91.0	88.7	8.0	−11.0	1.264	1.144
F10	75.9	73.7	23.1	4.5	1.256	0.941
Government mills, Bouaké:						
F1	44.6	99.5	42.4	−21.8	1.112	1.307
F10	31.7	87.1	55.3	−9.4	1.114	1.133
Government mills, Abidjan:						
F1	48.1	104.4	38.9	−29.2	1.160	1.429
F10	34.8	92.7	53.2	−17.6	1.169	1.260
F2	39.7	101.2	47.3	−26.1	1.235	1.433
F11	33.7	105.1	53.3	−30.0	1.289	1.531
F3	56.6	118.8	30.4	−43.7	1.160	1.641
F17	36.6	119.7	50.4	−44.6	1.219	1.736
F5	50.9	111.6	36.1	−36.5	1.216	1.608
F12	28.2	99.0	58.8	−23.8	1.265	1.412
F14	29.5	110.9	57.4	−35.8	1.263	1.670

NOTE: For an explanation of the concepts used, see Appendix A.
[a] For explanation of F numbers, see Table 2.1.
[b] Because the savings of domestic resources that occur when imported rice is no longer distributed are deducted from the costs in this table, these figures are not simply the sum of costs in Tables 2.7 and 2.8.

highest, often more than double the profitability in the small-scale sector. This result occurs because of the high government support price for paddy maintained by direct subsidies to government milling and distribution. As a consequence of these subsidies, the private cost of rice processed in state-operated mills is as much as 40 CFA francs/kg lower than that incurred in the sector using small-scale hullers. The difference in private profitabilities is less than the difference in costs because the small-scale sector is able to sell its rice at a premium above government rice. As a result of these subsidies, the profitability of government milling and shipment to Abidjan is about 20 CFA francs/kg rice higher than hand pounding for home consumption and about 30 CFA francs higher than milling by small hullers for village consumption. Government transfers are high enough to offset the lower quality of government rice, the higher assembly and milling costs, and the higher cost of shipment to Abidjan.

Results in Social Prices

Farm costs. The social costs of farm paddy production are about 40–80 CFA francs per kg of paddy, but most techniques have social costs in the range of 45–55 CFA francs (see Table 2.7). The cheapest paddy is produced traditionally in the savannah, whereas the most expensive relies on irrigation, modern inputs, and mechanization in the forest. The gap is almost 45 CFA francs, or more than a 100 percent cost increase.

In general, for the same type of production, costs are lower in the savannah than in the forest, although the differences between the two regions are less in social than in private costs. For traditional upland, costs in the forest are over 20 percent higher than in the savannah. The introduction of new techniques by the government also tends to diminish the cost advantage of the savannah. Within the same ecological zone, traditional techniques have lower social costs than those that use selected seeds, fertilizers, insecticides, and extension. The only exception is upland forest production using modern inputs, which has slightly lower social costs than traditional production. In this case, costs per mt fall because modern inputs appear to increase yields sufficiently to offset the additional costs per ha.

All upland techniques have lower social costs than the irrigated techniques, regardless of climate, inputs, or mechanization. The difference—when inputs other than irrigation are similar—often exceeds 10 CFA francs/kg paddy, or 20 percent of production costs. Even in the more arid savannah, the higher yields with water control are insufficient to offset the high costs of investment and operation for irrigation.

The use of modern inputs appears to reduce social costs only in the forest zone, and there the gain is only 2 CFA francs/kg paddy. In the savannah, modern inputs actually increase social costs per kg paddy, because the increase in yields is not sufficient to compensate for the high input costs per ha. The government input package is too expensive, given possibilities for upland production in the savannah. The use of irrigation in the savannah, which does raise yields, increases social costs more.

Results for labor-saving techniques vary. The use of power tillers on improved forest lowlands lowers social costs by over 5 CFA francs.[23] The use of oxen on savannah uplands with modern inputs also lowers social costs by over 4 CFA francs.[24] This method of farm production is the least costly of all techniques except traditional savannah production. In both cases, the gain is due both to the increase in yields and to savings of high wage labor.

---

[23] Without the increase in yields (0.5 mt paddy/ha) assumed to accompany the use of tillers, this technique would raise social costs by 0.5 CFA francs/mt paddy, even though labor costs are lower.

[24] Without the increase in yields (0.3 mt paddy/ha), the use of oxen would raise social costs by 2.7 CFA francs/mt paddy despite lower labor costs.

Unlike small-scale mechanization, the use of tractors increases social costs. Large-scale mechanization on savannah uplands, despite higher yields than the manual technique, remains the most costly of all upland techniques. Tractor cultivation on forest lowlands irrigated by pumps (F9—the Yabra experiment) has resulted in lower yields and is extremely costly—nearly double the cost of oxen cultivation. Even if yields could be raised to the levels in manual production, the mechanized technique at Yabra remains 30–50 percent more costly than manual irrigated production. Therefore, coupled either with expensive irrigation systems or with costly mechanical land clearing, the use of tractors and harvesters has been ineffective in lowering social costs. The use of large-scale mechanization has merely substituted more costly capital and imported inputs for labor, and hence social profitability has fallen.

Post-harvest costs. The social costs of assembly, milling, and marketing are given in Table 2.8. The most striking aspect of these costs is the variation in the social costs of milling—ranging from 4 CFA francs/kg rice to over 30 for hand pounding. But because milling costs vary in part because of differences in the related techniques of paddy assembly and rice distribution, comparisons are best made using the sum of post-harvest costs. These are given below for the forest zone (in CFA francs/kg rice).

Hand pounding, farms	30.1	Government mills, Bouaké	32.9
Small-scale hullers, villages	18.4	Government mills, Abidjan	34.2

Despite savings in assembly and distribution, hand pounding is extremely costly in social, as well as private, prices. The high costs exist mainly because of the large input of labor and the high interest charged on traditional capital tied up in on-farm storage of paddy. Compared to small hullers, there is also a substantial premium on government milling and marketing, amounting to more than 10 CFA francs/kg rice.[25] Finally, it costs almost 2 CFA francs/kg more to ship rice from the savannah than from the forest.

Social profitability. Only two of the Ivorian rice-producing activities have positive net social profitability. These two techniques—on-farm and village consumption of traditional savannah upland production—account for less than 20 percent of national paddy output. However, other activities—of which the most important is hand-pounded traditional production on forest uplands—have a slightly negative NSP (greater than a mi-

[25] This premium may be overestimated because the social costs of small-scale milling are based on the private market charges prevailing in 1975–76, which rose by 50–100 percent in the subsequent two years. These recent market adjustments may have been lagged responses to the high inflation between 1972 and 1976, when prices rose over 60 percent (6, Table A.11). As a result, charges in 1975–76 may have been abnormally low in real terms.

nus 10 CFA francs per kg rice), and these produce about 40 percent of domestic output. The remainder, about two-fifths of domestic rice production, clearly causes the inefficient use of national resources. All of the activities involving government milling and distribution have significant negative net social profitability. These unfavorable results occur because the social cost of government mills is significantly higher than that of small-scale mills and because rice in the government distribution channel is stored longer, incurs greater losses, and is transported over greater distances.[26]

Because the price of imported rice increases with the distance from the port, production and consumption that are located farther inland will be more profitable in social prices. Hence, production and consumption in the savannah are favored over the forest, which is closer to Abidjan. It also means that small-scale hullers, which produce for local consumption, have a higher NSP than government mills, which ship rice to Bouaké and Abidjan, where consumers are closer to the port. But these differences in the degree of natural protection can explain only a small share—no more than about 5 CFA francs—of the variation in NSP.

Divergences Between Private and Social Costs

A comparison of private and social costs helps illuminate the impact of present government interventions in the rice sector. The ranking of farm production by private and social costs changes considerably, as shown in Table 2.7. In particular, government programs have made dam irrigation and the use of tractors relatively more attractive to farmers. Conversely, traditional production in both regions, improved upland production in the forest, and the use of power tillers have received relatively little or no encouragement from government policies. Sources of these divergences between net social and private profitabilities are illustrated in Table 2.10 for paddy purchased and milled by the government and shipped to Abidjan. Divergences are greatest when farm production is combined with these post-harvest techniques.

The average increase in the NPP over the NSP for seven production activities using modern inputs is about 83 CFA francs/kg rice.[27] Assuming the output is processed through the government channel, the subsidy paid to the milling sector—that is, the domestic price support policy—is the most important source. It represents over 60 percent of the total divergence. Subsidies given to farmers for modern inputs and land development are second in importance, but they average only 17 CFA francs/kg rice, or one-fifth of the total. The third most important source of the di-

---

[26] For example, delivery to Abidjan of government rice rather than local consumption of privately hulled rice decreases the NSP by roughly 15–20 CFA francs per kg rice.

[27] The results are based on the following farm production techniques: F2, F3, F5, F11, F12, F14, and F17. The average is unweighted.

TABLE 2.10. *Explanation of Divergence Between Social and Private Profitability for Production Milled by the Government and Shipped to Abidjan*
(CFA francs/kg rice)

Production technique[a]	Divergence between net social and private profitabilities		Trade effect[b]	Social minus private factor price[c]	Net subsidies[d]			
	Value	Rank			Farm	Assembly	Milling	Distri-bution
F1	68.1	10	11.9	−0.3	−0.2	−1.6	51.4	6.9
F2	73.4	7	11.9	−3.9	8.7	−1.6	51.4	6.9
F3	79.1	6	11.9	−5.6	16.1	−1.6	51.4	6.9
F5	72.6	8	11.9	−11.4	15.4	−1.6	51.4	6.9
F9	125.2	1	11.9	−0.8	57.4	−1.6	51.4	6.9
F10	69.8	9	11.9	−0.3	−0.2	−1.6	51.4	8.6
F11	83.3	4	11.9	−0.6	13.6	−1.6	51.4	8.6
F12	82.6	5	11.9	−0.5	13.0	−1.6	51.4	8.6
F14	93.3	3	11.9	−0.5	23.6	−1.6	51.4	8.6
F17	95.0	2	11.9	−6.8	31.5	−1.6	51.4	8.6

[a]For explanation of F numbers, see Table 2.1.
[b]The trade effect equals the domestic minus the relevant import price. The domestic price used here is 87 CFA francs per kg and the import price 75.12.
[c]These figures measure the effect of shadow prices. Negative values indicate that private factor prices exceed social values.
[d]Positive values denote a subsidy, negative ones a tax.

vergence is the increase in the domestic price caused by trade control. By restricting imports, the government causes 12 CFA francs to be shifted from consumers to producers for every kg of rice grown and marketed. The other major cause of the increase in NPP shown in Table 2.10 is the subsidy on government shipments of rice to Abidjan. The subsidy amounts to 7–9 CFA francs and results directly from the rationing required by the policy to equalize domestic rice prices in all markets. Unlike the other divergences that increase the NPP by a constant amount, farm level subsidies vary widely—depending on the type of production—from less than 9 to more than 30 CFA francs per kg rice.

Two sources of divergence reduce the net subsidies of the other policies. The assembly activity is slightly taxed, a result caused by indirect taxes on transportation. Land rents, which are viewed here as a private but not a social cost, also reduce the divergence between social and private profits. They are important only for the irrigated techniques.

Apart from the project at Yabra (F9), which is an experiment with pump irrigation and large-scale mechanized production, the largest divergences exist for mechanized cultivation on savannah uplands and for dam irrigation in the north. Other than transfers to the milling sector, subsidies given directly to farmers constitute the bulk of these large divergences. The smallest divergences are for traditional production. Those that do exist stem almost entirely from trade protection and from subsidies to the milling and distribution activities, not from subsidies to farmers.

For improved production, divergences are consistently larger in the savannah than in the forest for similar techniques.[28] Three factors explain why government policies favor the northern region. First, the policy of equal producer prices throughout the country means that rice from the savannah must be more heavily subsidized to reach Abidjan. Second, although the government input package actually costs slightly more in the savannah than in the forest, yields are lower in the savannah, which increases the value of input subsidies per kg of rice produced by 50 percent compared to the forest. Third, investment in dams in the savannah is more costly than irrigation in the forest. The heavy subsidization of these dams makes this type of irrigated production privately but not socially profitable.

These divergences in profitability change only slightly if different assembly techniques are used. Small-scale milling is taxed (0.8 CFA francs per kg rice) rather than subsidized, which eliminates a substantial share of the total divergence. Nongovernment distribution is also taxed, depending on the distance. For rice that is not milled and distributed by state agencies, therefore, the government's major influence on the divergence is through the trade effect and subsidies on farm inputs.

The trade effect depends on the relevant import price, which is lowest at the port, and on the domestic price, which is greater for on-farm consumption and for hulled rice that sells at a premium. The trade effect is thus considerably more important for the private than for the public milling and distribution sector, because of the higher natural protection on home consumption by producers and the ability of private merchants to sell local rice at a premium over imports. For rice consumed on the farm and for locally hulled rice sold in villages, the trade effect is roughly twice as great as that shown in Table 2.10.[29]

It is clear from the results that a major portion of rice grown in the Ivory Coast has negative social profits. Production occurs, especially with nontraditional farm techniques and with government milling, only because government policies cause private profits to exceed social profits. If social costs are taken as a guide to the techniques that should be encouraged, government interventions have had the opposite effect. The more costly

---

[28] Private land rents are higher in the forest than in the savannah. The treatment of land rents as private but not social costs decreases the divergence between NSP and NPP, thus inflating the difference between forest and savannah.

[29] The following figures illustrate the magnitude of this trade effect under different assumptions (in CFA francs/kg rice):

Sector and location	Domestic price	Import price	Trade effect
Government, Bouaké	87.0	77.7	9.3
Private hullers, forest	99.0	77.8	21.2
Private hullers, savannah	99.0	78.3	20.7
Private hullers, Abidjan	104.0	75.1	28.9
Hand pounding, forest	102.6	79.9	22.7
Hand pounding, savannah	102.6	80.4	22.1

farm production techniques receive the greatest subsidies, and the least costly receive the smallest. As a result, the most inefficient techniques are most strongly encouraged.[30]

Government policies that give strong incentives to rice production can be evaluated by comparing these efficiency results with those of other agricultural crops. The comparisons must be made by using the resource cost ratio because of differences in units. Using a slightly different methodology, Stryker obtained the following coefficients for major agricultural activities (18):[31]

Crop	Resource cost ratio	Crop	Resource cost ratio
Coconuts	0.4	Maize, cotton, and	
Cocoa and pineapples	0.45	peanuts	0.8–1.0
Coffee and oil palm	0.5	Rice, savannah	1.5–2.5
		Rice, forest	2.0–2.5

Clearly, tree crops, fruits, oil crops, fibers, and other cereals are a more efficient use of domestic resources than is rice, even in the savannah. For other food crops, coefficients are subject to greater uncertainty owing to the lack of reliable technical and economic data. But preliminary estimates suggest that production of both cassava and banana plantains are more efficient than rice production in providing calories to substitute for rice imports. On the other hand, yams appear to be no more efficient than is rice.[32]

In order for rice to compete efficiently with these alternatives, the costs of new techniques of rice production must fall significantly below the current costs of domestic production. However, most of the new methods of rice production promoted by the Ivorian government have higher, not lower, social costs of production.

Sensitivity Analysis of Results

The results reported above rely on several major assumptions. The most important are the yields of paddy per ha, the shadow prices for pri-

[30] The ranking of the ten paddy production techniques by the divergence between net social and private profitabilities (Table 2.10) can be compared with the ranking by social costs (Table 2.7) through the use of the statistical test of rank-order correlations. The coefficient for these two rankings equals −0.515, which is statistically significant at a level of 10 percent. This coefficient means that the incentives created by government rice policies are strongly and negatively correlated with the social costs of production.

[31] Both the methodology and the cost assumptions used in Stryker's analysis differ from those used in this study. However, the rankings of different crops are unaffected so long as relative prices remain the same.

[32] These estimates use world prices based on the world price of rice and the proportion of calories in these foods relative to rice. Other costs are based on data similar to those used in this analysis.

mary domestic factors, the world price of rice, and the milling outturn. Increases in yields increase NSP, whereas increases in shadow prices have the opposite effect. Increases in the world price of rice and in milling outturn raise the NSP. The sensitivity analysis is based on estimated elasticities of changes in NSP as values of these major variables change.[33] The effects of changes in these values are presented in Table 2.11.

The level of yields is one of the most important variables in the analysis. Small changes in yields have a large effect on NSP, and the scope for reducing the gap between actual farm yields and feasible yields is large. Data in Table 2.11 show an average increase in NSP of 0.6–0.8 CFA francs per kg rice for each percentage improvement in yields. This increase is higher for irrigated than for upland rice. The range of observed yields suggests that the best farmers using improved irrigated techniques easily obtain 30 percent greater yields than those used in the analysis. If all other variables remained constant, such yield increases could raise NSP by almost 25 CFA francs, which would give most irrigated techniques a positive NSP. The scope for improvement of rainfed production is less, but yield increases of 15–20 percent would be sufficient to give positive NSPs.

However, yield increases are also associated with increases in unskilled labor costs, particularly for harvesting. These increases could amount to 15–20 percent of total unskilled labor costs for rice grown on improved lowlands, milled by the government, and shipped to Abidjan. Based on data in Table 2.11, the NSP would fall by about 0.55 CFA francs for each percentage increase in labor costs in the forest and by about 0.45 in the savannah. As a consequence, the advantages from a 30 percent increase in yields would be offset by 10 CFA francs, or 40 percent. The effect caused by increased labor costs would be almost as large for rainfed production.

The costs of both unskilled labor and capital have relatively large effects on net social profitability. If wage rates were assumed to exclude important in-kind labor costs and equal only money wages,[34] most of the traditional production, representing some two-thirds of national output, would become socially profitable. In no case, however, does domestic produc-

---

[33] In interpreting these elasticities it is important to note that the relative magnitude of the elasticity of any given shadow price depends on the importance of that factor in total costs. As a result, unimportant factors have low elasticities. The absolute magnitude of the elasticities also increases as the NSP approaches zero. Therefore, elasticities should be directly compared only across techniques with similar net social profitabilities. The estimates are point elasticities, and they are probably valid only for small changes in yields and factor costs. Finally, elasticities for capital are based on changes in the value of capital in production costs rather than on changes in interest rates. Since annuities, which comprise the bulk of capital changes, are not a linear function of the interest rate, these results cannot be used to interpret the effects of changes in the interest rate. A detailed listing of the elasticities for yields, labor, capital, and land is available in 3, Tables N1, N2, N3, and N4.

[34] These low wages would be equal to 350 and 300 CFA francs per day in the forest and savannah, respectively.

TABLE 2.11. *Changes in Net Social Profitability Resulting from One Percent Increases in Yields, Social Costs of Primary Inputs, and Milling Outturns*
(CFA francs/kg rice)

Production technique[a]	Yield	Unskilled labor	Skilled labor	Capital	Land	Milling outturn
F1	0.71	−0.60	−0.11	−0.26	−0.00	0.85
F2	0.68	−0.45	−0.18	−0.24	−0.03	0.82
F3	0.86	−0.56	−0.19	−0.31	−0.05	1.00
F5	0.78	−0.43	−0.19	−0.34	−0.09	0.92
F9	1.25	−0.22	−0.33	−0.71	n.c.[b]	1.38
F10	0.58	−0.52	−0.12	−0.21	−0.00	0.71
F11	0.70	−0.42	−0.23	−0.22	−0.01	0.84
F12	0.64	−0.36	−0.21	−0.25	−0.00	0.78
F14	0.76	−0.18	−0.26	−0.45	n.c.	0.90
F17	0.84	−0.42	−0.20	−0.44	−0.06	0.98

[a] For explanation of F numbers, see Table 2.1. These paddy production techniques are combined with assembly to government mills, where paddy is milled, and shipment of the rice to Abidjan by the government.
[b] "n.c." means not calculated.

TABLE 2.12. *Break-Even Prices, c.i.f. Abidjan, for Positive Net Social Profitability*
(U.S. dollars/mt)

Production technique[a]	Post-harvest activity			
	Hand pounding, on-farm consumption	Small hullers, village consumption	Government mills, consumption in Bouaké	Government mills, consumption in Abidjan
F1	329	356	399	418
F2	n.a.[b]	342	387	405
F3	n.a.	422	461	481
F5	n.a.	386	428	447
F9	n.a.	583	576	536
F10	288	296	349	372
F11	n.a.	347	398	420
F12	n.a.	322	373	396
F14	n.a.	371	422	444
F17	n.a.	410	457	479

NOTE: These prices can be converted into ones that are equivalent to Thai 5 percent brokens, f.o.b. Bangkok, by the following formula: $P_t = (P_a - 50)/0.7.$, where P_t is the price, f.o.b. Bangkok, for Thai 5 percent brokens, and P_a is the price, c.i.f. Abidjan, for imports of 25–35 percent brokens.
[a] For explanation of F numbers, see Table 2.1.
[b] "n.a." means that the production technique is generally not associated with the respective post-harvest activity.

tion for consumption in Abidjan become socially profitable. On the other hand, if wage rates were to rise by 50 and 100 CFA francs per day in the forest and savannah respectively, no production would be socially profitable. Given past trends, an increase in labor costs is more likely than a decrease.

The effects of changes in capital costs are difficult to assess in terms of interest rates alone. Capital costs would fall if investments were less expensive, if equipment and irrigated land were more fully utilized and had longer lives, and if interest rates were lower. If capital charges fall by 25 percent, results change only slightly because increases in NSP usually amount to only about 8 CFA francs, which is insufficient to offset the highly negative initial results.

Since government mills already achieve fairly high outturns, an improvement to 0.68 kg of rice per kg of paddy would increase net social profitability by less than 3 CFA francs. On the other hand, more efficient small-scale hullers (attaining 68 percent outturns) would raise NSP by 7–8 CFA francs, which would be sufficient to make all traditional production that is hulled and consumed in the producing areas socially profitable.

The world rice price also has a major impact on NSP. Table 2.12 summarizes the effects in terms of Abidjan import prices (c.i.f., 25–35 percent brokens). If the long-run value of imported rice were to increase 25 percent, all traditional and improved upland production either consumed on-farm or hulled for local consumption would become profitable. But government milling and distribution would remain unprofitable for most farm techniques. Conversely, if the price were to fall 25 percent, no domestic production would be profitable.

The following tabulation summarizes the changes in net social profitability resulting from different assumptions. These values are difficult to establish precisely, but they offer an order of magnitude for evaluating the sensitivity of the results.

	Changes in NSP (CFA francs/kg rice)	
New assumptions	Forest	Savannah
Yield increases:		
Irrigated rice (30 percent)	14	16
Rainfed rice (20 percent)	7	8
Increase in wages (to 500 and 450 CFA francs/day, in forest and savannah, respectively)	−5	−11
Reduction in capital charges by 25 percent	8	8
Increase in land values to market rents:		
Irrigated rice	−6	−6
Rainfed rice	−4	0
Increase in world rice price by 25 percent (to $375/mt, c.i.f., Abidjan)	20	19

Effects on Objectives

As discussed by Humphreys and Rader (7), increased domestic rice output has been viewed mainly as a vehicle for furthering secondary government objectives, notably import substitution (or balance-of-payments equilibrium) and a more nearly equal regional distribution of income. But because the primary Ivorian economic objective is a high rate of growth of national income, it is appropriate to examine first the effect of government rice policies on this goal.

Replacement of rice imports with domestic production reduced national income by perhaps 2.7 billion CFA francs in 1975–76, which implies an average NSP of −8.4 CFA francs per kg rice, as shown in the accompanying tabulation.[35] The only category that does not suffer a net economic loss is on-farm consumption of traditional production, and the net gain from this category occurs only because the lower costs of savannah production offset the higher production costs (and negative NSP) in the forest. Losses to the economy from irrigated production are nearly 14 times as great as from improved upland because of the larger quantity of rice produced by irrigation. Losses stemming from forest production are about four times greater than from savannah production, owing to the larger share of rice grown in the forest.

Type of production and consumption	Thousand mt rice	Average NSP (CFA francs per kg)	Economic gain (billions CFA francs)
Total production	318	−8.4	−2,658
Traditional, on-farm consumption only	167	0.8	133
Improved, all consumption	51	−31.1	−1,584
Improved, Abidjan consumption only	23	−39.2	−906
Improved upland, manual or oxen	13	−18.6	−242
Lowland or dam-irrigated	29	−32.9	−957
Forest, all	220	−9.5	−2,079
Savannah, all	98	−5.9	−579

These estimates are, of course, subject to error. But even with the more optimistic assumptions summarized in the preceding section, the initial conclusion holds: improved rice production in the Ivory Coast reduces national growth. Traditional production, even for consumption on the farm, is barely socially profitable.

[35] Theoretically, the concept of net social profitability measures the marginal social cost of producing the next unit of output. But because cost data are average costs—at least over a wide range—and because fixed input-output coefficients preclude factor substitution within the same technique, the marginal cost can be assumed to equal the average cost for each technique. The aggregate supply curve consists of several discrete steps, each representing a different and increasingly costly production technique. Total economic costs thus equal the sum of the costs of each technique. Aggregate net social profitability equals the product of the NSP and the quantity of rice produced, summed over all the techniques. Since rice production is a relatively small share of the economy in the Ivory Coast, large changes in rice output are unlikely to affect factor prices.

Rice policy affects the government objective of a more nearly equal regional income distribution in three ways. First, the policy of national price equalization implicitly benefits savannah farmers, since transport costs are greater from the northern region. The magnitude of this differential is about 1.7 CFA francs per kg rice, as shown in Table 2.10. Second, farm-level input subsidies, summarized in the accompanying tabulation, are as much as 50 percent larger for savannah farmers (in CFA francs per kg paddy). Investment subsidies are also higher in the north, especially for dam irrigation. In addition, subsidized government tractor services and mechanical land clearing are more widely used in the savannah than in the forest.

	Forest		Savannah	
	Investment	Inputs	Investment	Inputs
Upland, modern inputs, manual	0.0	6.5	0.0	10.0
Upland, modern inputs, oxen	—	—	1.6	9.4
Improved lowland, modern inputs, manual	6.5	5.9	7.6	6.0
Dam irrigation, modern inputs, manual	—	—	19.3	5.3
Tractor services	—	—	11.2	4.0

The third impact of government rice policy on regional income distribution occurs through the price support for paddy, which appears as the milling subsidy in Table 2.10. Roughly half of this subsidy goes to farmers who sell paddy to the government, which means that the paddy price support is about 17 CFA francs per kg.[36] Without this transfer, farm paddy prices could not be much greater than 48 CFA francs. Farmers using the least costly farm production methods benefit the most from this transfer. For the more costly techniques, like upland and irrigated production using modern inputs in the forest, the price support may be used largely to defray higher costs. To the extent that production is cheaper in the savan-

[36]The milling subsidy is divided into the following categories:

	CFA francs/ kg rice	CFA francs kg paddy
Reimbursement for milling costs	14.3	9.4
Paddy price support	25.8	17.0
Net surplus or profit	11.3	7.5
TOTAL	51.4	33.9

The value of the paddy price support equals the difference between total costs (paddy purchase price and milling costs) and the selling price set for mills. The value of net surplus or profit is net of other indirect taxes and subsidies. At costs used in these calculations, it represents a net profit to the government mills. The excess exists because subsidies were set in an earlier period when milling costs were higher and outturns lower. This excess may now be used to offset other losses by the rice agency.

nah, the price supports will increase incomes of savannah rice farmers compared to those in the forest. But because most paddy is produced in the forest, the major share of these price subsidies has been paid in the forest zone (7).

In summary, government input, investment, and distribution subsidies give poorer savannah farmers an extra 5–14 CFA francs per kg paddy, compared with forest farmers using similar improved techniques.[37] The margin is greatest for irrigated techniques and smallest for manual upland cultivation. These policies, therefore, have contributed to the government's goal of more equal regional income distribution, whereas output price supports—for which more money has been spent—have not.

The other secondary government objective is substitution of domestic production for rice imports, with the goal of saving foreign exchange. Estimating the impact on the balance of payments requires more analysis than simply calculating value added in world prices, which is the difference between the value of rice imports and the cost of imported inputs used in rice production. Although these results are positive for all Ivorian rice techniques, they do not guarantee a positive net effect on the balance of payments. The impact of government rice policies on this objective is best evaluated by analyzing the efficiency of rice production in converting domestic resources into foreign exchange. An indicator of this efficiency is the resource cost ratio (RCR) given in Table 2.9. If this ratio is greater than unity, the value of foreign exchange saved is worth less than the value of domestic factors used in the process of saving it. For all the activities described—except traditional savannah production that is consumed on-farm or locally—the ratio exceeds unity. Most improved paddy production techniques coupled with government milling and shipment to Abidjan have ratios of 1.5 or greater, implying that at least 15 francs worth of domestic resources must be expended to save 10 francs worth of foreign exchange. Alternatively, at least 50 percent more foreign exchange could be saved or earned if the domestic factors used in rice production were diverted to other activities, such as coffee, cocoa, cotton, corn, or coconuts, which have RCRs less than unity. Because there are alternative uses of domestic resources that are more efficient than rice production in saving or earning foreign exchange, policies that cause these resources to be

---

[37] These figures are based on the following differences in subsidies between forest and savannah (CFA francs/kg paddy) (the techniques for irrigated rice are improved lowlands for the forest and dams for the savannah):

Subsidy	Irrigated rice	Upland rice
Investment	12.8	0.0
Input	−0.6	3.5
Distribution	1.7	1.7
TOTAL	13.9	5.2

used in producing rice have a negative impact on the balance of payments. For the Ivory Coast, this impact has been strongly negative.

Virtually all of the government interventions in the rice sector have used rather than saved foreign exchange. The benefits of trade and domestic price policies have gone mostly to the forest zone, where the RCR is higher by 15 percent. The use of modern inputs, encouraged by large subsidies, have raised the RCR by as much as 20 percent when compared with traditional production in the savannah. Investments in irrigation, even more heavily subsidized, have increased the ratio by another 15 percent compared with upland techniques using modern inputs. The effect of using tractors has been almost as large. Hence, government policies have exacerbated the inability of rice production to replace rice imports efficiently and thereby to save foreign exchange.

Lessons and Conclusions

An increase in the production of rice is a false indicator of the success of government rice programs because a significant share of the production of rice in the Ivory Coast is socially unprofitable with existing technology and prices. Compared to the price of rice imports, the costs of domestic production are too high, and each kg of imported rice that is replaced by domestic production results in a waste of resources and a potential loss of foreign exchange. General policies—trade protection and domestic price supports—to promote increased domestic production have caused a significant misallocation of resources and reduced overall growth potential.

Increases in the use of fertilizers and improved seeds, expansion of irrigated areas, and greater reliance on tractors are also false indicators of the success of government policies because these changes in the structure of Ivorian rice production have usually been less profitable and less efficient in saving foreign exchange than traditional techniques. Government subsidies of modern inputs have raised the social cost of production and lowered profitability of upland rice, except in the forest. Irrigation facilities have increased the social cost of rice production by 20 percent above upland production. Large-scale mechanization with subsidized hire services has increased production costs relative to the costs of farms employing manual techniques and using oxen and small-scale motorization. Output subsidies that entice the delivery of paddy to government mills have increased the costs of processing domestic rice and idled capacity in the small-scale sector.

This failure of government policy has two main causes. Most important, the rice sector—even with traditional production techniques—cannot compete efficiently with imports. Policies to expand rice output have been largely ineffective in changing the resource constraints that make Ivorian rice production much more costly than imports.

With its low population density and its unused arable land, the Ivory Coast is clearly land-surplus and labor-scarce. As a result of these factor proportions, wage rates are relatively high. The share of labor in total costs of manual paddy production is also high, ranging from 40 to 80 percent. Thus, a reduction in the costs of Ivorian rice production requires that labor become more productive to offset its large share in these costs.

Government policies have not ignored the need to save labor. Labor days per mt rice decrease with the use of irrigation and modern inputs compared with traditional techniques, although these savings are smaller for irrigated production. The promotion of mechanized techniques has also reduced labor inputs. Unfortunately, this increase in the physical productivity of labor does not often coincide with an improvement in the productivity of labor in economic terms, especially when inputs and outputs are valued in social prices. These reductions in labor times have generally been paid for with increases in social costs. However, three new techniques—the use of modern inputs on manually cultivated forest uplands, power tillers, and oxen—have succeeded in lowering social costs while saving labor. But none of these techniques has lower costs than traditional production in the savannah, and the gains from the mechanized techniques depend on increases in yields.

The poor performance of new techniques can be explained by reviewing factor prices. In an economy like that of the Ivory Coast, where the wage-rental ratio is high, new techniques that increase the productivity of land more than of labor are unlikely to lower the total costs of production. Increasing the yield on land that has a zero shadow rent does nothing to reduce costs.[38] Cost reductions occur only by raising the productivity of other inputs, like labor. Because improved seeds, fertilizers, and irrigation are land-saving innovations, they will be effective in Ivorian conditions only if they are successful in producing high yields per labor day.

However, these new innovations, which rely heavily on imported intermediate inputs and capital, are usually expensive compared to the increased output they furnish. Low population densities and widely dispersed farms increase the costs of delivering inputs. Porous soils and rolling topography increase the costs of irrigation investment. Low and irregular rainfall lowers yields of upland rice and makes multiple-cropping more problematical if water control is only partial. Capital and im-

[38] The ability of land-saving inputs to reduce social costs by raising yields depends on the shadow rent for land. In order for modern inputs on savannah upland and on irrigated land in both regions to give social costs of production that are no greater than those of traditional production, the social value of land would have to be about 20,000 CFA francs/ha/crop. This value is four times the observed private rent for forest upland.

ported intermediate inputs are costly, in part because the Ivorian market is limited and there are few suppliers.[39] The regular repair of mechanical farm equipment is hampered by the lack of an adequate service infrastructure, reducing service life and raising capital costs.

High costs also exist because inputs are used inefficiently. Fertilizer application rates have not been tailored to local conditions, thereby decreasing the marginal return to fertilizer. Yield potentials of the improved seeds have not been fully realized.[40] Improved seeds are unnecessarily renewed each year, raising input costs. Irrigation networks are inadequately maintained, which shortens their productive lives and increases their annuities. Cropping intensities are low on irrigated lands, and abandoned fields are numerous. The delivery of mechanical services is inadequately organized, limiting the number of ha that are cultivated by a set of equipment.

The share of capital and imported inputs in total costs of the new techniques increases from less than one-quarter in traditional production to 35–67 percent for improved techniques. If the ratio of wages to the price of capital is low, as it probably is given the high costs and inefficient utilization of these new inputs, it is unlikely that new techniques will lower costs substantially by substituting these inputs for labor. For cost savings to occur in the production of improved rice, either the cost of these inputs must be lowered or their physical productivity raised.

Virtually none of the Ivorian policies to promote rice production has confronted the problems of increasing the economic productivity of high-wage labor, of decreasing the costs of expensive capital investments and imported inputs, and of raising the physical productivity of the new inputs. Trade controls and domestic price supports draw more resources into rice production at higher cost. Subsidies on inputs and investments encourage the adoption of new inputs, but they mask the fact that these new techniques are often more costly than traditional ones.

The reorientation of Ivorian policies suggested by this analysis of costs is threefold. If economic growth and increasing foreign exchange are strong national priorities, the rice sector—both traditional and improved—should no longer be encouraged. Second, if the growth of domestic food pro-

[39] Several examples illustrate the high cost of inputs used in rice in the Ivory Coast. Although the long-run price of urea, delivered West Africa, has been estimated at 43 CFA francs/kg (4), the price, c.i.f. Abidjan, prevailing in the Ivory Coast in 1975–76 was 60 CFA francs, or 40 percent higher. Power tillers usually sold for 750,000 CFA francs, although some suppliers offered them for 100,000 CFA francs less. The cost per ha of herbicides (2-4-D) in the Philippines appears to be only about one-fourth the cost in the Ivory Coast (4).

[40] For example, current experimentation by the Institut de Recherches Agronomiques Tropicales et des Cultures Vivrières (IRAT) indicates that the selected rainfed varieties presently used (Moroberekan, Iguapé Cateto, and Dourado Précoce) have yield potentials of 4–5 mt paddy per ha. Varieties under development (IRAT 10 and IRAT 13) may have a potential of 5.5–6.0 per ha (14).

duction is desired, crops other than rice should be promoted. Third, if the commitment to raise rice output remains unchanged, the government must engage in efforts to bring about significant technical improvements that can lower the social costs of production. Given that the government cannot alter the prices of labor and land in the short term and that both are likely to rise in the long run, efforts must be made to reduce the inputs of labor and assure that capital and other inputs have low social costs and high physical productivity. Such efforts will require a significant commitment to basic research, to the adaptation of cheaper foreign materials to local conditions, and to improvements in farm management.

Besides the use of modern inputs in the forest and of small-scale mechanization for land preparation, the analysis suggests two specific alternatives that might save farm labor efficiently. Small machines designed to save labor at harvesting currently appear to reduce costs, although only slightly.[41] Herbicides conserve weeding labor (about 20 percent in irrigated production) and might raise yields if they enable more timely weed control. Empirical results indicate that, even without yield increases, herbicides can possibly lower social costs of irrigated rice production in both forest and savannah.[42] In addition, methods to raise yields through improved farming practices—like more timely planting—can also increase the productivity of new inputs if such changes are relatively cheap.

Even without improvements in paddy production, the competitiveness of Ivorian rice production could be enhanced by changes in government policies to increase the use of small-scale milling, particularly to supply village markets. Not only do hullers appear less expensive to operate, but their small scale permits reductions in both assembly and distribution costs.

Unfortunately, the evidence presented here indicates that savings must be large—up to 30 CFA francs per kg rice—if Ivorian rice production is to

[41] The social costs of improved lowland production in the forest using modern inputs are tabulated below by different types of small-scale mechanization (in CFA francs/kg paddy):

Power tiller only (F5)	51.9
Power tiller with motorized thresher (F6)	52.1
Power tiller with mechanized cutter-binder (F7)	51.2
Power tiller with thresher and binder (F8)	51.5

[42] The social costs calculated for the use of herbicides (propanil and 2-4-D applied at the rates of 7.5 and 1.25 liters/ha, respectively) on different types of rice production are compared below with the social costs for manual weed control. The results for dam irrigation are estimated from the analysis of production on improved lowlands in the savannah. Data are in CFA francs per kg of paddy.

	Manual weed control	Herbicides
Upland, manual, modern inputs, forest (F2)	45.1	46.2
Improved lowland, manual, modern inputs, forest (F3)	57.6	55.6
Upland, manual, modern inputs, savannah (F11)	46.6	49.6
Dam irrigation, manual, modern inputs, savannah (F17)	56.2	56.0

become socially profitable for consumption in Abidjan. None of the changes studied offers a large cost reduction, highlighting the need for redoubled efforts to develop other, more efficient production and milling techniques.

In conclusion, existing techniques, together with current factor costs and the world price of rice, make much of the rice production in the Ivory Coast socially unprofitable. Government rice policies have reduced, not improved, social profitability. Despite some improvement in the regional distribution of income, policies have depressed national output and decreased the capacity to earn foreign exchange. Even improvements in income distribution have been achieved inefficiently (7). A continuation of past efforts to expand rice output can be justified economically only if they are coupled with technical innovations that increase the social profitability of producing Ivorian rice. Hence, government policies should be reoriented to develop, adapt, and apply the mechanical, chemical, and biological technologies that can save labor without incurring offsetting increases in other costs. In the interim, savings would result if policies were changed to increase the role of the small-scale private sector in marketing and milling.

Citations

1 Dana Dalrymple, *Development and Spread of High-yielding Varieties of Wheat and Rice in the Less Developed Nations*. Foreign Agricultural Economic Report No. 95, U.S. Department of Agriculture, Washington, D.C., Sept. 1978.

2 Walter P. Falcon and Eric A. Monke, "The Political Economy of International Trade in Rice," *Food Research Institute Studies*, Vol. 17, No. 3, 1979–80.

3 Charles P. Humphreys, "Data on Costs of Ivorian Rice Production," Stanford/WARDA Study of the Political Economy of Rice in West Africa. Food Research Institute, Stanford University, Stanford, July 1979.

4 Charles P. Humphreys and Scott R. Pearson, "Choice of Technique, Natural Protection, and Efficient Expansion of Rice Production in Sahelian Countries." Report submitted to the U.S. Agency for International Development, Food Research Institute, Stanford University, Stanford, June 1979.

5 Charles P. Humphreys and Patricia L. Rader, "Field Surveys of Farmers and Laborers." Gagnoa, Ivory Coast, 1977.

6 ———, "Background Data on the Ivorian Rice Economy," Stanford/WARDA Study of the Political Economy of Rice in West Africa. Food Research Institute, Stanford University, Stanford, July 1979.

7 ———, "Rice Policy of the Ivory Coast," Stanford/WARDA Study of the Political Economy of Rice in West Africa. Food Research Institute, Stanford University, Stanford, July 1979; Chapter 1.

8 Ivory Coast, Government of, Banque Nationale du Développement Agricole, *Rapport Annuel 1975–76*. Abidjan, 1976.

9 ———, Ministère de l'Agriculture, Direction des Statistiques Rurales, *Commercialisation des Produits Agricoles*. Abidjan, monthly.

10 ————, Ministère de l'Agriculture, Société pour le Développement de la Riziculture (SODERIZ), personal communications. Abidjan and Touba, 1977.

11 ————, Ministère de l'Economie et des Finances, Direction de la Statistique, *Bulletin Mensuel de Statistique*. Abidjan, monthly.

12 ————, Ministère du Plan, *Etude Générale de la Region de Man*. Vol. 3, *Rapport de Synthèse Agricole*, by Vo Quang Tri, Bureau pour la Développement de la Production Agricole. Abidjan, 1966.

13 ————, Ministère du Plan, *Région du Sud-Est—Etude Socio-Economique*. Vol. 3, *L'Agriculture I*, by Société d'Etudes pour le Développement Economique et Social. Abidjan, 1967.

14 ————, Ministère de la Recherche Scientifique, Institut de Recherches Agronomiques Tropicales et des Cultures Vivrières (IRAT), *Fiches des Variétés de Riz*, by B. Leduc. Bouaké, Dec. 1974.

15 John M. Page, Jr., and J. Dirck Stryker, "Methodology for Estimating Comparative Costs and Incentives," Stanford/WARDA Study of the Political Economy of Rice in West Africa. Food Research Institute, Stanford University, Stanford, July 1979; Appendix A.

16 Patricia L. Rader, "Ethnicity and Agricultural Production in the Ivory Coast," Stanford/WARDA Study of the Political Economy of Rice in West Africa. Food Research Institute, Stanford University, Stanford, July 1979.

17 Abdoulaye Sawadogo, *L'Agriculture en Côte d'Ivoire*. Presses Universitaires de France, Paris, 1977.

18 J. Dirck Stryker, "Economic Incentives and Costs in Agriculture," West Africa Regional Project. World Bank, Washington, D.C., April 1977.

19 J. Dirck Stryker, John M. Page, Jr., and Charles P. Humphreys, "Shadow Price Estimation," Stanford/WARDA Study of the Political Economy of Rice in West Africa. Food Research Institute, Stanford University, Stanford, July 1979; Appendix B.

20 Bastiaan A. den Tuinder, *Ivory Coast—The Challenge of Success*. World Bank and Johns Hopkins University Press, Baltimore, 1978.

21 U.S. Department of Agriculture, Agricultural Marketing Service, "Rice Market News," Federal-State Market News Service, San Francisco, weekly.

22 West African Rice Development Association (WARDA), Development Department, "Rizières à Perméabilité Elevée dans le Nord de la Côte d'Ivoire." Monrovia, June 1976.

PART TWO

Liberia

3. Rice Policy in Liberia

Eric A. Monke

Liberia has achieved rapid rates of economic growth in the post-war period. Monetary-sector gross domestic product (GDP) increased sixfold between 1950 and 1969—an annual rate of 10 percent. This expansion was due primarily to the growth of concessions in iron ore, rubber, and, more recently, timber, a pattern that left the vast majority of the population unaffected. With changes in political power and the stagnation of the economy in the late 1960's, however, the government began to look in other directions for sources of growth. As part of this search, agricultural development and increased attention to the rural population became important considerations in government policy.

Rice, the staple food and focal point of traditional agriculture, was one of the most obvious targets for government planners. On the production side, interest has centered on the possibilities of large-scale concessional operations and small-farmer development programs, although little actual investment has been undertaken. Trade policy, on the other hand, has been used vigorously, and the recent implementation of a variable levy system has served to maintain the high domestic prices established during the world price rises of 1973–74. Government trade policy reflects both a desire to encourage domestic production to substitute for imports and a strong constraint on the government budget, with the result that consumers rather than the government provide subsidies to producers.

The initial sections of this paper deal with aggregate characteristics of the Liberian rice economy and delineate trends in production, marketing, and consumption. The intent is to highlight the constraints on policy that are created by existing economic and physical conditions. The remainder of the paper is a discussion of the manner in which the choice of policy instruments and tools has affected the rice economy. The third section concentrates on the historical evolution of policy under the Tubman government, and the final section discusses new policies of the 1970's.

Economic Geography

Population

Rice plays an important role in the provision of two-thirds of total employment and involves roughly half of the population in its production.

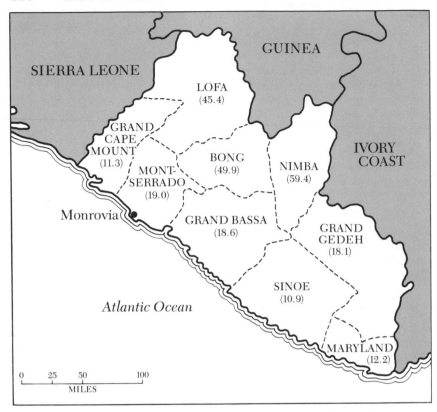

MAP 3.1. The Republic of Liberia. The numbers in parentheses following the county names indicate 1975–76 paddy production in thousand metric tons. From Republic of Liberia, Ministry of Agriculture, *Production Estimates of Major Crops, 1976*, Monrovia, 1977.

The annual national survey of crop production estimates the traditional agricultural population at 826,000, or 54 percent of the total population of 1.5 million.[1] Rice is a staple crop of 90 percent of agricultural households. Employment data from the 1974 Population Census indicate that 70 percent of the total employment of 430,000 emanates from traditional agriculture. Both agricultural population and rice production are concentrated in the three interior counties of Bong, Lofa, and Nimba (see Map 3.1). During the three crop years between 1974 and 1977, these counties were responsible for 64 percent of rice production. Fifty-eight percent of

---

[1] This figure may be somewhat underestimated because the household size found in the survey, 5.4 persons, is below the estimates of the microeconomic research of Currens, van Santen, and McCourtie.

agricultural households are located in these areas, and they have received the majority of the government's rice development efforts.

Land and Climate

The topography of Liberia features an undulating plateau, cut frequently by steeply sloped valleys. Valley bottoms often contain streams and small swampy areas. Almost every type of soil is used for rice cultivation. The combination of topography, soil type (predominantly latosols), and high rainfall (2,500–4,600 millimeters per year) means that upland soils are particularly low in nutrients, because the silica and humus are washed out of the thin topsoil. The subsoil is a distinct hardpan owing to the accumulation of iron and aluminum, further increasing the hazards of runoff and erosion. These characteristics suggest a potentially high payoff from the use of artificial fertilizers, while the fragility of the upland soil structure limits the potential for the introduction of continuous mechanized cultivation techniques. Nutrient retention in the swamps is greater than in the uplands since drainage is poor, and fallow periods are shorter. Continuous cultivation is possible only in the swamps, and resistance to iron toxicity is a requirement for new seed varieties. The small size, the lack of hardpan, and the unequal depth of the swamps make mechanized irrigation technologies impractical in most areas.

Production

Table 3.1 lists the major crops of Liberian agriculture, which together occupy about 9 percent of the total arable land area of 4.5 million hectares. Rice production techniques are described in Table 3.2. Upland cultivation is the single most important technique, accounting for 92 percent of total production. Traditional swamps produce 6 percent of output, and improved swamp and irrigation techniques account for 2 percent.

Key characteristics of existing rice production techniques are presented in Table 3.3.[2] For traditional techniques, labor and seeds are the major inputs, and the capital stock is limited to a few hand tools. No chemical inputs are used. Uplands are cultivated with slash-and-burn techniques and fallowed for ten or more years following the single rice crop. Swamp cultivation follows a similar technology, although transplanting may be substituted for broadcasting. Swamplands are cultivated for several consecutive years, with much shorter fallow periods than upland areas.

Government planners have been interested in establishing permanent as opposed to shifting forms of cultivation and in remedying the dearth of improved technological packages for upland rice. The interest in perma-

---

[2] More detail concerning the technologies of the rice economy is contained in the companion analysis to this essay (8).

TABLE 3.1. *Major Crops*

Crop	Hectarage, 1976–77	Production (mt)	Pct. of farmers growing 1971–72	Pct. of farmers growing 1976–77
Rice	200,000	245,000	88%	90%
Cassava	30,000	136,000	57	61
Coffee	23,300	4,600	20	25
Cocoa	21,100	2,900	18	21
Sugarcane	8,200	92,000	6	17
Rubber	119,100	82,400	8	6

SOURCE: *17.*

TABLE 3.2. *Characteristics of Rice Production*

Production technique	Hectarage (000 ha) 1974–75	Hectarage (000 ha) 1975–76	Hectarage (000 ha) 1976–77	Average size of holding 1975–76	Average size of holding 1976–77	Production (000 mt paddy) 1974–75	Production (000 mt paddy) 1975–76	Production (000 mt paddy) 1976–77
Upland	—	181.0	190.0	1.4	1.5	—	209.4	221.4
Swamp	—	8.0	8.0	0.6	0.6	—	13.8	13.8
Improved swamp	—	2.0	2.0	0.3	0.3	—	5.7	5.6
TOTAL	201.0	191.0	200.0	1.4	1.4	249.0	229.0	241.0

SOURCES: *15, 17.*

nent cultivation has resulted in a concentration on the development of improved swamp and irrigation technologies. A number of techniques have been contemplated, ranging from the introduction of water control in small swamps to the establishment of concessional large-scale, fully mechanized operations patterned after American and Australian production. Major efforts to date have focused on project areas of at least 50 ha in size, with mechanized land development. These projects developed 900 ha by 1977, although only 360 ha were actively cultivated at this time. In recent years, increased attention has been given to labor-intensive smallholder swamp development, and this technique forms the basis for most of the current and planned production investment. The Integrated Rural Development projects of the World Bank are the most prominent developments of this type. For upland rice, efforts begun in 1976 have centered on a pilot program to distribute an improved seed (LAC-23) and fertilizer package.

In economic terms, rice production techniques appear attractive for home consumption purposes but not for cash cropping. Wage rates for unskilled labor ($1.25/day in cash and in-kind payments in 1976), strong off-farm labor demand, and more profitable alternative crops have made the commercialization of traditional production a relatively minor phe-

TABLE 3.3. *Key Characteristics of Rice Production Techniques*

| Production technique | Type of water control | Source of power | | Improved seeds | Fertilizer | Recurrent farm labor (man-days/ha) | Annual land development cost ($/ha) |
		Land preparation	Harvest				
Traditional upland	none	manual	manual, knife	no	no	226	0
Traditional swamp	none	manual	manual, knife	no	no	243	8
Improved upland	none	manual	manual, knife	yes	yes	253	0
Improved swamp, labor-intensive	bunded swamp	manual	manual, knife	yes	yes	362	118
Improved swamp, partially mechanized	bunded swamp	tractor, power tiller	manual, sickle	yes	yes	222	237
Fully mechanized	dam irrigation	tractor	combine	yes	yes	n.a.	n.a.

SOURCE: Chapter 4.
NOTE: A single crop per year is produced with all techniques except "fully mechanized," which produces two to three.

nomenon. Preliminary studies (8) suggest that available new technologies, while more efficient than traditional techniques for home consumption, are not attractive as cash crops. Yield increases are not sufficient to offset increased labor costs in the labor-intensive swamp techniques or to offset the increased input and capital costs associated with the mechanized swamp and improved upland techniques.

Marketing

Virtually all farm households market some produce. Vegetable and fruit products are sold extensively, and recent national agricultural surveys indicate that 50 percent of households sell some specialty crop, such as avocados, bananas, plantains, palm oil, and okra. Producing crops wholly for sale is a relatively recent phenomenon, however, because transportation constraints limited the establishment of major market outlets until the late 1950's. Since that time, coffee, cocoa, and sugarcane have emerged as major cash crops, each grown by 20–25 percent of all agricultural households. In general, cash cropping remains a complementary activity to food crop production, and very little complete specialization has occurred (1, 7).

Rice has remained largely outside the market sphere. Data from three agricultural censuses, presented in Table 3.4, indicate that the Liberian rice economy remains dominated by a pattern of production for home consumption. Half of all producing households are self-sufficient, although market participation has increased markedly in recent years. Table 3.5 shows the distribution of marketing households by county. Market participation rates vary widely across counties, but about 70 percent of all sellers are located in the interior counties of Bong, Nimba, Lofa, and Grand Gedeh.

Both private and government marketing channels exist for rice. Government purchases for 1976–77 were only 3,000 mt paddy, less than 8 percent of total marketings. In the private marketing channel, farmers generally control rice through the processing stage. Data from the 1971 Agricultural Census revealed that only 4 percent of marketing farmers sold produce on the farm (22). Of the great majority who traveled to market, about half walked the entire distance and the remainder used motor transport. Head-loading capacities thus act as a constraint on the size of individual sales. The first transaction is made at the market, usually by a woman from the farm family. Average daily sales per seller are between 5 and 20 kg.[3] Seasonal variations are pronounced at the large regional markets, since both the number of sellers and the quantities per seller are

[3] Current estimated sales of 5 kg/seller for the Lawalaza market in Lofa County (1, p. 159); van Santen estimated sales of 14 kg/seller for the Kolahun market, one of the largest of Lofa County.

TABLE 3.4. *Market Participation*

| Year | Percent of holdings | | Producing sufficient rice for home consumption |
	Buying rice	Selling rice	
1971	30%	16%	70%
1975	51	24	54
1976	57	29	46

SOURCES: *26, 15, 17.*

TABLE 3.5. *Market Patterns by Area*

| County | Percent of holdings selling rice | | Number of holdings selling rice | | Percent of selling holders | |
	1975	1976	1975	1976	1975	1976
Bong	32%	27%	8,700	6,858	27%	17%
Grand Bassa	12	38	1,620	5,206	5	13
Grand Cape Mount	32	9	1,664	468	5	1
Grand Gedeh	29	40	2,697	3,800	8	10
Lofa	12	5	2,568	1,110	8	3
Maryland	26	28	2,132	2,324	7	6
Montserrado	5	10	650	1,320	2	3
Nimba	31	47	9,517	14,711	30	37
Sinoe	32	49	2,592	4,018	8	10
Liberia	24%	29%	32,256	39,730	100%	100%

SOURCE: *17.*

increased in the immediate post-harvest months of November-February.[4] Increased volume during the post-harvest months is due to bulking of larger quantities for shipment to Monrovia, the primary urban market. During the low season, sales are made for local consumption at the large regional markets and at the smaller weekly markets, where vehicle access is often limited. Local demand is strongest during the pre-harvest months, and thus the small markets show a sales pattern opposite to that for the larger markets.

Processing

Most processing is done by hand pounding, and rice is threshed and pounded daily in quantities sufficient for home consumption. Rice mills are generally used for marketings, and the farmer has paddy custom-milled before transporting it to markets for sale. Small-scale mills are the dominant type. Some 220 mills are scattered throughout the

[4] Kolahun, a major Lofa market, for example, averaged 102 traders and sales of 2.3 mt rice/day during the peak season, and only 69 traders and 0.8 mt rice/day during March-October (*10*).

country, principally in the northern and interior counties. Rubber roller mills make up about 75 percent of the total; the remainder are steel cylinder mills and simple hullers, primarily of English manufacture. Mill capacities are rated at 0.25 mt paddy per hour, but problems with poor quality of paddy (the presence of stones, high moisture content, insufficient threshing) and the need for occasional shutdowns of the system to allow the diesel engine to cool limit actual capacity to 65 percent of the technical maxima. These mills process 31,000 mt of paddy per year. This is about 35 percent of yearly milling capacity, although demands for processing in the post-harvest months result in utilization rates double the annual average. The government operates three two-ton-per-hour mill-market outlets in the interior counties of Bong, Lofa, and Nimba, where paddy is purchased for $0.264/kg, processed, and sold for $20/cwt bag ($0.44/kg) throughout the year. Paddy purchases are small (about 3,000 mt in 1976–77), and mill utilization rates are about 8 percent of capacity.

The small-scale mills are able to operate at much lower costs per ton of rice milled than the larger mills. Lower fixed costs enable them to function profitably at lower capacity utilization rates. Large mills incur additional costs in the form of hired labor, storage, and increased losses (equal to about 7 percent of rice output). Finally, the need to attract paddy at a greater distance from the mill and a reliance on commissioned agents for paddy delivery result in higher collection costs for the large-scale mills than for the small-scale mills.

Consumption

Per capita consumption of rice averages 110 kg/yr. Cassava is the next most important staple at an annual rate of 87 kg per capita. There is some variation across regions and between urban and rural areas. Surveys in major rice production areas suggest a per capita consumption of 135 kg, with adults consuming as much as 225 kg per capita. Cassava consumption is about 45 kg per capita in these areas (*1, 11, 12*).

Most imported rice is consumed in urban areas, and historical import patterns have followed the growth of urban areas. Imports were zero in 1945, when the population of Monrovia was about 15,000 (*16*). But as the urban population increased, marketings of domestic production stagnated. By 1965, annual import volume reached 33,000 mt, and it increased to as much as 54,000 mt over the next decade. Annual totals demonstrate a sawtooth pattern over this period, reflecting year-end carryover and the six-month storage capacity in Monrovia. (The only exception is 1971, when large imports may have been the result of a shortfall in 1970–71 domestic production.) The United States is the major source of supply. Relatively minor shipments originate from Asian countries and from Egypt. Prior to 1975, parboiled rice was imported almost exclusively, but a policy decision to encourage consumption of nonparboiled domestic rice resulted in the elimination of imports of parboiled white rice during the 1975–77 period.

Little is known about either income or price elasticities. However, price changes would be expected to have a larger impact in urban areas than in rural producing areas because the income and substitution effects work in opposite directions in the rural areas (as long as income elasticities are nonnegative). Urban consumers do appear responsive to price changes. The 1974–76 price increases and varietal switch from parboiled to raw rice resulted in reductions in average annual imports from 48,000 mt in 1970–73 to 34,000 mt in 1974–76. Some, but not all, of this difference was made up by domestic rice; flows to urban areas increased by about 9,000 mt of rice during this period. Based on production estimates and import data, average per capita rice consumption appears to have declined by about 8 percent between the 1970–73 and 1974–76 periods. Nominal rice prices in Monrovia increased by 56 percent, while the nonfood CPI increased by 48 percent between these two periods. This amounts to a 17 percent increase in real prices and is probably an underestimate of the effective price change, since it does not allow for the varietal change from parboiled to white rice. The real price of cassava also increased, while the real price of bread and plantains fell, suggesting that both substitution effects between staple commodities and own-price effects were responsible for declines in imports and consumption.

History of Rice Policy

The history of rice policy, and of economic policy in general, does not really begin until the post World War II era with the accession of W. V. S. Tubman to the Presidency. Prior to the war, government intervention in the interior regions was limited to the delineation of national boundaries and the establishment of political control over the interior. No internal road network existed, and, although cash cropping began in the 1930's and 1940's, rural trade was limited to goods that could be head-loaded and was directed toward the French colonies of Guinea and the Ivory Coast. By the time Tubman assumed the Presidency in 1944, the economy was beset by inflation and shortages of consumer goods, and the government had financial difficulties (revenues at this time were less than $2 million)(6).

The major economic policy established in this period was termed the Open Door Policy, which attracted foreign investment and skilled manpower through concessional grants for iron ore mining and rubber development. This policy was a continuation of historical patterns of Liberian development. The Americo-Liberians who settled the country were artisans and traders, not agriculturalists, and their interest lay in an export economy rather than in agricultural capitalism. Moreover, the concessional model had already proved successful, as evidenced by the tenfold expansion in rubber exports during World War II. The results of the Open Door Policy were dramatic, and Liberia exhibited sustained economic growth until the late 1960's. GDP increased from $48 million in 1950 to $366.6 million by 1969. During this same period, exports increased from $31.0 million to $196.0 million, and government revenue increased from $3.9 million to $61.8 million. The resultant economy was heavily export-oriented and dominated by foreign interests. Half of GDP, 80 percent of government revenue, and 70 percent of wage employment came from the concessional sectors.

Tubman launched the Unification Policy in 1954. This policy was intended to ensure a smooth transition to modernization for rural society. A comment by Tubman concerning the numerous coups d'état in Africa provides a succinct summary of the objectives of this program (6): "There would have been no coup if the poor man has an opportunity of eating smoked fish while the leader eats ham and eggs. But if the poor man has nothing to eat while you eat ham and eggs, he will surely do something against you."

A National Unification Council was established to oversee the integration of the people of the interior into the national society and economy. Attempts were made to reduce abuses of the original inhabitants by government officials and soldiers. Liberians from the interior were brought into government administration. Public relations officers were appointed to maintain cordial relations with various groups in the interior, and by 1967 the government was spending $1 million annually on this activity. The completion of a basic transport system to the interior in 1958 resulted in an increasing orientation of commerce toward Monrovia rather than toward the French colonies. During the 1960's, elementary schools and medical units were established in most of the major towns.

Although there was an interest in the political incorporation of the interior, little attention per se was devoted to agriculture—the major economic activity of the rural inhabitants. In 1950 the income of the traditional agricultural sector was $24.0 million, including $1.8 million in cash crops, or about 40 percent of national GDP. By 1960 the value of agricultural production had increased to only $25.4 million in nominal terms ($7.3 million in cash crops), and its share of GDP had fallen to 15 percent. Agriculture was also the least favored sector in terms of government expenditures, receiving less than 2 percent of total expenditures throughout the period. Over 70 percent of the Department of Agriculture's expenditures were on general administration and services; nothing was spent on materials or machinery (6). Major activities in agriculture during this period involved a national soils survey in 1951, the establishment of a Central Agricultural Experiment Station in Suakoko in 1953, and the creation of a National Extension Service in 1960.

In the 1960's government attention toward the rural economy began to increase, spurred in part by marked increases in rice imports and balance-of-payments crises in 1961–63. In 1963 President Tubman launched Liberia's first agricultural development program, Operation Production, Priority Number One (9): "We of this country must realize, know, and understand that we must produce to survive; that increased production is analogous to prosperity, progress, development, happiness, self-sufficiency, patriotism, national pride, and even Godliness."

The objective of the program was to double agricultural production through the establishment of local committees to work with farmers. A Rice Extension Program was begun in which experimental farms were established by the United States Department of Agriculture and the Food and Agriculture Organization of the United Nations to encourage rice production with swamp techniques. Exhortation rather than financial aid appeared to be the major method used to implement the plan. A two-man staff was charged with the administration of the program, but no public appropriations were made for its coordinating activities. Rather, the program was run in a manner similar to the Public Relations Program, as the President made spontaneous grants to honorary county chairmen. The relationship between these grants and agricultural development was often obscure. In one area, for example, $5,000 of a $14,900 grant was allocated for the purchase of an automobile for the local chairman (6). Although the impact of the program was negligible and demonstrated the central government's inexperience with agricultural development, the program was significant in that it represented the initial step in reorienting government development policy toward the agricultural sector.

A second attempt, the Crash Program for Agricultural Development, was launched in 1968. Tubman called for a program which would (14)

[serve] as an official guideline for the agricultural development of the country . . . [and] as a stimulant in getting agriculture moving in every field and on every level of the national endeavor. . . . It is expected that every citizen and foreigner . . . will earnestly undertake and relentlessly pursue some production effort, no matter what the size or extent, until the total national potential has been fully regimented, and the planned goals attained.

Program goals included "making the country as nearly self-sufficient in the production of food crops and particularly rice as soon as possible." The program emphasized a long-term commitment to assist small farmers, largely through the establishment of cooperatives. Cooperatives were to replace existing marketing channels, since it was believed (without evidence) that "local traders are generally inefficient and exploitative" (14). The cooperatives would provide a marketing outlet for commodities, supply inputs, and implement a system of standardized grades, weights, and measures.

In addition, the plans called for the establishment of 90 one-acre swamp rice demonstration plots and the development of 1,500 acres per year in large-scale projects with mechanical land clearing. Small farmers would be organized by the cooperatives to cultivate these large developments. Finally, a Rice Committee was formed to oversee the importation of rice. Initially, the Committee gave import rights to one company to import Egyptian brown rice and to mill it into white rice. Problems were encountered as the rice spoiled and developed odor problems, and the project was quickly disbanded (5). A tender system was then instituted in which the Committee would accept competitive bids from any prospective importer.

Although the Crash Program produced few tangible results, it was important for rice policy because it laid the foundation for much of the subsequent government involvement with the rice economy. Small swamp development projects were continued and reached fruition some ten years later as World Bank projects, while large-scale, partially mechanized developments formed a key component of government production efforts in the 1970's. Finally, the Rice Committee actions and government control of imports formed the cornerstone of trade and price policy and enabled the government to intervene effectively in the rice market.

Modern Rice Policy

By the end of the 1960's much of the growth potential of the iron ore and rubber sectors was exhausted. Although the two-product, export-oriented economy brought substantial gains in GDP and a strongly positive balance of trade, a desire to sustain economic growth dictated a search for new activities. Attention turned increasingly to the agricultural sector. With the succession of President Tolbert in 1971, a new economic policy—Total Involvement for Higher Heights—was formulated. Under this plan increased emphasis was placed on rural integrated development and diversification of agricultural production. The introduction of cash cropping concentrated on production of coffee and cocoa for export and production of rice for import substitution. Self-sufficiency in rice continued to be a goal of the plan, as it had been during Tubman's regime.

Accordingly, government expenditures on agriculture began to increase. Whereas expenditures were only $2.6 million in 1971 (4 percent of total expenditures), by 1974 the total budget for agriculture was $8.2

million (8 percent of the total), and by 1975, $12.8 million (*22*). About 44 percent of the 1975 total came from government revenues, the remainder from foreign donor organizations. Increased budgetary outlays reflected primarily increases in agricultural sector investment. Expenditures were primarily on research and on rice and tree crop production, and increased from $1 to $10 million between 1971 and 1976.

Government Investments in Rice Development

Government involvement in rice production followed the precedents set in the 1968 plan. Two special divisions were created within the Ministry of Agriculture to deal with rice production. The first, known as the Special Projects division, was concerned with the development of water control in large swamp areas (50 ha or more) and with the organization of small farmers to work these developments, resettling the farmers when necessary. Cooperatives were established to act as market outlets and sources of inputs and credit. Swamp rice was only one component of these projects, and tree crops, primarily coffee and cocoa, were to be planted in upland areas. Shifting slash-and-burn practices were to be abandoned. Mechanization was also a major component of this program. Agrimeco, an autonomous government corporation involved with agricultural mechanization, was established to provide mechanical services for land clearing and development.

Whereas plans called for the development of 1,500 ha per year beginning in 1971, actual developments fell far below this target, and by 1976 a total of only 900 ha had been developed (*20*). Although eight sites were involved, 650 ha were located in the Zlehtown Project in Grand Gedeh County and in the Foya Project in Lofa County. Total membership in the cooperatives associated with the eight projects numbered 2,460.

A number of serious bottlenecks were encountered in the Special Projects. Equipment down-times, poor management of the cooperatives, input shortages, and lack of farmer enthusiasm hampered development (*13*). The projects were also expensive to implement. By 1976, though the number of projects had been reduced to five and cultivated area had declined to 360 ha, subsidization costs for the Ministry of Agriculture were $0.3–0.4 million annually. Additional subsidies were provided to Agrimeco for capital equipment, and the Taiwanese government supplied 20 consultants (*18*). By 1977, it was decided to phase out the Special Projects. The experiment with partial mechanization had failed.

The experience at Gbedin, a project in Nimba County, illustrates the difficulties encountered by the Special Projects. Originally begun in the 1950's by the USDA as a swamp rice demonstration farm, the project was reactivated in the 1960's by the Ministry of Agriculture with the assistance of the Taiwanese Agricultural Mission and was incorporated into the

Special Projects program in the 1970's. Objectives of the project were to develop 1,200 ha of swamp and to relocate and train 600 families in swamp cultivation. By 1968 only 70 farmers were resettled and 75 ha cultivated. By 1976, in spite of a staff of 35 people, only 63 farmers were active in the project, of whom only three were original settlers. Cultivated area was 69 ha. Annual subsidy costs for the project were $43,000, and the cooperative required an additional subsidy of $10,000 per year.

Problems in the Gbedin project were manifold. The economic attractiveness of the production technique depended on government subsidies, and ineffective input delivery negated the effect of subsidy programs. Farmers complained about the nonavailability of fertilizer and water. Machinery breakdowns were frequent because Agrimeco had purchased used equipment, spare parts were in short supply, and machinery operators were inadequately trained. This resulted in delays in land clearing and preparation which upset cropping schedules. The administration was poorly trained and morale was low, partly because of delays in the receipt of wages from the government. Farmers did not understand clearly their obligations to the cooperatives to pay for services with portions of the harvest. Whereas management wanted to market rice, farmers themselves wanted to decide between home consumption and marketing. Management was also late in the reimbursement of farmers for rice sales. Finally, the experts from Taiwan were hampered by language problems, with the result that they often worked by themselves rather than with farmers.

The second type of project development, the Expanded Projects, involved the improvement of small swampy areas scattered throughout the country. Swamps were generally less than 1 ha in size and cultivated by individual households. An advance team was to identify areas to be improved and enlist farmer interest in development. Following this, a technical assistance team would arrive and assist with actual swamp preparation. The project began in 1972 with two sets of teams, and by 1975 four teams were working in the interior countries. About 60 employees were involved, with leadership provided by 20 Taiwanese advisers. By 1975, roughly 1,000 ha had been identified, but only half of this amount was developed, an annual rate of 125 ha (13). Expenditures approximated $0.2 million annually, excluding the costs of the Taiwanese assistants. Seventy-five percent of Ministry costs went for personnel. As with the Special Projects, organizational difficulties hampered the dissemination of the new technology. Lack of transport facilities and delays in procurement of equipment were frequently cited as constraints in team performance (18). In 1977, when the Taiwanese were replaced by advisers from the People's Republic of China, the program was disbanded.

By the mid-1970's government efforts to introduce new production technologies had not achieved much success. Improved techniques accounted for only 1 percent of cultivated area. Only 2 percent of farmers used improved seeds and 1 percent used fertilizer (15). Farmers had not demonstrated a strong affinity for the new technologies. In part, disinterest occurred because the new techniques were not so economically attractive to farmers as planners had assumed. Whereas the new techniques were more complex and involved increased use of intermediate inputs and more management and labor organization, without government subsidies the net returns per man-day were not substantially higher than for the traditional techniques. Even with subsidies, rice had limited attraction as a cash crop owing to alternative opportunities in tree crops and sugarcane (8).

Off-farm constraints were equally important. Although the new techniques appeared economically attractive for home consumption, organizational deficiencies in the delivery of essential supporting services made it difficult for farmers to find a reliable supply of improved inputs in most areas. Farmers were expected to maintain their own pure seed stock and had to travel to Monrovia to obtain fertilizer, a 200–300 km trip from most production areas. Shortages of planning and extension personnel caused a pervasive lack of technical assistance to the traditional farmer. Only 4 percent of agricultural holdings reported having received advice from extension workers (26).[5]

Although the new techniques required greatly increased annual out-of-pocket expenditures by the farmer, little access to seasonal credit was provided through government or private channels. Since capital inputs are very limited in traditional techniques, credit is usually used for nonagricultural purposes, and creditors are generally relatives or village officials rather than traders. Commercial banks do not operate in rural areas. Government credit for rice-related activities were provided through the Agricultural Credit Division. About $2 million was spent annually, primarily for the support of Agrimeco and the Expanded Rice Programs. Some seasonal credit was dispensed through four cooperatives—Interfowar and Gbandi in Lofa County, Amenu in Grand Gedeh County, and Dokadan in Nimba County—but these cooperatives comprised less than 1 percent of farm families.

[5] The extension service was created in 1960 to "provide technical assistance and help the small-holder." In 1977 there were 85 agents and aides, or one worker per 1,600 households. In general, the quality of workers is low. Average schooling is less than nine years, since past openings were often filled on the basis of political strength rather than ability (6). Few receive organized training from the Ministry of Agriculture. In addition, lack of transportation and support funds (84 percent of the extension service budget is spent on salaries), ineffective program design (the Improved Upland program, for example, makes farmer visits for one year only), late salary payments to extension workers, and logistical problems of obtaining necessary inputs at appropriate times have further hampered the morale and effectiveness of the extension service.

Fulfilling farmer needs for more technical information, improved credit availability, and modern inputs is made even more difficult because of meager infrastructure. Environmental factors such as the level and intensity of rainfall and the presence of numerous valleys and streams have made the construction of roads both expensive and logistically complicated, and transportation facilities have developed slowly. In 1945, there were 300 km of roads, none of which were hard-surfaced. By 1962, 1,300 km of surfaced roads had been built, and this increased to 2,000 by km by 1974. In addition, there were 3,000 km of dry-weather and secondary laterite roads. Of this 5,000 km total, 1,100 km was built in and by concessions.[6] The result is one of the lowest road densities in West Africa—0.02 km/km² for all-weather roads and about the same density for seasonal roads. Family plots and villages are often isolated, and access to markets and to supplies of inputs is costly. The use of foot trails and head-loading clearly limits the scale as well as the speed of transport operations, and, consequently, transportation problems affect government policies on both inputs and outputs. On average, farms are 30–60 minutes' walk from home villages. In addition, more than an hour is required to reach markets, and usually a combination of walking and motor transport is utilized (17).

By 1977 new development strategies emerged in response to these constraints. Success with the concessional model in rubber production and iron ore mining invited its application to the rice sector, and proposals were developed for large-scale, fully mechanized rice plantations of at least 1,000 ha in size. A pilot study was completed in 1978 for a project of 1,800 ha in Grand Gedeh County. This operation was to be fully mechanized, modeled after the techniques used in much of the United States and Australia, with negligible labor employment. Government planners were attracted to this method because of centralized control and the ability to direct production conveniently to the Monrovia market. In addition, this approach largely avoided the problems of inadequate transportation infrastructure, deficiencies in administrative abilities, and an ineffective extension service that hampered the small farmer-oriented projects. The key to the effectiveness of large-scale rice plantations hinged on economic considerations, however, and when preliminary studies suggested that substantial government subsidization would be necessary to sustain production, the proposal was discarded.

Two new types of programs for small farmers have emerged. The first, the Integrated Rural Development projects under the aegis of the World Bank, is essentially a reorganization of the small-scale swamp approach

[6] See the *National Socio-Economic Development Plan* (29). Transportation data for 1973 indicated a total of 23,000 vehicles, including 9,200 buses and trucks and 4,600 taxis. The transport business is dominated by the single-vehicle owner-operated variety.

advocated under the Expanded Rice Program, but with increased emphasis on the removal of infrastructural constraints which hampered previous programs. Modeled after similar projects in Sierra Leone, programs are planned for Bong and Lofa counties. Besides the development of 4,000 ha of swamp rise and 11,000 ha of improved upland production, the projects emphasize the development of health and educational facilities, the establishment of input distribution centers, and road construction to improve farmer access to markets. Only 50 percent of the $37 million project cost is intended for rice.

The final program currently contemplated by the government involves the extensive dissemination of an improved seed-fertilizer package to upland producers (19). Like the swamp projects, the improved upland program recognizes the importance of infrastructural and institutional constraints, and intends to increase the size of the extension service, the number of input distribution centers, and credit availability. The seed-fertilizer package is intended to reach 80 percent of upland producers. An eightfold increase in the size of extension force will lower the ratio of extension workers to holdings to 1 : 50. Each holder will be visited for one year, leaving the farmer responsible for maintaining his own seed stock and obtaining fertilizer. The plan also calls for the establishment of 20 regional centers to distribute seeds and fertilizers. The three-year progam cost is currently estimated at $32 million. It is unclear whether distribution facilities or extension service will be maintained after project completion.

A pilot version of the program was begun in 1976 in the southern counties of Grand Gedeh, Sinoe, and Maryland. In total, 1,369 farmers received seed and fertilizer, with an extension worker per 60 holdings. Although the rice was popular with consumers, it has a harder husk, and hand pounding was reported to require 35 percent more time than traditional rice. Seed repayment rates were 67 percent, although the purity of the seeds is doubtful, since farmers often mixed seeds and varieties during planting. Dilution of seed purity will have an obvious impact on potential output increases in successive years. Furthermore, the injection of increased variation in grain shape and size has major implications for the potential quality of rice that can be marketed. With the LAC-23 seed, which is harder to mill, the result after processing is overmilled traditional rice or undermilled LAC-23 rice. This implies a reduced output quality relative to imported rice.[7] Other problems that surfaced in the second year involved delays in the distribution of fertilizer, and new farmers in the program were required to delay planting. Finally, yield

---

[7] Similar problems have been encountered in Asian countries, where varietal innovation has increased the range of grain shapes and sizes. Examination of 18 milled rice samples in Liberia revealed only five to contain pure varieties.

effects appeared substantially less than the 87 percent increases promised in the program and achieved on experimental plots. Although no thorough yield studies were made, reports of extension workers suggest a 35–50 percent increase. Reasons for this difference are not known.

The upland package is seen by the government as an intermediate step toward ultimate technological change, since upland cultivation is viewed as wasteful of both land and labor. Since it will take time to convince farmers to cultivate swamp rice, the short-run approach is to improve yields of rice on the upland (16). At the present time, however, there is little evidence to suggest that upland production is inherently less efficient than more permanent forms of cultivation. Upland techniques are more land-extensive than irrigated techniques, but in most areas land is not a scarce resource and in general commands a market price close to zero. Labor and capital, on the other hand, are the more critical potential resource constaints, and upland and irrigated technologies do not appear inherently different in their utilization of these resources per unit of output. The improved swamp technique provides an output per man-day of 40 percent higher than the improved upland (10.6 versus 6.8 kg paddy/man-day), but capital costs are twice as large ($0.06 versus $0.03/kg paddy) (8). Given the dearth of expenditures on research and development of upland relative to swamp technologies, substantial scope may exist for improving the relative efficiency of the upland techniques.

The new technologies are improved relative to their traditional counterparts in terms of economic criteria and appear highly profitable for home consumption. The improved swamp technique, for example, offers a private profitability of nearly $200/mt rice and is nearly socially profitable (−$4/mt rice) (8). Therefore, the improved swamp and upland techniques increase the efficiency of resource utilization within the agricultural sector and contribute to income generation by freeing resources, primarily labor, for cash-cropping activities. The profitability of the new technologies, however, remains uncertain. Table 3.6 presents data concerning incentives for cash-cropping rice. Private profitability is negative for the improved upland program and only slightly positive for the labor-intensive improved swamp. Tree crops and sugarcane remain more attractive in terms of relative profitability, with returns above $2.00 per day for farmers who have access to the necessary capital inputs (primarily seedlings) and fertilizer. Furthermore, the levels of private profitability that do exist in the new techniques depend on subsidies. Government production subsidies for the new techniques are $20/mt rice (this effect is partially offset by taxation of postproduction activities), whereas subsidies from consumers of about $145/mt rice are provided through the imposition of a variable levy on imports. Thus the social profitabilities of the new techniques are strongly negative for import substitution.

TABLE 3.6. *Incentives for Cash-Cropping Rice in Liberia*
(Dollars/mt rice, delivered to Monrovia)

Production technique	Private		Social		Labor as percent of total costs		Private returns/ man-day
	Costs	Profit	Costs	Profit	Private	Social	
Upland	554	−96	545	−231	79%	81%	0.93
Improved upland	520	−62	533	−219	65	64	1.01
Unimproved swamp	464	−6	455	−141	75	77	1.22
Improved swamp, labor-intensive[a]	416	42	428	−114	55	53	1.50
Improved swamp, partially mechanized[b]	349	109	488	−174	53	38	2.34

SOURCE: Chapter 4.
NOTE: These calculations do not consider the economic incentives of home production–home consumption systems. In this case, the new techniques appear attractive relative to the traditional techniques, and farmers may find the new technologies attractive for this purpose. See (*10*).
[a] Based on a different fertilizer mix than that contemplated for the World Bank projects.
[b] Representative of the now defunct Special Rice Projects, and included for comparative purposes.

The importance of labor in total costs is highlighted by the last two rows in Table 3.6. As a source of cash income in agriculture, rice must compete with tree crops and sugarcane, while local crafts and the rubber and mining concessions also contribute to a strong rural labor demand and 1976 wage levels of about $1.25/day. The new technologies substitute intermediate inputs for labor and thus reduce the relative importance of labor in total costs, but input substitution effects are not sufficient to make the new technologies socially profitable. Consideration of economic efficiency suggests that it is less costly to the economy to continue to rely primarily on imports to meet market demand in urban areas and that both farmers and national income will benefit more by concentration on alternative crops for cash income.

Government Price and Trade Policies for Rice

Although government policy of the post-Tubman era achieved little with respect to new technologies, major innovations were introduced in price and trade policy. Policy makers soon realized that the exhortations of the 1960's were not sufficient to induce increased production and that prices and profit incentives were necessary to bring about the desired responses. Whereas it was recognized that subsidization was necessary to increase production, the choice between government and consumers as the source of subsidy has influenced the relative use of input versus output price incentives. To date, governmental budgetary constraints appear to have dictated substantial emphasis on consumers as the source of subsidization.

At the producer level, an attempt was made to support paddy prices through the establishment of a purchasing system at each of the three government mills located in the interior countries. A purchase program was begun in 1973–74 at a price of $0.11/kg, but drew little farmer response. By 1976–77, prices had been raised to $0.26/kg, but purchases remained small, at about 3,000 mt paddy, or about 1 percent of total production. In part, the limited farmer response was due to the organization of the purchase program. The Liberian Produce Marketing Corporation (LPMC) attempted to collect rice largely through licensed agents and cooperatives, who received a handling allowance and a commission for their efforts. However, cooperatives did not grade rice and generally made deductions from the support price without resort to moisture tests or other objective criteria. Licensed agents generally paid on a volume basis, and effective prices were as low as $0.16/kg.[8] The result was farmer suspicion and disinterest in the program. Finally, implementation of the paddy purchase program proved costly to the government. Losses on government account in 1976 were about $80/mt rice purchased.

A more important factor that limited the effectiveness of the paddy purchase program was the structure of the price system. LPMC maintained a constant purchase price of $0.394/kg milled equivalent, which was attractive to farmers in the post-harvest months of December–February, when retail rice prices in rural areas were about $0.33/kg. A constant selling price for rice of $0.44/kg was also maintained (for rice sold in lots of 40 kg). This price became attractive to rural consumers and retailers during the pre-harvest months of May–September, when rural retail prices became as high as $0.66/kg, with the upper bound established by the price of imported rice in rural markets. As a result, 78 percent of total paddy purchases by LPMC occurred between December and February, whereas 57 percent of sales to consumers took place between May and September. During most of the year, farmers preferred to process and sell rice on the private market.

Although price policy at the farmer level was ineffective, government controls over import prices had quite the opposite experience. The Rice Committee, by virtue of its control over imports, became the most effective policy-making group of the modern era. Established in 1968 to control imports and to avoid supposed shortages and surpluses that resulted

---

[8] High transport costs reduce rapidly the farmer returns from the price support system. But in Voinjama (Lofa County) farmers have organized truck-chartering and own-delivery systems, and currently deliver 75 percent of the paddy to the mills, thus bypassing all middlemen. The result is that the Voinjama mill is responsible for 50 percent (1,500 mt) of LPMC paddy purchases. On the other hand, at Ganta this system has not developed. Farmers directly deliver only 20 percent of total purchases of 700 mt; the farmers that do deliver directly are relatively large producers whose average transaction volume is 1.9 mt. Transactions with agents averaged 11.2 mt, suggesting that lower reported prices to farmers may reflect increased bulking costs rather than lack of competition in the rice market. These figures are based on October 1976–January 1977 survey data.

under free trade conditions, this group experimented with a number of different policies. In 1968, the government began a cost-of-marketing study with the intent to establish ceiling prices on imported rice at the wholesale level. Market shortages resulted as importers held up orders of rice, and the program was quickly terminated. Equally short-lived measures included the decision to import partially milled rice and an attempt to encourage marketing of domestic production through quantitative restrictions and rationing of imported rice in 1972 (1). The latter policy was terminated upon the appearance of a black market, which circumvented the limitations placed on retail rice prices.[9]

Currently, imports are regulated through a tender system. Three categories of rice are recognized: commercial rice in bags of 40 kg or more; luxury rice in small packages; and concessional rice, used as a wage good by the rubber, mining, and forestry industries. Only the first category is regulated by the Committee. The Committee monitors stock levels in Monrovia, and when supplies fall to 3,000 mt, tenders are invited. The Committee specifies the quantity (usually 3,000 mt), and prospective importers submit sealed bids, which stipulate time of delivery, price, and quality of rice. No attempt is made to regulate annual quantities of imports. The rice is sold at government-regulated prices of $21.55/cwt bag ($0.475/kg) through officially licensed rice stores located throughout Monrovia.[10] A variable levy is used to absorb the difference between the c.i.f. price plus distribution costs and the wholesale price.

Government control is simplified because the import market is dominated by two companies, one of which is government-owned, and by the foreign concessions, which sell rice at subsidized prices to their workers. These groups are responsible for about 90 percent of total imports; the concessions take about a 16 percent share.[11] Imports of rice are consumed almost exclusively in urban areas and compose less than 20 percent of total rice consumption. However, rice imports make up nearly 60 percent of total marketings. Consequently, the ability of the government to control the price of imports is a key influence on market price behavior in both rural and urban areas.

The price of local rice is less directly controlled than the price of imports, but the high degree of substitutability among varieties makes additional price control unnecessary. In Monrovia, LPMC makes local rice available to the rice stores at the same price as imported rice. However,

[9] Rationing is still practiced at the wholesale level when Monrovia supplies become low owing to late arrivals of rice.

[10] Official prices of $0.484/kg are also established at the retail level, but these are not strictly enforced. A standard 16-ounce blue plastic cup is distributed to market women, but *dashing* (overfilling to provide a small gift) and shortening of the cup by scraping it on the sidewalk make standardization difficult.

[11] Based on import data for 1973 (92 percent) and 1976 (90 percent).

demand is limited, since local traders supply most of the domestic rice and undersell LPMC. Domestic rice is sold at a discount relative to imported rice because of its inconsistent quality and the presence of impurities.[12] In rural areas, imported rice appears only sporadically, at a 1976 price of about $27/ cwt bag ($585/mt) in the interior counties. As in urban areas, the price is constant throughout the year.

The decision to implement a variable levy system on imported rice was the most significant policy of the Rice Committee in the 1970's. Both the easing of supply conditions on world markets after 1975 and the change in the quality of imported rice meant declining world prices and increases in the size of the variable levy. In 1975, c.i.f. prices were $444/mt, and the levy was $43/mt. By 1977, c.i.f. prices were $281/mt with tax revenues of $145/mt, equivalent to a tariff rate of 52 percent. The revenues generated by the variable levy are collected in the Agricultural Development Fund, which is to be used for investments in tree crops and rice and as a price stabilization fund in the event that world prices rise above the domestic price. No funds had been expended by mid-1977, and disbursements are controlled by the Ministry of Finance rather than the Ministry of Agriculture. The variable levy had become an effective generator of government revenue; by 1977 the levy amounted to $5 million annually, equal to one-third of total governmental expenditures on agriculture. Furthermore, this policy indicated a strong preference on the part of government to have subsidization for domestic producers come from consumers rather than from government coffers.

Examination of the time series of prices for Monrovia demonstrates some important changes between pre- and post-1973 price behavior. Before 1973, domestic rice sold at the same price or at a premium over imported rice. This price relationship reflected both a thin market and the sporadic flow of rice from rural to urban areas, since only a few thousand metric tons of domestic rice were sold in urban areas. However, the sharp price rise of 1973 resulted in a reversal of domestic-imported price relationships. Retail prices for imported rice rose 85 percent between 1972 and 1976, from $275/mt to $508/mt. Prices of domestic rice, on the other hand, rose less than 60 percent, from $308/mt to $488/mt. The Monrovia nonfood consumer price index increased 64 percent during this period, suggesting a real increase in the price of imported rice and a real decline in the price of domestic rice.[13]

---

[12] The discount in the annual average price series in Monrovia is probably underestimated because the prices are monthly averages, whereas the distribution of quantities is heavily biased toward the months of December–March. From December 1974 to March 1975, for example, local rice sold in Monrovia for $.459/kg and imported rice for $.55/kg. A year later, prices were $.446/kg and $.506/kg, respectively.

[13] The total CPI increased by 72 percent over this period. These numbers may differ from those referred to elsewhere in the text because of different year coverage.

The effect of the increase in the price of imported rice was amplified by the decision to import lower-cost white rice rather than American par-boiled rice. This policy was undertaken because it was felt that a major constraint to the adoption of local rice was that consumers were not accustomed to its taste. By changing the quality of imports, it was believed, consumer tastes could be altered so as to prepare for the adoption of domestic rice. The alternative policy of introducing parboiling into the rural production process was believed impractical. Finally, it was hoped that exclusive use of white rice would reduce speculation and black markets when supplies ran low, since parboiled rice could be stored for a longer period than white rice. Regardless of the above considerations, a definite effect of this policy was to increase the impact of the variable levy policies. Effective nominal price increases exceeded 100 percent.

Changes in price relationships caused major alterations in urban and rural marketing patterns. Wholesalers and importers noticed the appearance of marked seasonal patterns in urban rice sales. During the post-harvest months of December–March some 9,000 mt of domestic rice flowed to urban areas,[14] and sales of imported rice were only 40 percent of the monthly averages of the remainder of the year. Domestic rice actually outsold imports for these months—2,200 vs. 1,400 mt/month. During April–November, domestic rice sales in urban markets were only 500 mt/month.

These changes in urban marketing patterns reflected increased linkages to the rural producing areas. Before 1973, the amount of rice marketed was small, since rice made up only 10–12 percent of the value of cash sales per household (about 6 percent of the value of household rice production).[15] Average sales were 150 kg paddy/holding, suggesting total

[14] This estimate is based on monthly sales data for imported rice for the Abi-Djaoudi firm and LPMC for the 1975–76 year. These two firms accounted for 70 percent of total imports. Average monthly sales were 57,621 cwt bags. Average monthly sales for April–November were 70,248 cwt bags, and this is used as the consumption standard for urban areas. (Imported rice is consumed almost exclusively in urban areas, and only in years of extreme shortfall are there any movements of imported rice to rural markets. Imported rice is transported to rural areas only in small quantities, resulting in high transport costs. Rural prices of imported rice are about $0.15/kg greater than in Monrovia.) Average monthly sales in urban areas for the December–March period are 28,926 cwt bags. Since prices are constant throughout the year, these figures suggest that 165,300 bags of imported rice are "replaced" by domestic rice. This figure is then adjusted for the fact that the sales figures apply only to 83 percent of total imports, suggesting that 8,800 mt of rice flows from rural to urban areas in the post-harvest season. The remaining 17,000 mt of total domestic marketings is distributed during the rest of the year among rural and urban areas, although shipments to urban areas during April–November are limited, owing to high market prices for rice in rural areas vis-à-vis urban areas.

[15] Van Santen's results (*11,12*) found average sales of 103 kg of rice per holding. A national survey during 1970–71 found that less than 2 percent of farmers reported selling more than 50 percent of their rice crop. In contrast, the same survey found that 46 percent sold more than half of their cassava output (*7*).

marketings for 1971–72 of 17,000 mt paddy. Since only 16 percent of producers sold rice, average sales were about 1 mt paddy per seller.[16] By 1976–77, market participation among rice farmers had increased substantially. Twenty-nine percent of farmers reported selling rice, nearly double the total five years earlier. The number of small-scale rice mills, an additional indicator of market activity, doubled to over 200 mills during this period. Total marketings were now on the order of 39,000 mt paddy, or about 16 percent of total production.[17] This amounts to average sales of 0.3 mt paddy/holding, and 1.0 mt paddy per seller. Thus the substantial increases in marketings in the 1970's were not due to increases in marketed output per farmer, and the expansion of the market economy had taken an extensive rather than intensive course. The number of participants in the market increased, as a result of both population increase and the creation of additional profitable opportunities arising from the increase in rice prices.[18]

Although price data are scarce for rural producing areas, the pattern of monthly price movements was different from that in the urban areas. Until 1972, annual average prices in rural areas seemed relatively stable in nominal terms. Rice sold in cwt bags for $8.50 ($0.19/kg), and in rural markets retail prices averaged $0.22/kg, with seasonal ranges from $0.17 to $0.28 (10,11). This was below prices for domestic rice on the Monrovia

[16] This is equivalent to the output from 1 ha of upland rice or 0.6 ha of swamp rice, and implies that the larger upland farms and individually controlled swamp holdings were primary sources of marketed output. Data from one study (7) found that 8 percent of upland farms, representing 9,000 upland rice holdings, were greater than 5 ha in size. These large farms were usually operated by village elders who had claims on voluntary labor or by farmers with the financial ability to hire. Another survey (2) found that rice from personal farms run primarily by women was a major source of marketed output. Swamp rice holdings totaled 10,000.

[17] Microeconomic production data were not available for 1976–77, necessitating the use of less direct indicators of marketings. Rubber roller mills compose 73 percent of total small-scale mills, and sales of rubber rollers for 1976–77 were 640 pairs. With a capacity of 35 mt/paddy pair, and assuming that nonrubber roller mills process the same amount of paddy as rubber roller mills, annual small-scale mill throughput is 31,000 mt paddy. Government mills purchased 3,000 mt paddy. Hand-pounding sales are estimated from government purchase data. Government mills purchase hand-pounded rice for $0.44/kg throughout the year, but purchases occur only in December and January. For 1976–77, 200 mt of hand-pounded rice were purchased, or about 15 percent of government paddy purchases during these two months. Hand-pounded sales can thus be estimated at 5,000 mt of paddy, giving total rice marketings of 39,000 mt paddy, or about 26,000 mt rice.

[18] The number of total holdings increased by 35 percent over this period. The number of upland rice holdings increased by 34 percent, from 96,000 to 129,000, while swamp holdings increased by 37 percent, from 9,500 to 13,000. Increases in marketings would also result if a substantial number of farmers began sales on a very small scale, perhaps substituting cassava for rice in their own consumption patterns. But cassava prices have increased at roughly the same pace as rice prices, thus negating substitution incentives. Retail rice prices increased by 88 percent in Monrovia between 1972 and 1976, from 12.9¢/lb to 24.2¢/lb, while cassava prices increased by 119 percent, from 3.2¢/lb to 7.0¢/lb. During the 1971–76 period, rice prices increased by 74 percent, cassava prices by 79 percent.

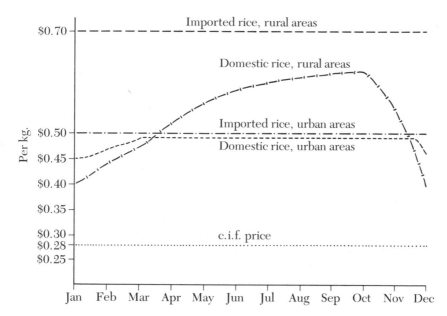

FIGURE 3.1. Retail Rice Prices During 1977 (U.S. $/kg).

market, which averaged about $0.30/kg during this period. By 1976 prices for domestic rice in rural areas had increased sharply. Rice in cwt bags now sold for $20 ($0.44/kg), with a seasonal range of $15–25 ($0.33–0.55). Prices at the retail level were in the range $0.45–0.75/kg, with the magnitude of the seasonal price movement varying from year to year.

Hence, by 1977, rice markets had become substantially more integrated, since rural prices were now more in line with urban prices. In fact, for much of the year and probably on an annual average basis, retail rice prices in rural areas were *above* prices in urban areas, a reversal of historical relationships. Figure 3.1 summarizes these price relationships. Prices for imported rice ran the pricing system. Constant throughout the year, they promoted flows of domestic rice to urban areas only in the postharvest months. Imported rice prices acted as a ceiling on rural area prices by limiting potential increases in the price of domestic rice on rural markets.

For rural producers, rice prices had increased by about 150 percent in nominal terms, perhaps 75 percent in real terms, and marketed supply increased by 135 percent. But in the aggregate, the magnitude of the supply response was small. Since population growth accounts for 6,000 mt of additional marketed rice production, the maximum supply elasticity, resulting if all the increased output came from expanded acreage and there was no substitution in farmer consumption patterns (presumably between cassava and rice), is only 0.1. The response was indicative of the low levels of average private profitability for cash-crop production with the traditional technologies. The traditional upland techniques had an average profitability of −$96/mt rice for Monrovia delivery, whereas swamp rice offered −$6/mt rice. If two-thirds of swamp and irrigated production is assumed marketed, this leaves about 20,000 mt of paddy marketed from upland farms, or about 10 percent of upland production. For the majority of upland rice farmers, rice was not an attractive source of cash income.

A low supply elasticity does not imply that farmers lack price responsiveness. Relative prices of rice did increase, but the traditional two-factor technology (land and labor), the lack of access to profitable new technologies, and severely restricted capital markets can be important constraints on the shape of an aggregate supply curve. Coffee and cocoa prices increased by 30–40 percent, reducing the incentives to shift resources from alternative cash crops into rice. Finally, nominal rice price movements were partly offset by an 80 percent increase in the nonfood price level during the 1970–76 period. The result of price policy under these conditions reinforces inflation rather than increasing real output, and rural unskilled wage rates doubled during this period, while land remained an essentially free resource.

Government Policy and Income Distribution

Numerous references have been made to the disparities between rural and urban income levels (3, 29). But in contrast to historical patterns, government policies of the 1970's benefited rural areas. Rural areas made substantial gains in income relative to the urban sector during the 1972–76 period.

Information contained in the National Development Plan and publications on national accounts estimation (23, 27, 28, 29) can be used to specify, in general terms, the nature of the rural-urban income linkages, and thus to estimate changes over time in rural-urban income differentials. The rural population's contact with the monetary economy comes primarily through the sectors of Mining and Commercial Agriculture and Forestry, which serve as major sources of off-farm employment. The mining sector is a foreign enclave, connected to the rest of the economy through taxes (5–7 percent of sector GDP) and wage payments (15 percent of sector GDP). The Commercial Agriculture sector includes plantations of rubber, coffee, cocoa and oil palm, forestry concessions, export fishing firms, and large-scale poultry farms. Rubber production alone accounts for 60 percent of the GDP of this sector; wage payments absorb 40 percent of GDP originating in rubber. All wage payments of the min-

TABLE 3.7. *Income Distribution*
(Million dollars, except as indicated)

Distribution	Urban		Rural	
	1972	1976	1972	1976
Monetary sector:				
Agriculture	2.7	5.6	15.3	32.2
Mining	9.9	12.9	21.3	27.7
Other sectors	191.9	260.0	—	—
Traditional sector	—	—	71.9[a]	143.9
TOTAL	204.5	314.5	108.5	203.8
Income per capita (*dollars*)	686	888	95[b]	168
Real income per capita, 1972 prices	686	516	95	98

SOURCES: 23, 27.

[a] Differs from official estimates, owing to revisions in the estimation of the value of rice production. Based on data presented in this study, 1972 rice output is valued at $31.5 million rather than the $19.7 million presented in official estimates.

[b] C. E. van Santen, in two micro-surveys of Foya and Gbarnga, estimated per capita incomes at $54 and $83, respectively (*11, 12*). Van Santen evaluated rice at a farm gate price of $132/mt rather than at rural retail prices of $220/mt. If the higher price is used, van Santen's estimates become $70 for Foya and $103 for Gbarnga.

ing sector and 40 percent of the commercial agriculture GDP are allocated to the rural sector. This assumption overestimates rural income somewhat, since some of the wage payments in mining accrue to urban dwellers and rubber is more labor-intensive than the other industries in the commercial agriculture sector, and thus rural-urban income differentials will be underestimated. Finally, absolute urban incomes are distorted by the inclusion of concessional income in urban income figures. Concessions (primarily timber, rubber, and iron ore) provide about 50 percent of monetary GDP. The urban population benefits from this income primarily through taxes (about 7 percent of sector GDP), the remainder accruing to expatriates and foreign sources. Only tax revenues are attributed to urban incomes.[19]

Results of these calculations are presented in Table 3.7. Real incomes in urban areas fell by 25 percent between 1972 and 1976, while real incomes in rural areas remained constant. As a result, urban-to-rural nominal income differentials declined from 7 : 1 to 5 : 1 during this period. Thus, in a period of strong inflation between 1972 and 1976, when the economy experienced a 72 percent increase in the CPI, government policy was able to maintain rural sector real incomes, and shift income distribution significantly in favor of rural areas.

Rice price policy played an important role in maintaining the purchasing power of rural inhabitants. Forty percent of the increase in rural in-

[19] Including concessional income in the estimation of urban income leads to per capita estimates of $1,126 and $1,415 for 1972 and 1976, respectively.

TABLE 3.8. *Impact of Rice Price Policy on Real Incomes, 1972–76*
(Million dollars)

Type of transfer	Total transfer	Type of transfer	Total transfer
Income transfers:[a]		Consumer welfare losses:[e]	
Urban consumers–rural producers[b]	1.5	Rural consumers	6.4
Rural consumers–rural producers	22.1	Per capita (dollars)	5.0
Urban consumers–government[c]	2.4	Urban consumers	0.6
Rural consumers[c,d]–government	0.6	Per capita (dollars)	2.0

SOURCES: Data are from the text, 4, 25, and Eric A. Monke, "Government Intervention and International Trade of Rice," Ph.D. dissertation, Stanford University, Stanford, 1980.

[a] Transfers for 1976 are estimated under the following estimates of consumption patterns:

	Urban	Rural
Per capita consumption (mt)	0.107	0.107
Imports (mt)	29,000	8,000
Domestically produced (mt)	9,000	122,000

[b] Real producer prices estimated to have increased by 75 percent above 1972 prices of $220/mt.
[c] Based on a 1976 tariff of $82/mt.
[d] Rural consumption of imports comes primarily through subsidized sales by the rubber and mining concessions. Concessional imports are about 15 percent of annual imports.
[e] Excess-burden consumer welfare losses are calculated in terms of the Hicksian equivalent variation measure. Price elasticity of demand is assumed as −0.9, the income of elasticity is assumed as 0.5. The price elasticity is the most important parameter in empirical estimation. The more price-inelastic the demand, the lower the value of the welfare loss. These estimates are not included in Table 3.6.

come during this period was due directly to the decision to maintain high rice prices through the variable levy system.[20] Table 3.8 presents some summary estimates of the impact of rice price policy on real incomes of the rural and urban sectors. Most of the income transfers occurred between rural consumers and rural producers; this accounts for the stability of rural real incomes observed in Table 3.7. Real income transfers from urban consumers were directed primarily into government coffers rather than toward rural producers because imports continued to dominate urban consumption patterns. In 1977, with the variable levy at $144/mt due to declines in world prices, the difference between urban-government and urban-rural transfers on rice expenditures increased even further, and total consumer-government transfers were about $5 million.

A final area of welfare loss resulting from the rice price policy is the impact on consumer welfare of a real price change. These are not incorporated in the estimates of Table 3.7, owing to the uncertainty of estimation of the critical parameter—the price elasticity of demand. But even at the relatively large elasticity of −0.9, the equivalent income effect of the changes in rice prices is relatively small; estimates of real per capita income presented in Table 3.7 fall from $98 to $93 for rural inhabitants and from $516 to $514 for urban dwellers.

[20] The value of rice production during this period rose from $31.5 million to $70.8 million (evaluated at rural retail prices of $220/mt and $488/mt, respectively). These figures underestimate the actual effect of rice policy on incomes, since they neglect the effects of rice price increases on unskilled wage rates (rice, as the staple food, acts somewhat as a wage good), and thus on the costs of other commodities produced in the rural economy.

Conclusion

The ability of the government to control price levels in both rural and urban areas through trade restrictions has been the major success of Liberian rice policy. Policy has been successful in encouraging a much closer integration of rural and urban rice markets and has effectively transmitted real price increases to rural producers. Although rice as a cash crop is still economically unattractive to a majority of producers, market participation nearly doubled in only five years. In the face of severe technological and infrastructural constraints, the price response of farmers was impressive.

Implementation of the variable levy system for imported rice has allowed the government to encourage rice production through transfers from consumers to producers rather than through direct government subsidization of producers, thus sidestepping potential constraints on the government budget that might result from direct production subsidies. Although government finances are relatively secure in Liberia, the government has largely avoided policy alternatives that are likely to require subsidization, such as production projects, and preferred policies that result in revenue generation, such as the variable levy. This suggests that budgetary considerations are of critical importance to government planners.

The Liberian experience with rice policy demonstrates that prices are necessary but not sufficient conditions for transforming agriculture. Increases in rice production since 1974 have been limited in the aggregate because the effects of rising input prices have eroded much of the impact of rice price increases. Government attempts to introduce technological change in rice production have met with very limited success, as direct interference in rural production has proved to be far more complex and demanding of government resources (both financial and human) than import price manipulation. The inability to relieve the institutional constraints of input delivery and credit availability as well as the erratic delivery of technical and mechanical services have done little to create enthusiasm among farmers. In addition, the basic problems of transportation availabilities to encourage marketing and of the size and quality of the extension service to provide adequate follow-up services remain unsolved. But, most important, the new technologies are on average less privately profitable than alternative cash crops. To date, therefore, there has been little incentive for the farmer to adopt much more complex, and, under Liberian conditions, often riskier, rice technologies.

The future development of the rice sector depends on three factors. The first, and most critical, centers on the creation and dissemination of economically attractive new technologies, i.e. the ability to increase production efficiency. The improved swamp technologies appear more efficient than the traditional techniques and are nearly socially profitable when produced for home consumption. The results imply that the adoption of the improved techniques for home consumption should not require the heavy taxation of consumers that results from the variable levy system. But the new techniques are not socially profitable for delivery to Monrovia and thus represent inefficient utilization of agriculture's resources. Even with heavy subsidies, average private profitabilities are not high, and the impact of the new technology on the level of rice imports appears uncertain. New technologies are needed that can efficiently substitute capital for labor. Relative to irrigated technologies, rainfed technologies have received little research and development, and the potential for new upland technologies is largely unknown.

The second critical factor involves future investment in infrastructure. Improved transportation facilities will undoubtedly increase rice marketings as more farmers become integrated into the commercial sphere. The adoption of new technologies will require both an effective extension service and improvements in financial capital markets to allow farmer access to improved inputs. These developments are likely to benefit the more profitable cash-crop alternatives of coffee, cocoa, and sugarcane, as well as rice production. The new production programs recognize the importance of these constraints, but the investment requirements are not insubstantial. The $27 million programs of the World Bank, for example, will reach at most 9 percent of total rice farmers.

The final area of importance involves the future behavior of domestic rice prices. It is unclear whether the relative gap between domestic and world rice prices will be maintained over time or be allowed to erode as world prices rise in nominal terms. The relationship between domestic prices of rice and inputs will be an important determinant of future production response. Failure to maintain the tax on rice imports will further shift production incentives for cash-cropping away from rice and increase the reliance of the urban sector on rice imports.

Modern rice policy has been most affected by, and often the result of, a search for new sources of economic growth. That policies should result in benefits to farmers and thus improve income distribution was not central to the government's decision-making process, as evidenced by the willingness to explore technologies of any scale, regardless of distributional impact. Rather, the emergence of small-scale techniques has been based on economic superiority relative to their large-scale counterparts. But as the companion paper (8) to this analysis suggests, these new techniques will affect rice production largely through increases in the efficiency of production for home consumption. This will increase rural incomes by freeing resources for cash-cropping activities, which, under current conditions, will not include much marketing of rice. Opportunities for growth of the Liberian agricultural economy lie with small farmers, but in the absence of more efficient new technologies, import substitution for rice appears an unlikely source of growth in real income.

Citations

1 Gerald E. Currens, "The Loma Farmer: A Socioeconomic Study of Rice Cultivation and the Use of Resources among a People of Northwestern Liberia." Ph.D. dissertation, University of Eugene, 1974.

2 ———, "Women, Men, and Rice: Agricultural Innovation in Northwestern Liberia," *Human Organization*, 35 (1976): 355–65.

3 International Labour Office, "Total Involvement: A Strategy for Development in Liberia." Geneva, 1972.

4 Wesner Joseph, "Liberia," in John Caldwell, ed., *Population Growth and Socioeconomic Change in West Africa*. Columbia University Press, New York, 1975.

5 Liberian Produce Marketing Corporation, "Annual Report." Monrovia, 1970–73.

6 Martin Lowenkopf, *Politics in Liberia: The Conservative Road to Development*. Hoover Institute Press, Stanford, 1976.

7 W. D. McCourtie, *Traditional Farming in Liberia*. University of Liberia, Monrovia, 1973.

8 Eric A. Monke, "The Economics of Rice in Liberia." Stanford/WARDA Study of the Political Economy of Rice in West Africa, Food Research Institute, Stanford University, Stanford, July 1979; Chapter 4.

9 C. B. Ufondu, "Small Scale Swamp Rice Farming in Liberia—Bottlenecks and Recommendations." Cuttington College, Suakoko, 1969.

10 C. E. van Santen, "Rural Markets in the Kolahun District, 1971–72." Ministry of Agriculture, Monrovia, 1974.

11 ———, "Smallholder Farming in the Foya Area, 1971." Ministry of Agriculture, Monrovia, 1972.

12 ———, "Smallholder Farming in the Gbarnga District—Bong County, 1972." Ministry of Agriculture, Monrovia, 1973.

13 Republic of Liberia, Ministry of Agriculture, *Annual Reports—Ministry of Agriculture*. Monrovia, various years.

14 ———, *Crash Program for Agricultural Development, 1968–71*. Monrovia, 1968.

15 ———, *National Rice Production Estimates, 1975*. Monrovia, 1976.

16 ———, "Policy Objectives, Strategies, Programs and Projects of the Ministry of Agriculture Along with Major Achievements Since 1971." Monrovia, 1976.

17 ———, *Production Estimates of Major Crops, 1976*. Monrovia, 1977.

18 ———, "Progress Report of Development and Recurring Projects." Monrovia, various years.

19 ———, "Second Revised Short Term Self-Sufficiency Plan." Monrovia, 1977.

20 ———, *Statistical Handbook: Republic of Liberia*. Monrovia, 1975.

21 Republic of Liberia, Ministry of Planning and Economic Affairs, *Demographic Annual of the Population Growth Survey, 1971*. Monrovia, 1973.

22 ———, *Economic Survey*. Monrovia, various years.

23 ———, *Estimates of Domestic Product at Current and Constant 1971 Prices 1964–1973*. Monrovia, 1975.

24 ———, *External Trade of Liberia*. Monrovia, 1968–75.

25 ———, *1974 Census of Population and Housing*. Monrovia, 1976.

26 ———, *1971 Census of Agriculture: Summary Report for Liberia (Preliminary)*. Monrovia, 1973.

27 ———, *Quarterly Statistical Bulletin of Liberia*. Monrovia, various years.

28 ———, *Sources and Methods of Estimation of National Product, 1970–73*. Monrovia, 1975.

29 Republic of Liberia, National Planning Council, *National Socio-Economic Development Plan: July, 1976–June, 1980*. Monrovia, 1976.

4. The Economics of Rice in Liberia

Eric A. Monke

Rice is the primary staple food of both urban and rural areas in Liberia, and almost the entire Liberian population produces or consumes it. In rural areas rice remains the focal point of traditional agricultural life. Yet as urbanization has fueled market demand for rice and as the agricultural sector has become increasingly integrated into the market economy, domestically produced rice has remained largely outside the market. Imports of rice have satisfied increases in market demand, and coffee, cocoa, and sugarcane have emerged as dominant cash crops. Government attempts to encourage marketings of domestic rice through increases in prices and the introduction of new technologies have had only limited effects.

The primary objective of this Chapter is to analyze the economic structure and performance of the rice sector at both the production and post-production levels. The chapter is divided into five parts. The following section describes the existing and potential techniques of rice production, processing, and transport. Emphasis is on the technological and environmental constraints on performance rather than on specific cost incentives, which are the subject of the rest of the paper. The second section discusses government incentives and infrastructural and institutional constraints, and the third section is a discussion of factor markets and shadow prices. The fourth section analyzes costs and incentives from both a private and social perspective, and the fifth section subjects the results to sensitivity analysis to test for the importance of changes in underlying assumptions. The chapter concludes with a brief summary of principal results.

Delineation of Techniques

Production

Major characteristics of Liberian rice production techniques are highlighted in Tables 4.1 and 4.2. Upland rice production is the dominant

The author would like to thank the members of the Rice Project, particularly C. P. Humphreys, S. R. Pearson, and J. D. Stryker, for many helpful comments.

TABLE 4.1. *Key Characteristics of Rice Production Techniques in Liberia*

Production technique	Area, 1976–77 (ha)	Paddy yield (mt/ha)	Paddy production, 1976–77 (mt)	Improved seeds	Fertilizer	Pesticides
Upland	190,000	1.05	199,500	no	no	no
Improved upland	—	1.57	—	yes	yes	no
Unimproved swamp	8,100	1.55	12,555	no	no	no
Improved swamp I	2,000	3.50	7,000	yes	yes	yes
Improved swamp II	included in improved swamp I	3.50	included in improved swamp I	yes	yes	yes

SOURCES: Production data are from 23 and 24. For description of the techniques, see 16, 17, 18, 19, and 25.
 NOTE: A single crop per year is produced with all the techniques listed, and water control is a factor only in the two "improved swamp" techniques. The land is prepared and the crop harvested manually in every case except "Improved swamp II," in which tractors and power tillers are used in preparation and sickles are used in harvesting (instead of the knives used in every other case).

technique, accounting for over 90 percent of the annual national output of 210,000 metric tons of paddy. Rice is produced with a slash-and-burn technology and is intercropped with numerous fruit and vegetable crops, including corn, pepper, bananas, plantains, cassava, and okra. Rice is usually grown for only one year, since low soil nutrient levels and weed competition make rice production impractical in succeeding years. Cassava, peanuts, or sugarcane is occasionally planted in portions of old fields. The land is fallowed from seven to 50 years, depending primarily on the degree of population pressure. In general, younger bush is preferred by farmers and as yet there is little evidence of pressure on the fallow cycle.[1] Farm size is thus primarily a function of consumption needs, the health of the household head, and the availability of labor. Area per household planted to upland rice averages 1.4 hectares.

Farming begins in February or March with brushing, when the undergrowth is cutlassed and left to dry.[2] Trees are felled with an axe, and toward the end of the dry season (April–early May) the field is set on fire.

[1] See Currens (3) and *Production Estimates of Major Crops (24)*. The age of bush is estimated by the type of vegetative growth that predominates. In addition to population pressure, the age and health of the farmer are also important determinants of the fallow cycle, since older bush requires more effort in land clearing (1, 3). Pressures on the fallow cycle in Liberia appear less severe than in Sierra Leone, where a survey by Spencer (11) found average fallow times of 8.6 years for the southern region.

[2] The crop calendar in this discussion applies to the northern part of the country, where 70 percent of production occurs. Timing of rice production in the southern areas is one or two months ahead.

TABLE 4.2. *Quantities and Costs Per Hectare of Major Inputs in Liberia*

Production technique	Recurrent farm labor (*man-days*)	Seeds (*kg*)	Annual Land develop-ment cost (*dollars*)	Extension service cost (*dollars*)
Upland	214	50	0	0
Improved upland	231	50	0	24.67
Unimproved swamp	243	65	7.99	0
Improved swamp I	331	45	118.38	49.26
Improved swamp II	235	45	237.38	134.58

SOURCES: *3, 5, 8, 9, 11, 16, 17, 18, 19, 21, 22, 24*, and *25*. Interviews in Lofa, Nimba, and Grand Gedeh counties were also used to evaluate conflicts in data from the various sources and to complement existing data (e.g. the collection of 1975–76 input prices).

NOTE: 42 kg of fertilizer (nitrogen, phosphorus, and potassium) were applied in each of the three improved techniques; no fertilizer was used in the others. Fertilizer levels are those used in the Expanded Projects. The new World Bank projects in Lofa and Bong counties intend to use different proportions: 40–90 kg N, 40 kg P, and no K. The improved upland areas will receive 20 kg N, 20 kg P, and no K.

Stumps are left in the ground. These tasks are influenced by weather conditions, and early rains can cause a higher labor input per hectare or lower yields as clearing and weeding are made more difficult. The importance of the weather factor was demonstrated in 1975 when heavy rains in March caused a 10 percent decline in national production. In the succeeding year, 47 percent of farmers reported being able to make larger farms, and production returned to its 1974 level (*24*).

When the rains begin, rice seeds are broadcast and the land is lightly tilled with a hand hoe. Planting may commence at any time between mid-April and mid-July and continues for several weeks. There is large variation in the timing of planting although excessive delays result in lower yields, owing to the flushing of nutrients that occurs if the ground is left unoccupied during the early rains. Seeding rates are about 50 kilograms per hectare (kg/ha).[3] *Oryza sativa* is the dominant species and is represented by numerous varieties. *Oryza glabberima* is also present in most areas and comprises 1–10 percent of the plant population (*1*). Maturation times are 150–80 days. The presence of numerous varieties of rice and the long planting period result in an extended harvest period between September and mid-December, with most rice harvested at the end of the period. The genetic diversity probably also aids in resistance to pests and diseases.

The interim period is devoted to a number of tasks. Wooden fences with traps are erected around the field as protection from rodents, which

[3] Increases in the seeding rate do not effect yields, nor does changing the form of application (some southern counties use seed drilling). See Carpenter (*1*).

may take as much as 30 percent of the crop (1). An on-farm shelter and a scaffold to aid in birdwatching are also built. Weeding begins 6–8 weeks after planting and is normally done only once because the long fallow period effectively suppresses weed growth. Although highly variable, bird watching may be necessary for two weeks following seed germination and for four weeks before the harvest. Bird watching is usually performed by children and less able adults. The most severe damage occurs on the early- and late-maturing crops, thus placing additional constraints on the timing of farm operations.

Rice is hand-harvested with small knives. The stalks are gathered into bunches and stacked near the on-farm shelter for eventual transport to the village. Paddy is stored in a loft in the "rice kitchen." The area below the loft is used for cooking, with the smoke from the fires acting to inhibit mold, bacteria, and insects. The rice is threshed immediately prior to processing by drawing the stalks between the feet or flailing. While hand harvesting is a time-consuming process, the flexibility in the timing of planting and the presence of varieties with different maturation times allow a temporal dispersion of labor requirements and make the introduction of labor-saving techniques such as sickle-harvesting difficult. Harvest and post-harvest activities consume 36 percent of the total expenditure of 214 man-days/ha.

Pesticides, herbicides, and chemical fertilizers are almost nonexistent in traditional practice. The burning process acts as a partial substitute for chemical application. Because of the abrupt change in environment from bush to rice, losses due to disease and insect pests are low, although stem borers are occasionally reported. The success of the burn and the age of the bush are inversely related to the extent of weed development, since high forest discourages weeds and successful burns kill weed seeds. Finally, most of the nutrients available to rice are deposited on the soil as soluble ash after burning, with nitrogen provided by the accumulation and decomposition of leaf litter during the fallow. Most upland farms exhibit signs of nutritional stress in the latter half of growth, however, owing to nitrogen and phosphate deficiencies (1). Hence the quantity of vegetative cover that is felled and burned is an important factor in yield determination. Average yields are about 1,100 kg/ha.

The improved upland technology considered in this chapter involves the adoption of a seed-fertilizer package. Weed control with herbicides is not included in the package because of the uncertain effects on intercropped vegetables. The improved variety, LAC-23, selected from locally grown varieties, is red-grained and long-panicled, and has yielded 87 percent above traditional varieties on experimental plots. On-farm results are not yet available, although preliminary indications from pilot schemes suggest that a 50 percent increase in yields is a more realistic estimate of the impact of the package. It offers limited savings in labor input per unit of output because harvest and post-harvest requirements increase along with yield increases. Total labor requirements increase by about 8 percent above the traditional technique to 231 man-days/ha.[4]

The cultivation of small swampy areas scattered throughout the country is the second most important production technique, accounting for 15,000 mt of paddy. Introduced in the late nineteenth century from Guinea and Sierra Leone, swamp cultivation generally plays a supplementary role in subsistence production. Swamps are usually cultivated by women, who may have a claim on the upland farm rice supply but desire their own supply for cash purposes or social reasons such as the support of relatives on the periphery of the household organization. Average holding size is 0.6 ha (23, 24).

Swamp cultivation techniques differ little from those of the upland. Improved nutrient availability and retention of the swamp soils make fallowing necessary only once every four to five years. Fallow periods are only one or two years. Initially, small trees and undergrowth are removed. This operation does not have to be repeated as long as the swamp is cultivated. In July, after the upland fields have been planted, grass is cut or pulled from the swamp. When the fields are wet, the seed is pregerminated and then broadcast. Transplanting may be practiced in poorly drained swamps that remain flooded for most of the year. Transplanting results in a reduction of labor times for weeding but increases planting labor demand. With either seeding method yields average about 1,600 kg/ha, although broadcasting requires a higher seeding density of 65–110 kg/ha. Compared with upland production, swamp rice suffers less from rodent damage, but weed competition is severe. Bird watching is necessary for four to six weeks before harvest, and the small size of fields means increased labor requirements for this task on a per hectare basis relative to upland production. Rice is harvested in December and January.

The introduction of improved swamp technologies with increased water control, fertilizers, pesticides, and improved seeds has received the most attention from government planners, who envision that this type of production will eventually dominate rice production. Currently, improved swamp and irrigated techniques are responsible for 7,000 mt of paddy per year, or 3 percent of national production. Two of these techniques are considered here—the improvement of small swamps and the development of larger swamp areas of at least 100 ha each.

The first technique involves the improvement of traditional small

---

[4]The actual impact on labor usage may be greater than this, since LAC-23 is a hard-grained variety and thus more difficult to process than traditional varieties. Small mills, for example, charge a 40 percent premium in the one area marketing LAC-23. Hand pounding is also reported to be more difficult.

swamp areas with labor-intensive methods. This technique was introduced through the Expanded Rice Projects, in which teams of experts assisted villagers in the identification and development of suitable areas, and it is similar to the technique intended for dissemination through the World Bank Integrated Rural Development Projects. By 1975–76 about 500 ha had been developed with this approach. Development begins with the removal of tree and bush cover, including the stumps. This process is followed by the excavation of major drainage and irrigation canals, the separation of fields with small bunds, leveling of the fields, and finally the creation of tertiary canals and drains to supply individual fields. Leveling is the highest cost component in land development and the most critical as well, since otherwise areas of higher elevation will become weed-infested and rapidly have their fertilizer flushed into low-lying areas, while in the low areas the water will be too deep for adequate tillering and healthy root growth.

Recurrent operations begin with land preparation, which includes killing of weed cover. The field is then lightly flooded and puddled to reduce water and nutrient seepage. Seeds, usually South or Southeast Asian varieties, are grown in a nursery for three weeks before transplanting.[5] Seeding rate are 45 kg/ha. Plants are set out in lines for easy weeding and a basal dressing of fertilizer is applied. Weeding is done by hand. Fertilizer is applied twice more during the 120-day growth period and bird watching is necessary for a month before harvest. Yields average 3,500 kg/ha.

Labor inputs are high, on the order of 330 man-days/ha. This level is about 90 man-days/ha greater than labor requirements for the traditional swamp technique. About 65 percent of this difference arises from increased labor demand in land preparation and seeding. The remainder is due to increased harvest and post-harvest demands, since yields are more than double traditional levels.

The second major class of government projects, the Special Rice Projects, are partially mechanized. These projects are larger developments than the small swamps discussed above, occupying as much as 300 ha each. The scarcity of large swamp areas limits the potential of this technique, and only 750 ha have been developed. Land development is mechanized, a power tiller is employed in land preparation, and sickles and a pedal thresher aid in harvesting. Although two crops per year are possible, sufficient water is rarely available in the dry season unless costly dam

---

[5] Varietal development remains incomplete, and there is a need for a variety that is resistant simultaneously to blast and iron toxicity. Resistance to diopsids and stem borers is also desirable, particularly for continuous cultivation of a given land area (1). The Liberian experience with continuous cultivation is limited, and long-term potential pest-rice relationships are not known.

and pumping systems are introduced. In addition, the increased incidence of blast disease and heightened bird problems make rice an unattractive second crop. Yields approaching 6,000 kg/ha have been achieved, but the average is 3,500 kg/ha. Labor times are reduced to 235 man-days/ha, and, except for sickle harvesting and pedal threshing, recurrent operations are similar to those for the labor-intensive swamp technique.

Milling

Four milling technologies are prevalent in Liberia (see Table 4.3). Hand pounding is the dominant technique and is used almost exclusively to process rice for home consumption, although a small proportion of hand-pounded rice is marketed. Two types of small-scale mills with technical capacities of 0.25 mt paddy/hr are also widespread. The Japanese-made Yanmar rubber roller is the more common type, accounting for 160 of an estimated 220 small mills. The remainder are steel cylinder mills and simple hullers, primarily of British manufacture. Only the rubber roller type is locally marketed. Steel cylinder mills are mostly purchased in Sierra Leone. Finally, there are three government-operated Japanese Kyowa large-scale mills. They use the rubber roller principle, and each has a technical capacity of 2 mt paddy/hr.

Hand pounding is the traditional method of processing. Paddy is placed in a wooden mortar and pounded with a long pestle. After pounding, the rice is winnowed and the operation repeated, depending on the degree of milling desired. Three poundings are generally adequate. Outturns for home consumption are estimated at 65 percent. Asian data suggest 50 to 60 percent is a more appropriate range for marketable quality, and 60 percent is assumed in the estimation of commercial channel costs. Hand-

TABLE 4.3. *Key Characteristics of Rice-Milling Techniques in Liberia*

Milling technique	Projected full capacity (mt paddy/year)[a]	Quality of output (percent brokens)	Milling ratio	Rice milled in 1976 (mt milled rice)	Social cost ($/mt milled rice)
Hand pounding	7/person	40–50%	0.65	126,200	78.60
Small-scale rubber roller	64,000	25–35	0.67	16,000	53.37
Small-scale steel cylinder	24,000	35–45	0.65	6,000	51.07
Large-scale rubber roller	12,000	25–35	0.67	1,200	118.72

SOURCES: Data are from a survey of 25 small-scale mills and three government mills that was conducted by the author in 1976. Hand-pounding data are adopted from *12*.

NOTE: All techniques resulted in by-products used for feed except hand pounding, which had none.

[a] Estimates of actual capacity/shift are based on operating records and survey results. Full capacity assumes 250 days/year of one-shift work (10-hr shifts) for hand pounding and the small-scale mills. Two 8-hr shifts per day are assumed for the large-scale mills.

pounded rice contains about 40 percent brokens (12). Rice is processed daily for home consumption at a rate of 4.6 kg paddy/hour,[6] normally by women. It is difficult to determine the number of hours of pounding that constitute a full labor day because no large-scale commercial pounding operations are in existence. It is assumed below that pounding is done for six hours per day and that the remainder of the work-day is occupied with supporting activities such as bagging.

Steel cylinder mills remove the hulls from the rice by the abrasive action of a fluted steel blade inside a rotating steel cylinder. The rubber roller mills operate by passing the rice between a pair of rubber-sheathed rollers turning at different speeds. Pressure is applied across the grain width rather than the length, where the kernels are more uniform in size, thus improving outturns and reducing breakage relative to the steel cylinder mill. Rollers are spring-loaded, which provides a degree of self-control to the system and reduces the risk of damage both to the rice and to the machine. All mills evaluated in this study are equipped with polishers.

Overheating of the diesel engines and impurities in the paddy necessitate constant shutdowns, and only rarely are the technical capacity rates approached. More realistic rates of operation are 0.15–0.20 mt rice/hr. This is not an important constraint at current levels of throughput. Additional limitations on operations result from the consistent misalignment of the engine and the mill, causing excessive belt wear and machine breakdown. Mill owners also report great difficulty in obtaining spare parts and repair services. Down-times are on the order of two months per year. The steel cylinder mills are more resilient, since the hulling mechanism is more resistant to damage from impurities.

The large-scale government mills operate similarly to the small-scale rubber roller mills, except that larger roller size results in increased processing capacity/hr. The mills are equipped with sifters to remove stones and thereby to reduce the likelihood of damage to the polishing screens. Mills are electrically powered and suffer from occasional generator breakdowns. As with the small mills, spare parts and repair services are difficult to obtain. Down-times averaged 68 percent in recent years.

Small-scale milling is done almost exclusively on a custom basis, and thus only minimal storage capacity—3 or 4 mt—is available. Buildings are of minimum size, with walls made of local materials, a cement floor, and a sheet-metal roof. More substantial structures are required in the large government mills because of the increased scale of operation. Most of the paddy purchases occur at harvesttime. Since sales of rice are roughly uniform throughout the year, storage is required. Storage facilities of 400–500 mt are adequate for current levels of operation.

[6] See 12. This is close to the rate for Indonesia of 3.9 kg/hr (14).

Although small mills operate with two workers, the large mill requires seven. Four or five additional workers are required to operate the purchasing and warehouse operation. Shift work for the small mills is not generally practiced, since electricity is rarely available. The large mills operate multiple shifts sporadically.

Outturns average 67 percent for both small- and large-scale rubber roller mills and 65 percent for the steel cylinder mills, although the variance is substantial—from 50 to 70 percent. Outturns are primarily a function of operator skill and of the quality of the paddy, which varies with differences in grain size, content of impurities, and moisture content. Although there are great variations in both operator skill and paddy quality, paddy quality is the more serious problem. Bran outturn varies from 1 to 2 percent to as high as 7 percent, although the last figure probably indicates a large proportion of rice flour rather than true bran.

Transportation

Two final items—the transportation of rice from the farmer to the mill, and the movement of rice from the mill to Monrovia, the major point of consumption—complete the rice production and marketing scheme. The isolation of family plots from villages is an important constraint in obtaining inputs and marketing paddy. Because Liberia has only 2,500 kilometers (km) of all-weather roads, marketing generally requires head-loading the paddy from the farm either to the village or to the nearest road, on average a one-hour round trip (24). Then a pick-up truck is used to transport the paddy to the mill.[7] Rice that is hand-pounded at the village may be transported to a bulking point, such as a local market, and then shipped to Monrovia. Farmers generally control output through the processing stage.

Distance to small mills is probably not greater than 15 km for the majority of rice-marketing farmers. Transport charges average $0.12/km/person. Transport to the large mills is three to four times more costly, owing to the increased average distance from farm to mill. Much of the collection of paddy for these mills is done by commissioned government agents, although direct farmer transport and truck chartering is increasing in some areas.

Two techniques to transport rice from the mill to Monrovia are considered below. Both involve transport by 8-ton trucks. The cost of the technique that applies to small mills and hand pounding is greater than that for large mills because of bulking necessitated by the limited capacity associated with small-scale techniques.

[7] The National Rice Survey revealed that 56 percent of farmers who went to market used the above technique; the remainder walked the entire distance. Eighty-six percent of the farmers spent less than $0.90/person to reach the market. Marketing data specific to rice are not available.

Government Incentives and Sectoral Constraints

Government intervention in the rice economy has been limited.[8] Weak private price incentives, lack of information, and institutional and infrastructural constraints have hindered the expansion of new technologies. Attempts to draw the agricultural sector into the money economy must first relieve these constraints. The new production programs have been designed accordingly, and about half of project expenditures are directed to infrastructural and institutional improvements. Milling is a partial exception to the above generalization as private small-scale mills have spread rapidly through the countryside since 1973, but poor transportation infrastructure and the resultant segmentation of markets have contributed to low-capacity utilization rates and high costs relative to other West African countries.

Until recently, average returns per man-day for commercial rice production were below both market wages for unskilled labor and the returns earned in the production of the major alternative cash crops of cocoa, coffee, and sugarcane. As a result, annual marketings of domestic rice production were less than 20,000 mt paddy, about 10 percent of national production. In order to control the domestic price of rice and to stimulate production, the government instituted a variable levy system in 1975 to maintain rice prices at 1974 levels in spite of subsequent price declines on the world market. By 1977, the levy reached 50 percent of the c.i.f. price. Marketed output more than doubled during this period, to 39,000 mt paddy, but this figure represented only a small increase in total production. Response to the price increase was limited in part by a doubling of unskilled wage rates during the same period and concomitant increases in the world prices of coffee and cocoa (10).

Input price policy has been largely passive. Hand tools and agricultural chemicals are exempted from import duties, but subsidization policies on inputs are limited to the few hundred farmers participating in government projects. There has been no encouragement of the establishment of input marketing channels, and to obtain purchased inputs other than the traditional hand tools farmers must travel to Monrovia—on average a trip of 300 km (20). This situation has not proved attractive to farmers. The 1975 National Rice Survey revealed that only 2 percent of farmers used improved seeds and only 1 percent used fertilizer. With the exception of the World Bank project areas, no sales outlets for agricultural chemicals exist in rural areas.

The transportation and credit systems have been important constraints on the introduction of improved inputs and thus on the implementation of

[8] This section provides only a summary view. A fuller discussion is presented in 10.

government input subsidization policies. In transport the difficulty lies not in road transport costs per km, which are comparable to other West African countries, but in the availability of roads. Liberia has one of the lowest road densities in the region (0.04 km/km²), half of which are seasonal roads, and obtaining and delivering inputs to the farm involve substantial amounts of walking. Some rough calculations are illustrative of farmer access to roads. The interior counties of Lofa, Bong, and Nimba are responsible for 70 percent of national rice production and have 1,500 km of roads. If one assumes that land use is 20 ha per household and that two-thirds of the land is used for agriculture (thus probably underestimating road scarcities), at most 10,000 households, or 12 percent of the total, are within 1 km of a road.[9]

Credit plays only a limited role in the traditional farming systems. Increases in operating capital availability are an essential feature of the introduction of new production technologies, which imply the need for hired labor and purchases of chemical and mechanical inputs. Results presented in Table 4.7 highlight the importance of the credit constraint. The improved techniques require nonlabor expenditures per kg of paddy that are three to six times greater than those of traditional techniques. Expenditures per farmer must increase even more if both production and marketings per holding are to increase.

By 1977, only minor amounts of production credit had been made available to small farmers through government programs. Government credit was distributed through four cooperatives comprising less than 1 percent of farmers. About $2 million was spent annually on rice, primarily for fixed capital formation in government projects. Little was lent for annual input purchases, owing in large part to the inability of the cooperatives to obtain repayment from farmers. The result was a short lifespan for revolving credit funds.

A final institution affecting the rice economy is the extension service, which was established in 1960. By 1977 the service consisted of nine county agents, seven assistants, and 76 aides, or about one worker per 1,600 households. Service was limited primarily to larger farmers who requested assistance and to farmers participating in government projects. The 1971 Agricultural Census found that only 4 percent of agricultural holdings had received advice from extension workers.[10] Training was also a problem, since extension workers had on average less than nine years of education.

[9] The land use assumption is based on a ten-year fallow cycle and 2.0 ha/farmer/year, of which 1.5 is for rice. These calculations are based on county production data presented in the *Production Estimates of Major Crops* (24).

[10] The worker/household ratio is slightly higher, 1:1,400, for the four major rice-producing counties of Bong, Lofa, Nimba, and Grand Gedeh. The proportion of households getting advice, however, was 3.2 percent, slightly below the national average.

Finally, the lack of both a coherent policy and supporting funds to disseminate information further hampered its delivery to farmers.

This discussion indicates the need for caution in the interpretation of the cost analysis of the following sections. None of the improved techniques examined in this chapter has been widely disseminated and most are only in the pilot stages of development. Optimal seed varieties and levels of fertilizer application, for example, are still under study, and hence both the practices and the input levels for the improved techniques may be less reliable measures than those for the traditional techniques. Furthermore, time will be required to implement the improved techniques. The importance of a limited rural infrastructure and institutional constraints are clearly recognized by the new production programs, but the capability and cost of delivering new inputs and information to small farmers, particularly those in isolated communities, remain uncertain.

This is not to say that farmers are unresponsive to profit incentives. The relative profitabilities of rice, cocoa, coffee, and sugarcane and substitution in consumption between rice and cassava influence marketing decisions. Rice, as the staple crop, is also clearly influenced by noneconomic factors. Social obligations to kin and friends and thus demands for consumption tend to increase, for example, with increases in production. Nevertheless, the social structure has shown a remarkable ability to adapt to profitable opportunities. National production surveys found that the percentage of rice producers who market rice increased from 16 to 29 percent between 1971 and 1976.[11]

Liberian agriculture has in the recent past demonstrated the ability to adapt to new cash-cropping opportunities; the rapid adoption of coffee and cocoa are cases in point. Cash cropping was uncommon until the 1950's, and by 1960 coffee and cocoa exports were still each less than 1,000 mt per year. But the completion of a basic road network improved the linkages between Monrovia and the interior, and in 1967 the government began to implement an effective floor price scheme. By 1975, coffee exports were 4,128 mt and cocoa exports were 3,175 mt. Sugarcane has also become an important cash crop. In 1971, only 6 percent of farmers grew cane, but by 1976 this level had increased to 17 percent, and sugarcane had become almost as important as coffee and cocoa (10). Production surveys (3, 6, 9, 17, 18, 21, 22) have found that these crops offer substantially higher returns per man-day than rice.

Production incentives provided by government projects to participating farmers are described in Table 4.4. The partially mechanized technique, representative of the Special Projects, received the largest sub-

[11] Currens (4) has noted the recent development of personal rice fields where one person has control over the disposal of output, and he suggests this is a major source of expanded marketed output.

TABLE 4.4. *Farm Input Subsidies and Charges in Liberia*
(Dollars/ha)

Input subsidies[a]	Production technique		
	Improved upland	Improved swamp, labor intensive	Improved swamp, partly mechanized
Intermediate inputs			
Seeds	0.05	—	0.19
Fertilizer	1.24	1.24	1.24
Mechanical services			
Land preparation	—	—	2.00
Land development	—	—	−212.43
Extension	−24.67	−49.26	−134.58
TOTAL	−23.88	−48.02	−343.58
Net input subsidy in $/paddy	−15	−14	−98
Net input subsidy in $/mt rice	−22	−20	−146

SOURCES: Liberia, Republic of, *Tariff Schedules of the Republic of Liberia*, 1974. Other sources are the same as in Table 4.2.
   [a] Positive figures denote a net tax; negative figures denote subsidies.

sidy. Both the total cost and the level of subsidization of extension services are greater for this technique. Land development costs were almost entirely paid by the government, whereas the farmer bears these costs with the labor-intensive swamp technique.

In addition to input subsidies, protection against imports is also provided through a variable levy. This amounts to $144/mt rice and equals the difference between the 1976 Monrovia wholesale price for domestic rice of $458/mt and the c.i.f. equivalent price of $314/mt.[12] Table 4.5 indicates the dominance of the use of output rather than input price incentives and of higher prices to consumers rather than government subsidization to encourage production. This policy mix reflects both the importance of governmental budgetary constraints and the relative difficulty of implementing input versus output price incentives. Except for the partially mechanized projects, almost all protection is provided through the price differentials maintained between the c.i.f. and domestic selling prices for rice. Hence, effective protection coefficients are close to the value of the nominal protection coefficient on output of 1.46.

   [12] An equivalent c.i.f. price for domestic rice must reflect the quality differential between domestic rice and current imports. The 1975 monthly average retail price for domestically produced rice in Monrovia was $0.506/kg, whereas the average price for imported rice was $0.546/kg. Marketing margins equaled $0.048/kg. These were calculated on the basis of import price data as the difference between the retail price and the sum of c.i.f., port charges, tariffs, and the variable levy. The domestic retail price less the marketing margin gives a price for domestic rice delivered to the wholesaler of $0.458/kg. A comparable c.i.f. price for domestic rice was calculated by discounting the 1976 c.i.f. price (assumed a "normal" price year) by the ratio of domestic to imported wholesale prices. Thus an equivalent c.i.f. price for domestic rice was $314/mt, or $30/mt less than actual import prices.

TABLE 4.5. *Net Taxes (+) or Subsidies (−) of Production Systems*
(Dollars/mt rice)

Production technique	Total	Inputs (government)	Output (consumer)
Upland	−135	+9	−144
Improved upland	−157	−13	−144
Traditional swamp	−135	+9	−144
Improved swamp, labor-intensive	−156	−12	−144
Improved swamp, partially mechanized	−282	−138	−144

SOURCES: Same as for Tables 4.2–4.4.
    NOTE: These costs include paddy collection, processing, and delivery to Monrovia.

Factor Markets and Shadow Prices

Labor is the most commonly mentioned constraint in traditional agriculture. Opportunities for off-farm employment on rubber, timber, and iron ore concessions, in the larger cities as government or construction workers, and in local crafts and trade, has fostered a strong demand for off-farm labor. The 1971 Agricultural Census found that 17 percent of holdings had a member of the household engaged in a nonagricultural occupation, representing about 4 percent of the agricultural work force.[13] Microlevel studies suggest that this is not a static group. Rather, off-farm jobs are taken for a temporary period of a few months to a few years. As a result, over an extended time period the proportion of families participating in the off-farm market is higher, probably between 35 and 45 percent.[14]

In addition, the margin for variability in the timing of upland farming tasks has allowed for the redistribution of labor time in the crop calendar as opportunities for cash cropping have developed. Additional time is also occupied with other food crops. Labor resources have thus been nearly fully exploited. Farmer complaints about loss of leisure time and six-day work weeks (3) also suggest that at current returns there is little available surplus labor.

The substantial use of hired labor, a recent phenomenon coinciding with the increased importance of cash crops, completes the picture of a well-integrated labor market. Hired labor is not generally employed di-

[13] McCourtie's national survey of 1,281 farmers in 1970–71 found that 19 percent of the farmers had off-farm jobs (9). Seventy percent of these were engaged in local trades and crafts. In the Agricultural Census (24), the agricultural work force was estimated at 562,000, or 68 percent of the agricultural population. Some 24,800 people reported working off-farm.
[14] Carter (2) found that 47 percent of the male residents in Zolowo, a town in Lofa County, had worked off-farm. Van Santen (17), in another Lofa survey, found that 50 percent of the holders had worked off-farm. A national survey by McCourtie (9) yielded a figure of 35 percent.

rectly on cash crops but substitutes for family labor in staple food production, thus releasing family labor for cash crop production. Whereas hired labor was uncommon in the 1950's, by 1975, 35 percent of farm households used hired labor at some time in the production cycle, primarily for the initial upland tasks of brushing and clearing.[15] About 10 percent of the total labor input comes from hired sources.[16] Hired workers are usually migrants, although local work groups of women or men are also available in most areas.

Given the existence of a tight labor market, the question arises as to what determines the wage rates for hired labor. Assuming constant returns to scale, labor is paid the value of its average (and marginal) product. But returns are often lower in the subsistence than in the cash crop portion of output, owing perhaps to the willingness to accept lower returns in exchange for increased security of food supply.[17] If so, the farmer divides his own labor between nonrice farming tasks and rice farming until returns in rice farming plus the value attached to risk reduction equal market wages. The net demand for hired labor is largely determined by cash crop–related demands and the substitution of labor for leisure time on the part of the farm family. With competitive demand, hired labor is paid the average (and marginal) value product of cash crop production. Thus earnings in the subsistence sector plus a risk discount (which will tend toward zero as the willingness to rely on market sources increases) equal returns in cash cropping and also the market wage for unskilled labor.[18]

In order to induce consistent cash cropping of rice, returns to the farmer must be equal to the returns he can earn in other cash crop activities. Cash crops, not subsistence crops, are the foregone alternatives in

---

[15] Data from the National Rice Surveys (23, 24) are presented below. The use of hired labor on subsistence crops rather than cash crops reflects both the relative preferences of family labor and the need for family supervision of cash crops to prevent theft.

Task	Percent of households 1975	1976	Task	Percent of households 1975	1976
Brushing	35	35	Weeding	9	14
Clearing	28	25	Birdwatching	2	5
Planting	21	21	Harvesting	14	15

[16] McCourtie (9) found that 38 percent of farmers used hired labor at an average intensity of 35 man-days/ha, or 10 percent of labor requirements. Further data are provided by Currens (3), who found that 12 percent of total farm labor comes from hired or nonreciprocated sources. The remainder comes from the household and cooperative or reciprocal work groups.

[17] This does not disallow the possibility for substitution in consumption among crops to maximize income in the face of price changes that occur between the time of the decision to plant and the harvest.

[18] This view of risk differs from others which predict that risk aversion will result in expanded planting area and marketings beyond the level that would be predicted by competitive price considerations alone. The implications of both portrayals for wage rates are the same.

the decision to market rice.[19] To the extent that government intervention causes domestic prices to be below world prices for cash crops,[20] returns to cash cropping are artificially low and thus the shadow price of labor exceeds the market wage. Time-series data for producer and f.o.b. prices for cocoa and coffee were used to estimate the effective tax on output. Averages for 1963–66, when there were no taxes, establish a "free market" relationship between the two prices, with the margin representing transport and handling costs. For cocoa, local prices were 83 percent of f.o.b. prices; for coffee the figure was 74 percent.[21] During the 1973–74 period, production was aggressively taxed and local prices fell as a proportion of export prices for both coffee and cocoa. For cocoa, local prices were 58 percent of export prices; for coffee, 54 percent. Government policies are thus able to maintain an artificially low return per man-day in the cash crop sector and to depress wage rates. Assuming that substitution possibilities in production result in an equalization of returns in the cash crop markets (cocoa, coffee, and sugarcane), a shadow wage rate 40 percent above market wages would be reasonable.[22]

[19] This assumes that the farmer has access to the complementary resources (land and capital) necessary for cocoa, coffee, or sugarcane production.

[20] Government purchases through the Liberian Produce Marketing Corporation began in 1966. Purchases are not necessarily made directly from farmers; traders may act as intermediaries. In 1975–76 an average of 52 percent of farmers sold some or all of their crop to traders (24). Data for coffee and cocoa purchases are presented below.

	Coffee (mt)		Cocoa (mt)			Coffee (mt)		Cocoa (mt)	
Year	LPMC purchases	Exports	LPMC purchases	Exports	Year	LPMC purchases	Exports	LPMC purchases	Exports
1963		3,674		1,043	1970	5,105	4,944	2,102	1,633
1964		7,847		1,542	1971	5,284	5,534	2,526	2,767
1965		3,175		726	1972	3,287	5,570	2,659	3,175
1966	8,162	8,890	1,490	1,542	1973	4,605	6,940	2,962	2,404
1967	4,060	4,173	1,710	1,451	1974	3,815	3,402	3,074	3,266
1968	5,552	4,672	1,827	2,268	1975	4,161	4,128	2,550	3,175
1969	4,623	4,264	1,102	1,905					

[21] Price data are drawn from McCourtie (9) and the *Quarterly Statistical Bulletin of Liberia, 1975 Summary* (28).

	Coffee (¢/lb)		Cocoa (¢/lb)			Coffee (¢/lb)		Cocoa (¢/lb)	
Year	Local	Exports	Local	Exports	Year	Local	Exports	Local	Exports
1963	15.9	19.1	15.3	18.8	1970	17.0	20.3	18.0	27.8
1964	19.0	23.7	15.6	17.6	1971	17.5	33.0	18.0	20.6
1965	14.8	21.1	11.9	15.3	1972	15.0	36.9	15.0	20.9
1966	20.0	29.5	13.2	15.6	1973	19.6	33.1	21.2	35.5
1967	18.0	26.2	16.0	19.6	1974	26.6	52.6	30.2	59.1
1968	18.0	27.6	16.3	24.3	1975	27.0	51.2	36.0	55.0
1969	16.0	27.0	17.5	37.6					

[22] This estimate is based on the following formula: $P + tP + T = $ f.o.b., where $P = $ local price, $t = $ export tax, $T = $ transport and trader margins, and f.o.b. = f.o.b. price, Monrovia. For cocoa, $T = (0.17)$ (f.o.b.) and $P = (0.58)$ (f.o.b.). So $1 + t = 0.83/0.58 = 1.43$ and $t = 43$ percent. For coffee, $T = (0.26)$ (f.o.b.) and $P = (0.54)$ (f.o.b.). So $1 + t = 0.74/0.54 = 1.37$ and $t = 37$ percent.

However, an offsetting adjustment should be made to account for domestic prices of rice being above world levels. In Liberia high rice prices are maintained through a specific tax on imports of $11.01/mt and a variable levy to maintain a constant wholesale price for imported rice (35 percent brokens) in Monrovia of $475/mt. Using 1976 as a "normal" year, when c.i.f. prices were $344/mt, an ad valorem tariff equivalent is 40 percent. On balance, then, price policies in the rice and cash crop sectors appear to have largely offsetting effects on wage rates, and the market wage of $1.25/day is a good approximation of the shadow price used in this analysis. [23],[24]

In general, there is no pressure on land resources. Upland fallow cycles are 7–30 years, depending on farmer preferences (3, 6). Land may be purchased from the government for $2.50/ha, but most is held communally by villages, with permission to farm granted by the village authorities. Less than 5 percent of farmers report land as a constraint to establishing a larger farm (9). Near major roads and some of the larger towns there is pressure on the fallow cycle, reflecting site value and a constraint in the form of territorial boundaries of towns. The problem of local population pressure is often overcome, however, by leasing land from neighboring villages for a minimal fee. Since swamp areas are in equally abundant supply, both the market and shadow prices of land equal zero. [25]

Capital markets are poorly developed in agriculture because of both supply and demand factors. On the supply side, institutional lending has

[23] The use of a single wage rate disguises regional variations. Reported wages were as high as $1.50/day around areas of strong demand such as mines and were as low as $0.75/day in the more remote areas. The assumption here is that these differences are offset by transport cost differences in the collection and delivery of rice to Monrovia. Similarly, there are undoubtedly differences in daily wages between men and women. But the similarity of charges for contract labor regardless of sex and recent Sierra Leonean survey data suggest that this difference is due to differences in hours worked per day rather than productivity differences. As long as labor estimates are in terms of man-days, there is no estimation problem.

[24] A potentially important distortion in these conditions for the Liberian rice economy that will raise the opportunity cost of labor involves the elements in the social structure that discourage the marketing of rice. There are, for example, social restrictions against the use of cooperative labor to produce rice for sale. An additional hindrance is the aforementioned tendency for producer obligations to increase with output. These factors imply that profit incentives may have to be higher than shadow wage rates suggest. To the extent that such social factors differ among ethnic groups, a regional bias in marketing is encouraged. Such biases already appear—in Lofa County an average of 8 percent of households reported selling rice during the 1974–76 seasons. For Nimba County, the average was 39 percent, for Grand Gedeh 45 percent, and for Bong 31 percent. Finally, the assumption implicit in the calculation of shadow wages is that increases in output prices result in similar proportional increases in the marginal value products of each factor. Given limited substitutability of factors in traditional techniques over the price range considered here and the predominance of labor costs in the traditional technologies, this seems a reasonable assumption.

[25] This assumes that the land is a completely renewable resource—i.e. that there are no long-run problems of erosion and the potentially marketable stands of timber that are destroyed in the older forest can be fully reestablished.

been almost exclusively for mechanized land development, and this portion of government expenditure is to be phased out by 1980. Bank lending is not available in the interior. Farmers rely primarily on relatives and village elders for loans. Traders appear to be only occasional sources of credit (17, 18). Interest is generally paid in kind and effective interest rates are not known.

In the commercial capital market, nearly 20 percent of outstanding loans of $15.3 million by banks have gone to agriculture and about 90 percent of these have been made to rubber and forestry concessions. No loans are extended to rice-related activities. Interest rates for term loans greater than six months range between 8.5 and 15 percent, with a mode of 12 percent (26). Bank officials suggest that loans made to small-scale farmers would require interest rates substantially above this level.

On the demand side, fixed capital needs are minimal in traditional agriculture and are limited to a few hand tools. Draft animals are not used because of disease problems, and mechanical implements are used only in larger government project areas, currently comprising 750 ha. Tree crops represent the most substantial form of capital investment in traditional agriculture, and returns are above 40 percent.[26] Though the marginal value product of capital appears high, the response of this rate of return to increases in aggregate small farmer investment (i.e. the elasticity of demand for financial capital) is not known. The rate targeted for annual loans by cooperatives and small farmer development projects is 15 percent, and this figure is used in this study as the real opportunity cost of capital.[27]

Private and Social Costs and Benefits

Production

Five production techniques are compared in Tables 4.6 and 4.7. All production costs are grouped with the costs of small-scale rubber roller processing and delivery to Monrovia, the dominant market route in Liberia. This categorization also allows for an analysis of the specific effects of the various production techniques on total costs. It is important to note that these are largely hypothetical constructs, since very little rice actually flows from rural areas to Monrovia.

Of the two traditional techniques that currently dominate production,

[26] See 17, 18, 21, 23. Ministry of Agriculture calculations show internal rates of return (IRR) of 40–45 percent, at official producer prices. At world prices, the IRR is substantially higher. The calculations assume annual labor inputs of 50–60 days/ha, 500 kg/ha of fertilizer, and yields of 1,100 kg of output from years 9–25 of production.

[27] This is consistent with the view of opportunity cost as the return that can be gained in the best alternative investment. Thus the determination of opportunity cost is based on an estimate of internal rates of return rather than on the supply price of capital.

TABLE 4.6. *Indicators of Private and Social Profitability in Liberia*
(Dollars/mt milled rice)

Technique	Private cost	Social cost	Net private profit-ability	Net social profit-ability	Effective rate of protection	Resource cost ratio
Monrovia delivery:[a]						
Upland	554	545	−96	−231	1.46	1.78
Improved upland	520	533	−62	−219	1.62	1.99
Unimproved swamp	464	455	−6	−141	1.46	1.48
Improved swamp, labor-intensive	416	428	42	−114	1.52	1.44
Improved swamp, partially mech-anized	349	488	109	−174	1.70	1.69
Home consumption:[b]						
Upland, hand pounding	520	520	42	−128	1.43	1.33
Improved swamp, labor-intensive, hand pounding	378	398	184	−4	1.47	1.01

SOURCES: Same as for Tables 4.2–4.4.

[a]Activities listed under Monrovia delivery represent average costs for each production technique combined with the costs of collection, small-scale rubber roller milling and delivery to the Monrovia wholesaler. Private profitability is based on a Monrovia wholesale price for domestic rice of $458/mt. Social profitability is based on a c.i.f. domestic-equivalent price of $314. These are essentially hypothetical constructions, since actual deliveries are very small under current conditions. Thus a negative private profitability implies that the technique is not attractive to the average producer.

[b]Activities listed under home consumption exclude costs of collection and delivery. Hand-pounding processing is assumed. Private profitability is based on rural market retail prices of $525/mt plus delivery costs to the farm gate. Social profitability is based on a price of $394/mt for farm-gate delivery of imported rice. This includes the c.i.f.-equivalent price for domestic rice, transport costs, and retailer margins of $24/mt. Imported rice, when available, sells at a price of $595/mt in rural producing areas.

the swamp technique appears the most attractive as a source of cash income. Yields in traditional swamps are 48 percent greater than those on the uplands, but labor requirements increase only 14 percent as increased weeding, harvest, and post-harvest requirements are somewhat offset by lower land preparation and seeding times. The results support the observation that the relative importance of swamp rice in marketed output exceeds its relative importance in total output (4). Even so, traditional swamp farming does not offer on average a positive return for Monrovia delivery. Thus negative average private profitability for Monrovia delivery of rice produced with traditional techniques is consistent with observed commodity flows.

The preceding paragraph does not imply that producers who market rice are earning negative profits. Farmers able to achieve above-average yields or use less than the average quantity of labor inputs will attain above-average profitability. In statistical terms, profitability is a concept that has a distribution (variance) as well as a mean value. The variance properties and aggregate production statistics from Table 4.1 can be used

TABLE 4.7. *Production Data*

Technique	Output/ man-day (kg paddy)	Private returns/ man-day ($/day)	Farmer nonlabor expenditures[a]	
			$/ha	$/kg paddy
Monrovia delivery:				
Upland	4.9	0.93	33.95	0.03
Improved upland	6.8	1.01	127.73	0.08
Unimproved swamp	6.4	1.22	38.00	0.02
Improved swamp, labor-intensive	10.6	1.50	266.00	0.08
Improved swamp, partially mechanized	14.9	2.34	228.74	0.06
Home consumption:				
Upland	4.9	1.39	33.95	0.03
Improved swamp, labor-intensive	10.6	2.56	266.00	0.08

SOURCES: Same as for Tables 4.2–4.4.
[a]These do not necessarily represent cash expenditures, but include some imputed values for items such as seeds and working capital.

to elaborate the relationships between profitability and observed marketing patterns. If profitability is assumed to have a normal distribution, then half of traditional swamp production, or about 4,200 mt rice, appears profitable for Monrovia delivery. If half of the improved swamp production (2,300 mt rice) is assumed profitable for Monrovia delivery, these two techniques account for over half of total Monrovia marketings of 12,000 mt rice (10). Upland marketings can be estimated as a residual of 8,000 mt paddy, only 4 percent of upland production. This pattern is consistent with the strongly negative private profitability of −$96/mt for Monrovia delivery. In spite of substantial tariff protection, cash cropping of rice for Monrovia delivery was not a profitable activity for the vast majority of upland farmers.

At the level of private incentives, both of the improved swamp rice techniques show positive net profitability for delivery to Monrovia, ranging from $42/mt to $109/mt rice. Relative to upland production, these techniques have higher yields/man-day, and this, combined with the incidence of input subsidies, more than offsets higher nonlabor input costs for fertilizer, land development, and extension services. As Table 4.7 shows, private returns/man-day for commercial marketing are greater than market wages ($1.25/day) for the improved swamp techniques, and significantly so for the partially mechanized technique. The difference in returns to swamp production is due entirely to the incidence of subsidized government services. Without the subsidies, private returns/man-day fall to $0.88 for the partially mechanized technique and $1.35 for the labor-intensive technique.

All of the improved techniques provide greater private production incentives than their traditional counterparts. Thus successful dissemination of the new technologies and replacement of traditional production should result in increased marketings of rice for Monrovia delivery and thus contribute to the goal of self-sufficiency. Two techniques, the labor-intensive improved swamp and the improved upland, are receiving the greatest emphasis in current programs. The former technique appears the most promising in terms of profitability, since average returns are $42/mt rice. The improved upland program offers higher returns than traditional upland cultivation, but average earnings remain below market wage levels. Hence, contrary to government expectations, the adoption of the improved upland packages is not likely to have a substantial impact on marketed supplies.

The achievement of self-sufficiency with the new technologies appears to have a substantial opportunity cost. All of the systems discussed here have negative net social profitabilities (and thus resource cost ratios greater than one), ranging from −$114 to −$231/mt rice. The least attractive techniques in terms of social costs are the upland and the partially mechanized swamp techniques. Table 4.8 provides a breakdown of social costs by input category. The savings in labor costs attained by the improved techniques relative to their traditional counterparts result from the yield-increasing effects of the packages, which spread the labor input for some activities such as land preparation over a greater output. Capital and tradable input costs, primarily fertilizer, land development, and extension service costs, however, are increased.[28] The labor-intensive improved swamp technique, for example, reduces labor costs by $122/mt (including costs of processing and delivery to Monrovia) relative to the traditional swamp technique, but it increases capital and tradable input costs by $95/mt. The social profitability of import substitution remains negative, suggesting that further technical change would be desirable. The adoption of sickle harvesting, simple mechanical threshing, and increases in the efficiency of fertilizer use may be areas where significant cost reductions can be introduced.

The difference between private and social costs is captured by the net input subsidy. The post-production techniques are taxed, thereby reducing the effect of subsidies given to production. On balance the traditional techniques are taxed at a rate of $9/mt rice, whereas the improved techniques are subsidized from $12 to $138/mt (3–28 percent of social costs). It is difficult to determine the limits to governmental willingness to sub-

[28] Data for this calculation are contained in the production budgets available from the author. In the partially mechanized technique, down-times for mechanical equipment also cause increased costs. But even if 100 percent utilization is assumed, rice production costs are reduced by only $14/mt rice, and net social profitability rises to −$160/mt rice. The ranking of the techniques does not change.

TABLE 4.8. *Social Costs*
(Dollars/mt rice)

Technique	Labor	Capital	Tradables	Total
Delivery to Monrovia, machine-processing:				
Upland	440	87	18	545
Improved upland	340	100	93	533
Traditional swamp	350	87	18	455
Improved swamp, labor-intensive	228	147	53	428
Improved swamp, partially mechanized	184	242	62	488
Home consumption, hand pounding:				
Upland	488	32	0	520
Improved swamp, labor-intensive	269	93	36	398

SOURCES: Same as for Tables 4.2–4.4.

sidize rice production. But the recent decision to terminate the expansion of the partially mechanized swamp package suggests that an input subsidy of $100/mt paddy exceeds these limits and thus may serve as an upper bound on the budgetary constraint. An increase in production of 40,000 mt of rice (equal to the volume of 1977 imports) would involve annual government subsidies of $6 million in the partially mechanized approach and at least $1.2 million in the labor-intensive swamp technology (assuming that a 10 percent return on total costs is sufficient to induce farmer participation). Total recurrent expenditures in agriculture were $6.2 million in 1975.

The replacement of imports with expanded domestic production is, of course, much different from the replacement of domestic production with imports. This reflects the role of transport costs and the shift in the consumption point from Monrovia to the farmgate. In terms of social costs, the border price of rice is increased relative to the Monrovia price by the costs of transport and distribution. Furthermore, since the farmgate is the consumption point, total costs are reduced by the same transport cost margin that affects the border price of rice. These effects are summarized in the last two lines of the Tables 4.6, 4.7, and 4.8. Given existing tax/subsidy policies, production for on-farm consumption is privately profitable. Even though hand pounding is more costly than the mechanical milling technique used in the Monrovia delivery systems, private profitabilities are increased by $140 per mt relative to Monrovia delivery. Thus farmers find it more profitable to produce for their own needs than to rely on market sources. This finding conforms to observed production patterns, since 85 percent of production is home-consumed and over 50 percent of rice producers are self-sufficient (*10*).

All of the improved techniques increase private and social profitability relative to the traditional techniques and thus contribute to increases in agricultural sector income. The labor-intensive improved swamp, for example, increases profitability for home consumption by over $140/mt relative to the traditional upland. Returns per man-day are double market wages, and labor productivity is more than twice that of the traditional upland technique (10.6 versus 4.9 kg paddy/man-day). Adoption of this technique should increase the proportion of self-sufficient rice producers and offer scope for increasing rice consumption, rice marketings, or production of other cash crops.

But given that the improved techniques are more efficient than traditional techniques and that social profitability is nearly positive for home consumption, increases in the level of home consumption and the release of farmer resources for additional income-generating opportunities appear attainable without the substantial income transfers from consumers to producers that result from the variable levy policy. The only effect of the variable levy and other agricultural tax policies, described in the preceding section, is to influence the flow of resources into cash crop activities. Dissemination of improved rice production techniques and maintenance of subsidies on rice and taxes on alternative cash crops will bias the flow of resources toward rice, and thus reduce reliance on rice imports. But this set of policies, while redistributing income from consumers to producers, does not serve to maximize national income. Furthermore, these policies may not maximize real farmer incomes either, since the encouragement of alternative cash crops (through the elimination of coffee and cocoa taxes, for example) may result in returns to farmers equal to or greater than those offered by marketing rice.

Post-Production

Post-production technologies examined in this section are described in Tables 4.9 and 4.10. They are distinguished by the milling technique, but differences also exist in collection and distribution, since both are influenced by the type of processing. All techniques are combined with labor-intensive improved swamp production to facilitate comparison of delivering domestic rice to Monrovia vis-à-vis the alternative of importation. Post-production costs are 20–25 percent of total production costs for the small-scale processing systems. The range for total private costs is $416–511/mt and for social costs $428–517/mt. The large-scale rubber roller mills, which are government-operated, have private and social post-production costs roughly double those of the small-scale techniques. The ranking of small-scale post-production costs differs between Table 4.9 and 4.10, since the effect of differences in outturns of farm-level costs outweighs the small differences in post-production operating costs. Owing to

TABLE 4.9. *Indicators of Private and Social Profitability, Monrovia Delivery of Improved Swamp Production*
(Dollars/mt milled rice)

Technique	Costs		Profitability		Effective rate of protection	Resource cost ratio
	Private	Social	Net private	Net social		
Hand pounding	449	484	9	−170	1.52	1.64
Small-scale rubber roller	416	428	42	−114	1.52	1.44
Small-scale steel cylinder	423	436	35	−122	1.52	1.47
Large-scale rubber roller	511	517	−53	−203	1.56	1.83

SOURCES: Same as for Tables 4.2–4.4.

high outturns, the rubber roller is the least expensive system, followed by the steel cylinder and the hand-pounding systems.

In contrast to the farm sector, government incentives have little effect on the costs of post-production activities. On balance there is slight discrimination against the sector. The small-scale techniques are taxed between $5/mt and $8/mt rice, reflecting tariffs on tradable inputs. The large-scale milling technique exhibits a slightly greater divergence between private and social costs of $16/mt, owing largely to the government policy of paying unskilled labor at wages 60 percent above market rates.

For the small-scale techniques, transportation and bulking costs are about equal in magnitude to milling costs. In the government channels the ratio is 0.75. Hand-pounding offers the lowest collection costs because of the savings that result from the transportation of milled rice rather than of paddy. Rubber roller mill costs are lower than steel cylinder mill costs as a result of higher outturns. The government collection system is nearly twice as costly as the small-scale techniques, owing to the use of commissioned agents and the higher transport costs involved in moving paddy over greater distances to reach the mill. Costs of delivery to Monrovia are lower for the government mills, since some bulking costs are avoided at this stage. In total, however, the more centralized collection and distribution system is more costly.

In milling, the small-scale techniques are again roughly similar, with costs of $53–56/mt. The mechanical mills are cheaper than hand pounding and demonstrate a profitable substitution of capital and purchased inputs for labor. The steel cylinder mill is slightly lower-cost than the rubber roller mill because the lower outturn is more than offset by lower capital charges on equipment owing to lower initial costs and longer equipment life, as well as lower spare parts costs. Outturns from the steel cylinder mill contain a higher percentage of brokens than those from the

TABLE 4.10. *Costs of Post-Production Activities by Production Technique Used*
(Dollars/mt rice)

Technique	Collection		Milling		Delivery to Monrovia		Total	
	Private	Social	Private	Social	Private	Social	Private	Social
Hand pounding	40	39	78	78	24	20	142	137
Small-scale rubber roller	46	44	56	54	24	20	126	118
Small-scale steel cylinder	48	45	53	52	24	20	125	117
Large-scale rubber roller	73	72	130	118	20	17	223	207

SOURCES: Same as for Table 4.3.

rubber roller mill, but this does not effect the price of output on the market.

Milling costs of the government mills are more than double those of the small-scale techniques, with major differences occurring in labor costs ($51/mt vs. $32/mt) and capital costs ($66/mt vs. $21–24/mt). Technical inefficiencies in the large mills account for some of this difference. Payment of official wages by the government mills is responsible for $9/mt of the difference in labor costs. Since hiring practices make labor a largely fixed cost, capacity utilization and operating down-times seen in Table 4.11 also contribute to high labor costs. Differences in capital costs result partly because the capital charges on buildings and equipment are spread over a smaller throughput in the government mills.

The small mills appear profitable to operate even at low annual average rates of capacity utilization, and they run at much higher utilization rates than government mills. But the economic advantages of the small-scale systems extend beyond capacity-utilization factors. Although increasing utilization to 100 percent reduces large mill costs by $61/mt and small mill costs by $28/mt, this is not sufficient to offset the existing differences in the costs of the systems.[29] Losses, for example, are 250 percent greater for the government mills relative to the small mills, which rely on on-farm storage. Finally, these cost factors are an important offset to the technical advantages claimed by proponents of large-scale mills, who argue that higher outturns of rice can be attained than under the small-scale technologies present in Liberia. An increase in outturns to 70 percent, for example, reduces effective farm production costs per mt rice by only $18–

[29] For the large mills, unskilled labor is assumed to increase by only 60 percent (based on an operation of two shifts of the mill crew and one shift of the warehouse crew). Skilled labor requirements remain unchanged. Data for the small-scale mills are based on the rubber roller technique. Unskilled labor costs are doubled and skilled labor costs remain constant.

TABLE 4.11. *Milling Capacities*
(Mt rice/hr)

Capacity	Small-scale mills	Large-scale mills
Technical capacity[a]	0.25	2.0
Actual capacity[b]	0.16	1.0
Average production[c]	0.06	0.15
Capacity utilization (*percent*)[d]	35	7.5

SOURCES: Same as for Table 4.3.
 [a]Manufacturer's estimate.
 [b]Based on reports of millers and allows for shutdowns to cool the diesel engine and to remove impurities from the paddy.
 [c]Small mills: based on average down-times of 2 months/year, and 6 days/week operations the rest of the year (250 days). Large mills: based on daily records for 16 months of operation for the three mills between Oct. 1976 and Apr. 1977.
 [d]Based on 1 shift/day (10 hr) for small mills and 2 shifts/day 8 hr/shift) for large mills. Multiple-shift work is not normally practiced in small mills, since lack of electricity and storage facilities and bookkeeping and administrative skills necessary for night work are not generally present. Large government mills are equipped to handle multiple-shift work. Of 124 operating days covered in this survey, 16 days had double shifts and two days had triple shifts.

25 for the techniques studied here. This is not sufficient to offset the differences in costs between large- and small-scale processing systems.

The technical superiority of the large mills has no economic value because market prices for rice do not vary with the four processing techniques studied here. Consumers, at least in Liberia, do not strongly discriminate among the range of qualities of raw rice represented by the outputs of these techniques. Technical superiority at increased costs without increased benefits is, in economic terms, inappropriate. Although price differences do exist between domestic and imported rice, these price effects are determined by pre-processing considerations—paddy quality and the presence of impurities, fermented or immature grains—and these problems are not rectified with the processing technologies considered here.

Sensitivity Analysis

The sensitivity of the results to changes in assumptions is discussed within an elasticities framework. The values presented in Table 4.12 represent the percentage change in social profitability that results from a 1 percent change in the value of the parameters listed as column headings. Only the production techniques from Table 4.6 are presented, since differences among post-production activities have little influence on the elasticity values. Yields and unskilled labor are the most important parameters. But as Table 4.13 demonstrates, values for man-day inputs, wage rates, and yields that would be required individually to cause social profitability to be positive differ substantially from estimated values. This suggests that the conclusions presented here are relatively insensitive to changes in estimates and assumptions. Skilled labor and capital have little

TABLE 4.12. *Elasticities of Net Social Profitability with Respect to Yields and the Social Cost of Primary Inputs*

Technique	Yields	Social costs, unskilled labor	Social costs, skilled labor	Social costs, capital
Monrovia delivery:				
Upland	1.83	−1.76	−0.14	−0.38
Improved upland	1.82	−1.30	−0.22	−0.43
Traditional swamp	2.37	−2.23	−0.24	−0.62
Improved swamp, labor-intensive	2.70	−1.64	−0.36	−1.30
Improved swamp, partially mechanized	2.12	−0.78	−0.28	−1.40

SOURCES: Calculated from data in Tables 4.1, 4.2, and 4.8.

TABLE 4.13. *Yields, Inputs of Unskilled Labor, and Wages, Estimated Levels, and Levels Required for Positive Net Social Profitability*

Technique	Yield (mt/ha)		Unskilled labor (man-days/ha)		Wages ($/man-day)	
	Estimated	Required	Estimated	Required	Estimated	Required
Upland	1.05	1.63	214	92	1.25	0.54
Improved upland	1.58	2.43	243	56	1.25	0.29
Traditional swamp	1.55	2.20	243	134	1.25	0.69
Improved swamp, labor-intensive	3.50	4.80	331	129	1.25	0.49
Improved swamp, partially mechanized	3.50	5.14	235	< 0	1.25	< 0

SOURCES: Same as for Table 4.12.

influence on net social profitability. Real skilled wages must be negative and real interest rates less than 1 percent in order for net social profitability to be positive.

The approach used here is useful to simulate additional changes that might occur in the input-output coefficients of the new technologies as the producers move from production for home consumption into the commercial sphere. For example, one of the major difficulties with the introduction of sickle harvesting into upland production is the traditional use of multiple varieties with varying maturation times. Under commercial cropping, however, decision criteria may change and allow the cultivation of pure stands with uniform maturation times. Introduction of sickle harvesting into the improved upland package could reduce unskilled labor input by 10 percent and raise social profitability by 13 percent. Introduction of sickle harvesting into the labor-intensive swamp package reduces labor input by 15 percent and increases social profitability by 25 percent.

These changes are not sufficient to induce positive profitability for delivery to Monrovia, but social profitability for home consumption increases to $96/mt rice.

The elasticity values of Table 4.12 suggest that new techniques or improved seeds that can increase yields with a lesser relative impact on increased input requirements in other factor categories (especially unskilled labor) offer the most promising means to reduce social costs. Improvements in the adaptation of Asian rice varieties to Liberian environments, better techniques of vertebrate pest control, intermediate mechanical aids in harvesting and processing, such as sickles and threshers, and improvements in fertilizer response and input distribution systems may provide sufficient cost reductions in the new swamp technologies to induce positive social profitability. But in the upland environments, where social costs of the improved upland package are $105/mt greater than those of the improved swamp, the magnitude of the effects of a new package must be large. With the improved upland package, for example, even if the yield increases achieved on experimental plots (87 percent above traditional yields) are attained, net social profitability remains strongly negative (−$123/mt). Thus the results of Tables 4.12 and 4.13 highlight the need for further development of upland production technologies.

TABLE 4.14. *Net Social Profitability in Relation to the World Market Price of Rice*
(Per mt)

Technique	$200	$250	$300	$350	$400
Monrovia delivery:					
Upland	−341	−291	−241	−191	−141
Improved upland	−348	−298	−248	−198	−148
Traditional swamp	−250	−200	−150	−100	−50
Improved swamp, labor-intensive	−228	−178	−128	−78	−28
Improved swamp, partially mechanized	−288	−238	−188	−138	−88

Technique	$450	$500	$550	$600
Monrovia delivery:				
Upland	−91	−41	11	61
Improved upland	−91	−41	2	52
Traditional swamp	0	50	100	150
Improved swamp, labor-intensive	22	72	122	172
Improved swamp, partially mechanized	−38	12	62	112

SOURCES: Same as for Table 4.12.

Finally, the elasticities emphasize the value of good land or management ability that enables the farmer to attain yields that deviate positively from the mean within each of the techniques. A traditional swamp farmer, for example, who can obtain yields of 2,000 kg/ha reduces social costs by $350/mt.[30] Unskilled labor costs increase by only $60/mt, resulting in a net social cost of $140/mt. This is highly profitable at both domestic and world prices for rice.

The final parametric variation involves the world price of rice. Results are presented in Table 4.14. Only at prices above $430/mt (for 35 percent broken white rice) do any of the techniques become socially profitable. Because social prices of all techniques are discounted from actual c.i.f. costs by 10 percent to account for quality differentials between domestic and imported rice, this value represents a social cost greater than any historical import cost level except for 1975, when c.i.f. prices were $485/mt (an equivalent value for domestic rice of $443/mt).

Conclusion

In spite of the importance of rice in Liberia, the picture that emerges is one of a subsistence-oriented agriculture only peripherally involved with the market economy. But this portrayal reflects the paucity of rural infrastructure and the limited efforts in agricultural development before the mid-1970's rather than disinterest of farmers in market opportunities. Nearly all farmers market some produce, and the rapid spread of sugarcane, coffee, and cocoa over the last 25 years without government encouragement and the shift response (though limited in quantity) to government-induced changes in rice prices indicate a strong interest by farmers in the generation of cash income. Moreover, the spread of small-scale mills (for all types of processing) without government support in the face of an underdeveloped infrastructure and a limited credit market suggests that substantial entrepreneurial potential exists.

The Liberian experience with new production techniques serves as an example of the technology transfer problem that results when relative factor abundance differs markedly between donor and recipient countries. The Asian approaches and Asian technologies were developed under conditions of scarce land and abundant labor supplies, whereas in Liberia the supply conditions are reversed. As a result, the Asian technologies are more costly when placed in a West African context and do not offer promise of efficient import substitution in the major consumption center of Monrovia. Self-sufficiency in rice and increases in cash cropping are thus possible only with production subsidies from consumers or the government budget.

[30] This yield level was 30 percent above the mean and comprised less than 1 percent of the holdings (24).

These considerations, however, do not negate the importance of the new production technologies in the development of Liberian agriculture. The new technologies are more efficient than the traditional techniques, and dissemination of these new techniques may be the most rapid way to generate increasing incomes in rural areas. Over 90 percent of the agricultural population produces rice, and 85 percent of total production is consumed on-farm. Thus improvements in production efficiency can have a major impact on farm incomes by increasing the amount of resources available for producing other cash crops. Given that the labor-intensive improved swamp technique appears socially profitable with only minor improvements in technology (such as sickle harvesting), these resources can be freed without imposing taxes of $145/mt on consumers.

The economic costs of the new techniques arise in the context of import substitution, and the goal of self-sufficiency in rice appears attainable only by increasing rice prices well above their c.i.f. value. Whereas such a policy redistributes income from urban to rural areas, the effect is achieved only at a loss in potential national income. To avoid these losses, one option for the Liberian government is to try to improve further the efficiency of the new production technologies and thus to create comparative advantage in rice production. The labor-intensive improved swamp technique appears the most cost-efficient and thus may offer the greatest promise in this regard. Further development of improved varieties resistant to blast and iron toxicity, improvements in methods of vertebrate pest control, and the development of low-cost mechanical aids in harvesting, threshing, and land preparation may reduce costs sufficiently to allow efficient cash cropping of rice. The potential of upland production remains a major unknown, and increases in basic research may provide improvements in production efficiency.

A second option for the government involves looking at opportunities in the production of other agricultural commodities for export and continuing to rely on rice imports until more efficient rice production technologies emerge. Although the policies of the 1970's have increased farmer incomes through increases in rice prices, it does not follow that such policies have maximized potential farmer incomes. The experience in Liberia and neighboring countries with a myriad of agricultural cash crops, such as coffee, cocoa, sugarcane, palm oil, bananas, pineapples, and rubber, most of which are heavily taxed, suggests substantial potential for increases in farmer incomes through increased efforts to disseminate new technologies for these profitable crops and to reduce levels of taxation affecting them.

The post-production techniques, although relatively unimportant in total costs, demonstrate two important characteristics of the Liberian rice economy. First, high costs of transportation and distribution make rice production for home consumption much more attractive than production for markets. Second, the small, privately run milling sector appears to operate at far lower costs than the large-scale government mills. This is due to several underlying economic factors, including the lack of large increasing returns to scale, fuller utilization of capital equipment by small mills, their better adaptation to market conditions and the constraints imposed by a limited infrastructure, and the lack of economic value for the technically better performance of the large-scale mills.

Replacement of imports through major increases in absolute levels of commercial marketings is a stated government objective. Whether it can be achieved at a tolerable cost by expanding the technologies discussed here remains open to question. The results of this study provide arguments against import substitution through an expansion of available production techniques and in support of the development and dissemination of new technologies of rice production for home and local consumption. Introduction of more efficient rice production techniques would also free resources for cash cropping. But in the absence of significant technological changes in rice production, greater potential economic gains in cash-crop agriculture lie in the production of other commodities, such as tree crops. Whether agricultural development follows an import-substitution or export-oriented strategy, the realization of agricultural potential depends critically on improvements in extension services, infrastructure, and credit and input availabilities.

Citations

1 Alan J. Carpenter, *Draft Terminal Report on Rice Farming in Liberia*. Republic of Liberia, Ministry of Agriculture, Monrovia, 1975.

2 Jeanette E. Carter, *Household Organization and the Money Economy in a Loma Community, Liberia*. Ph.D. dissertation, University of Oregon, Eugene, 1970.

3 Gerald E. Currens, *The Loma Farmer: A Socioeconomic Study of Rice Cultivation and the Use of Resources among a People of Northwestern Liberia*. Ph.D. dissertation, University of Oregon, Eugene, 1974.

4 ———, "Women, Men and Rice: Agricultural Innovation in Northwestern Liberia," *Human Organization*, 35 (1976).

5 R. E. Figueroa, "Small Swamp Development: Method and Cost Study." Third National Conference on Development Objectives and Strategy, Ministry of Planning and Economic Affairs, Monrovia, 1973.

6 K. Harteveld, "Terminal Report of Farm Management." Farm Management and Production Economics Working Paper Number II, Ministry of Agriculture, Monrovia, 1975.

7 International Labour Office, "Total Involvement: A Strategy for Development in Liberia." Geneva, 1972.

8 A. W. Kannangara and S. Pillai, "Report to the Government of Liberia on Swamp Rice Production." Food and Agriculture Organization of the United Nations, Rome, 1970.

9 W. D. McCourtie, *Traditional Farming in Liberia*. University of Liberia, Monrovia, 1973.

10 Eric A. Monke, "Rice Policy in Liberia." Stanford/WARDA Study of the Political Economy of Rice in West Africa, Food Research Institute, Stanford University, Stanford, July 1979; Chapter 3.

11 Dunstan S. C. Spencer, "The Economics of Rice Production in Sierra Leone—I, Upland Rice." Bulletin No. 1, Department of Agricultural Economics and Extension, Njala University College, University of Sierra Leone, Njala, 1975.

12 Dunstan S. C. Spencer, I. I. May-Parker, and F. S. Rose, "Employment, Efficiency and Income in the Rice Processing Industry of Sierra Leone." African Rural Economy Paper No. 15, Department of Agricultural Economics, Michigan State University, East Lansing, 1976.

13 Hisamitsu Takai, L. Ebron, and B. Duff, "Nature and Characteristics of Farm Level Paddy Storage in Luzon, Philippines." IRRI Saturday Seminar Paper, June 1978.

14 C. Peter Timmer, "Choice of Technique in Rice Milling on Java," with comment by W. L. Collier, *Bulletin of Indonesian Economic Studies*, 9, 2 (1973).

15 Z. Toquero, C. Maranan, L. Ebron, and B. Duff, "Assessing Quantitative and Qualitative Losses in Rice Post-Production Systems." Paper No. 77-01AE Manila; IRRI 1977. Same authors, "An Empirical Assessment of Alternative Field-Level Rice Post-Production Systems in Nueva Ecija, Philippines." Paper No. 76-03 AE, Manila; IRRI 1976.

16 C. B. Ufondu, "Small Scale Swamp Rice Farming in Liberia—Bottlenecks and Recommendations." Cuttington College, Suakoko, 1969.

17 C. E. van Santen, "Smallholder Farming in the Foya Area, 1971." Ministry of Agriculture, Monrovia, 1972.

18 ———, "Smallholder Farming in the Gbarnga District—Bong County, 1972." Ministry of Agriculture, Monrovia, 1973.

19 ———, "A Tentative Economic Appraisal of Rice Production—Foya Project." UNDP Development of Rice Cultivation Project in Liberia, Monrovia, n.d.

20 J. G. Vianen, "Self-Sufficiency in Rice in the WARDA Region: A Model for Specialization in Paddy and Rice Production." *Socioeconomic Aspects of Rice Cultivation in West Africa: Seminar Proceedings*, 3, WARDA; Monrovia 1975.

21 Liberia, Republic of, Ministry of Agriculture, "Cost and Returns for Bringing One Acre of Cocoa into Production." Monrovia, 1976, mimeo.

22 ———, "Cost and Returns for Bringing One Acre of Coffee into Production." Monrovia, 1976, mimeo.

23 ———, *National Rice Production Estimates, 1975*. Monrovia, 1976.

24 ———, *Production Estimates of Major Crops, 1976*. Monrovia, 1977.

25 ———, "Second Revised Short Term Self-Sufficiency Plan." Monrovia, 1977.

26 Liberia, Republic of, Ministry of Planning and Economic Affairs, *Economic Survey*. Monrovia, various years.

27 ———, *External Trade of Liberia*. Monrovia, 1968–75.

28 ———, *Quarterly Statistical Bulletin of Liberia*. Monrovia, various years.

PART THREE
Sierra Leone

5. Rice Policy in Sierra Leone

Dunstan S. C. Spencer

The purpose of this chapter is to review and analyze both historical and recent government policies affecting the production, consumption, and trade of rice in Sierra Leone. An attempt is made to identify the major objectives of government and the main constraints on policy implementation. The policy instruments used by the government in the colonial, immediate post-independence, and more recent periods are analyzed to determine the success of different policies in furthering objectives.

The chapter is divided into six sections. The next section provides a general description of the physical, demographic, and socioeconomic conditions of the country. This is followed by a summary of the techniques of rice production and a review of output levels during the last two decades. The fourth section contains a brief description of the marketing system and consumption patterns. An analysis of the shifts in government rice policy is presented in the fifth section. Three periods are examined—the colonial period up to 1961, the immediate post-independence period (1961–67), and the period since 1968. In the sixth section a detailed analysis of trade, tax and subsidy, and investment policies is provided. The final section contains recommendations for future policies.

Setting

Sierra Leone, a former British colony which attained independence in 1961, has an area of 72,600 square kilometers and a population estimated at three million in 1978. National population growth is around 2.1 percent per annum and annual urban growth is currently estimated at 5.9 percent. The urban population comprises 32 percent of the total population. Freetown, the capital and largest city, has about 300,000 people, a third of the urban population.[1] Gross domestic product (GDP) in 1976 was reported at Le 613 million (U.S. $613 million) and annual per capita income at Le 216 (10).

Agriculture is the mainstay of the Sierra Leone economy. It employs

[1] The urban population includes towns with more than 2,000 inhabitants, more than 50 percent of whom are engaged in nonfarm activities.

about 70 percent of the working population and produces about one-third of GDP. During the 11-year period between the population census of 1963 and 1974, employment in agriculture, forestry, and fishing declined from 77 to 72 percent of total employment, but the actual number of people employed in agriculture increased by 5.5 percent (10). The agricultural sector, a major provider of foreign exchange, has become relatively more important in recent years with the decline of the mining industry. In the early part of the twentieth century, agricultural exports accounted for about 90 percent of all exports from Sierra Leone. With the start of iron ore and diamond mining in the 1930's, its share fell to between 15 and 25 percent. But in the last four years there has been a dramatic rise to around 40 percent. This was the result of the closure of the country's only iron ore mine, a decline in diamond mining, and small increases in the tonnage of agricultural exports, coupled with the large rise in the world market prices for coffee and cocoa, Sierra Leone's major agricultural exports.

Rice, the staple food and most important crop in Sierra Leone, is grown by over 85 percent of the country's farmers. Of the roughly 465,000 hectares cultivated in 1976 about 72 percent contained rice in mixed or pure stands.[2] Coffee and cocoa ran a distant second, together occupying 15 percent of the cultivated area (6,7). The development of the rice industry is therefore of crucial importance to the economy of Sierra Leone.

Rice Production

Rice is grown throughout Sierra Leone. Average annual rainfall varies from 4,000 millimeters in the southwest to 2,250 millimeters in the northeast, has a unimodal distribution, and is sufficient to allow at least one crop of rainfed rice to be cultivated during the rainy season (May–October). A brief description of the main system of rice production follows. A more detailed description and analysis of the systems are provided in Chapter 6 (23).

Upland rice cultivation, practiced with shifting cultivation on well-drained land not subject to flooding, is the major system of rice production in Sierra Leone. Because of better rainfall and soils, yields are higher in the south than in the north. Starting in 1976/77 an improved system of upland rice cultivation was introduced by the Integrated Agricultural Development Projects (IADP). Farming practices are the same as in traditional upland rice cultivation except that improved seeds and fertilizers are used.

Mangrove swamps are located along the coast where tidal action causes

---

[2] The fact that only 9 percent of the country's land area is cultivated should not be interpreted in itself as showing a land surplus situation. Shifting cultivation, which is practiced on about half of this area, requires a fallow period of at least ten years, given present levels of technology. Most uncultivated uplands are therefore under fallow.

inundation at high tides and drainage at low tides. Salty water floods the land during the dry season because there is no tidal control in Sierra Leone. Owing to differences in land preparation, transplanting, and other practices, yields are lower in the south than in the north.

Bolilands are low, saucer-shaped, swampy grasslands located in central and northern Sierra Leone. They are flooded up to 1.5 meters for periods varying from three to six months. Fertilizer use is common. Bolilands are farmed either completely by hand or by using the government's tractor-hire scheme for land preparation.

Extensive riverain grasslands are located in the southern coast where silt deposited by two rivers has formed grassy plains that flood up to four meters in the rainy season. Hand cultivation is uncommon because of high labor demand due to heavy weed infestation and because of a shortage of labor in the area. Mechanical land preparation using the government's tractor-hire scheme is widespread.

Inland swamps are found throughout the country wherever depressions occur in the rolling upland. Traditionally the swamps are cultivated for a number of years before being fallowed. Transplanting is usual but broadcasting is not uncommon. No water control is practiced and only one crop of rice is taken each year. Yields are higher in the south than in the north, because of better soils and rainfall. To improve the traditional systems of inland swamp cultivation, the swamps are completely stumped and partially leveled, dikes and contour bunds are constructed to provide partial water control, and improved seeds and fertilizers are used.

The key characteristics and area and yield data for each of the subsystems are summarized in Chapter 6 (23). Improved systems of rice cultivation in which fertilizer, improved seed, and tractor plowing are used account for less than 10 percent of total rice production. Yields are higher in swamps than in uplands. Improved inland swamps yield twice as much as improved uplands.

Table 5.1 shows the growth of area and production of rice during the last two decades. Because there is no systematic annual crop survey in Sierra Leone, the figures are extrapolations based on the 1965/66 and 1970/71 agricultural sample surveys (6,7), and on smaller surveys by Njala University College in 1971/72 and 1974/75 (21, 26). The table shows that by the mid-1970's area under rice had increased about 50 percent over the average of the early 1960's, and that tonnage produced had roughly doubled. Average national rice yields have increased somewhat during the same period.[3] Sierra Leone's present average yield of 1.4 tons per

---

[3] The reported average yield of 1,055 kg/ha for 1960–64 is probably an underestimate. There is no reasonable explanation for the 30 percent increase in yield between the averages of 1960–64 and 1965–69, except an improvement in statistics. Yields after 1965–66 are much more reliable, since they are based on extrapolations of the 1965–66, 1970–71, and other objective sample surveys.

TABLE 5.1. *Average Annual Rice Production, Imports, and Consumption in Sierra Leone*

Category	1960–61—1964–65	1965–66—1969–70	1970–71—1974–75	1975–76	1976–77
Area planted (*000 ha*)	295.8	325.8	423.4	434.6	463.6
Paddy yield (*kg/ha*)	1,055.0	1,357.0	1,369.0	1,401.0	1,385.0
Production (*000 mt*)	312.0	442.0	580.0	609.0	642.0
Milled equivalent (*000 mt*)[a]	206.0	292.0	383.0	402.0	424.0
Imports (*000 mt*)	16.2	21.3	34.3	0.0	3.5
Rice consumption (*000 mt*)[b]	185.2	265.9	336.8	326.0	355.7
Nat'l population (*000*)	2,155.0	2,367.0	2,622.0	2,786.0	2,843.0
Per capita consumption (*kg*)	85.9	112.5	128.5	117.0	125.1
Self-sufficiency (*percent*)[c]	91.3	92.0	91.6	98.8	94.1

SOURCE: 29.

[a] 66 percent recovery.

[b] Domestic production less allowance for losses and seed, plus imports.

[c] Domestic production the preceding year as a percent of domestic consumption. Although there was no importation in 1975–76, the ratio was not 100 percent because of imported stock carry-over from previous years.

hectare is equal to the West African regional average but much less than the world average of 2.4 tons per hectare (29).

Since 1965 average national rice yields have remained more or less constant only because of the faster annual growth in the area under swamp culture (6.3 percent) compared with the growth in area under upland cultivation (2.4 percent). Swamp rice yields, even under traditional cultivation, are generally higher than upland rice yields. Over the last two decades there are indications that the national upland rice yield has been declining an average of 1 percent per annum because the length of the bush fallow has been reduced. On the other hand, the average swamp rice yield has been increasing at about 3 percent per annum because of the gradual adoption of improved varieties and cultural practices under the stimulus of various government programs.

Rice Consumption and Trade

Despite the virtual doubling of rice output during the past two decades, imports of rice have continued. Table 5.1 shows that the annual quantity imported increased from an average of about 16,000 metric tons in the early 1960's to about 34,000 tons in the early 1970's.[4] In 1974 and 1975 high rice prices forced down consumption and increased domestic production, and as a result Sierra Leone was temporarily self-sufficient in 1975. Imports resumed the following year with 3,500 tons, rising to 16,500 metric tons in 1977 and 18,000 tons in 1978. But Sierra Leone has

[4] Imports have usually been average-quality rice containing 20–40 percent brokens, although about 30 percent of the imports in 1974 was of higher quality (less than 5 percent brokens).

been 90–95 percent self-sufficient in rice during the last decade, and there are indications that complete self-sufficiency might be achieved in the 1980's (27).

Per capita consumption of rice, 123 kilograms (kg), is the highest in West Africa and about the same as in many Asian countries. Detailed consumer surveys estimating the calorific contribution of different food items in Sierra Leone are not available. Some rough estimates indicate that close to 90 percent of all the calories in the rural diet are supplied by rice (11).

In a recent survey of rural household consumption in Sierra Leone, Byerlee and King (11) estimate that average propensity to consume rice is about 0.4, whereas that for all food items is 0.7. Average income elasticity of demand for rice was estimated at 0.97. They found only a small drop in income elasticity with increasing incomes compared to a substantial drop for other staples, indicating that average per capita rural rice consumption might still grow despite the high levels already attained. This is probably due to shifts from other foodstuffs to rice. Although Byerlee and King's estimate of average income elasticity might be too high, the estimates of close to zero by Levi (12) and Snyder (19) are certainly too low. Both authors use data from the 1964 household consumption survey conducted in Freetown by the Central Statistics Office (8). Those data were collected during only two months of the year and were too aggregative to provide accurate estimates of income elasticities.

In the absence of official measures of rice marketings, it is estimated that annual marketing of domestically produced rice amounts to about 105,000 tons, or 35 percent of annual domestic production.[5] As discussed below, virtually all of this rice is marketed by the private trade. The government-owned Rice Corporation annually accounted for less than 10 percent of the domestic crop marketed throughout its 25-year life.[6] Imported rice, which was handled exclusively by the Rice Corporation, was sold primarily to licensed wholesale merchants who took delivery in Freetown, the main port.

[5] In 1974–75 Byerlee and King estimated that 79 percent of rural household expenditure on rice was subsistence consumption. A maximum of 20 percent of total urban expenditure on rice might also be subsistence expenditure. (Urban areas include settlements of as few as 2,000 inhabitants in which the proportion of subsistence rice consumption could approach 50 percent. On the other hand, the proportion could be virtually zero in large urban areas like Sefadu and Freetown.) Table 5.1 shows that out of the 337,000 metric tons consumed annually in the 1970's, about 303,000 tons were produced domestically. Virtually all imported rice is consumed in urban areas, and, assuming a per capita urban and rural consumption figure of 123 kg, about 71,000 tons of domestic rice are consumed in urban areas. If one assumes that 30 percent of urban consumption and 21 percent of rural consumption are local rice that is not home-grown, annual marketings of domestically produced rice amount to 105,000 tons, or 35 percent of annual domestic production.

[6] The Rice Corporation was liquidated in April 1979. Since it handled all imported rice, the Rice Corporation was responsible for up to 30 percent of all rice that moved in commercial trade in its more active years, for example 1974/75.

Domestic paddy is assembled in villages by resident village and itinerant merchants who transport the rice to larger towns, where they sell to wholesalers or the Rice Corporation, or process it before resale. Rice processing involves parboiling and milling in one of the 350 or more small village mills existing in the country. Milled rice is then transported and sold to retailers or wholesalers in urban areas.[7]

Government Rice Policy

A review of government rice policy in Sierra Leone can be conveniently divided into three periods: the long colonial period preceding independence in 1961, the immediate post-independence period prior to the onset of military rule in 1967, and the period since 1967. Following these reviews of rice policy in each period, an attempt is made in the next section to analyze the effectiveness of policies adopted by the government to further its objectives.

Colonial Rice Policy

By the fourth quarter of the nineteenth century, the colonial presence in the rural areas of Sierra Leone was well established. The Department of Agriculture was created in 1911 to coordinate the government's efforts in developing the agricultural sector. But explicit goals for the agricultural sector were not published until 1961. These goals were the conservation and improvement of lands and forest, the attainment of self-sufficiency in all foodstuffs that could be produced in Sierra Leone, the expansion of exports to pay for imports, and the improvement of methods of agricultural production (3).

The colonial government adopted a policy of minimum intervention in agriculture throughout the long period of colonial rule. Direct government participation in production was ruled out, and no attempt was made to displace traditional small-scale farming. In fact, contrary to its policy in East Africa and Asia and unlike the policy of other colonial governments in West Africa, the British government took active steps to prevent the development of large-scale plantations under European settler control in Sierra Leone and in the rest of British West Africa. Foreigners were prevented from owning land and requests from them for permission to establish plantations were refused (18).

Consistent with this policy, the Department of Agriculture concentrated its efforts in two directions. Attempts were made, first, to increase farmers' production by providing advisory or extension services backed

[7] For a more detailed description of the marketing and processing systems, see Spencer (20), Spencer, May-Parker, and Rose (24), and Mutti, Atere-Roberts, and Spencer (15).

by a rudimentary research and education network and, second, to regulate the marketing of agricultural produce by exercising supervision and providing direct assistance. Toward the end of the colonial period, input subsidy policies were introduced on a small scale. As discussed in the following section, most of the colonial schemes were failures mainly because of poor planning and underfinancing.

The Immediate Post-Independence Period, 1961–67

The onset of independence saw a rather dramatic shift in agriculture policy in Sierra Leone. Driven by a desire to get agriculture moving quickly and encouraged by the large reserve funds of the export crop-marketing agency, the Sierra Leone Produce Marketing Board (SLPMB), the government changed its approach and decided to go into direct agricultural production. The government's "white paper" stated (3): "Hitherto the Department (of Agriculture) has confined its activities to research and advisory services. It is now felt that while the vital research work must be continued and expanded, much more money must be spent on switching the emphasis from advisory services to actual direct supervision of productive effort."

Most of the direct production schemes were undertaken in the export crop sector by the SLPMB. But the government-owned Rice Corporation for the first time cultivated about 590 hectares on its own account in 1966.

The experiment in direct production by government was short-lived. The schemes were poorly planned, very unwisely located, generally staffed with unqualified personnel, and, consequently, yields were extremely low. For example, the Rice Corporation reported average yields of 438 and 795 kg per hectare for its mechanically plowed riverain and inland valley swamps in 1966, less than half of traditional farmers' yields. Furthermore, since the projects were poorly, if at all, documented, it was impossible to secure external financing for them and so the Rice Corporation and SLPMB had to dig heavily into their reserve funds. By the onset of military rule in 1967, they could no longer perform even their traditional marketing operations, and instead of paying farmers in cash, they gave them I.O.U.s which were not redeemed for months. The military government closed the production divisions of the Rice Corporation and SLPMB and confined them to traditional marketing activities, thus ending a short but tumultuous chapter in the history of Sierra Leonean agricultural policy.[8]

Throughout this short post-independence period other government policies, including marketing and price policies, were virtually unchanged. The only other notable occurrence was the establishment of a

[8] It is difficult to estimate exactly how much money was lost by the two corporations in this experiment, but it is likely that the SLPMB lost at least Le 6 million and the Rice Corporation at least Le 1 million.

university college to train agricultural staff at the middle level (agricultural instructors) as well as at higher levels (agricultural officers).

Rice Policy Since 1968

Following the total failure of the direct production projects, policy makers after 1967 went back to the drawing boards to redesign agricultural development policy. Emphasis shifted to encouragement of small-scale agriculture. An agricultural planning team was set up in 1967 with United Nations assistance. Starting in 1966, Njala University College began to turn out increasing numbers of middle- and higher-level graduates to staff the extension service. The level of government expenditure on agriculture increased to levels that for the first time could be regarded as more than token.

Agricultural investment policy also assumed a different form. In contrast to the approach during the colonial era, when the complex of interrelated problems facing the farmer was attacked in a piecemeal fashion,[9] an integrated approach began to be adopted. In 1967/68, a pilot scheme financed by the Diamond (Marketing) Corporation was introduced in the eastern province. Farmers were provided with a cash grant to cover the costs of improved seed and fertilizer as well as part of the cost of hiring labor for land development. There was also heavy extension input. Participating farmers cleared, developed, and brought new inland valley swamps into production. In 1970/71 the Ministry of Agriculture took over the scheme and expanded it into a nationwide scheme. The subsidy was increased from Le 35 to Le 74 per hectare.[10]

The scheme had many problems including poor organization, inadequate supervision, dishonesty on the part of some officials, and a poor input delivery system (30). It was closed in 1972 because of lack of funds to pay the subsidies and the decision to replace it by a scheme that had a more limited geographical coverage. By then 2,500 hectares of new inland swamps had been brought into cultivation although yields were only a little above traditional swamp levels (30). Although its success was limited, the scheme provided very useful lessons for future projects.

Starting in December 1972 a series of new Integrated Agricultural Development Projects (IADP) have come into operation while others are in advanced stages of planning. The main features of these integrated projects are that they cover a limited geographical area, are usually mainly financed by foreign concessional loans and grants (World Bank, African

[9] One project would emphasize drainage, another in a different area would stress seed distribution, while yet another would focus on credit.

[10] The subsidy was intended to cover only part of the total cost of development. Although it was recognized, for example, that labor costs exceeded Le 125 per hectare, only Le 14.50 was provided in the subsidy.

Development Bank, World Food Program, European Development Fund, Chinese Government, Dutch Government), provide a heavy extension input with extension agent/farmer ratios between 1:40 and 1:70 compared with a current national average of 1:1,200, supply improved planting materials, improved tools, fertilizer, and chemicals on credit to farmers, provide low-interest development and seasonal loans to finance farmer operations including hiring of labor, make provision for infrastructural items such as feeder roads and village wells, and utilize highly qualified, usually expatriate, personnel for senior management positions.

The integrated projects usually try to improve the cultivation of more than one crop, but the development of an improved system of inland-valley swamp rice cultivation has been the focus so far. About 8,000 hectares had been brought under improved cultivation by the end of 1977. The newer integrated projects also include upland, boli, and riverain rice cultivation, and it is expected that mangrove swamp development will be included in future projects.

Analysis of Rice Policies

The objectives of agricultural policy in Sierra Leone have remained more or less those established during the colonial era. They were restated in the 1974/75–1980/81 development plan as follows: to stimulate development from the traditional subsistence type of production to a more productive system of commercial agriculture; to achieve self-sufficiency in staple foodstuffs and other products; to diversify agricultural production with emphasis on food and cash crops in suitable areas; to increase the productivity, incomes, and living conditions of the rural population; to maximize foreign exchange earnings through expansion of export crops and import substitution; to increase rural employment through stimulation of private investment in various agricultural enterprises; and to improve human nutrition and to conserve the fertility of the soil and other natural resources (9).

This list contains elements of what Pearson, Stryker, and Humphreys (16) have categorized as fundamental objectives—generation of income, distribution of income, and security (the probability of obtaining income)—as well as what they regard as proximate objectives—achievement of self-sufficiency in staple foodstuffs, conservation of land resources, and increasing agricultural exports.

Although there is no published order of priority for Sierra Leone's stated objectives for rice, observation of the implementation of government policies since independence leads one to believe that of the three fundamental objectives income distribution has been given top priority, followed by income generation. Security of income has received minor emphasis.

Rice policies used in achieving these objectives can be categorized into three groups—investment in research, education and extension, trade and price policies, and input subsidies. These policies are analyzed with the aim of determining how successful they have been in helping the government advance toward its objectives. The analysis concentrates on the effect of the different policies on the fundamental economic objectives, but reference is also made to the other stated objectives of the government.

Investment in Research, Education, and Extension

Research, education, and extension policies are the oldest of Sierra Leone's agricultural policies. The Department of Agriculture established a series of agricultural experiment stations between 1920 and 1940 to carry out research and facilitate extension work. Each station emphasized one or two crops. Rokupr Rice Research Station was established in the Scarcies mangrove swamp area in 1934 with funds from the Colonial Welfare Fund. This station developed slowly, and its laboratories were completed only around 1950. It was converted into a West African regional rice research station in 1953, and it functioned as such until independence. Unfortunately, during the colonial period the station concentrated almost exclusively on research on mangrove swamp rice, a system that was already quite well developed under traditional conditions and that contributed only a small proportion of Sierra Leone's total rice output. However, some work was done in the late 1950's with inland swamp and deep-flooded rice. In general, although some useful experimental work was done on mangrove soil chemistry, plant pathology, and rice agronomy, and some improved rice varieties were selected, the station was understaffed and inadequately funded. It therefore had very little measurable impact on Sierra Leone's rice production during the colonial period.

To carry out its extension activities the Department needed a cadre of trained workers. An agricultural school was established in 1924 at Njala, the headquarters of the Department. But like much of the Department's other activities at that time, it was underfunded. In fact, funds were cut off from the school during the depression of the 1930's, and by 1944 probably fewer than 50 lower- and middle-level agricultural instructors had been trained. Moreover, by independence in 1961 fewer than ten Sierra Leoneans had been trained abroad for senior agricultural positions in the Department. Agricultural training and research were given a much needed boost after independence, when Njala University College was established in 1964 as a degree-granting institution. An average of about 30 middle-level certificate holders and seven B.S. graduates have been produced annually since 1968.

Table 5.2 shows the extension staff position and levels of government expenditure on agriculture, including research. The Department, later

TABLE 5.2. *Extension Staff and Government Expenditure on Agricultural Research and Extension in Sierra Leone, Selected Years*

Year	Number of extension workers		Government expenditure on agriculture from different sources (000,000 Le)			Government expenditure on agriculture as a percentage of total expenditure
	Senior	Intermediate and junior	Agric. Dept.	Educ. Dept.[a]	Total	
1922	3	5	n.a.	n.a.	n.a.	0.88%
1929	7	29	n.a.	n.a.	n.a.	3.71
1934	6	n.a.	n.a.	n.a.	n.a.	2.31
1938	11	11	n.a.	n.a.	n.a.	2.85
1944	15	75	n.a.	n.a.	n.a.	3.35
1950	18	n.a.	0.12	0.01	0.13	2.98
1955	17	n.a.	0.65	0.13	0.78	3.13
1960–61	19	16	0.62	0.23	0.83	2.59
1966–67	19	40	1.08	0.33	1.40	2.78
1970–71	49	129	1.99	0.60	2.59	4.12
1976–77	53	230	10.68	1.20	11.90	7.10

SOURCES: 1920–60 data are from *14*; 1960–77, *4* and *9*.
[a]Appropriations for agricultural education are estimated at 50 percent of total cost of Njala Training Center and University College.

Ministry, of Agriculture has been heavily understaffed and underfinanced throughout its history in relation to the job it was expected to do. The ratio of farmers to extension workers, recently about 1,200:1, has been extremely high.

The proportion of total government expenditure devoted to agriculture has been much lower than warranted by the importance of the sector in the national economy. Table 5.2 shows that less than 4 percent of total development and recurrent expenditure was annually spent on agriculture throughout the long colonial period. In the past decade the proportion of expenditure spent on agriculture has increased modestly, reflecting the government's heightened interest in agricultural development, but is still unsatisfactory considering that agriculture contributes one-third of gross national product.

The proportion of government expenditure on agriculture that went to the rice industry is not known, but it is likely that it is over 50 percent. Although it is not possible to quantify the results of this spending, there is ample evidence that the investments on research extension and training have had some stimulating effects. The recent increased level of activity in rice development projects has been partly made possible by the larger pool of trained manpower available for planning and executing the projects. Post-independence investment in rice research and training has been increasing the supply of manpower and available technology, thus easing one of the constraints on agricultural development in Sierra Leone.

Input Subsidy Policies

Input subsidy policies have been widely used in Sierra Leone to encourage producers to use inputs that the government feels would increase farm production and incomes. During the long colonial era, the government's input subsidy policy concentrated on land development. As has been pointed out, swamp rice yields are higher than those of upland rice. The colonial government believed that the goal of land conservation could be achieved simultaneously with that of increased rice production and incomes if there was a general switch from upland to swamp rice production. In 1938, therefore, the government started giving interest-free loans to farmers to help them finance the heavy labor costs of clearing and bringing mangrove swamps into cultivation in the southern province. The scheme was ended in 1951 when it proved difficult to recover loans. Official records showing that about 3,230 hectares were brought under cultivation under the scheme are exaggerated, since that figure implies that all the presently cultivated mangrove swamp area in the south was cleared under the scheme (17).

Starting in 1940, attempts were made, with very little success, to encourage communal clearance of inland swamps. In 1941 the Agriculture Department formulated an ambitious and peculiar scheme to drain 222,000 hectares of land in the Scarcies River Basin. Saylor (18) reports that because of technical miscalculations and delays only 385 hectares had been drained by 1944, when the scheme lapsed. Subsequent irrigation and drainage schemes for the mangrove swamps also failed for similar reasons. Furthermore, in the few cases where some land was drained, it was either not cultivated by farmers because of land tenure problems or, more important, abandoned after a few years because of weed and soil problems.

In the modern period, emphasis has shifted to the inland valley swamps. Farmers participating in integrated agricultural development projects receive subsidies to cover part of the costs of clearing new swampland, building partial water control structures, and purchasing other inputs. It has been estimated that about 48 percent of the annual user costs (depreciation plus interest charges) of improved inland swamp development is currently subsidized (23).

As part of its land development policy, which attempted to encourage swampland development, the colonial government started contract plowing of previously uncultivated riverain grasslands and bolilands in 1949, using government-owned equipment. Starting with 20 hectares, the pre-independence peak was reached in 1956 when 4,900 hectares were plowed. Thereafter, there was a drop in area as difficulties arose with fee collections. Costs have always exceeded the revenue collected. During the colonial period, the scheme was subsidized at around 50 percent. In the post-independence period, the costs and rate of subsidy have increased. The subsidy rate was recently about 77 percent (23), and because of this heavy subsidy, the government has found it difficult to budget sufficient amounts to provide the service to farmers consistently. Funds are not always available for purchase of spare parts, fuel, or replacement tractors. Consequently, area cultivated in the post-independence period has fluctuated widely.

Tractor plowing usually results in about a 40 percent drop in labor use per hectare (21), but because the scheme is concentrated in the riverain grasslands and bolilands—areas with unutilized land and low population density—it has had little adverse effect on aggregate rural employment. In fact, there is evidence that the plowing scheme has encouraged rural to rural migration, increased farm sizes resulting in higher rice output, and therefore generated higher incomes per family (23, 26). By concentrating the scheme in the northern bolilands during the last ten years, the present government apparently has also been trying to achieve its income distribution goals by shifting resources to the poorer northern province. Unfortunately, in the northern province the technology appears to be having an adverse effect on the sexual division of labor within the family. The work load of women is increased as they are called on to weed and harvest larger areas, while that of men, replaced by the tractor, falls (25). Although net social returns at the national level have remained positive, social profitability is lower for mechanized than for hand-cultivated bolilands (23).

Prior to the start of the pilot Rolako mechanization project by the Chinese in 1975, little serious effort was made to encourage farmers to adopt yield-increasing practices at the same time as tractor plowing. The experience in the Rolako pilot project, where the average yield per hectare of 190 farmers cultivating 315 hectares has risen from 1.2 tons in 1975 to 2 tons in 1977, indicates the possibilities. The government's tractor-plowing services in the riverain grasslands have recently been taken over by the Torma Bum project, jointly financed by the Sierra Leone Government, the African Development Bank, and the Dutch Government. It is hoped that this project will have yield-increasing results similar to those of the Chinese project but on a larger scale, while reducing costs, so that social profitability will be increased.

Agricultural credit policy, except that tied to land development, was not a feature of colonial policy until the 1950's, when rudimentary efforts were made. A Development Industries Board, established in 1946 to give credit to industry and agriculture, gave virtually nothing to agricultural enterprises. A Cooperative Loan Scheme was conceived in the 1950's but was just getting off its feet at independence. It was designed to give low-cost institutional credit to cooperative societies to finance their marketing and other activities. Loans guaranteed by the government were obtained from the commercial banks by the Registrar of Cooperatives and passed on to cooperative societies. The societies loaned the money to their members or used it to invest in society projects such as purchase of tractors, rice mills, and weighing scales, boat building, and construction of rice stores.

The cooperative movement in general and the credit scheme in particular enjoyed a measure of success in the early 1960's. By 1965–66 there were 56 registered rice-marketing societies, about 30 of which owned small rice mills and seven owned tractors purchased with loans received from the scheme. The loan fund stood at over Le 400,000 and repayment rates exceeded 60 percent. Unfortunately, reflecting developments in other areas of national life, politics intruded increasingly into the operation of the scheme. Unqualified staff of doubtful integrity were recruited as cooperative officers, and loans were disbursed to people with political connections who regarded them as gifts. Moreover, produce collected by cooperative societies was sometimes stolen by employees, and what was delivered to the SLPMB or Rice Corporation was not paid for on time. As a result, cooperative members were completely disillusioned. Because of massive defaults the commercial banks refused to advance any more credit and called on the government to repay outstanding loans. The loan scheme collapsed in 1967 and the whole cooperative movement almost went into oblivion.

In the post-1968 era, institutional credit to agriculture has been better handled. In the integrated agricultural development projects, supervised credit is given in kind to farmers adopting yield-increasing practices. Cash loans to cover part of the cost of hiring labor for land development activities and seasonal credit in kind are given for inputs such as improved seeds and fertilizer. Subsidized interest rates are about 8 percent for 4-10-year development loans and 10 percent for seasonal loans. Total loans advance by the longest-established projects in the eastern and northern provinces amounted to over Le 1 million in 1978. Repayment rates, which were close to 100 percent in the early years, have now stabilized at around 65 percent.

In addition the National Development Bank (NDB), a government-owned institution, disbursed Le 562,000 (51 percent of its total loans in 1978) to medium-size farmers. The Cooperative Development Bank, established in 1971 to finance cooperatives, advances about Le 50,000 annually to marketing cooperatives. But the commercial banks still make only negligible advances to agricultural enterprises. The above figures show the steady increase in institutional supervised credit for agricultural production after the debacle of the immediate post-independence period.

With the expansion of agricultural projects financed by foreign concessional loans, the increase in foreign funds available to the NDB, and the planned development of rural banks, it is likely that the availability of well-supervised, low-cost institutional credit to farmers will increase in the near future.

Colonial rice policy did not include supply of fertilizers. After research revealed the need for fertilizer use in the bolilands if farmers were to obtain reasonable yields, a subsidized fertilizer distribution scheme was started in the early 1960's (1). Only about 3 percent of Sierra Leone farmers use the 5,000–8,000 tons of fertilizer imported each year. Subsidy rates are currently estimated at about 57 percent for farmers in the IADP and 66 percent for others (23).

In order to encourage widespread distribution of improved seed selected by Rokupr Rice Research Station, a seed loan scheme was started in the Scarcies area by the colonial government in 1936. Pure seed was issued to farmers, who repaid with interest in kind after harvest. When the scheme closed in 1947 because of high administration costs, a total of 2,500 tons of seed had been distributed. Subsidized distribution of improved seeds started again on a national scale in 1977 with the establishment of the National Seed Multiplication project, financed mainly by a grant from the German government. About 300 tons of pure seed were supplied to the IADP and other development projects in 1979.

What has been the effect of the government's input subsidy policy on income generation and distribution? Inputs used in improved systems of swamp rice cultivation (bolilands, riverain grasslands, and inland valley swamps) have received the greatest subsidy. This is in line with the government's belief that swamp rice cultivation is more attractive than upland cultivation from the national point of view. It was not until 1977 that the IADPs started to encourage upland rice farmers to adopt an improved farming system in which improved seed and fertilizer are used.

Because they employ the widest range of subsidized inputs, farmers cultivating improved inland swamps receive the highest amount of subsidy—over Le 95 per hectare. Because of relatively higher yields, the subsidy per ton of rice produced, valued at about Le 50, is lower than that provided to farmers using tractor plowing (Le 70–120 per ton). Traditional systems of cultivation in uplands and swamps receive virtually no input subsidies. Swamp rice cultivation is more socially profitable than upland cultivation (23), and the tendency for government input subsidy policy to favor swamp rice is also consistent with the proximate goal of land conservation, since there is much less danger of soil erosion in the swamps. Moreover, the policy is in theory neutral to the regional income distribution goal, since inland swamps are widely distributed throughout the country. But in practice, by concentrating the IADPs in the poorer northern province, the government is trying to reduce the regional income disparity.[11]

Trade and Price Policies

Rice trade and price policies have long featured prominently in agricultural policy in Sierra Leone. The stated objectives of such policies of the colonial government were to ensure adequate supplies of good-quality food and fiber to both domestic and foreign (imperial) markets and to provide stable prices to producers. To further these objectives in the rice sector, the government created a rice mill division in 1936. During the Second World War this agency was converted to a Supplies Department. In its early years, especially during the war, the Department mainly purchased rice for government employees in Freetown.

During the war the transactions were at controlled prices. Otherwise, until 1952, the Department operated alongside private (usually foreign) traders, and prices were determined solely by market forces. The Department usually purchased husk rice in the Scarcies area. A large rice mill was installed in Freetown in 1936 and a second at Mambolo in the Scarcies area in 1951.

In 1952 there was a change in the free market policy when the government accepted the following Rice Committee recommendations (5):

That in order to encourage the planting of rice a guaranteed price to producers should be fixed annually before clearing of farms. . . . That in order to implement the guarantee . . . the government must be prepared to buy a considerable proportion of the crop offered for sale to the present capacity. . . . That Government should hold adequate reserve stocks of rice not only in Freetown but also in the large urban centers in the Protectorate.

The Committee was set up "to inquire into, and to report on the production and marketing of rice in Sierra Leone, to make recommendations on the methods to stimulate production, to regulate prices, and to regulate speculation in the best interest of producers and consumers." Coinciding with the start of the diamond mining boom around 1950, rice short-

---

[11] Since the establishment of the first IADP in the eastern and southern provinces in 1972, two others have been established and two more are in advanced stages of negotiation for finance, all located in the northern province. The northern province contains the poorest farmers in Sierra Leone with fewer alternative possibilities for profitable agricultural activities (11, 26). Hence, this policy is consistent with the objective of more equitable distribution of income. However, the policy could be faulted from the point of income distribution within a given geographical area because it concentrates on a system of rice farming practiced by less than one-third of Sierra Leone's farmers. Farmers adopting the improved technology have the potential of widening the income gap between them and nonadopters in the same village. The improved upland rice farming system, which could theoretically be adopted by virtually all Sierra Leone farmers, therefore, has greater potential for minimizing rural-rural income differences while increasing rural incomes.

ages had developed in Freetown and other urban areas. Prices had shot up because perhaps 100,000 farmers had left their farms and migrated to the diamond areas. Market demand increased although total national production remained nearly constant.

The Rice Department established a price of Le 69.45 per metric ton (rice equivalent) in 1952. This was raised 60 percent to Le 111.10 in 1954 and another 5 percent to Le 116.67 in 1955 in an attempt to secure more domestic supplies. At the same time the Department imported over 30,000 tons of rice in 1954, marking the change from self-sufficiency to Sierrà Leone becoming a rice-deficit country (see Table 5.3).

The Rice Department quickly ran into trouble in its trade in domestic rice. Purchases averaged about 7,000 tons and the limited available storage space was soon filled, indicating that in 1954–55 the Department's producer price was attractive enough to command supplies.[12] The two rice mills averaged more than one shift operation. Nonetheless, the Department was still losing money on its domestic account. These losses, amounting to about Le 125,000 in 1955, were officially claimed to be subsidies to producers. In fact, they represented the Department's rice milling costs, its storage losses (which were substantial), and other losses due to its inefficient operation. Moreover, although there are no precise data to indicate whether farmers were actually receiving the Department's announced producer price, the impression of people active in the field at that time was that farmers in general received less than the announced buying price because traders took a larger share than their officially allowed margins.[13]

Plagued by its losses on the domestic rice account, the Department progressively reduced its producer price to about Le 90 per ton in the immediate post-independence years. Consequently, it was able to purchase very little domestic rice (Table 5.3).[14] The Department (later Rice Corporation) was therefore not able directly to defend its official producer price for paddy. Private traders, using small mills to process paddy, were able to compete effectively with the Corporation. Although there are no

---

[12] Purchases were much lower in subsequent years. Unfortunately, there is no information on the exact amount of storage space owned by the Department at that time, but it was probably insufficient for 12,000 tons. After considerable expansion, it was about 23,000 tons in 1966 (15). Of course, a large percentage of the available capacity was used for storing imported rice.

[13] This does not imply that traders earned excess profits. It is quite likely that actual marketing costs were higher than those in the Corporation's officially allowed margin. Existing evidence indicates that rice-marketing margins in the private trade have not been excessive in Sierra Leone (15).

[14] Much of the domestic rice purchased in the early 1960's, when prices were around Le 90 per ton, was sold to the Department by Cooperative Societies under pressure from the Cooperative Department.

TABLE 5.3. *Rice Equivalent of Quantities of Domestic Paddy Purchased and Rice Imported, Rice Prices of the Sierra Leone Rice Corporation, and Urban Free Market Retail Prices, 1954–77*

Year	Domestic rice[a]		Imported rice				Urban retail price[e] (Le/ton)
	Quantity purchased (000 tons)	Producer price[b]	Quantity[c] (000 tons)	C.i.f. price (Le/ton)	Wholesale price[b]	NPC[d]	
1954	n.a.	111.10	4.59	126.50	n.a.	n.a.	n.a.
1955	3.37	116.67	21.06	91.91	n.a.	n.a.	n.a.
1956	4.95	116.67	36.80	89.67	n.a.	n.a.	n.a.
1957	6.67	116.67	31.05	92.04	n.a.	n.a.	n.a.
1958	6.67	111.10	21.78	94.29	n.a.	n.a.	n.a.
1959	5.54	111.10	43.31	92.23	n.a.	n.a.	n.a.
1960	12.61	106.50	28.59	86.53	117.80	1.34	n.a.
1961	12.30	100.90	4.11	102.24	117.80	1.13	n.a.
1962	7.26	95.30	26.83	101.32	117.80	1.14	172.70
1963	12.08	89.70	20.82	90.43	117.80	1.28	151.10
1964	6.36	89.70	0.54	173.11	n.a.	n.a.	174.20
1965	2.14	89.70	18.72	97.52	n.a.	n.a.	202.80
1966	0.94	89.70	34.55	99.83	120.60	1.18	191.80
1967	2.87	112.15	23.85	99.05	130.90	1.30	193.60
1968	2.21	112.15	16.88	148.91	130.90	0.85	187.10
1969	3.63	112.15	12.68	119.73	159.60	1.31	210.00
1970	0.53	112.15	49.36	112.23	198.00	1.74	249.90
1971	1.85	127.90	26.93	100.11	198.00	1.95	204.20
1972	4.75	127.90	6.63	132.24	198.00	1.47	235.40
1973	0.46	127.90	43.72	136.25	198.00	1.43	239.40
1974	3.27	177.90	45.02	388.73	198/412[f]	0.48/1.03	366.10
1975	7.17	277.75	0	—	412.50	1.03	447.00
1976	0.53	250.00	3.50	343.00	412.50	1.18	457.60
1977[g]	0.75	277.75	16.50	n.a.	412.50	n.a.	n.a.

SOURCE: Rice Corporation, Ministry of Labor, and Central Statistics Office.

[a] Paddy converted to rice equivalent using 66 percent recovery rate.
[b] Official government prices; wholesale rice prices were the same for domestic and imported rice.
[c] Usually medium-quality parboiled rice with 12–35 percent brokens.
[d] Nominal protection coefficient, defined as the sum of c.i.f. and implicit tariff (wholesale price c.i.f. plus landing costs) divided by c.i.f. Landing costs were estimated at Le 9.00 per ton, or 2.62 percent of c.i.f price in 1976.
[e] Free market domestic rice in Freetown.
[f] Price raised progressively from Le 198 to Le 412.50 during the first half of the year.
[g] Provisional.

precise data to that effect, observations lead one to believe that farmers received prices from traders that were higher than the official producer price. In 1974/75, when the official producer price was doubled and for the first time was substantially higher than the price in the 1950's, traders once more delivered noticeable quantities to the Corporation.

The Corporation was not able to control producer prices directly by its trade in domestic rice. But its restrictive rice import policy and price setting for imported rice affected the level of producer prices indirectly through the effect it had on domestic rice prices. As pointed out earlier, imports of rice have been handled exclusively by the Rice Department/ Corporation since 1954. The Board of the Corporation each year decides what quantities are to be imported.[15] Since there is no annual crop survey or production forecast in Sierra Leone, decisions on quantities to be imported have been made on what could be described as nothing more than inspired guesses of what available domestic supplies would be. Wholesale prices for imported rice are also set by the Corporation's Board. Table 5.3 shows that these prices have almost always been fixed so as to allow the Corporation a large market margin. The nominal protection coefficient (NPC) dropped below 1.0 only in 1968 and the early part of 1974. In those years sharp increases in the c.i.f. price of imported rice were not immediately passed on to consumers, who were therefore subsidized for a short period. In all other years consumers of imported rice have been taxed by the Corporation's pricing policy.

Once imported rice left the Corporation's stores, it was traded in a free market. There is no available retail price series for imported rice. The available series for domestic rice, presented in Table 5.3, shows that there is a strong correlation between the wholesale price for imported rice and the retail price for domestic rice. Domestic rice was usually not purchased from the Corporation, indicating that the Corporation's import price policy had an effect on domestic rice prices.

The evidence was strongest in 1974/75. In that year there was a big increase in the Corporation producer and consumer prices. Anticipating a shortfall in domestic rice production, the Corporation imported 45,000 tons of rice in 1974. There was no concrete evidence to support this anticipation. In fact, the rather unusual event of a stock carry-over of about 3,000 tons of imported rice at the end of 1973 might have pointed to the need for less imports in 1974 compared with 1973. But seeing the skyrocketing world market prices, resulting mainly from world-wide crop failures and inventory reductions in 1972 through 1974, the government overreacted. Consequently, a record 45,000 tons of rice were imported in 1974 at almost Le 390 per ton, three times the 1973 price.

The government at first tried to subsidize consumers by fixing a wholesale price less than the c.i.f. price, but soon abandoned this attempt because of its severe budgetary impact. The consumer price was gradually raised to Le 412.50 per ton by mid-1974, allowing the Corporation an unusually small, but still positive, profit margin.

At the same time the guaranteed producer price was more than doubled, rising from Le 128 per ton (rice equivalent) at the end of 1973 to Le 278 in October 1974. This rise in producer price had the effect of making the Corporation competitive in the domestic market. Hence, in 1975 the Corporation was able to purchase enough domestic rice to fill all its avail-

---

[15] The Board includes representatives of the Ministries of Agriculture, Finance, Trade, and Industry, as well as people representing farmers and private rice traders.

able stores, about 10,800 tons of paddy, or 7 percent of all domestic crop marketings. As shown in Table 5.3, domestic retail prices increased over 80 percent between 1973 and 1975.

Domestic retail rice prices have remained high in Sierra Leone since then, despite the fact that world rice prices have dropped substantially since 1975. This has been partly because the Corporation has maintained its wholesale prices at the level reached in 1975. Faced with a drop in consumer demand, full stores, and all its operating capital tied up in stock, the Corporation imported virtually no rice in 1975 and 1976 while maintaining its high prices.[16,17]

In summary, the effect of the government's rice trade and price policy during the past 25 years has been to tax consumers and protect domestic rice producers. Retail prices have been generally kept at a higher level than would have prevailed if imported rice had been allowed to flow freely into Sierra Leone.

Summary and Prognosis

This paper has presented a review and analysis of Sierra Leone's rice policies. Rice is the staple food of the country's 3 million people, two-thirds of whom live in the rural area and grow rice. Upland rice production, using the traditional bush fallow method, is still the most important system, occupying 75 percent of the land under rice and producing 55 percent of national rice output. But the higher-yielding swamp rice systems, in which rice is grown in standing water, are gradually increasing the importance. The inland valley swamps in particular are being developed with a higher level of technology, using water control, improved seed, and fertilizers under the sponsorship of government development projects.

During the colonial period the government made only minor attempts to increase the pace of agricultural development. The Agriculture Department, established in 1911, concentrated its marginal efforts on research and extension and made only limited attempts at land development, staff training, and supply of inputs to farmers. In the last years of colonial rule a rice-marketing organization was established to stimulate domestic production, using price policy. It soon had to abandon its attempts to maintain a guaranteed producer price for paddy and concentrated instead on trading imported rice. Because of poor organization, inadequate staffing, and underfunding, all colonial rice

[16] Per capita consumption is estimated to have declined from 132 kg in 1973 to 117 kg in 1975.

[17] The Corporation had substantial stock carry-over—30,000 tons at the end of 1974 and 25,000 at the end of 1975.

schemes with the exception of the research program ended in varying degrees of failure.[18]

In the immediate post-independence years, 1961–67, there was a dramatic shift in government policy as the government attempted to increase agricultural output by directly establishing and managing state farms. Because of grave errors in concept, planning, and management, the experiment was a complete failure and incurred losses of several million Leones.

The present era, starting in 1968, has seen a more cautious approach to agricultural development, with the emphasis shifting to small farm development. A series of integrated projects has been launched, with rice as a major component. These projects have a high ratio of extension worker to farmer (about 1:70 compared with a national average of 1:1,200) and are attempting to encourage farmers to adopt yield-improving technology (improved seed, fertilizers, water control measures), while providing the villages with needed infrastructure (feeder roads, wells). In the last five years government expenditure in agriculture has shown a slight tendency to increase relatively faster than expenditure in other sectors.

In recent years rice policy has aimed at increasing rural incomes by increasing rice production. Another major objective has been to improve regional distribution of rural incomes, whereas increasing security of income has not been an important goal.

The government has implemented a series of input subsidy measures that have provided fertilizers, improved seed, land development, water control, extension, tractor hire services, and credit at subsidized prices to farmers participating in agricultural development projects. These subsidized inputs have been mainly available to farmers cultivating swamp rice. Between 1970 and 1977 about 8,000 hectares of improved swamps were brought under cultivation under the stimulus of the scheme. Since yields are 30–50 percent higher than in traditional swamp rice systems and more than double those on traditional upland farms, domestic rice production in Sierra Leone has been increased by government input subsidy and investment policies.

Government agricultural investment policies are currently funded mainly by external funds that are obtained on long-term, highly concessional rates. This situation contrasts sharply with that in other sectors for which the government has had to obtain short-term loans at very high in-

---

[18] Success of the research program was mainly in generating knowledge. Although some improved varieties were released for mangrove swamp cultivation, the net effect on Sierra Leone rice production was minimal. The only activity in which colonial agricultural production policy in Sierra Leone had some success was the establishment of the export tree crop industry (coffee and cocoa). Even here, compared to the success in Ghana and Nigeria with cocoa or in Kenya with coffee, the Sierra Leone experience could only barely be described as successful.

terest rates. Therefore, whereas developments in most other sectors of the Sierra Leone economy are restricted because of the ever tighter government budgetary constraint, developments in the agricultural sector are less affected.[19]

By concentrating rice project investment in the northern province, the government has also been trying to advance its regional income distribution objectives. Farmers in the northern province are poorer and have fewer alternatives for profitable agricultural production than those in the southern and eastern provinces. But rice production systems in the north have lower social profitability than those in the south and east. Consequently, the policy of emphasizing developments in the north results in a lower gain in social profitability. This lower social profit is the cost in terms of reduced efficiency of the government's regional income distribution policies.

Rice trade and price policies have also aimed at encouraging increases in rice production by providing stable and remunerative prices to producers while ensuring adequate supplies of good-quality rice to consumers. Through its restrictive import policy the government has succeeded in protecting rice producers by keeping domestic rice prices higher than they would have been without such policies. The rice trade and price policies have therefore had some effect in stimulating domestic production. Since the policy affected all rice production systems in the same way, it has had little effect on regional income distribution.

The net effect of the present government input subsidy and trade policies is to make net private profit exceed net social profit for all rice production systems in Sierra Leone (23). This indicates that producers are on the average subsidized by government policy at the expense of consumers and the government treasury. This conclusion is contrary to one of the major conclusions recently arrived at by Levi (13).[20]

What should be the direction of government rice policy in the future? Since Sierra Leone has a comparative advantage in producing rice to replace imports as well as to export to neighboring countries, efforts to increase domestic rice production should continue. Net social profitability is highest in the inland swamps and mangrove swamps in the south and in

---

[19] The budgetary constraint has had some effect on agricultural policies in the last few years. The tractor hire scheme has been particularly affected by the inability of the Ministry to provide capital and operating funds. Also, attempts are being made to reduce the level of subsidy on fertilizers for the same reason, although foreign funds are being brought in to relieve the situation. The tractor-plowing scheme in the southern riverain grasslands has been taken over by a foreign-financed project, which hopes to run the scheme efficiently enough to make subsidization unnecessary.

[20] Levi claims that a consumer-oriented, cheap rice policy has been pursued in Sierra Leone (13). He arrived at this conclusion because he failed to take proper cognizance of the effects of input subsidy policies and of the producer price support effect of the restrictive rice import policy.

manually cultivated bolilands in the north. But exclusive reliance on their development, although yielding highest social returns, could worsen regional income distribution. The essential elements of a rice production strategy in the 1980's, which would give Sierra Leone a reasonable chance of efficiently achieving self-sufficiency or an export capability and of improving reasonable income distribution, include increasing the area under improved inland swamp cultivation at a rate of perhaps 10 percent per annum, doubling the area under mangrove swamp in the southern province by 1990, improving yields and efficiency of mechanical cultivation in the bolilands and riverain grasslands, and bringing about 80,000 hectares of upland rice under improved management (22).

Two specific changes in present practice would facilitate rice development. First, steps need to be taken to improve the morale of extension staff, particularly intermediate- and lower-level staff. It is presently difficult to hire and keep high-caliber staff in the Ministry of Agriculture because of the unattractive salary for what is a difficult job. Second, a marked feature of policy making in the past has been the taking of crucial decisions on prices, subsidy levels, and so forth with little or no empirical information. It is not surprising that such policy making has sometimes led to mistakes (e.g. importation of too much rice in 1974 and of too little in 1961 and 1964). The gathering and processing of agricultural statistics have improved during the last decade, but there is still one major gap—the provision of regular production statistics. Presently, statistics on rice and other agricultural output are the result of guesses. More accurate policy making calls for a more systematic crop-reporting system.

Since all but one of the rice production systems in Sierra Leone have positive social profitability, government protection and subsidy policies are not necessary so long as world rice prices (c.i.f. Freetown) do not drop below Le 225 per ton for 25 percent broken rice, the quality imported by Sierra Leone. Both protection and input subsidies could therefore be progressively reduced in the immediate future while efforts are made to stimulate domestic production.

The level of subsidies on fertilizers, credit, tractor hiring, and improved seed should continue to be reduced. The government is presently attempting to reduce subsidies on these inputs by raising the farmer price and trying to reduce the government costs of providing tractor hire services and improved seed by more efficient operation of the agencies providing these inputs. Removal of these subsidies would help bring Sierra Leone rice production in line with world market conditions and reduce the strain on the government budget. Given the high cost of land development (construction of water control structures in swamps) and extension input (due to the need to use expatriate personnel to supplement scarce local manpower) and the availability of foreign funds to finance such subsidies, it might be desirable to continue to provide subsidies on these two inputs for a longer period.

Although the restrictive trade policies that have raised domestic rice prices have not recently been necessary to protect local rice production, it would appear unwise to remove them completely and immediately, given the unpredictable nature of world prices. Government agencies that restrict imports should strive to reduce their costs of operation so that the implicit tariff could be reduced. Moreover, the government should ensure that the private trade, which has efficiently processed and transported domestic rice in the past, is encouraged to continue to do so. As import substitution proceeds, milled domestic rice from private traders could supply institutional and other urban customers, and the inefficient government rice-processing operations could then be phased out.[21] By carrying out this set of policies, the government could promote self-sufficiency or exports of rice through expansion of efficient production systems whose increased incomes are mainly distributed to Sierra Leone's poorer farmers.

Citations

1 R. Q. Crawford and A. J. Carpenter, "Partial Mechanization of Rice in Sierra Leone," *World Crops*, 20, No. 1 (1968).

2 J. Davies, "Development Plan of the Agricultural Services of Sierra Leone." N.p., 1964.

3 Government of Sierra Leone, "White Paper on National Resources Policy." Freetown, 1961.

4 ———, "Annual Estimates of Income and Expenditure." Freetown, 1958–78.

5 ———, "Report of the Commission Appointed to Enquire into and Report on Matters Contained in the Director of Audit's Report on the Accounts of Sierra Leone for the Year 1961/61; and the Government Statement Thereon." Freetown, 1963.

6 ———, *Agricultural Survey of Sierra Leone, 1965/66*. Central Statistics Office, Freetown, 1967.

7 ———, *Agricultural Statistical Survey of Sierra Leone, 1970/71*. Central Statistics Office, Freetown, 1975.

8 ———, *Household Survey of the Western Area*. Central Statistics Office, Freetown, 1968.

9 ———, *National Development Plan, 1974/75–78/79*. Central Planning Unit, Ministry of Development and Economic Planning, Freetown, 1974.

[21] As noted above, the Rice Corporation was dismantled in April 1979. Two of its three large rice mills (Mambolo and Torma Bum) have been handed over to two agricultural development projects. The third (in Freetown) as well as all other assets of the Corporation are apparently being passed over to the export crop marketing corporation (SLPMB). What its policy for rice marketing will be is as yet unclear. Imports of rice are apparently to be handled by private firms that will be licensed and will submit bids for the import quota.

10 ———, *National Accounts of Sierra Leone*. Central Statistics Office, Freetown, 1978.

11 R. P. King, and D. Byerlee, "Factor Intensities and Locational Linkages of Rural Consumption Patterns in Sierra Leone," *American Journal of Agricultural Economics*, 60, No. 2 (1978).

12 J. F. S. Levi, "The Demand for Food in the Western Area of Sierra Leone, An Analysis of the 1967 Household Expenditure Survey," *Bank of Sierra Leone Economic Review*, 7, No. 1 (1972).

13 ———, "African Agriculture Misunderstood: Policy in Sierra Leone," *Food Research Institute Studies*, XIII, No. 3 (1974).

14 ———, *African Agriculture: Economic Action and Reaction in Sierra Leone*. Commonwealth Agricultural Bureau, Slough, 1978.

15 R. J. Mutti, D. N. Atere-Roberts, and D. S. C. Spencer, "Marketing Staple Food Crops in Sierra Leone." University of Illinois and Njala University College, 1968.

16 Scott R. Pearson, J. Dirck Stryker, and Charles P. Humphreys, "An Approach for Analyzing Rice Policy in West Africa." Stanford/WARDA Study of the Political Economy of Rice in West Africa, Food Research Institute, Stanford University, Stanford, July 1979; Introduction.

17 G. A. Petch, "Report on the Oil Palm Industry of Sierra Leone." Freetown, 1958.

18 R. G. Saylor, *The Economic System of Sierra Leone*. Duke University Press, Durham, N.C., 1967.

19 D. W. Snyder, "An Econometric Analysis of Household Consumption and Savings in Sierra Leone." Unpublished Ph.D. thesis, Pennsylvania State University, University Park, 1971.

20 Dunstan S. C. Spencer, "Rice Production and Marketing in Sierra Leone, in I.M. Ofori, ed., *Factors of Agricultural Growth in West Africa*. ISSER, Legon, Ghana, 1971.

21 ———, "The Efficient Use of Resources in the Production of Rice in Sierra Leone: A Linear Programming Study." Unpublished Ph.D. thesis, University of Illinois, Champaign-Urbana, 1973.

22 ———, "Rural Development in Sierra Leone: Improvement in Production Systems for Development in the 1980s," *Sierra Leone Agricultural Journal*, forthcoming.

23 ———, "Rice Production in Sierra Leone." Stanford/WARDA Study of the Political Economy of Rice in West Africa, Food Research Institute, Stanford University, July 1979; Chapter 6.

24 Dunstan S. C. Spencer, I. I. May-Parker, and F. S. Rose, "Employment, Efficiency and Income in the Rice Processing Industry of Sierra Leone." African Rural Economy Paper No. 15, Department of Agricultural Economics, Michigan State University, East Lansing, 1976.

25 Dunstan S. C. Spencer and D. Byerlee, "Technical Change, Labor Use and Small Farmer Development: Evidence from Sierra Leone," *American Journal of Agricultural Economics*, 58, No. 5 (1976).

26 ———, "Small Farms in West Africa. A Descriptive Analysis of Employment, Incomes and Productivity in Sierra Leone." African Rural Economy Working Paper No. 19, Department of Agricultural Economics, Michigan State University, East Lansing, 1977.

27 West Africa Rice Development Association (WARDA), "Prospects for Intra-

Regional Trade of Rice in West Africa." WARDA/77/STC.7/9, Monrovia, Liberia, 1977.

28 ———, "Classification of Types of Rice Cultivation in West Africa." WARDA/78/STC.8/11, Monrovia, Liberia, 1978.

29 ———, "Rice Statistics Yearbook (Abstract)." Monrovia, Liberia, 1978.

30 L. Weintraub, "Introducing Agricultural Change: The Inland Valley Swamp Rice Scheme in Sierra Leone." Unpublished Ph.D. thesis, University of Wisconsin, Madison, 1973.

6. Rice Production in Sierra Leone

Dunstan S. C. Spencer

Sierra Leone is the leading rice-producing and -consuming country in West Africa and the third most important in Africa (after Egypt and Malagasy Republic). Rice is the staple food and most important crop grown in Sierra Leone. Per capita consumption is over 120 kilograms (kg), and about 465,000 hectares (ha) of rice are cultivated each year by 85 percent of Sierra Leone farmers. Rice production generates about 15 percent of gross national product. Although domestic rice output has doubled during the past two decades, imports of rice have continued (17). The government of Sierra Leone believes that rice importation unnecessarily uses scarce foreign exchange and, like all other West African governments, has adopted a goal of eventual self-sufficiency in rice.

Government policies have had differential impacts on the many rice production and post-production systems in Sierra Leone. The objective of this analysis is to examine the structure of costs and benefits in the principal systems in order to estimate comparative advantage in domestic rice production and to determine the varying effects of government policies.

The chapter is divided into nine sections. Section 2 contains descriptions of the main rice production techniques in Sierra Leone, classified on the basis of soil and water regimes and of technology. Private costs and returns in each of the production systems are analyzed in Section 3. Section 4 provides a description of the post-production systems, which is followed by an analysis of their private costs in Section 5. In the rest of the chapter the analysis concentrates on social costs and returns. Derivation of shadow prices and estimation of net input subsidies are discussed in Section 6. Private and social profitability for each production and marketing channel are contrasted in Section 7, and sensitivity analyses to determine the effect of changes in input and output values on the results are discussed in Section 8. The final section contains a summary and conclusion.

Rice Production Systems

Rice is grown in the uplands and lowlands of Sierra Leone. Using WARDA's classification system (16), one can demarcate four basic types of

TABLE 6.1. *Key Characteristics of Rice Production Techniques in Sierra Leone*

Production technique	1975 area (000 ha)	Paddy yield (mt/ha)	1975 output (000 mt)	Improved seeds	Fertilizer	Planting method[a]	Weeding practice[b]
Traditional upland, south	152.8	1.30	198.6	no	no	B	M
Traditional upland, north	179.2	0.81	115.2	no	no	B	M
Improved upland, south	—	—	—	yes	yes	B	H
Improved upland, north	—	—	—	yes	yes	B	H
Mangrove swamp, south	3.1	1.74	5.4	no	no	T	H
Mangrove swamp, north	24.3	3.15	76.5	yes	no	T	N
Boliland (manual), north	12.3	.96	11.8	yes	yes	BT	L
Boliland (tractor plow), north	5.9	1.13	6.7	yes	yes	B	L
Riverain (tractor plow), south	5.5	1.84	10.1	no	no	B	L
Traditional inland, south	20.9	2.65	55.4	no	no	BT	L
Traditional inland, north	36.0	2.20	79.2	yes	yes	T	L
Improved inland, south	3.5	3.98	13.9	yes	yes	T	L
Improved inland, north	2.0	3.30	6.6	yes	yes	T	L

NOTE: Only improved inland techniques had partial water control; all other techniques had no water control. All harvesting was done manually.
[a] B = broadcast, T = transplant, BT = combined broadcast and transplant.
[b] H = heavy, M = medium, L = light, N = none.

rice cultivation—upland (hill rice and flatland rice), mangrove without tidal control, freshwater without water control, and freshwater with partial water control. On the basis of location, planting method, weeding and harvesting practices, source of power, and use of fertilizer and improved seeds, these four basic types can be further stratified to give the 13 systems of rice cultivation described and analyzed in this paper. The key characteristics of these systems are summarized in Table 6.1.

Upland Rice Cultivation

Upland rice cultivation is practiced on well-drained land not subject to flooding, where rain is the only source of water. The crop does not draw water from a high groundwater table. This type of rice cultivation occurs on hillsides (hill rice) and fairly flat areas (flatland rice) in Sierra Leone, but for the purposes of this analysis, there is no need to make a distinction between the two.

Upland rice culture is the major system of rice production in Sierra Leone. As shown in Table 6.1, it occupies about 75 percent of the area under rice and accounts for about 55 percent of domestic rice production. It is a traditional method of rice cultivation in which soil fertility is maintained by the bush fallow method. Generally, the land is brushed (i.e. the forest vegetation is cut and allowed to dry), burned, and cleared between January and April. With the onset of the rains the land is slightly plowed, seeded by broadcasting, and harrowed with a short-handled hoe. Traditional varieties are normally planted, and intercropping is common, usually with cassava, maize, benniseed, or broad beans. Hand weeding is time-consuming, but has a great effect on yields. Hand harvesting with a small knife usually takes place between August and October. Because of better soil and rainfall, yields are slightly higher in the center, east, and south of Sierra Leone than in the north. Consequently, in this discussion upland rice cultivation in the north is differentiated from that in the south.[1]

Starting in the 1976–77 crop season, an improved upland rice cultivation system was introduced into Sierra Leone by the Integrated Agricultural Development Projects (IADPs) located in both the north and the south. Cultural practices are the same in this system as in the traditional upland rice system except that improved seeds and fertilizer are used. Less than 1,000 ha were under this system of cultivation by the end of 1978. Therefore, in this chapter upland rice cultivation is subdivided into four systems: traditional upland, south (System 1), traditional upland, north (System 2), improved upland, south (System 3), and improved upland, north (System 4).

Mangrove Rice Cultivation Without Tidal Control

Mangrove swamps are located along the coast where tidal action causes inundation at high tides and drainage at low tides (Map 6.1). The mangrove soils are acid sulphate or cat clays.

Cultivation of swamp rice in Sierra Leone probably started about 1880 in the mangrove swamp areas around the Great and Little Scarcies rivers in northern Sierra Leone. Mangrove swamps gave such good results that by 1900 practically all the better swamps in the Scarcies were under cultivation following clearing of the native mangrove forests, a very difficult and expensive task. Clearing and cultivation of mangrove swamps in the southern coastal belt, especially along the banks of the Ribbi River, developed later and more slowly. Most of the currently unutilized mangrove swamps (about 50,000 ha) are in this southern coastal belt.

Tidal mangrove swamps are continuously cultivated from year to year. Transplanting takes place in July and August after the rains have pushed the salt tongue out to sea and leached the salt from the soil. Empoldering to protect crop land from the intrusion of saline water is not practiced in Sierra Leone. Weeding is uncommon, and hand harvesting takes place in December and January. Because of differences in land preparation and transplanting practices, yields are different in the northern and southern mangrove swamps. Consequently, two systems of mangrove swamp cultivation without tidal control—mangrove swamp, south (System 5), and mangrove swamp, north (System 6)—are distinguished in this paper.

---

[1] South is defined here to include the southern and eastern provinces of Sierre Leone. The north comprises the northern province. Annual rainfall averages about 2,700 mm in the north and about 3,430 mm in the south.

MAP 6.1. Sierra Leone, Showing Swamp Areas Important in Rice Cultivation.

Freshwater Cultivation Without Water Control

There are three types of freshwater rice cultivation systems without water control in Sierra Leone: bolilands, riverain grasslands, and inland valley swamps.

Bolilands. Bolilands are low, saucer-shaped swamp grasslands associated with the Rokel River and its tributaries in central and northern Sierra Leone (see Map 6.1). It is estimated that 30,000 ha of bolilands are suitable for rice cultivation, and about 60 percent of this area is cultivated. Boliland swamps (bolis) are dry throughout the dry seasons, but are flooded up to 1.5 meters for periods varying from three to six months as a result of rain water accumulation and river flooding. The soils are acidic and very low in phosphorus. Without the use of phosphatic fertilizers, yields are very low. Consequently, even under semitraditional conditions

(System 7) superphosphate is used in this area. In fact, almost all of the fertilizer consumed in Sierra Leone was used in the bolis until recently.

Because of the flat topography, mechanized plowing using a government-provided and heavily subsidized tractor-hire scheme has been introduced into the area (System 8). In 1975–76 about 6,000 ha in the bolis were mechanically plowed. When mechanical plowing is utilized, seed is broadcast in May. Under hand cultivation, broadcasting takes place in May or transplanting in June and July. Yields are higher with transplanting. Weeds are a problem in the bolis and hand weeding is necessary. Hand harvesting occurs in December and January.

Riverain grasslands. Small patches of riverain grasslands are scattered throughout the country. Extensive areas are located only in the south where two rivers, the Waanje and the Sewa, are prevented from flowing directly into the sea by a raised sand bar (see Map 6.1). The silt deposited by the rivers has formed extensive grassy plains, which flood up to four meters in the rainy season, necessitating the use of floating rice varieties. Hand cultivation of these areas is extremely labor-intensive because of the heavy infestation of weeds. Because of very low population density in the area, there has been little hand cultivation of these soils.

Mechanical cultivation under a government scheme started in 1949 and has been quite popular (System 9). A substantial number of the users are absentee farmers. Area cultivated fluctuates with the size of the government tractor fleet, declining to virtually zero in some years and rising up to 6,000 ha in others.

It is estimated that over 50,000 ha are suitable for rice cultivation in the riverain grasslands, and only about 10 percent of this area is cultivated. Seed is usually broadcast in April and May so that initial growth has started before deep flooding takes place. Weeding is essential if the crop is not to be choked by weeds. Hand harvesting among the tangled mat of lodged straw takes place in December and January after the floods have receded. Production in the riverrain grasslands is unpredictable because of the uncontrolled nature of flooding, the weed problem, and labor shortage in the area.

Inland valley swamp. Inland valley swamps are found throughout the country wherever depressions occur in the rolling upland. It is estimated that there are about 300,000 ha of inland swamps in Sierra Leone, of which only about 65,000 ha are currently under cultivation. Under traditional culture, inland swamps are cultivated for a number of years before being fallowed. The swamps are not completely stumped and there is no water control. Transplanting is usual, but broadcasting is not uncommon. Only one crop is planted a year in pure stands. As shown in Table 6.1, yields are nearly double those on uplands, and yields in the south are higher than in the north, the result again of better soil conditions. Consequently, two types of inland swamps without water control—traditional inland, south (System 10), and traditional inland, north (System 11)—are distinguished here.

Freshwater Rice Cultivation with Partial Water Control

Freshwater rice cultivation with partial water control involves stumping, partial land leveling, and the construction of dikes and contour bunds in inland valley swamps. The dikes and bunds allow some control over the submersion and drainage of plots. This system was introduced on a pilot basis in 1966–67. Presently it is being promoted by various development projects through which farmers are encouraged not only to adopt the water control system but also to use improved seeds and fertilizers. In 1976–77 about 8,500 ha were under these improved inland swamp systems comprising improved inland, south (System 12), and improved inland, north (System 13). Yields are generally 30 to 50 percent above those in the corresponding traditional systems.

Private Costs of Rice Production

Table 6.2 shows the private cost per hectare of the 13 rice production systems. Physical input-output data were collected in a detailed farm management survey conducted in 1974–75 (*11*). Prices have been adjusted to 1976 prices.

Land

Land development (or investment) costs apply only to inland swamp rice systems where land is cultivated for a number of years after forest vegetation has been cleared.[2] In the traditional system tree stumps are not completely removed. Hence, costs are lower than in the improved systems, where stumping is more complete and bunds and water canals are constructed. Land development costs are mainly labor costs, since the sluice gates are made out of palm logs.

Labor inputs per hectare into inland swampland development are about 47 and 60 man-days, respectively, for traditional swamps in the south and north, the difference arising from better stumping in the north. For improved systems land development takes about 185 man-days per hectare. Costs are prorated over ten years. For traditional systems an estimated real market rate of interest of 40 percent is used.[3]

[2] Forest vegetation is also cleared in upland rice farming systems. However, since the land is only cropped for one year, such costs are not regarded as land development (or investment) costs.

[3] This figure is based on Linsenmeyer's (*4*) estimate that fishing households paid about 43 percent interest on short- to medium-term loans after defaults were taken into account. In the absence of alternative empirical evidence, this rate is also used for farming households.

Farmers adopting the improved system receive a five-year loan of Le 172.90 per hectare at 8 percent. This loan covers the land development costs (185 man-days at Le 0.73 in the south and Le. 0.65 in the north) and costs of farm tools. To calculate the private cost of the land development, the annuity on the five-year loan is discounted at the market rate of interest (40 percent) to obtain the discounted present value of the loan. This is then prorated over the estimated life of the investment (ten years), using the market interest rate to give the annual private cost of land development.

Annual fees for land use are paid by some farmers in Sierra Leone to persons who control use of the land.[4] Although such fees are paid by less than a quarter of all farmers (*11*), they are used in Table 6.2 to reflect the rental value of land in calculating private cost. Fees paid are usually higher for the more productive swamplands.

Farm Labor

Labor use in rice production systems ranges from 68 to 445 man-days per hectare. Low labor use figures are found in systems in which seed is broadcast on grassland farms, especially when tractor plowing takes place. High labor utilization occurs in manually cultivated swampland farms where there is transplanting. Improved systems of production use additional labor for harvesting the increased yield as well as for changes in cultural practices including shifts from broadcasting to transplanting, better weeding, and application of fertilizers and chemicals.

Rural wage rates in Sierra Leone vary by region, sex, season, and task, reflecting the active nature of the rural labor market (*11, 13*). The daily wage rates shown in Table 6.2 are weighted average annual wages per man-day. They are based on data collected in a detailed farm management survey in 1974–75, adjusted upward in proportion to increases in the official minimum wage rate since 1975.[5] The wages include cash payments as well as the value of payments in kind. Generally, agricultural wages are higher in the south, where there are ample employment opportunities in more profitable tree crop production (coffee, cocoa, and oil palm). They are also higher in swamp than in upland rice farming systems because of the arduous nature of work in swamps.

The quantities of labor used to produce rice in Sierra Leone are similar to those used in neighboring Liberia (*7*), but are higher than in the Ivory

[4] In the communal system of land tenure practiced in Sierra Leone, the direct controller of land is usually the head of the extended family, but in some chiefdoms the chief exercises this control (*5*).

[5] There is evidence that wages in general in Sierra Leone started increasing rapidly in early 1974. In the absence of direct empirical measurements and rural cost-of-living indexes, the rate of increase was assumed the same as that for the official minimum wage. This approximation is used in spite of the fact that the official minimum wage rarely affects actual rural wages.

TABLE 6.2. *Quantities and Private Costs of Major Inputs into Rice Production Systems in Sierra Leone, 1976*
(Leones per hectare unless otherwise indicated)

Production technique	Land		Farm labor			Fertilizer N-P-K		Seeds		Annual tool cost	Interest on working capital	Plowing fee	Total private costs	
	Devel-opment	Annual fee	Man-days	Daily wage	Total cost	Qty. (kg)	Cost	Qty. (kg)	Cost				Per hectare	Per ton paddy
Traditional upland, south	—	3.51	205	0.70	142.88	—	—	54	10.53	2.23	4.62	—	163.77	140.0
Traditional upland, north	—	3.51	238	0.52	123.76	—	—	52	10.71	1.76	4.16	—	143.90	177.6
Improved upland, south	—	3.51	225	0.70	157.50	50–50–0	21.34	56	16.40	2.23	6.63	—	207.61	110.9
Improved upland, north	—	3.51	258	0.65	167.70	50–50–0	21.34	56	16.40	1.78	6.88	—	217.59	149.2
Mangrove swamp, south	—	6.84	220	0.75	165.00	—	—	92	17.94	3.11	5.91	—	198.80	114.5
Mangrove swamp, north	—	20.40	445	0.80	356.00	—	—	150	29.25	5.86	11.83	—	423.34	151.0
Boliland (manual), north	—	1.73	112	0.60	67.20	9–9–0	1.98	70	14.42	7.07	3.51	—	95.91	99.6
Boliland (tractor plow), north	—	5.19	68	0.60	40.80	13–13–0	2.96	60	12.36	2.10	5.30	24.70	53.41	82.5
Riverain (tractor plow), south	—	4.20	91	0.80	72.80	—	—	48	9.36	3.67	5.23	24.70	119.96	65.8
Traditional inland, south	14.21	12.96	274	0.73	200.02	—	—	56	10.92	7.13	6.09	—	251.33	107.7
Traditional inland, north	16.15	14.72	356	0.65	231.40	8–8–0	1.90	105	21.63	7.21	8.13	—	301.14	140.6
Improved inland, south	28.52	12.96	336	0.73	245.28	53–53–0	22.70	67	19.63	9.62	9.15	—	347.86	114.6
Improved inland, north	25.40	14.72	390	0.65	253.50	53–53–0	22.70	67	19.63	9.62	9.36	—	354.93	127.4

NOTE: Le 1.00 = $1.00.

Coast (3), and some Sahelian countries (6,15). This result is due to the fact that cultivation in heavy rain forest areas is more labor-demanding than in open savannah regions and thinner rain forests such as exist in the Ivory Coast.[6] On the other hand, rural wages rates are lower in Sierra Leone than in neighboring Liberia, reflecting the smaller return in alternative rural employment opportunities compared with those in the timber and mining concessions in Liberia. This lower wage rate is also reflected in the lower per capita gross national product in Sierra Leone than in Liberia. Labor is the single most important cost item, accounting for 44–61 percent of total private costs in partially mechanized systems and over 70 percent in manual systems.

Fertilizer

Fertilizer is used by farmers in improved systems of inland swamp and upland rice cultivation and in partically mechanized bolilands. It is also used by otherwise traditional farmers who manually cultivate bolilands and inland swamps in the northern province. These farmers have apparently learned about fertilizers from neighbors participating in the government tractor-hire scheme in the bolilands. As shown in Table 6.2, average fertilization levels are quite low on such farms.

Seed

Farmers provide their own seed in all systems except in the improved upland and inland swamp, where they are supplied with improved seed from seed multiplication farms run by the development projects.[7] Seed rates are higher in swamp rice systems in which transplanting takes place, particularly in northern mangrove swamps where long transportation distances and damage by pests result in much seedling loss (9).

Farmer-provided seed costs 12–18 percent more than the average paddy producer price, reflecting the expected price increase between harvest and planting dates. Seed supplies by the development projects costs about 29 cents per kilogram.

Tools

The annual user cost (depreciation plus interest) of hand tools is a small proportion of total costs. When the investment is financed out of the farmers' own resources, the actual acquisition cost is depreciated. When

---

[6] Average annual rainfall in the rice-producing areas of Sierra Leone is about 3,200 mm, and it is 2,000 mm in Liberia, 1,400 mm in Ivory Coast, 1,200 mm in Senegal, and about 620 mm in Mali.

[7] With the recent establishment of a specialized seed multiplication project, this situation is changing. Farmers in and out of development projects are now to be supplied with certified seed from central seed multiplication farms and certified seed growers.

financed by a loan, the discounted present value of the medium-term loan is depreciated.[8]

Working Capital

The cost of working capital is calculated using the market interest rate of 40 percent. Working capital is tied up in the labor input for an average of three months, since work in the rice fields spans the six-months average rice-growing season. For seed, fertilizer, and plowing fees, it is tied up for an average of six months, since these items are invested in at the start of the growing season.

Plowing Fees

A fixed plowing fee of Le 24.70 per hectare is paid by farmers using the government tractor-hire service in the bolilands and riverrain grasslands.

Total Costs

Total private costs per hectare vary from Le 93.41 in tractor-plowed bolilands to Le 423.34 in manually cultivated mangrove swamps in northern Sierra Leone. The adoption of improved technology within a production system generally leads to increased private costs per hectare, except where the level of subsidy is high, as with tractor-plowed bolilands. By contrast, with one exception, costs per ton of paddy produced are reduced by adoption of the improved technologies, since resulting yield increases more than make up for increased costs per hectare. For the same reason comparisons across systems show that there is a much smaller range in private costs per ton of paddy (Le 111.70) than in private costs per hectare (Le 329.93). Private costs of production per ton of paddy are lowest in the heavily subsidized, partially mechanized systems and highest in the manual systems, which receive few or no subsidies.

Rice-Marketing Systems

About 105,000 metric tons, or 35 percent, of annual domestic rice production are marketed in Sierra Leone (13). Assembly, processing, and distribution of this rice are performed by private intermediaries and, until recently, by the government-owned Rice Corporation.[9]

[8] As stated earlier, farmers adopting the improved inland swamp system receive a five-year loan of Le 30.00 at 8 percent for tools. In calculating the discounted present value of the loan, the annuity at the loan rate (8 percent) is discounted at the market rate (40 percent) and summed over the life of the loan (five years).

[9] The Rice Corporation was dissolved in April 1979, leaving the domestic rice trade completely in private hands, although the government apparently intends to continue to attempt to control prices.

Assembly

Resident village and itinerant merchants (who might also be farmers) purchase directly from the farmers. These private merchants handle virtually all the rice sold by farmers. Only a small proportion of rice marketings is sold through cooperative societies or directly to consumers in the small towns and villages. Farmers head-load their produce to the merchant's place of business in the village. Village or itinerant merchants transport their produce, using boats in the Scarcies and riverain grasslands and trucks elsewhere. They typically have sold less than 5 percent of the produce to the Rice Corporation and the bulk to private wholesalers.

Processing

Rice processing involves parboiling and milling. Parboiling consists of saturating paddy with water and raising the temperature to that required to gelatinize the starchy endosperm. In the most common village method of parboiling rice, a mixture of paddy and water is boiled in large iron pots or in 44 gallon drums, or part of a drum, until the grains are slightly swollen and soft and some of them burst. The paddy is then removed and spread out in the sun. Paddy is parboiled in this way by farmers themselves or by itinerant, village, or wholesale merchants, who use either their own family labor or hired workers. Commercial parboiling involves passing wet steam through grain that has been soaked for a few hours, followed by mechanical or sun drying. It is used only in the large rice mills. About 40 percent of rice consumed in rural areas is parboiled (60 percent in the northern province and 20 percent in the south). In urban areas the proportion is as high as 80 percent (10).

There are three basic types of rice-milling techniques in Sierra Leone—hand pounding with small wooden mortars and pestles, small rubber roller or steel cylinder mills processing about 0.2 tons of paddy per hour, and large mills processing 0.75–3.0 tons of paddy per hour. Hand pounding is used by farmers to prepare their rice for subsistence consumption and for sale in small village markets. In addition, over 350 privately owned small mills are concentrated mainly in small towns in the major rice-producing areas. The mills operate at about 50 percent of capacity. Finally, there are four large rice mills in the country. The Rice Corporation owned three of these mills, with a total nominal capacity of 6 tons per hour. They have been run at an average of less than 25 percent of capacity during the last ten years. Throughput increased to 44 percent of capacity in 1975–76. The privately owned large mill has a capacity of 0.75 tons per hour, but has rarely operated since it was set up in 1975.

Milling outturn is about 67 percent for hand processing and small mills, which produce rice with 20–40 percent brokens. The large mills produce rice with less than 10 percent brokens and have recovery rates averaging around 64 percent.[10]

Between 60 and 70 percent of the 160,000 tons of paddy milled annually in Sierra Leone is processed in the small rice mills. Virtually all the rest is hand-pounded and goes to supply small towns and large villages. Large mills process less than 5 percent each year.

Distribution

Before April 1979, imported rice was handled exclusively by the government-owned Rice Corporation. The Corporation sold its imported rice primarily to licensed wholesale merchants, who took delivery usually in Freetown. The rice was sold to retailers in Freetown or trucked by the wholesale merchants to other towns. Secondary channels for imported rice were direct sales to retailers and consumers by the Rice Corporation from its depots in Freetown and the major towns. The Corporation similarly distributed the small proportion of domestic rice that it handled.[11]

The privately marketed domestic rice is trucked to urban areas either by itinerant merchants or by large wholesalers who have taken title in the small provincial towns and villages. In the urban areas the rice is sold to retailers who sell in the public markets, using volume measures such as the cigarette or milk tin.

Private Costs of Rice Marketing

Table 6.3 shows the costs of marketing rice produced with the three alternative means of rice processing. Assembly costs consist of the annual user cost of sacks as well as transportation and handling charges. Sacks last two years. Transportation costs involve head-loading of paddy for the first five kilometers followed by trucking.

Milling costs are highest in hand pounding because of the high labor input, and small mills have the lowest milling cost. The relatively high cost of milling using the large disc sheller mills of the Rice Corporation was due to the poor physical condition of these mills and the low average milling ratio of 64 percent. But distribution costs to Freetown, the capital city, are lowest for the large mills, since they are in general located closer to Freetown than the average distance for hand-processing and small mills.

Total assembly, milling, and distribution costs are lowest for the small

[10] Most of the rice that is commercially milled in Sierra Leone is parboiled. For further details of rice processing in Sierra Leone, see Spencer, May-Parker, and Rose (*10*).

[11] The Corporation handled less than 5 percent of the domestic rice crop that was marketed annually between 1965 and 1975.

TABLE 6.3. *Annual Private Costs of Rice Marketing Systems in Sierra Leone*

(Leones per ton of clean rice)

Marketing costs	Hand processing	Small mills	Large mills
Assembly:			
Cost of sacks	7.85	7.85	7.85
Head-loading[a]	0.00	7.48	7.48
Trucking	0.00	0.00	1.37[b]
Handling	0.00	1.32	1.77
Total assembly cost	7.85	16.65	18.47
Milling:			
Unskilled labor	30.60	0.02	2.37
Skilled labor	0.00	0.97	3.16
Rent	0.00	0.31	1.15
Equipment cost[c]	0.30	5.33	11.92
Electricity, fuel, and oil	0.00	3.25	3.91
Repairs and maintenance	0.00	4.62	2.16
Others	0.00	0.00	0.40
Total milling cost	30.90	14.50	25.07
Distribution to Freetown:			
Head-load	7.48	0.00	0.00
Trucking[d]	14.48	14.48	13.26
Handling	4.40	2.90	2.65
Total distribution cost	26.36	17.38	15.91
Total marketing cost	65.11	48.53	59.45

SOURCE: *10*

NOTE: Assume mills operate at 67 percent of maximum capacity, i.e. 20 hours per day, 200 days a year. This rate is used for consistency among mills, although it is higher than the rates achieved up to 1975. For large mills, maximum capacity is only 65 percent of manufacturers' stated capacity because of the poor physical condition of the mills.

[a] Le 0.40 per 68 kg bag of paddy for 5 km.

[b] 15 km.

[c] Depreciation and interest. Market interest rate of 20 percent for longer-term capital investment.

[d] Average cost of Le 0.061 per ton per km.

mills channel. As stated earlier, this channel handles 60–70 percent of domestic rice marketing and over 90 percent of the domestic rice that moves to Freetown. To simplify the analysis that follows, the small mills channel is used exclusively.

Social Costs and Government Subsidies

Because of government policies and market imperfections, private costs do not reflect the true social costs of rice production. In order to calculate social costs, returns, and profitability, it is necessary to use shadow prices for domestic factors and to estimate the amount of subsidies and taxes on tradable inputs and outputs.

Shadow Prices

Appendix B provides details of the procedures used in estimating shadow prices in this study. These shadow, or social accounting, prices reflect the value of income forgone by using scarce resources in the different rice production and marketing systems rather than in the most profitable alternative activities.

The shadow price of Sierra Leone rice, the price of imported rice, 10–25 percent brokens (wholesale buying) at Freetown, is estimated at Le 309.00 per metric ton. This figure is based on an expected c.i.f. price of Le 300.00 per ton plus landing costs of moving the rice from ship to wholesale warehouses of Le 9.00 per ton. This price level is consistent with a projected long-run f.o.b. price of Thai 5 percent brokens of about $350.00 per metric ton (2).

Shadow wage rates for unskilled labor are taken to be the same as market wages. There is ample evidence of the existence of an active rural labor market with minimum distortion of wages (12).

For the cost of capital, a market interest rate of 40 percent is used when traditional sources of credit finance the investment in tools and working capital. A rate of 20 percent is used for three- to five-year investments in rice mills usually financed by traders in larger towns.[12] During the last five years, Sierra Leone has been successful in obtaining concessional credit for financing its agricultural development projects. Because this foreign aid is expected to continue during the next ten years, a shadow interest rate of 3.5 percent is used for government investment capital in the Integrated Agricultural Development Projects (for improved inland swamp and upland development). A higher interest rate of 8 percent is used for government investment in mechanical cultivation, reflecting the fact that the government employs foreign commercial credit for this activity. All interest rates have been adjusted for the effects of inflation, which averaged about 12 percent in 1975.

Since land is not in short supply (and site value plays no role in this analysis), the shadow price of land is assumed to be zero. This assumption is supported by the earlier observation that private land costs are low and account for only a small proportion of total private cost.

Government Subsidies

In order to encourage farmers to adopt improved methods of rice cultivation, the government has subsidized some inputs and has supported domestic rice prices through restrictive trade policies (13). Subsidies are

---

[12]This rate is actually applicable only to small mills, but in the absence of more complete information it is also used for large mills in this analysis.

provided on land development, fertilizers, plowing, and extension service costs. The net subsidy on each input is shown in Table 6.4.

The subsidy on land development for improved inland swamp systems arises because farmers receive loans at a subsidized interest rate of 8 percent. The subsidy is the difference between the annual cost of total investment at the market rate of interest and the actual private cost at the subsidized interest rate. Forty-nine percent of the annual user cost of land development is subsidized.

Extension advice, defined as the teaching of new skills to farmers, is traditionally provided free to farmers, and thus its cost can be regarded as a subsidy to farmers. Since the degree of extension input varies according to the system of production, the rate of subsidy also varies. In Sierra Leone only rice farmers using improved systems of upland and inland swamp rice cultivation receive any extension input.[13] Extension input subsidies are concentrated in the Integrated Agricultural Development Projects.

Extension has two phases. The first occurs during the initial three to six years of project development when farmers are being taught new skills and the extension input is heavy (the development phase), whereas the second takes place in subsequent years when extension effort is to maintain project achievements and introduce small changes (the maintenance phase). During the development phase, costs are treated as capital investment prorated over 20 years at the shadow government interest rate of 3.5 percent. During the maintenance phase, costs are projected annual costs. The subsidies shown in Table 6.4 are based on the costs of Integrated Agricultural Development Projects (Phase II Eastern Area and Phase 1 Northern Area).[14]

The government started to provide tractor-plowing services to Sierra Leone farmers in 1949. The government fleet plows, harrows, and sometimes seed harrows for farmers who pay a highly subsidized fee, Le 24.70 per hectare. This service actually costs the government about Le 108 per hectare, so that the subsidy rate is about 77 percent.[15] Because of the

[13] Farmers in the bolilands and riverain grasslands are in contact with government extension agents, but the input of the agents is not true extension education. Instead, extension agents provide tractor-plowing services or deliver fertilizers. In both instances the costs are treated as subsidies on the inputs handled.

[14] Costs of road improvements, well construction, and technical assistance for feasibility studies are not directly chargeable to project farmers and therefore are excluded. The average cost is Le 10.93 per hectare. Since the teaching of the techniques of the improved inland swamp system is more difficult than that of the improved upland system, inland swamp extension costs during the development phase are assumed to be 30 percent above the average, or Le 14.20 per hectare. Annual extension costs for upland rice systems are assumed to be half of those for inland swamp systems.

[15] The reported estimates, based on 1971 estimates by Due and Whittaker (1) that were updated in 1975 by the Ministry of Agriculture, have been recomputed using an 8 percent rate of interest.

TABLE 6.4. *Net Input Subsidies and Charges*
(Leones per hectare unless otherwise indicated)

Production technique	Land development	Extension service Development	Extension service Maintenance	Plowing service	Fertilizer	Farm tools	Working capital	Net input subsidy Per hectare	Net input subsidy Per ton rice[a]
Traditional upland, south	—	—	—	—	—	—	—	—	—
Traditional upland, north	—	—	—	—	—	—	—	—	—
Improved upland, south	—	7.10	8.34	—	28.71	—	1.08	45.23	36.06
Improved upland, north	—	7.10	8.34	—	28.71	—	1.08	45.23	46.30
Mangrove swamp, south	—	—	—	—	—	—	—	—	—
Mangrove swamp, north	—	—	—	—	—	—	—	—	—
Boliland (manual), north	—	—	—	—	3.94	—	—	3.94	6.10
Boliland (tractor plow), north	—	—	—	82.92	5.83	—	—	88.75	117.00
Riverain (tractor plow), south	—	—	—	82.92	—	—	—	82.92	68.00
Traditional inland, south	—	—	—	—	—	—	—	—	—
Traditional inland, north	—	—	—	—	3.80	—	—	3.80	2.64
Improved inland, south	27.43	13.96	16.67	—	30.53	9.26	1.21	99.06	48.73
Improved inland, north	24.42	13.96	16.67	—	30.53	9.26	1.21	96.05	51.48

[a] Paddy converted to rice, using 67 percent recovery rate.

heavy subsidy, it has been difficult for the government to import spares and new equipment as well as purchase fuel that would provide the service consistently to farmers. The area mechanically plowed has therefore fluctuated widely, reaching 11,250 hectares in 1971, dropping to 8,000 in 1973, increasing to 21,000 in 1974, and then dropping again to less than 10,000 hectares in 1977.[16]

Fertilizers are imported and distributed to farmers by the Ministry of Agriculture. Only about 3 percent of Sierra Leone's farmers use the 5,000 to 8,000 tons of fertilizers imported each year. Until the end of the 1974–75 crop season, farmers paid about 3 cents a kilogram for compound rice fertilizer (20-20-0), which cost the government about 20 cents a kilogram in 1974. In 1975–76 the price was raised to 8.5 cents to farmers in the Integrated Agricultural Development Projects and 4.5 cents to others. Subsidy rates were therefore between 57 and 66 percent.

Subsidies on farm tools and working capital are due to the subsidized interest rate on the loans given to farmers in the development projects. They are calculated in the same way as the land development subsidy.

Total input subsidies per hectare, shown in Table 6.4, are highest in improved inland swamp systems. But because of high yields obtained with those systems, the subsidy per ton of paddy is much lower than that provided to farmers using the tractor-plowing service.

Private and Social Profitability

Details of the methodology used in this analysis are given in Appendix A. Net private profitability (NPP) is the difference between the market value of output and the market costs of all inputs. It measures the incentive provided to private producers of the commodity. Net social profitability (NSP) is the profit calculated when all inputs and outputs are valued at their social opportunity costs. Net social profitability therefore measures the economic efficiency and comparative advantage of producing rice domestically to substitute for imported rice. The difference between private and social profit is a measure of the impact of government programs on domestic rice production.

The other measure of economic efficiency utilized in this chapter is the resource cost ratio (RCR), which is the ratio obtained by dividing the sum of all domestic factor costs valued at social opportunity cost by the value added in world prices. Unlike NSP, this ratio is independent of the unit of measurement and is therefore useful as a relative measure of economic

[16] At the peak of the mechanical cultivation scheme in 1974, the government spent about Le 1.74 million to subsidize tractor plowing. This was equivalent to about 55 percent of the annual recurrent expenditure of the Agriculture Division of the Ministry of Agriculture and Natural Resources.

TABLE 6.5. *Expected Yield and Private and Social Profitability Per Metric Ton of Rice Delivered to Freetown, Sierra Leone*

Production technique	Expected yield (kg/ha)	Net private profit		Net social profit		Resource cost ratio	Effective protection coefficient	Social cost of policies (Leones)
		Leones	Rank	Leones	Rank			
Traditional upland, south	1,170	80.00	10	55.00	8	0.811	1.022	25.00
Traditional upland, north	810	26.00	13	1.00	12	0.996	1.022	25.00
Improved upland, south	1,872	128.00	7	62.00	6	0.756	1.115	66.00
Improved upland, north	1,458	75.00	11	−4.00	13	1.018	1.147	79.00
Mangrove swamp, south	1,736	117.00	8	94.00	4	0.679	1.022	25.00
Mangrove swamp, north	2,803	64.00	12	48.00	9	0.837	1.022	16.00
Boliland (manual), north	963	147.00	4	108.00	1	0.622	1.044	39.00
Boliland (tractor plow), north	1,132	165.00	2	24.00	11	0.904	1.056	141.00
Riverain (tractor plow), south	1,820	192.00	1	96.00	3	0.651	1.024	96.00
Traditional inland, south	2,334	137.00	6	107.00	2	0.634	1.022	30.00
Traditional inland, north	2,142	92.00	9	58.00	7	0.802	1.031	34.00
Improved inland, south	3,034	158.00	3	65.00	5	0.757	1.080	93.00
Improved inland, north	2,785	140.00	5	45.00	10	0.831	1.086	95.00

NOTE: These normal expected yields are generally lower than 1975 yields shown in Table 6.1, since 1975 was generally a good year for most rice production techniques. The small mills channel was used for delivery to Freetown.

efficiency for making comparisons among different economic activities and international comparisons. When the ratio is less than unity, the activity is socially profitable and the country has a comparative advantage in its production. The lower the ratio, the more socially profitable the activity.

The effective protection coefficient (EPC) is the ratio of value added in domestic prices to value added in world prices. It measures the net increase in domestic value added permitted by trade and price policies over the value added in the absence of such policies. The higher the EPC, the higher the degree of protection. Trade and price policies include restrictions on imports (duties or quantitative controls), subsidies on exports, and domestic price support policies.

Table 6.5 shows private and social profitability as well as the resource cost ratio and effective protection coefficient for each rice production activity in Sierra Leone (using the small-mills marketing channel for delivery to Freetown). All of the systems have positive NPP, indicating that farmers have a positive incentive to produce rice for commercial trade. Private profits are highest for the systems using the government tractor hire scheme. They are also quite high in the improved inland swamp farming schemes and the improved upland system in the south, systems fostered by the Integrated Agricultural Development Projects. NPP is not very high in the improved upland system in the north, in part because of relatively high labor costs in areas around Makeni, where the system has been introduced.

Net social profitability is lower than NPP for all the systems examined, showing that all domestic rice production has been subsidized by government policies. Producers earned higher profits than they would have in the absence of such policies, i.e. there was a net transfer from the government budget and from consumers to producers. The social costs of policies, measured by the differences between NPP and NSP, are also given in Table 6.5. They range from Le 16–23 per ton for mangrove swamps to Le 96–141 for tractor-plowed bolilands and riverain grassland. This transfer to producers was effected by the government through input subsidies as well as through higher domestic market prices. For each technique the difference between the social cost of policies in Table 6.5 and the net input subsidies to producers in Table 6.4 is the net effect of trade policies that raise the domestic price of rice.[17] These differences range between Le 16 and Le 44 per ton of rice delivered to Freetown. The variations among systems were due to differences in input costs and yields as well as in the location of systems relative to Freetown resulting in different transportation costs. The greatest beneficiaries of government policies are the

---

[17]These trade policies included import restrictions, a tax on imported rice in the form of trading surpluses by the government Rice Corporation, and government purchase and storage of domestic rice (13).

improved systems. This result is further confirmed by the fact that EPCs are highest for improved uplands and swamps and for boliland systems, where farmers benefit from the full range of government input subsidies and protection policies. Traditional farmers only benefit from the protection provided by trade policies.

NSP is positive for all techniques except the improved upland system in the north, indicating that Sierra Leone has a comparative advantage in using 12 of its 13 rice production systems to replace imports in its major urban center.[18] This is also shown by the RCRs, which are less than unity for the 12 efficient systems. NSP is highest in the more traditional manually cultivated swamps (bolilands and mangrove swamps in the north as well as inland swamps in the south).

Since production for the Freetown market is socially profitable, it follows that production for closer urban markets and home consumption would also be socially profitable. Such production is accorded the further protection of transport costs from the port inland. In addition, all systems except three in the northern province—traditional and improved upland techniques and tractor-plowed bolilands—have a comparative advantage in exporting to Monrovia, the major urban center in neighboring Liberia, as shown in Table 6.6. The production systems in the north are further away from Monrovia than those in the south, and hence their social profit for supply to Monrovia is lower. Generally, social profit for sales to Monrovia is slightly lower than for sales to Freetown, reflecting the greater transportation distances.

The net effect of government incentives in rice production is illustrated by the relative ranking of the systems in terms of private and social profitability (Table 6.5). The changes in rankings are statistically significant, with a 90 percent confidence interval (based on Spearman's rank-order correlation coefficient, one-tailed test: $R_3 = 0.51$, $Z = 1.77$). However, the effect on three-fourths of the production systems is negligible (a change of one to three places). Some of the smallest changes occur in the more traditional and least privately profitable systems (1, 2, 11). The most dramatic effect occurs in the boliland swamp farms, which use the government tractor hire scheme. Because of the heavy input subsidies, one of the least socially profitable systems, ranking eleventh in NSP, is transformed into one of the most privately profitable systems, with a rank of second in NPP. Improved uplands in the north as well as improved inland swamps (north and south) and mechanized riverain grasslands are the other systems whose rankings are improved in NPP relative to NSP.

The net effect of government policies on production incentives is thus to transform some of the least socially profitable production systems into

[18] Virtually all imported rice is consumed in Freetown (13).

TABLE 6.6. *Private and Social Profitability Per Metric Ton of Rice Delivered to Monrovia, Liberia*

Production technique	Net private profit (Leones)	Net social profit (Leones)	Resource cost ratio
Traditional upland, south	58.00	40.00	0.867
Traditional upland, north	−8.00	−26.00	1.088
Improved upland, south	106.00	47.00	0.822
Improved upland, north	41.00	−32.00	1.129
Mangrove swamp, south	94.00	79.00	0.738
Mangrove swamp, north	16.00	5.00	0.983
Boliland (manual), north	114.00	80.00	0.724
Boliland (tractor plow), north	131.00	−3.00	1.012
Riverain (tractor plow), south	169.00	81.00	0.715
Traditional inland, south	115.00	92.00	0.694
Traditional inland, north	58.00	30.00	0.898
Improved inland, south	136.00	50.00	0.819
Improved inland, north	106.00	17.00	0.936

the most privately profitable systems. The experience of the development projects, which have had no difficulty getting farmers to adopt the improved systems of production, bears testimony to the fact that farmers are aware of the advantages provided by the government incentive structure. For example, the government tractor hire scheme has never been able to completely satisfy farmer demands for its services.

A question that arises from this analysis is why the Sierra Leone government chooses to protect and subsidize rice production when almost all techniques are socially profitable. First, it should be pointed out that the world price used in this analysis represents almost a 50 percent increase compared with the prices prevailing before 1972. Government protectionist policies that started in an era when most domestic rice production was socially unprofitable continued when increases in the world market price changed the situation. As is shown in the following section, a 33 percent fall in the world rice price would make most systems socially unprofitable. Furthermore, the Sierra Leone government desired to receive revenue from imports of rice, and the Rice Corporation made a profit on its trade in imported rice in most years (13).

Sensitivity Analysis

The empirical results discussed above are based on best estimates of the average values of several parameters. It is interesting to examine the effect of variations in these values on the empirical results. Such variations could reflect errors in data, heterogeneity among farmers, climatic effect on yields, or alternative social costs of inputs.

Elasticities of changes in NSP resulting from changes in parameter val-

TABLE 6.7. *Elasticities of Net Social Profitability Resulting from Changes in Yields, Milling Ratio, and Social Costs of Labor and Capital*

Production technique	Yields	Milling ratio	Unskilled labor	Skilled labor	Capital
Traditional upland, south	3.65	3.95	−3.66	−0.15	−0.48
Traditional upland, north	231.34	246.28	−229.64	−7.62	−26.84
Improved upland, south	3.16	3.43	−2.33	−0.23	−0.54
Improved upland, north	59.45	63.21	−44.00	−3.64	−8.48
Mangrove swamp, south	1.74	1.91	−1.74	−0.09	−0.28
Mangrove swamp, north	4.38	4.72	−4.43	−0.18	−0.32
Boliland, (manual), north	1.40	1.55	−1.20	−0.08	−0.37
Boliland, (tractor plow), north	9.58	10.26	−3.96	−0.35	−5.09
Riverain, (tractor plow), south	1.70	1.87	−0.93	−0.09	−0.85
Traditional inland, south	1.41	1.56	−1.33	−0.08	−0.33
Traditional inland, north	3.48	3.77	−3.17	−0.15	−0.72
Improved inland, south	2.98	3.24	−2.11	−0.23	−0.78
Improved inland, north	4.77	5.14	−3.41	−0.34	−1.16

NOTE: These are point elasticities measuring the effect of a 1 percent increase in parameter value. They are therefore sensitive to the absolute value of NSP. As the value of NSP approaches zero, the elasticity becomes very large.

ucs are given in Table 6.7. Since these are point elasticities, they are only valid for small changes. Furthermore, the values are dependent on the absolute value of NSP, becoming very large as NSP approaches zero, so that comparisons can only be made between input and output values within techniques. Such comparisons show that NSP is most sensitive to variations in yields and the milling ratio. Variations in unskilled labor costs also have an important effect on NSP, but changes in capital and skilled labor costs, both minor cost items, have minimal effect on NSP.

Table 6.8 shows the percentage changes in yields, world rice price, and social costs of labor and capital that are necessary to reduce NSP to zero. These are the percent changes that would cause Sierra Leone rice production techniques to lose their comparative advantage. Except for uplands in the north (traditional and improved) and tractor-plowed bolilands, it would take more than a 100 percent increase in the social opportunity cost of capital, other things being equal, for production techniques to lose their comparative advantage. The increases in skilled labor cost needed are also very large. Those for unskilled labor are smaller, but usually over 30 percent. On the other hand, yields and world market prices need to drop less than 30 percent for all but four of the systems to lose their comparative advantage.

Generally, one can conclude that the estimates of net social profitability are moderately sensitive to yield, the milling ratio, the cost of unskilled labor, and the world market price of rice. Nonetheless, most production techniques in Sierra Leone are likely to maintain their comparative advantage in replacing imports in the medium term, since over a 25 percent

TABLE 6.8. *Approximate Percent Change in Social Cost of Labor, Capital, World Price of Rice, and Yields Needed to Change Net Social Profitability of Rice Production in Sierra Leone to Zero*

Production technique	Unskilled labor cost	Skilled labor	Capital cost	World price of rice	Yields
Traditional upland, south	27	667	208	−20	−27
Traditional upland, north	0.4	13	4	−1	−0.4
Improved upland, south	43	435	155	−21	−32
Improved upland, north	−2	−27	12	+2	+2
Mangrove swamp, south	58	757	357	−33	−57
Mangrove swamp, north	23	556	192	−18	−23
Boliland (manual), north	83	1,250	270	−37	−71
Boliland (tractor plow), north	25	286	20	−10	−10
Riverain (tractor plow), south	108	1,111	118	−33	−59
Traditional inland, south	75	1,250	303	−37	−71
Traditional inland, north	31	667	139	−20	−29
Improved inland, north	47	435	128	−23	−34
Improved inland, north	29	294	86	−15	−21

increase in the real costs of most inputs or a similar fall in yields or world price of rice would be needed, other things being equal, for them to become inefficient.

Summary and Conclusion

Thirteen systems of rice production have been identified and analyzed in this chapter. They range from traditional upland rice cultivation in the northern province, with yields averaging about 800 kilograms of paddy per hectare, to improved systems of inland swamp cultivation using partial water control, improved seeds, and fertilizers, with yields averaging 3,000 kilograms of paddy per hectare. Private costs per hectare are between Le 93 and Le 423, and labor usually accounts for over 60 percent of the total costs of production.

Three post-harvest channels for rice moving to Freetown were described and analyzed. The estimated costs per ton of rice delivered to Freetown are about Le 49 for small mills, Le 59 for large mills, and Le 65 for hand pounding. The analyses of private and social profitability of replacing imported rice by domestic production concentrated on the channel using small mills, which is estimated to handle over 90 percent of domestic rice distributed to Freetown.

The Sierra Leone government provides subsidies on several rice production inputs. Net input subsidies on land development, extension service, plowing service, fertilizer, farm tools, and working capital range in the aggregate from zero in the traditional production systems to almost Le 100 per hectare in improved inland swamp rice farms participating in the

Integrated Agricultural Development Projects. Because of these input subsidies and trade policies that raise domestic rice prices, net private profitability exceeds net social profitability for all systems analyzed. This indicates that producers are on average subsidized by government policies, a continuation of past practice. But net social profitability has recently been positive for all except one system of production, the improved upland system in the north, and hence Sierra Leone now has a comparative advantage in producing rice to replace imports. Furthermore, ten of the thirteen production systems are also competitive in exporting to neighboring Liberia.

The net effect of government input subsidy and trade policy is to transform some of the least socially profitable systems of rice production into some of the most privately profitable. Tractor plowing in the bolilands in particular is given a big boost. As a result, the socially desirable pattern of production is distorted at a cost to society, and the government should review whether its objectives are well served by this set of policies.

Two other important conclusions emerge from this analysis. First, within the medium- to long-term framework, government policy should emphasize production in swamps rather than in uplands. Furthermore, manual traditional cultivation in swamps should be given priority over mechanized or improved production. However, the analysis in this chapter considers only efficiency and not other objectives of government, notably income distribution. These and other policy considerations are the subject of the previous chapter.

Citations

1 J. Due and V. Whittaker, "The Estimated Cost of Mechanical Cultivation of Rice in Sierra Leone and Suggestions for Cost Reduction," *Bank of Sierra Leone Economic Review*, 6 (1977).

2 Walter P. Falcon and Eric A. Monke, "The Political Economy of International Trade in Rice." Stanford/WARDA Study of the Political Economy of Rice in West Africa, Food Research Institute, Stanford University, Stanford, July 1979.

3 Charles P. Humphreys, "Rice Production in the Ivory Coast." Stanford/WARDA Study of the Political Economy of Rice in West Africa, Food Research Institute, Stanford University, Stanford, July 1979; Chapter 2.

4 D. Linsenmeyer, "Economic Analysis of Alternative Strategies for the Development of Sierra Leone Marine Fisheries." African Rural Economy Working Paper No. 18, Department of Agricultural Economics, Michigan State University, East Lansing, 1976.

5 I. I. May-Parker and S. S. Deen, "Land Tenure and Agricultural Credit—The Sierra Leone Case," *Proceedings of Seminar on Land Tenure Problems in Africa*. IITA, Ibadan, Nigeria, 1972.

6 John McIntire, "Rice Production in Mali." Stanford/WARDA Study of the Political Economy of Rice in West Africa, Food Research Institute, Stanford University, Stanford, July 1979; Chapter 10.

7 Eric A. Monke, "The Economics of Rice in Liberia." Stanford/WARDA Study of the Political Economy of Rice in West Africa, Food Research Institute, Stanford University, Stanford, July 1979; Chapter 4.

8 John M. Page, Jr., and J. Dirck Stryker, "Estimating Costs and Incentives." Stanford/WARDA Study of the Political Economy of Rice in West Africa, Food Research Institute, Stanford University, Stanford, July 1979; Appendix A.

9 Dunstan S. C. Spencer, "The Economics of Rice Production in Sierra Leone: II, Mangrove Swamp." Bull. No. 2, Department of Agricultural Economics, Njala University College, Njala, 1975.

10 Dunstan S. C. Spencer, I. I. May-Parker, and F. S. Rose, "Employment, Efficiency and Income in the Rice Processing Industry of Sierra Leone." African Rural Economy Paper No. 15, Department of Agricultural Economics, Michigan State University, East Lansing, 1976.

11 Dunstan S. C. Spencer and D. Byerlee, "Small Farms in West Africa. A Descriptive Analysis of Employment, Incomes and Productivity in Sierra Leone." African Rural Economy Working Paper No. 19, Department of Agricultural Economics, Michigan State University, East Lansing, 1977.

12 Dunstan S. C. Spencer, "Labor Market Organization, Wage Rates and Employment in Rural Areas of Sierra Leone," *Labour and Society*, 4, No. 3 (1979).

13 ———, "Rice Policy in Sierra Leone." Stanford/WARDA Study of the Political Economy of Rice in West Africa, Food Research Institute, Stanford University, Stanford, July 1979; Chapter 5.

14 J. Dirck Stryker, John M. Page, Jr., and Charles P. Humphreys, "Shadow Price Estimation." Stanford/WARDA Study of the Political Economy of Rice in West Africa, Food Research Institute, Stanford University, Stanford, July 1979; Appendix B.

15 A. Hasan Tuluy, "Rice Production in Senegal." Stanford/WARDA Study of the Political Economy of Rice in West Africa, Food Research Institute, Stanford University, Stanford, July 1979; Chapter 8.

16 West Africa Rice Development Association (WARDA), "Classification of Types of Rice Cultivation in West Africa." WARDA/78/STC.8/11, Monrovia, Liberia, 1978.

17 ———, "Rice Statistics Yearbook (Abstracts)." Monrovia, Liberia, 1978.

PART FOUR

Senegal

7. Rice Policy in Senegal

Kathryn Craven and A. Hasan Tuluy

For over a century, Senegal has exported peanuts and imported rice to cover its food deficit. That deficit has grown over time, and during the drought of 1968–73, Senegal imported large quantities of rice, sometimes at high world prices.

As a result, the government has increasingly emphasized the expansion of local rice production in its development plans. In part because of Senegal's unstable climate, the government has opted for more secure but costly irrigated systems of rice production. The expansion of production has been supported by parastatal land development agencies that distribute subsidized inputs as part of improved technological packages. But large capital investments plus high levels of modern input use have led to high production costs for rice, forcing the government to adopt protective trade policies. Although some success has been achieved in expanding local rice production, these increases have been insufficient to meet growing demand.

This chapter seeks to analyze the evolution of the political and economic influences on the rice sector in Senegal and to evaluate the effectiveness of government policies in furthering objectives. The physical setting for agriculture and the conditions of production, milling, marketing, and consumption of rice in the country are summarized in the succeeding sections. A description of historical changes in national economic policy in general and rice policy in particular sets the stage for an evaluation of major policies with respect to their impact on government objectives. A final section containing a summary and conclusion follows.

Setting

Economic Geography

Senegal has an area of 197,000 square kilometers and is divided into eight regions: urban Cap Vert in the west; Diourbel, Thiès, Louga, and Sine Saloum, which form the Groundnut Basin in the center-west; the northern Fleuve, which follows the Senegal River; the Oriental province in the east; and the verdant Casamance to the south of the Gambia (see Map 7.1).

Senegal's population was estimated at just over 5 million people in 1976

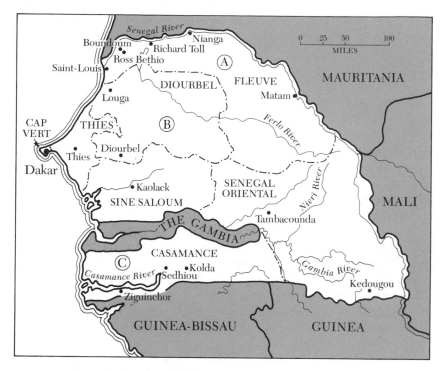

MAP 7.1. Senegal, Showing the Three Major Zones of Crop Production: (A) the Northern Crop Zone; (B) the Central Crop Zone (Groundnut Basin); and (C) the Southern Crop Zone.

and is thought to be growing at about 2.6 percent per annum (42). Population of the Cap Vert region, in which the capital city, Dakar, is located, has increased at 5 percent per year since independence in 1960 and now contains one-fifth of the country's population. The rural population is expanding at a rate of about 1.9 percent per annum (42).

Senegal has a good network of all-weather roads, which connect Dakar with the northern, central, and southern areas of the country although the southeastern part remains relatively isolated. Slow service and insufficient capacity at the Gambia River ferry have created a major bottleneck on the road connecting Dakar with the Casamance and have contributed to the continued isolation of this southern region.

Only about 13 percent of the total land area in Senegal was under cultivation in 1976, up from 10 percent at independence. It is estimated that nearly twice this amount is cultivable. Senegalese soils are generally poor—a condition that is aggravated in areas of high population density where the land is overcropped. The presence of marine salts in the richer

TABLE 7.1. *Area of Land Devoted to Principal Crops by Region, 1976*
(Thousand ha)

Region	Rice	Peanuts	Millet	Cotton	Total
Fleuve	7.7	9.2	91.7	—	108.6
Central Basin	4.0	1,119.8	676.6	5.9	1,806.3
Casamance	64.1	145.9	88.8	18.1	316.9
Eastern Senegal	5.4	46.8	94.2	19.8	166.2
Other	—	1.9	1.1	—	3.0
TOTAL	81.2	1,323.6	952.4	43.8	2,401.0
As percent of total cultivated area[a]	3.2%	52.0%	37.7%	1.7%	95.0%

SOURCE: Sénégal, Government of, Ministère du Développement Rural et Hydraulique, Direction Générale de la Production Agricole, *Rapport Annuel, Campagne Agricole, 1976/77*, Dakar, 1978.
    [a] 2,529,090 ha.

alluvial soils, which are found along the banks and in the mouths of the major rivers, restricts their full utilization for agriculture.

The absolute amount, seasonal distribution, and variability of rainfall are major physical constraints to Senegalese agriculture. Generally, the quantity of rainfall increases and its variability diminishes as one moves from north to south. The north Fleuve region, which lies in the Sahelian zone, receives 500 millimeters or less of rain during a three-month period. A Sudanese transitional zone extends from south of the Fleuve to just north of the Gambia and has a higher, more reliable rainfall of 700 to 800 millimeters a year. The major peanut-producing region lies in this middle zone. The southern, subtropical (Guinean) regions of the Casamance and part of Eastern Senegal have a four- to five-month rainy season with up to 1,800 millimeters of rain.

Because of climatic differences, Senegal's three major zones specialize in different patterns of crop production. As seen in Table 7.1, the northern Fleuve region, which is subject to a short, erratic rainy season, has traditionally grown short-cycle millets intercropped with cowpeas and, when possible, flood recession sorghum. Irrigated rice was first introduced into the region prior to independence, and the area is currently the country's second largest rice producer. The central Groundnut Basin, as its name implies, specializes in peanut (groundnut) cultivation while continuing to produce important quantities of millet and cassava.[1] Traditional swamp rice is cultivated in the Lower (western) Casamance, whereas in the upper reaches of this region and in Eastern Senegal, cotton, peanuts, and millet predominate.

Table 7.2 shows that during the past 40 years, the sum of the percent-

---

[1] Cassava is often consumed more as a condiment than as a staple food, but the quantities produced are nonetheless quite large (176,000 tons, 1969–73 average).

TABLE 7.2. *Percentage Distribution of Total Cultivated Area by Major Crop*

Years	Peanuts	Millet/sorghum	Rice	Total area culti-vated (*000 ha*)
1936–37	48%	45%	3.7%	1,411.1
1959–60	48	40	3.6	1,846.0
1976–77	52	38	3.4	2,529.1

SOURCES: Valy-Charles Diarassouba, *L'Evolution des Structures Agricoles du Sénégal*, Editions Cujas, Paris, 1968, pp. 122, 124; and Sénégal, Government of, Ministère du Développement Rural et Hydraulique, Direction Générale de la Production Agricole, *Rapport Annuel, Campagne Agricole, 1976/77*, Dakar, 1978.

ages of total cropped area devoted to peanuts and millet—the principal cash and food crops, respectively—has remained stable at about 90 percent. However, there appears to have been a shift away from the major cereal crop to peanuts. The peanut–food crop competition takes place mainly in the Central Basin, Upper Casamance, and Eastern Senegal through the demand for labor. There has also been some switching from rice to peanuts in both the Lower and Middle Casamance owing to the greater profitability of peanuts.

The Senegalese economy is highly dependent on agriculture and agricultural exports—particularly peanuts—for government revenue and foreign exchange earnings. Approximately 70 percent of the labor force works in activities directly related to agriculture, which provides about one-third on the gross domestic product (*43*).

Although rural per capita income grew during the early 1960's, low export prices for peanuts coupled with a series of droughts at the end of the decade led to a decline in the real value of rural income between 1961 and 1971. During this same post-independence decade, the importance of Dakar as the administrative center of the French West African colonial empire was diminished with the withdrawal of the French from the area. This shift, together with the decrease in agricultural income, led to a fall in the real value of per capita income earned in urban areas by 2.4 percent (*5*). In 1976 the average per capita income was approximately $400, and there was a wide disparity between urban and rural areas (9, p. 1).

With the exception of the central peanut regions, agricultural land is widely available.[2] Throughout the country, however, there is a seasonal shortage of labor. The peak in labor demand is closely tied to the timing and duration of a short rainy season. Once the rains begin, all crops must

[2] Although Senegalese agriculture has in the past generally been practiced using extensive techniques, in areas such as the central Groundnut Basin and southwestern Casamance there is some population pressure on the land, resulting in declining fallow periods, reduced soil fertility, and even feuds over prime agricultural land. But for Senegal as a whole, sufficient cultivable land is still available.

be planted almost simultaneously. In most areas, land preparation prior to planting cannot even start until the soil has been sufficiently softened by rain.

Production

There are two major rice-producing regions in Senegal. The bulk of production (65–70 percent) comes from the Casamance, where swamp and upland rice have been grown traditionally as a staple crop. Traditional rice cultivation depends heavily on the rainfall calendar. Land preparation is done with a few hand tools. Rice is transplanted in June or July and knife-harvested over an extended period beginning in November.

About 51,000 ha of rice are cultivated in the Casamance. With yields varying from 0.8 to 1.2 tons/ha, annual production from traditional farmers is about 50,000 tons of paddy. The basic production unit for this technique is the small family farm of between 4 and 5 ha, of which one-half to 2 ha are in rice (33, 36). This average figure includes the Lower Casamance, where rice is the predominant crop, and the Middle and Upper Casamance, where it is planted, along with millet, peanuts, and maize or cotton.

Recently, the government has begun extending improved rice techniques into this region, primarily aimed at swamp rice. Under the supervision of public agencies, modern inputs, including improved seeds, chemical fertilizers, and insecticides, have been introduced; more effective soil preparation, seeding, and weeding have been encouraged; and limited land improvements have been made. Following nearly a decade of extension work, substantially higher yields of 3 tons/ha have been obtained on the 13,000 ha of riceland that are under project supervision.[3] Thus far, irrigation has only been introduced on a small scale in the region, although tentative plans have been made to develop a large, mechanized irrigated scheme in the Upper Casamance. In addition, some smaller complexes of water control near the coast are being planned to control salt incursion on cultivable riceland.

Despite the preponderance of output from the southern region, most government investment in rice development has been concentrated along the Senegal River Valley in the north. There insufficient rainfall precludes rainfed rice cultivation so that only irrigated techniques can be employed. Polders were initially constructed to control flooding, but these soon proved inadequate. The rise and fall of the Senegal River, on which this type of irrigation depends, vary markedly from year to year, and reliance on natural flooding does not always assure sufficient inundation for a

---

[3] Planners anticipate that an additional 17,000 ha can be brought under this kind of supervision in the near future (61).

120-day rice crop. As a result, since independence—and particularly since the drought period of 1968–73—water security has been improved by pumping on leveled parcels, where high-yielding varieties can be used. Yields have increased from about 1 ton/ha to 3.5 or 4 tons/ha, and large interannual production variations are avoided in all but severe drought years, when there is little water in the river. At present, polders have varying degrees of water security, but it is planned that all will ultimately have total water control through leveling, installation of pumps, and construction of irrigation and drainage canals.

All rice production in the Fleuve is under the supervision of a large, parastatal organization—Société d'Aménagement et d'Exploitation des Terres du Delta (SAED)—which provides improved seed, chemical fertilizers, insecticides, and herbicides on credit to the project rice farmers. Short-stalked varieties, mainly I Kong Pao, are used on leveled parcels, and longer-stalked varieties, such as D52-37, are used on unleveled areas. In addition to these inputs, SAED provides machinery services for plowing and seeding on credit to farmers on the larger perimeters. On the smaller perimeters all of these operations are carried out by hand. Perimeter areas range from 2,000 to 3,000 ha in the Lower Delta to 15 to 20 ha in the Upper Valley around Matam. Average holdings of riceland are 1 to 2.5 ha per farmer on the large projects and 0.25 to 0.50 ha on the smaller projects. These smaller labor-intensive polders have the highest yields in Senegal, with nearly 5 tons of paddy per ha for a single rice crop.

Most polders are in the Senegal River Delta, where only one crop a year can be grown owing to saltwater incursion from the sea between March and July. Upriver, two rice crops are feasible, although maize and industrial tomatoes are usually preferred as the dry-season crop because of their lower water requirements and higher profitability. If too much water is drawn off for irrigation upriver, the ocean saltwater moves further upstream, making agricultural land adjacent to the river uncultivable. This saline incursion plus the general insecurity of water availability from the river have generated considerable interest in a downstream saltwater barrage (Diama) and an upstream storage dam (Manantali). Already, with only 3,000 ha under irrigation in the Middle and Upper Valley, upstream-downstream water management has become a critical issue. Future expansion of rice cultivation along the river will ultimately be linked to the decision to build these dams.

Marketing and Milling

Three-fourths of Senegal's total rice consumption is met by imports. The quantity imported each year varies according to the size of the domestic harvest, the world price, and the stocks on hand. In an unusually bad year for domestic rice production, such as 1974, imports may exceed 200,000 tons, whereas in favorable years, such as 1975, they decline to around 100,000 tons. Between 1969 and 1975, annual rice imports averaged 160,000 tons, compared with about 120,000 tons for each of the first five years of the 1960's.

The official rice-marketing structure in Senegal has two branches—one for imported rice, which constitutes over 95 percent of total official sales, and the other for locally produced paddy/rice, which handled only 8,000 to 12,000 tons of paddy per year. In addition to the operations of the government marketing agency, the Office National de la Coopération et de l'Assistance au Développement (ONCAD),[4] an illegal private market in domestic paddy/rice, also exists, although the area and extent of its activities are difficult to determine.[5]

Retail prices for imported rice are set in Dakar on the basis of the c.i.f. import price plus a variable levy. Small, fixed marketing margins are added to this base price, which allow wholesale and retail prices to differ regionally as a function of transport costs.

The government maintains a monopoly over most rice imports, though some imports of whole grain and packaged rice by licensed private traders are also permitted.[6] An average of 90 percent of the value of all imported rice is low-quality, often 100 percent brokens. Imports are delivered in sacks to Dakar, where they are released to wholesale-retail traders approved by the Ministry of Finance. ONCAD also maintains its own storage and distribution centers to supply traders in the outlying regions. Most small towns in Senegal have access to imported rice year-round both at government retail stores and from private stores that buy from the larger wholesale-retail traders.[7]

Official purchases of domestic rice come almost exclusively from government-supervised projects. SAED, for example, marketed an average of 40 percent of its annual production of paddy between 1969 and 1976. Seasonal farmer debts for inputs, machinery services, and water charges are collected in paddy at harvesttime and form the bulk of SAED pur-

[4] ONCAD has a monopoly on the purchase and processing of paddy and the selling of rice. At the official margins ONCAD has found transactions in local cereals unprofitable, however, so that rice and millet operations remain secondary marketing activities of the agency, which focuses mainly on peanuts.

[5] See SONED report (50, vol. 1 p. 103ff) for a description of the private market. One of the difficulties encountered in this study was the inadequacy of spatial and temporal data on prices and quantities marketed.

[6] The quantity of whole grain imports fluctuated in the 1969–75 period in response to relative changes in prices of whole grains and brokens. Fourteen percent of total imports in 1969 and 39 percent in 1971 were whole grains, yet in other years these imports amounted to less than 1 percent. See (56, p. 21).

[7] The Société Nationale de Distribution (SONADIS) maintains about 100 outlets throughout the country.

chases.[8] Perhaps 10 percent of total production in this area is sold on the private market.[9] In other rice projects, the amount of paddy sold through official channels is lower than in the Fleuve, probably because farmer indebtedness to the projects in these areas is less.[10]

In order to offset the differences between delivered costs of local and imported rice, a common official retail price is established through the operations of a stabilization fund, the Caisse de Péréquation et Stabilisation des Prix (CPSP).[11] The official price schedule in 1976 for locally produced rice from SAED is presented in Table 7.3. The total cost of rice to ONCAD at the warehouse has recently been 94 CFA francs per kilogram of rice.[12] This compares with the real cost of imported rice delivered to the ONCAD warehouse in St.-Louis of 61 CFA/kg in 1976, when the official retail price in St.-Louis was 82 CFA francs/kg.

At present, there are four large rice mills operating in Senegal with a combined rated hourly capacity of 17 tons, or 85,000 tons of paddy a year.[13] A two-ton per hour private mill in the Casamance works under government contract, whereas the other three mills are owned and operated by land development agencies. Given the low levels of official paddy purchases, all of these mills operate at below 20 percent of their rated capacities. The quality of milled outturn also varies greatly—from 90 percent brokens at the SAED mill in Ross Béthio to 50 percent brokens at the southern mills.

In certain areas in both the north and the south, small diesel-powered rice hullers operate despite the fact that they are officially discouraged.[14]

[8] Niaga, a mechanized SAED polder in the Middle Valley, marketed 690 tons of paddy, or 33 percent of production, in 1976. Eighty-five percent of this amount was for debt repayment.

[9] The gap between net production retained by farmers and current estimates of family rice consumption in the Fleuve is about 10 percent.

[10] In Eastern Senegal, the Société pour le Développement des Fibres Textiles (SODEFI-TEX) purchased only 21 percent of paddy production in 1976–77, half of which was debt repayment. Official purchases are even less than this in the Middle and Lower Casamance, reflecting lower yields, greater local consumption of rice, and more private trade.

[11] In May 1976 the official retail price in Dakar was established at 80 CFA francs/kg for 100 percent brokens. This price is set by an interministerial committee, the Comité Permanent Interministériel des Grands Produits Agricoles (CGPA), which each November also sets the producer prices for rice, peanuts, maize, millet, and sorghum (9, p. 39).

[12] The real cost is 96.97 CFA francs/kg because ONCAD only pays 85.44 CFA francs/kg instead of 87.98. The difference is covered by the stabilization fund (CPSP), which subsidizes SAED.

[13] A 6-ton/hour Schule (installed in 1971) and a 7-ton/hour Guidetti (installed in 1952) are located in Ross Béthio and Richard Toll in the north. Two 2-ton/hour Schule units are located in Séfa and Kedougou (installed in 1957 and 1975, respectively). Annual capacity has been calculated assuming a 20-hour day and 250 days of operation per year, or a total of 5,000 hours per year.

[14] These small machine hullers are illegal in the Fleuve, but their status is unclear in the rest of the country, where they are rare but operate openly.

TABLE 7.3. *Price Schedule for Domestically Produced Rice,*
SAED, 1976

Cost category	Official price (CFA fr/kg paddy)	Official price (CFA fr/kg rice)
Producer price	41.5	
Transport and handling	8.75	
Milling costs	7.70	
Subtotal/kg of paddy	57.95	
Subtotal/kg of rice (at 0.66 milling outturn)		87.66
Mill storage		0.85
Sales of by-products		0.53
Subtotal/kg of rice		87.98
Purchase by ONCAD, ex-mill		85.44
Delivery cost to St.-Louis		1.10
ONCAD charges		7.89
Total ONCAD cost at St.-Louis warehouse		94.43

SOURCE: Office National de la Coopération et de l'Assistance au Développement, "Barème du Riz Usiné par la SAED, 1975/77," Dakar, 1976, mimeo.

In the Fleuve, easy access to these machines has helped reduce transport costs, handling, and commercial margins and has made it profitable for an estimated 10 percent of production to flow into private milling and trade. In the Casamance, these hullers are used primarily during times of peak agricultural labor demand, when women working in the field do not have enough time to pound rice for the family meals.

Consumption

Rice is an important element in the Senegalese diet. Of a total average daily calorie consumption of 2,300 per capita, rice contributes 680 calories, or about 30 percent (55, pp. 481–85). The absolute amount of rice consumed annually remained fairly stable between 1965 and 1975, whereas per capita consumption declined. As Table 7.4 shows, however, these national averages mask highly uneven rice consumption patterns, especially between urban and rural consumers and among different regions.[15]

Urban centers, the largest consumers of rice, are supplied almost exclusively with imported rice.[16] The Cap Vert region has 20 percent of the total population and probably consumes between 100,000 and 130,000 tons of rice per year, or between 40 and 60 percent of net

[15] Derived consumption figures based on production and import estimates do not account for the flows that occasionally take place clandestinely between Senegal and the Gambia. Stocks of local and imported rice are also not accounted for in these figures.

[16] A Statistical Office study (44, pp. 12 and 21) found that 87 percent of all imports in 1973–74 went to urban areas, 61 percent of which was consumed in Cap Vert, Dakar.

TABLE 7.4. *Regional Population and Rural/Urban Rice Consumption Per Capita, 1973–74*

Region	Regional population			Rice consumption (kg/capita)	
	No. (000)	Pct. rural	Pct. urban[a]	Rural	Urban[a]
Cap Vert	936.2	0	100%	0	103
Peanut Basin:					
Thiès	664.5	75.3%	24.7	2	71
Diourbel[b]	801.3	85.1	14.9	8	70
Sine Saloum	958.0	85.3	14.7	17	110
Oriental					
(Eastern Senegal)	272.1	86.3	13.7	11	74
Casamance	700.2	82.7	17.3	58	130
Fleuve	502.4	78.1	21.9	30	80
TOTAL	4,834.7	70.4%	29.6%		

SOURCES: The number of inhabitants is estimated back from the 1975–76 population, assuming a constant 2.56 percent growth rate per annum. The 1975–76 population data and percent breakdown by rural and urban groups are from *42*. Rice consumption data are from *44*.
NOTE: Although consumption of rice in 1973–74 may have been influenced by the drought, that year is the only one for which such a breakdown is available.
[a] Urban refers to towns and cities of 10,000 people or more.
[b] The Department of Louga was split off from Diourbel to form an eighth region in 1976–77.

available rice.[17] As seen in Table 7.4, in other urban areas outside of Cap Vert, per capita consumption is well above the surrounding rural areas, mainly because of higher relative incomes and easier access to supplies of imported rice.

Average per capita rice consumption since 1968 has been below the average for the previous eight years.[18] This declining trend might be explained in part by falling incomes. As noted above, Senegal experienced an annual decline of 1 percent in real per capita gross national product between 1965 and 1975.

The government expects that future rice consumption will rise at least as fast as income growth.[19] Given the high proportion of rice consumed in urban areas, the rate of urbanization should also be considered as a factor in future consumption. Unfortunately, the lack of empirical data on demand elasticities for rice and cross elasticities between rice and other cereals means that consumption projects cannot be made with any reasonable degree of assurance.

Except in parts of the Casamance and the Fleuve, millet is the pre-

[17] According to a 1974 budget study conducted by the Institut Universitaire de Technologie (IUT) (*22*), average per capita consumption in Cap Vert was 132.5 kg/year, 29 percent above the estimates of the Statistical Office (*44*, p. 12).

[18] This observation seems to conflict with some government-planning assumptions that the national demand for rice has grown faster than population (*13*, pp. 10–11; *49*, p. 1; *57*, p. 1).

[19] A recent government cereals strategy statement (*38*) projected an income growth of 20 percent between 1977 and 1985 and demand for rice, if unchecked, of 284,000 tons in 1981 and 335,000 tons by 1985.

dominant cereal in rural Senegal. The fact that the government has chosen to emphasize rice production may be due to the lack of available technological innovations for increasing millet production—particularly of high-yielding varieties. The difficulty in increasing the domestic supply of millet has also been a major obstacle to greater urban millet consumption and to government attempts to encourage substitution of millet for both rice and wheat (in bread).

Historical Background to Senegalese Rice Policy

Peanuts were introduced into Senegal because European soap and vegetable oils manufacturers were seeking new sources of raw materials. Peanut cultivation spread rapidly in Senegal in the early 1840's and 1850's. By the end of the nineteenth century, nearly 100,000 tons were exported from the new French colony as thousands of virgin and food crop hectares were converted to production of the crop. Millet farmers rapidly adopted peanut cultivation because of its higher returns. This expansion was supported by private traders, who bought peanuts and sold food and other articles.

Rice was imported from French Indochina to sustain the growing urban areas as well as the peanut farmers during the "hungry season." By the eve of the First World War, Senegal was exporting nearly 300,000 tons of peanuts (7, p. 27) and importing about 26,000 tons of rice (47, p. 364). The demands for imported food in the countryside were made greater by the annual inflow of up to 70,000 peanut workers from neighboring countries. Because of the importance of this increased labor for Senegalese exports, the government provided migrants with reduced train fares and vaccinations and food on arrival (18, p. 224).

To aid peanut expansion, the government constructed roads, railways, and river wharves. Strong administrative support for peanuts was, in turn, encouraged by the French colonial policy that all colonies must strive for financial self-sufficiency; given Senegalese conditions, no other crop was so profitable as peanuts. The principal political as well as economic objective during this early colonial era, therefore, was to expand peanut exports as much as possible, and with them national income and budgetary revenue.

1930–59

With the collapse of the world vegetable oils market during the 1930's, farmers turned increasingly back to subsistence farming (29, p. 834). Paper money nearly ceased to circulate in the countryside (14, p. 119), and government revenues dropped sharply. In response, new concerns about dependence on a single commodity began to be voiced in government circles. Old suggestions that the entire Senegal River Basin be improved through water control so that other crops could be developed were taken down off the shelf,[20] and in 1934 the Mission d'Etudes du Fleuve Sénégal was formed in order to study the feasibility of dams and irrigation works for the area. This group was replaced in 1938 by the Mission d'Aménagement du Sénégal, whose primary aim was the installation of cotton schemes along the river. But lack of financing and the outbreak of the Second World War held up implementation of the project.

With the disruption of shipping and trade during the Second World War, Senegalese peanut farmers were forced to retreat to subsistence millet farming to an even greater extent than they had during the 1930's (18, p. 254). Rice imports dwindled from 70,000 tons per annum to a few thousand tons imported from outside West Africa and were supplemented by 8,000 to 10,000 tons from the French Sudan (now Mali) and Guinea (3, p. 344; 2, p. 33) and 20,000 tons of maize from Dahomey (51, p. 163).

Despite its flirtation with diversification, the government remained solidly committed to peanut production after the war. New policies aimed to enhance peanut production through seed selection, improvement of soil fertility, and the introduction of animal traction (29, p. 856). In order to assure a steady supply of vegetable oil and to respond to pleas for financial help from its colonies, France instituted preferential tariffs against all oilseed products originating in non-French territories. Administrators in Senegal also encouraged production by reducing head taxes, train tariffs, and the peanut export tax (29, p. 864).

World cereals markets recovered after the war, and in spite of preferential railroad tariffs, Sudanese rice could not compete with imports from other sources (53, p. 372). Moreover, supply difficulties from French Indochina forced Senegal to search for sources of rice supplies outside the franc zone. Scarcity of foreign exchange made this a difficult undertaking.

In late 1951, Senegal's main supplier—Indochina—suspended rice exports. The official retail price on the Dakar market rose to 40 francs/kg. Difficulties in finding other suppliers led to temporary shortages, which drove up retail prices temporarily to 100 francs/kg on the private market (1, p. 27). The government was under considerable pressure from trade unions and the press to lower the retail price of rice, but little could be done. After import quantities returned to normal, the government faced higher c.i.f. prices than it had in the past (1, p. 33). During this era, the trade unions also suggested the creation of a rice stabilization fund much like the one for peanuts (1, p. 33).

The problem of acquiring imported rice underscored the wartime con-

---

[20] A water storage facility had been under study in the Fleuve since 1925.

cern for domestic food self-sufficiency and resulted in inflows of foreign capital for new agricultural projects. Between 1947 and 1956, French public investment in all of French West Africa was twice as great as it had been in the previous 43 years (*18*, p. 280). About 20 percent of this investment found its way into agricultural development projects in Senegal, and most went to finance transport infrastructure and social services. A fully mechanized scheme for rice cultivation, covering 6,000 ha, was planned at Richard Toll in the Senegal River Basin. By 1948, a 120 ha experimental plot had been expanded to 600 ha, but equipment delivery delays and problems with pests resulted in unforeseen cost overruns and disappointing yields (*31*). In Sine Saloum and in the Lower and Middle Casamance, increased rice sales and better farm practices for rice cultivation were encouraged (*53*, p. 253). Nevertheless, by independence Senegal's annual food deficit was approximately 60,000 to 80,000 tons of grain, whereas local rice production had expanded by only 5,000 to 10,000 tons.[21]

1960–68

After independence in 1960, Senegal's new leaders continued to follow pre-war orientations in their economic policies. The basic economic objective remained income growth based largely on peanut-export expansion, with some new concern for expanded food production. Under the First Four-Year Plan (1960–64), initial investments continued to be made predominantly in social services, with agriculture receiving only 10 percent of total outlays (*15*, p. 452).

The new government also attempted to bring the economy under closer government control. The Office de Commercialisation Agricole (OCA) was created to handle domestic marketing of important crops, such as peanuts (but not peanut oil), rice, millet, and imported wheat, as well as to import and distribute agricultural inputs and equipment (*27*, p. 34). A number of other government organizations, working through the cooperative system, were set up to supply farmers with food, farm equipment, and credit. Financial assistance for the cooperatives' marketing operations came from the Banque Nationale de Développement du Sénégal (BNDS), which provided short- and medium-term credit. ONCAD, created in 1966, was responsible for the formation of cooperatives, buying cooperative products, distributing inputs, and collecting cooperative debts for the BNDS. The Office de Commercialisation Agricole du Sénégal (OCAS), which supplanted OCA, was given the responsibility for domestic and foreign marketing of all agricultural goods collected by ONCAD and became the sole importer of "essential" consumer goods such as rice and wheat (*27*, p. 58).

---

[21] Production figures prior to independence can only be estimated. The assumption here is that production was between 50,000 and 55,000 tons of paddy around the Second World War and had increased to about 60,000 tons by 1960.

Nevertheless, control over economic forces was not always within the new government's reach. Conditions developing outside of Senegal had a profound impact on the country's economic fortunes. France's accession to the European Economic Community (EEC) meant that it had to drop its preferential price supports for Senegalese peanuts by 1967.[22] The loss of this support, which amounted to approximately one-sixth of total export earnings, had serious implications for the entire economy (19, p. 3). The Senegalese government responded by reducing producer prices for peanuts by nearly 16 percent between 1963 and 1968 (19, p. 11) and by renewing efforts to develop crops that would either supplement peanut exports, such as cotton, or substitute for food imports, especially rice. The government was financially able to do this because the EEC and France had established sizable aid programs to help Senegal and other former colonies adjust to the new trading conditions.

Government efforts to participate directly in agricultural development were furthered by the establishment of land development agencies (LDAs) to facilitate receipt of this foreign aid. They were initially concerned with the promotion of one or two crops in a limited geographical area, but later they became more comprehensive in their approach to rural development within each region.[23] LDAs combined a number of different functions, including marketing, input delivery, credit, and extension. The government supported this approach because such concentrated investments were felt to yield faster, more tangible results than broader approaches to rural development. The government also instituted the Programme Agricole in 1964, which provided subsidies on fertilizers, farm implements, and machinery. Whereas cooperatives channeled these benefits to peanut farmers, the LDAs provided access to a few cotton and rice farmers. But unlike the cooperatives, the LDAs were also able to undertake large investments such as major irrigation works. In all these activities, the LDAs were aided by their access to foreign financial and technical assistance.

LDAs were among the first institutions to be made the object of explicit policies concerning rice production. In 1961 a number of public agencies were set up in the Fleuve region to develop rice projects.[24] In a con-

[22] In 1963 Senegal was informed officially that peanut support prices would be discontinued in 1966. The EEC pledged support during the phasing-out period until 1967 (25, pp. 507–9).

[23] In the Senegal River Valley, SAED's main function is to promote rice production. In the Peanut Basin, the land development agency is the Société de Développement et de Vulgarisation Agricole (SODEVA). In 1977–78, the Société pour la Mise en Valeur de la Casamance (SOMIVAC) was founded. In Eastern Senegal as well as in the Upper Casamance, the Société pour le Développement des Fibres Textiles (SODEFITEX) is engaged in cereals production in addition to cotton, its primary concern.

[24] The Société de Développement Rizicole du Sénégal (SDRS) took over the private operations at Richard Toll, and two sister organizations—the Organisme Autonome du Delta

certed attempt to reduce the nation's food deficit, these agencies constructed perimeteral dikes, sluice gates, and irrigation networks in order to expand rice cultivation on previously unfarmed land.[25] In addition to Richard Toll, 9,000 to 10,000 ha of rice land were developed in the Delta through the construction of controlled submersion polders. Although development costs of these polders were low, recurrent costs were high because mechanical services, other inputs, and free extension services were not covered by the projects' receipts. Furthermore, lack of control of the height and timing of river flooding meant that these polders were risky investments. To avoid this, water security had to be increased.[26] In contrast, apart from the partial transformation of an old peanut scheme into a rice scheme in 1964, rice production in the Casamance remained in the hands of the traditional farmers and outside of the influence of state projects and subsidies.[27]

During this era the government attempted to carry out a program of consumer price stabilization for rice. Stabilization operations were implemented by a variable levy on imports. Although large increases in the world price of rice were passed on to consumers, smaller, temporary fluctuations were absorbed by government taxes or subsidies. To generate revenues for the rice stabilization fund, consumer prices were usually set above the c.i.f. import price. In addition, because the cost of producing local rice was higher than the average c.i.f. import price, the higher retail price afforded protection to domestic rice producers.

(OAD) and Organisme Autonome de la Vallée (OAV) established rice perimeters elsewhere. SAED was set up later to do other work in the Delta and eventually took over all these operations.

[25] SAED was responsible for resettling hundreds of people into designated project areas in the Delta. Without the barrier dikes that had been constructed, there was little cultivable land in the Delta prior to the rice projects.

[26] This was accomplished in three stages: primary, diking and the construction of sluice gates; secondary, providing pumps and irrigation and drainage channels; and tertiary, leveling the fields within the diked area.

[27] The government's decision to invest in irrigated rice production in the Fleuve instead of in rice development in the Casamance is difficult to explain, but may be attributed to a number of factors. Plans to control the waters of the Senegal River for agricultural and navagational purposes go back to the beginning of the colonial era. Once some investments were actually made there in the 1930's and 1940's, money and personnel required for their maintenance were employed to carry out additional projects in the same area. In addition, the Fleuve's economic preeminance during the early slave and gum arabic trades has given it a certain prestige, and strong political connections still exist between Dakar and St.-Louis, the former capital. Moreover, since the Fleuve was not a traditional peanut-growing area and hence the investments in land and labor there would not directly jeopardize production of the country's leading export crop, it was thought to be an ideal area in which to emphasize food crop production. Finally, the Casamance was effectively cut off from the rest of Senegal until the completion of the trans-Gambian highway in 1958. The scale, the dispersion, and the highly evolved traditional techniques of its rice producers probably also dissuaded the government from attempts to introduce new techniques.

The 1968 drought occurred at a time of rising world food prices. The domestic production shortfall had to be made up with imports, the price of which rose 40 percent between 1966 and 1968. As a result, in 1968, 14 percent of total export earnings was used to purchase rice imports. Nevertheless, in 1967 and 1968 the government continued to pursue its policy of retail price stabilization by subsidizing rice consumption. These subsidies forced OCA to pay out 710 million CFA francs, which exhausted its resources and caused its dissolution in 1968. The coincidence of drought, increased quantities of rice imports, and high world prices led to the first serious budgetary crisis of independent Senegal.

1968–77

With the exception of one normal year, 1969–70, the drought lasted from 1968 through 1973. Output of the major cereals, millet and rice, fell to half of their pre-drought levels. As seen in Table 7.5, net domestic rice production, which had diminished to 31,200 tons in 1969, recovered to 85,900 tons with favorable weather the following year, but fell sharply in 1973 to 19,600 tons. In 1975, pre-drought levels were finally regained.

The duration of the drought plus the experience of paying high world prices, especially between 1972 and 1974, led the government to place much greater emphasis on income security than on income growth. As part of this new concern, government policies focused on food production security and on ultimate self-sufficiency in food.[28] The drought hit the northern and eastern parts of the country hardest and pointed up the need for more infrastructural and production investments to protect incomes in these areas. Plans were made to invest more resources in irrigation facilities, and LDAs were encouraged to focus more intensively on food crops in their extension efforts. This new policy orientation lent impetus to increased rice production under more secure conditions.

Production investment policies. Progress in developing more secure rice production was slow. In the Senegal River Valley, all public development efforts, including on-going rice projects, were centralized under SAED's control. The policy of establishing settlement schemes in the Delta, part of SAED's original charter, was abandoned. Because of the series of droughts, SAED's major aim became instead the extension of water control on existing polders rather than the expansion of production area. Therefore, efforts were begun to convert existing polders to total water control. In the absence of double-cropping, however, the yields on these rice polders were not sufficient to cover the high investment costs required. Double-cropping has not been possible, except on a limited

---

[28] At the beginning of the Fourth Plan (1973), the target date for self-sufficiency in rice was set for 1985 (16).

TABLE 7.5. *Net Availability of Rice in Senegal, 1965–76*
(Thousand tons)

Year	Production*a*	Imports	Net availability	Per capita availability*b*	Self-sufficiency ratio*c*
1965	60.6	179.2	239.8	62.3	0.25
1966	68.1	159.3	227.4	57.6	0.30
1967	69.8	153.4	223.2	55.1	0.31
1968	76.4	185.2	261.6	62.9	0.29
1969	31.2	145.9	156.8	41.6	0.18
1970	85.9	125.6	211.5	48.4	0.41
1971	49.1	186.8	235.9	52.6	0.21
1972	59.8	165.8	225.6	49.1	0.26
1973	19.6	192.5	212.1	45.0	0.09
1974	35.0	207.2	242.2	50.0	0.14
1975	65.2	102.1	167.3	33.7	0.39
1976	61.3	235.4	298.5	58.8	0.21

SOURCE: *61*, Table A.10.
a Net domestic production minus seeds and losses at milled rice equivalent (0.65).
b kg/capita.
c Production of local rice (net) divided by net availability.

scale in the Middle Valley, owing to lack of water during the low flood period and the problem of saltwater incursion in the Delta.

Small village-level perimeters were established in 1974 around Matam under the auspices of the Food and Agriculture Organization and the Société d'Aide Technique et de Coopération (SATEC), a French consulting firm. Although still under SAED's control, these perimeters are very different in design from the rice projects along the middle and lower river. The schemes at Matam have achieved high yields at low cost, and decision making for each polder remains at the village or farmer level.

In 1972, after nearly a century of proposals for developing the Senegal River Basin, Senegal, Mali, and Mauritania formed the joint Organisation pour la Mise en Valeur du Fleuve Sénégal. This international organization is currently considering construction of two dams. One, the Diama, is a saltwater barrage planned for the Delta near St.-Louis; the other, the Manantali, is a regulatory high dam in Mali on the Bafing—one of the Senegal River's main tributaries. Construction of these dams would permit double-cropping on a much wider scale.

Until the foundation of the Société pour la Mise en Valeur de la Casamance (SOMIVAC) in 1977, the Casamance did not have a regional development agency like SAED. Instead, a number of jointly funded agencies operated in each of the departments within the region. In the Lower Casamance, the European Development Fund financed a Dutch-supervised project (ILACO) in 1969 to improve mangrove swamps through saltwater control and the introduction of improved cultivation techniques to smallholder farmers. The scope of this project was limited and its efforts plagued with technical difficulties related to the drought—specifically, insufficient rainwater to permit desalinization of mangrove swamps. In 1974, the ILACO project merged with another group of projects to form the Project Interimaire pour le Développement Agricole de la Casamance (PIDAC). In 1978, PIDAC received major funding to revive and expand the old ILACO project, including the construction of a number of saltwater barrage/retention dams on the tributaries of the Casamance River.

In the Middle Casamance the World Bank–financed Projet Rural de Sédhiou has been operating as an integrated agricultural projects on 11,000 ha, of which about 5,700 ha are in rice. The project has been very successful in raising rice yields by introducing simple improvements such as better weeding, better timing of planting, the use of fertilizer, and animal traction. In Eastern Senegal, the Société de Développement des Fibres Textiles (SODEFITEX), the former French cotton development company, has introduced improved rice cultivation techniques and has had relative success by providing an efficient extension and collection service and timely input deliveries. In the early 1970's, SODEFITEX extended its operations to the Upper Casamance, where it took over several rice projects from SATEC.

A large portion of the rice development costs were met by the substantial foreign aid that flowed into Senegal during and after the drought. Between 1969 and 1976, rice projects received 20 percent of total development outlays and 10 percent of all foreign funds, much of which was in the form of grants. Hence, the country did not bear the full capital cost of these projects.

Input subsidy policies. For a number of years, subsidies on fertilizers were about 50 percent of the cost price, whereas those on seed, mechanical implements, and other chemical inputs were more modest. These input subsidies were not confined to rice producers and in fact probably had a greater impact on peanut and cotton production. In the Casamance, in particular, the problem of input distribution and the difficulties that traditional rice farmers have had in gaining access to the distributing agent, ONCAD, has lessened considerably the potential influences of the subsidies on rice output.

Producer price policies. Official producer prices for local paddy production were introduced in 1964. In 1968, just prior to the drought, the producer prices for millet, cotton, and peanuts were 21 CFA francs/kg, 37.7 CFA francs/kg, and 17.67 CFA francs/kg, respectively, while that for paddy was 21 CFA francs/kg. In November 1974, when the government raised the official retail prices of imported rice and peanut oil, the producer prices were raised as well. The paddy price was increased 66 percent to 41.5 CFA francs, equal to that of peanuts, while the millet price was raised 20 percent to 30 CFA francs/kg, and cotton prices by 37 percent to 47 CFA francs/kg. In practice, however, only the increase in the producer prices for the export crops—peanuts and cotton—were relevant for most farmers because these are the only crops widely purchased by ONCAD. The official price for paddy serves mainly as the means of evaluating government purchases in development projects.

Consumer price policies. While working to expand rice production, the government remained committed to maintaining adequate cereals supplies at stable consumer prices through its rice import policy. During the drought the government greatly increased cereal imports. Between 1970 and 1974, an average of 320,000 tons of grain were imported per year, or nearly one-third of total consumption requirements. Since the government had to pay much higher prices between 1972 and 1974, it was more difficult to maintain stable consumer prices than it had been in the past.

In the early 1960's, the variable levy on rice was maintained just above the c.i.f. landed price, ex-port. The small amount of tax that the government realized through the levy was used to smooth out modest fluctuations in price. In 1967, however, import prices rose above the consumer price level, and the government chose to subsidize rice prices in order to keep them close to former levels. Then in 1968 the government made an upward adjustment in the official wholesale and retail prices.

As can been seen from Figure 7.1, if the 1968 landed import price had remained the same through 1969, the government would have realized approximately the same level of net tax on imported rice that it had imposed during most of the 1960's. But the import prices fell, and the government made a larger net gain than it had in the past. When the import price continued to fall through 1971, the government decided to lower consumer prices, but this time it left a substantially larger margin between their costs and receipts.

Import prices shot up in 1973, and the government found itself once more in the position of subsidizing consumers. This time it reacted more rapidly, and by mid-1973 consumer prices were raised above the import price level. The period of price instability was not yet over, however, and in 1974 world prices rose to 235 percent over the average level of 1970 through 1972.

These fluctuations in import prices created financial strains on the Senegalese economy. Because of the increased requirements for cereals (and petroleum) imports at much higher prices, Senegal's international reserve position deteriorated rapidly. Total reserves fell from $38 million in 1972 to $6 million in 1974.[29] In that same year, the consumer subsidy on the

[29] To finance its growing trade deficit, Senegal borrowed on the Eurocurrency market at commercial rates. Its debt service/export earnings ratio doubled to 14 percent in 1974 alone (5, p. 57).

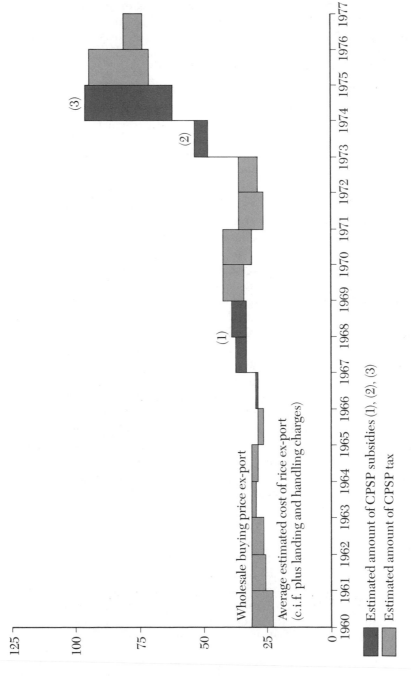

FIGURE 7.1. Estimated Variable Levy and Resulting CPSP Taxes or Subsidies in CFA Francs Per Kilogram of Imported Rice, 1960–77. Data are from the sources listed in Table 7.7.

rice account alone reached 5.5 billion CFA francs. As a result, the CPSP was unable to repay ONCAD for subsidies on input sales to cooperatives. ONCAD in turn ran behind in its repayment of short-term credit extended by the BNDS for marketing and input purchases. If high 1974 prices had persisted into the following year, the estimated subsidy on rice, sugar, and oils would have reached 27 billion CFA francs, which not only was much more than the funds available to the CPSP, but was also larger than the entire development budget of Senegal. Pressure to revise prices began to build within the government.

In November 1974, the government intervened with a major upward adjustment in price levels. The subsidy on rice imports was totally eliminated and subsidies on sugar, oil, and wheat were substantially reduced. The retail price of rice was raised from 60 to 100 francs; sugar and oil prices were raised 88 and 33 percent, respectively. As compensation, people subject to a fixed wage structure, such as those earning the minimum wage and some professional groups, were granted wage increases.

Because world rice prices fell after 1974, the adjustment of the domestic price levels led, temporarily, to a situation where the official consumer retail price was much higher than the c.i.f. price for imported rice. This significantly increased government revenue from rice imports. The taxes accruing through the variable levy gave the government a substantial buffer against future price rises. In May 1976, following the continued downward drift of the c.i.f. price, consumer prices were reduced 20 percent, and the tax margin was reduced accordingly.

Evaluation of Rice Policies

During most of Senegal's history, the government's major economic objective has been economic growth, based largely on peanut exports. Concentration on peanuts rather than on traditional food crops was facilitated by up-country traders who bought peanuts and sold food. Because the rapidly growing urban areas were not being fed entirely by the countryside, they had to import a considerable amount of their foodstuffs, primarily rice. This system worked reasonably well, although disruptions in peanut production, world markets, and shipping often made the country painfully aware of its dependence on imported food. New policies were implemented in an effort to cover some of the national cereals deficit through expanded domestic food production, but never directly at the expense of peanut production. With independence, the government attempted to bring all of agriculture—and peanuts in particular—under more careful supervision. A new government organization replaced the private peanut buyers and took over the provision of imported food to the countryside.

The effect of the loss of franc-zone supplies of inexpensive rice was compounded by the elimination of the French preferences on peanuts and the period of droughts and unstable world prices in the late 1960's and early 1970's. Until this time, cheap imported rice had supported the major economic objective of income growth. But during the period of instability, rice imports became a constraint on that objective. The large sums needed to pay for rice imports and price subsidies impinged on both Senegal's international reserve position and its domestic budgetary stability.

In the post-drought era, Senegal has tried to reduce the risk of income insecurity. The dependence of national income and government budgetary revenues on fluctuating peanut production and prices has been eased somewhat by the expansion of phosphate production and higher world phosphate prices. In order to stabilize farm family income, the government has tried to develop more reliable production methods, especially for food crops, and to introduce better farming techniques and modern inputs through regional development agencies. By emphasizing rice production systems based on water control and irrigation, the government has sought to avoid the impact of wide fluctuations in the prices and quantities of imported rice in recent years. This focus on domestic rice production can be seen as an effort to break an important perceived constraint to further national growth.

Senegalese policies directed toward the rice sector have primarily involved public investments in land development for producers and price stabilization measures for consumers. In both instances, the immediate goal of the policy instruments has been to reduce the instability of food supply in Senegal and thereby to improve food security. In the following sections each major type of policy is examined to assess its contribution in advancing government objectives.

Production Investment Policies

Because of Senegal's climatic instability, the only secure systems of food production are probably those that can assure the availability and distribution of water when needed. The need to control water is the main rationale behind the government's investments in irrigation systems for rice. This effort to increase the security of water delivered for production is wholly consistent with the major physical constraint in much of Senegalese agriculture. An examination of costs, production security, and productivity at the micro-level suggests that the strategy has been moderately successful. Virtually complete water control can be achieved, with reasonable assumptions about cost, at an increase of about 150 percent over the costs of current, secondary perimeters. The security of production would then rise from roughly two-thirds to well over 90 percent, and yields could increase by 500 to 600 percent.[30]

In the Fleuve, which has been the focus of these policies, progress has been slow, despite sizable investments. Although the first land improvements were begun in 1947, only 14,700 ha currently have any form of water control, and during 1971–77, only 4,500 ha were upgraded from rudimentary empoldering. A full record of costs is not available, but almost ten billion CFA francs were expended during 1974–78.[31] In addition to high costs, the policy has been impeded by difficulties of resettlement, double-cropping, and farmer incentives.

At the national level, production information suggests that the strategy has failed thus far to achieve either an increase in annual production or greater stability of this production. A comparison of annual averages for the two periods 1961–68 and 1975–78, prior to and succeeding the main drought years, shows that national yields have risen insignificantly, yearly paddy production and area planted in paddy have actually fallen slightly, and the variability of all three indicators has increased somewhat:

	1961–68	1975–78
Average paddy yields *(mt/ha)*	1.252	1.270
(Standard deviation)	(0.171)	(0.179)
Average national paddy production *(000 mt)*	103.5	101.7
(Standard deviation)	(25.2)	(26.6)
Average area in paddy *(000 ha)*	81.8	79.0
(Standard deviation)	(11.2)	(11.6)

Given the stagnation of production, imports have continued to rise, thwarting the government's goal to replace imports with secure domestic production. During the 15 years between 1961 and 1976, imports grew an average of 4,000 mt per year, or at an annual rate of 2.9 percent.[32] During the same period, population increased at an estimated 2.6 percent per year. On balance, imports grew faster than population when per capita consumption was either constant or falling. In contrast to very low imports in 1975, which resulted from large stock carryovers from 1974

[30] See *41*, p. 9. Security is measured as the percent of planted land that is harvested, and the estimate of two-thirds is based on data for the Delta in 1970–74 and 1976–77. Complete water control can never assure 100 percent security because of salt incursion during the years of low floods. Yield differences are based on 1976–77 SAED estimates.

[31] These figures include foreign funds, central budget funds for land development, and Treasury transfers to cover SAED operating deficits.

[32] These figures are based on the trend line regression of import quantities, the results of which are as follows: constant equals 123,745 mt; slope coefficient equals 4,188 mt per year; standard error of slope coefficient equals 1.970. The level of confidence for the slope coefficient exceeds 95 percent.

(56,000 mt) and a good harvest in 1974–75, 1976 imports were the highest in history.

The investment strategy of the government can be judged on two major counts—stability and profitability. The issue of stability depends on whether reliance on the world rice market is inherently more unstable than dependence on domestic rice production. An analysis of paddy production and rice imports for Senegal during the period 1961–76 gives no evidence to support the argument that the world market is less stable than Senegalese production. In fact, the opposite is true when the c.i.f. price per unit is compared to local production; domestic output is over twice as variable as the nominal price of imports.[33] Of course, the impact of the variability in the world price of rice will also depend on tonnages imported and the value of major export commodities. Empirically, these factors increase the variability of trade in rice for Senegal, but only slightly. The conclusion still holds that the government's investment strategy has not been justified on the grounds that import substitution for rice reduces the insecurity of food supplies.

Although the major irrigation investments in the past have been located in the Fleuve, the long-run stability of production in this region may well depend on the construction of the two proposed dams—the Diama and the Manantali. Recent production has suffered from the interconnected problems of low rainfall, the inability to pump or flood-irrigate, and saltwater incursion. Until these technical difficulties are resolved, the predictability of future production from this area will be tenuous. Because of the emphasis placed by the government on water security, more attention might be paid to the Casamance, because the frequency, reliability, and total amount of rainfall are considerably higher there than in the north.[34]

The profitability of the investment program focuses on two elements that influence costs of production—geographic location and choice of technique. The techniques of rice production associated with the large-scale irrigation schemes have had high capital costs, borne mainly by foreign aid, and large recurrent costs, which must be supported by the government budget. Recent evidence demonstrates that the small-scale, Matam-type model of rice production, as well as labor-intensive techniques in the Casamance, is relatively more efficient than the large-scale, heavily mechanized, and centrally directed schemes in the Delta and

---

[33] This relationship is statistically significant at a 90 percent probability level. All analyses were made using the variances of the trend line regression estimates, normalized to account for differences in units. Data and results are available upon request.

[34] For the drought years 1972, 1973, and 1974, when rainfall in the north of Senegal was 37 percent, 44 percent, and 61 percent of normal for the June–September rainy season, the rainfall in the Casamance was 69 percent, 92 percent, and 94 percent of normal for the same years (5, p. 27).

lower Fleuve (54, pp. 25a–26a). Despite the seasonal labor shortages in much of Senegal, the small-scale, relatively labor-intensive techniques are still more efficient than those utilizing heavy machinery. In the context of Senegalese conditions, large mechanization increases, rather than reduces, production costs.

Investments in rice development should be judged on their social profitability, that is, whether they contribute to or diminish national income. When investments are designed to replace imports with domestic rice, national income will fall if imports cost less than their domestic substitutes. At current costs, local rice production cannot be expected to replace imports in the capital city, except at great cost to the economy. The Fifth Four-Year Plan projects that the Fleuve region will produce 114,000 tons of paddy by 1980–81. At the social costs of production and world prices that prevailed in 1975–76, the attainment of this target would cost the Senegalese economy about 3.25 billion CFA francs ($13 million) in the loss of production efficiency alone.[35]

In areas far from Dakar, there is evidence that production and milling costs for certain rice production techniques are low enough and transport costs for moving imported rice from Dakar are high enough to permit domestic rice to compete with imports in the local market. In Matam, as well as in Kolda and some other areas of the Casamance, local production can profitably replace rice imports and lead to local self-sufficiency. Domestic rice will be able to compete with imported rice farther away from these local production centers and closer to Dakar as local production and milling costs are further reduced. Although some economies of scale and lower costs (especially in milling) might be realized through increased output alone, major cost reductions will probably come only from important technological innovations.

Input Subsidy Policies

In addition to large investments in infrastructure, the rice sector has enjoyed subsidies on agricultural inputs, with those on fertilizer by far the most important. As shown in Table 7.6, subsidies on composite fertilizers have averaged nearly 55 percent since 1970. Although a complete series of data is not available for urea, subsidy rates seem to be similar.

Except for agricultural extension, which is provided free by the LDAs, the other inputs—including insecticides, herbicides, oxen equipment, and mechanical services such as deep-plowing—are charged to farmers at

[35]This estimate assumes that 46 percent of the Fleuve production comes from highly mechanized operations in the Delta, 38 percent from the Middle Valley, also heavily capital-intensive, and 16 percent from the labor-intensive, small-scale perimeters near Matam (61, Table F-10). The social profitability of production is negative, averaging −45.8 CFA francs per kg of rice produced and delivered to Dakar. See (54, p. 25a).

TABLE 7.6. *Fertilizer Subsidies for Rice*

Crop year	Purchase cost[a] (CFA fr/kg)	Farm price (CFA fr/kg)	Per unit subsidy (CFA fr/kg)	Percent subsidy	Fertilizer applied to rice (000 mt)[b]	Estimated total cost of subsidy (000,000 CFA francs)[c]
1966–67	15.9	12	3.9	25%	1.0	3.9
1967–68	15.9	16	0	0	1.3	0
1968–69	20.0	16	4.0	20	1.2	4.8
1969–70	22.8	12	10.8	47	2.0	21.4
1970–71	24.0	12	12.0	50	0.5	5.9
1971–72	24.0	12	12.0	50	0.7	8.5
1972–73	26.3	12	14.3	54	0.8	11.3
1973–74	21.0	12	9.0	43	1.9	17.1
1974–75	34.4[d]	12	22.4	65	2.1[e]	46.2
1975–76	54.7	16	38.7	71	2.7[e]	104.9
1976–77	48.2	20	28.2	59	n.a.	n.a.
1977–78	n.a.	25	n.a.	n.a.	n.a.	n.a.

SOURCES: Prices are based on Sénégal, Government of, l'Office National de Coopération et de l'Assistance au Développement, personal communication, Dakar. Figures on fertilizer consumption are taken from Sénégal, Government of, Ministère du Développement Rural et Hydraulique, Direction Générale de la Production Agricole, personal communication, Dakar. Fertilizer usage for 1974–75 and 1975–76 is based on growth rates estimated in *24*.

NOTE: These price figures refer only to composite fertilizer ($N-P_2O_5-K_2O$) used on rice, primarily 16–48–0.

[a] Equals only the ex-factory price. Since it excludes the costs of transport, storage, financing, and delivery, the subsidy figures are conservative estimates.

[b] All fertilizers, both composite and urea. Tonnages are in gross fertilizer weight, not nutrient tons.

[c] Subsidies indicated are only estimates and are not confirmed by any actual budget figures. Although the subsidies present all fertilizers used, cost data apply only to composite fertilizers. It is not known if the cost and subsidy structure for urea is similar. Subsidies on transport, storage, financing, and delivery of fertilizers are excluded.

[d] Because the price for rice fertilizer is not available, the price of fertilizer for peanuts and millet has been used. In subsequent years, all fertilizers had the same purchase cost.

[e] Values have been estimated, based on assumed growth rates from 1973 to 1974.

cost or only slightly less. Selected seeds carry a somewhat higher subsidy at a rate of 15–20 percent. In addition, the input distribution system both subsidizes the delivery of most inputs to the farm and finances the working capital required for their purchase.

Although no particular agricultural crop seems to be favored by special subsidy rates on various inputs, total input subsidies paid to rice have been negligible. The main reasons are that riceland amounts to less than 10 percent of the land devoted to peanuts and cotton, which are virtually always fertilized, and that the majority of rice producers use few modern inputs. For example, during the period 1961–74, less than 3 percent of all fertilizers used were applied to rice.

Large variations in climatic conditions make it difficult to assess accurately the effect of the fertilizer subsidy on fertilizer use on rice. The available evidence shows no strong relationship—applications in 1974–75 were scarcely larger than in 1969–70. Despite a favorable price ratio of nitrogen to paddy in 1975 (approximately 8–10), distribution remains limited, in part because most LDAs have not been organized to reach widely dispersed, small-scale traditional farmers.[36]

As a result of the limited use of fertilizer on rice, the budgetary impact of this policy has not been large. During the ten years between 1966–67 and 1975–76, the total value of these subsidies amounted to less than a quarter billion CFA francs. Compared to investment policies for rice, which provided over 1.5 and 3.0 billion CFA francs to SAED alone in 1975–76 and 1976–77, respectively, input subsidies have been unimportant as a national policy.

Price and Trade Policies

Since independence the government has made an effort to stabilize consumer prices by adopting an official price of rice, which it has defended with large quantities of imports. By using a variable levy on the value of imports, the government has attempted to compensate for changes in world prices without altering official prices.

The government has not been able to defend official prices very effectively. During the ten years 1967–76, for which comparable data are available, the observed market price for 100 percent brokens in Dakar averaged 22 percent above the official price. The divergence between these prices does not seem to be correlated with shortfalls in either production or imports, but it does suggest that imports were insufficient in most years to defend the official price.

Given this failure to defend official prices, it is not surprising that Senegalese consumer price policy has failed to reduce the variability of official wholesale prices. As the following statistics show, both import and wholesale price series are very similar:

	C. i. f. price	Official wholesale price
Average price (*CFA francs/kg*)	36.80	42.35
Standard deviation	18.58	19.60
Coefficient of variation	0.50	0.46
Range	65.22	66.00

Domestic prices have basically tracked the c.i.f. import prices, with slight lags in adjustment, as shown above in Figure 7.1. In most years, imports have been taxed to a small degree. When world prices rose rapidly, imports were subsidized for one or two years, before domestic prices were brought into line with import costs. Large taxes accrued to the budget only when world prices fell rapidly and domestic prices were not

[36] These price ratios are based on the following assumptions: a paddy price of 41.5 CFA francs/kg; a fertilizer (16-48-0) price of 12 CFA francs/kg, which gives a price of N equal to 75 CFA francs/kg; and a response rate of 15–20 kg paddy per kg N.

TABLE 7.7. *Variable Levy on Rice*
(CFA fr/kg of imported rice, unless otherwise noted)

Year	C.i.f. price[a]	Est. landing margin[b]	Est. import cost[c]	Est. off. wholesale buying price[d]	Est. variable levy[e]	Est. ann'l budget rev. (bill. CFA fr)[f]	Annual CPSP rice import rev. (bill. CFA fr)
1960	21.8	1.4	23.2	30	6.8	0.6	n.a.
1961	24.6	1.4	26.0	30	4.0	0.4	0.2
1962	25.0	1.4	26.4	30	3.6	0.4	0
1963	27.3	1.4	28.7	30	1.3	0.1	0.3
1964	26.7	1.4[g]	28.1	30[g]	1.9	0.4	0
1965	25.0	1.4	26.4	28	1.6	0.3	0.1
1966	27.2	1.4	28.6	28	−0.6	−0.1	0
1967	35.9	1.6[g]	37.5	33	−4.5	−0.7	n.a.
1968	38.1	1.4[g]	39.5	33	−6.5	−1.2	n.a.
1969	32.0	2.0	33.4	43	9.6	1.4	n.a.
1970	28.0	2.0	30.0	43	13.0	1.6	n.a.
1971	24.7	3.0	25.7	37[g]	11.3	2.1	n.a.
1972	25.0	3.0	28.0	37[g]	9.0	1.5	n.a.
1973	50.0	3.0	53.0	48.5[g,h]	−4.1	−0.8	n.a.
1974	87.0	8.0[g]	95.0	63.3[g,i]	−31.7	−6.6	0
1975	59.3	11.7[g]	71.0	94.5[g]	23.5	2.4	n.a.
1976	68.5	5.7[g]	74.2	82.3[g,j]	8.3	2.0	n.a.

SOURCES: C.i.f. prices are taken from Sénégal, Government of, Ministère des Finances et Affaires Econo-miques, Direction de la Statistique, *Importations: Commerce Spécial*, Dakar, various years. Landing margins are taken from Sénégal, Government of, ONCAD, *Budget Provisionelle*, Dakar, various years, and *Bilan, Exercise, 1974–75*, Dakar. Wholesale prices are based on ONCAD, Direction de la Commercialisation, personal com-munication, Dakar, 1977, and on West Africa Rice Development Association, *Rice Statistics for Senegal*, Monro-via, 1976. Official prices are also regularly published in Sénégal, Government of, *Journal Officiel*, Dakar, weekly. CPSP revenue comes from Senegal, Government of, Caisse de Péréquation et de Stabilisation des Prix, Direction Générale, personal communication, Dakar, 1977.

[a] Average of brokens and whole grains.
[b] Includes unloading at the port, storage, financial charges, and administrative costs.
[c] Equals the c.i.f. price plus the estimated landing margin.
[d] Equals that price at which wholesalers purchase rice from ONCAD and is based on the official retail price minus official commercial margins.
[e] Calculated as the difference between the estimated wholesale buying price and the estimated import cost.
[f] This revenue equals the produce of total annual rice imports and the estimated variable levy. This figure is hypothetical.
[g] Actuals; others have been estimated on the basis of these figures.
[h] Average of 37, which prevailed through May, and 57, which existed from June on.
[i] Average of 57, which prevailed through October, and 94.5, which existed from November on.
[j] Average of 94.5, which prevailed through May, and 74, which existed from June on.

lowered accordingly. Based on estimates for 1960–76 (see Table 7.7), the average annual tax levied on rice imports amounted to just over 200 mil-lion CFA francs per year, or less than 4 percent of the average value of rice imports. It is significant to note that between 1961 and 1974, the cu-mulative budgetary gain on rice imports was negative, which means that rice imports had, on average, been slightly subsidized since indepen-dence. By 1977, this cumulative figure was somewhat positive, following two years of fairly high taxation of rice imports.[37]

[37] The figures for budgetary revenue used in the text are estimates. Actual revenue data, available only for 1961–66, are less than one-half the estimates for these six years. Conse-quently, the budgetary impact of rice imports is probably more strongly negative than indi-cated by the estimates in Table 7.7.

In addition to its impact on the consumer price level and stability, Senegalese price policy has also affected domestic rice production. The effect can be divided into two parts—trade protection and domestic producer price supports. As evidenced above, the trade protection provided by the variable levy has probably not been significant, although quantitative import restrictions may have caused the real market price to stay significantly above the world price in some years. The competitiveness of local production is hampered because Senegal usually imports inexpensive qualities of rice. Even with international shipping tariffs included, it is difficult for Senegalese rice to compete with 80 to 100 percent brokens from Southeast Asia, the quality most frequently imported. Consequently, government policy to purchase large quantities of 100 percent brokens, which are only lightly taxed, creates little incentive for local production to replace imports.[38]

The effectiveness and cost of producer price policy depend on the consistency of government efforts to maintain official paddy prices, the relation of these official prices to import and market prices, and the costs of production. As the following data show, official producer prices have recently exceeded official wholesale prices when both are expressed in equivalent units.[39] If these official prices are respected, government buying programs must subsidize the post-harvest sector. To the extent that government processing costs are higher than those used in estimating equivalent wholesale prices, the subsidies would be larger than implied above. The budgetary impact of these official prices is not large, however, owing to the small amount of official marketings.

Year	Official paddy price farmgate (CFA francs/kg paddy)	Wholesale equivalent of paddy price, Dakar (CFA francs/kg rice)	Official wholesale rice prices, Dakar (CFA francs/kg rice)
1972	21	53	37
1973–74	25	59	37
1974–75	41.5	84	57
1976	41.5	84	74

The market price of rice will also be associated with some paddy price, depending on costs of processing and marketing. As shown below, these

---

Moreover, the Senegalese government has not really followed an import price stabilization scheme. Revenues from the variable levy are often used to pay the subsidies on domestic rice milled by national agencies. During 1961–66, the Caisse de Riz showed a net deficit of over one-half billion CFA francs as a result of these expenditures.

[38] The selection of import quality may be strongly affected by the government's desire to supply cheaper, rather than more expensive, foodstuffs to consumers.

[39] The official farmgate paddy prices have been converted to rice equivalent, Dakar, by adding 14 CFA francs per kilogram of paddy for collection, milling, and distribution, and by converting to rice at a milling ratio of 0.66.

prices tended to be below estimated production costs until prices were increased in 1975.[40] As a result, there was little incentive to produce for shipment to Dakar, a conclusion consistent with empirical evidence. Even though market prices for rice usually exceed official prices, this additional incentive still appears insufficient to generate a large enough supply of rice to replace imports. The costs of production for most techniques and locations have been above prices in the Dakar private market.

	1973–74	1975	1976
Market wholesale price, Dakar *(CFA francs/kg rice)*	61/66	116	84
Estimated equivalent farmgate paddy price *(CFA francs/kg paddy)*	26/29	62	41
Estimated private farm-level production costs *(CFA francs/kg paddy)*			
Matam		26	
Delta		41	
Traditional swamp		72	
Improved swamp		39	
Improved rainfed with oxen		24	

In conclusion, producer price policy has been ineffective in the past because prices were set below costs of production. Moreover, producer prices appear to have been set with little regard for consumer prices, which have been closely linked to movements in import prices. As a result, implementation of official pricing policies has required subsidies to the post-harvest sector. Private producer prices, which are heavily influenced by official and private consumer prices, have probably been too low to cover costs of producing paddy and delivering it to Dakar.

Summary and Conclusion

Recent Senegalese rice policy has aimed at expanding domestic production under conditions of more secure water supplies. This policy has been implemented by LDA extension activities, government investments in irrigation projects, and input subsidies. At the same time, the desire on the part of the government to protect rice consumers and national income has reinforced the historical policy of importing the least expensive rice available on the world market, whenever possible. There is thus a policy conflict between the desires for income growth and for income security.

Price policy has clearly been tied to movements in world prices. Although consumers may have lost slightly as a result of government restrictions, producers have seldom gained much from price policy. On balance,

---

[40]The market wholesale price is estimated at 95 percent of the retail price. The equivalent farm price is calculated using information given in note 39. Production costs are from 54 and include all government taxes and subsidies.

the government budget seems to have realized small amounts of revenue from imports, which may have been transferred to handlers of domestic rice. Both the investment and input subsidy policies have made large demands on the government budget, with no apparent increase in stability of output, yields, or area of production. Rice imports have not been replaced significantly by domestic production. Real costs of production have not fallen, in part because the production policies seem often to have been focused on the wrong areas and techniques. As a result, potential national income has been reduced, without much compensating gain.

If the expansion of domestic rice production continues to be a focus of government policy, the methods need to be improved and the costs of production, milling, and marketing reduced so that local rice is at once the source of greater food security and a profitable food alternative. To this end, the choices of production technique and location are important. The country could promote a variety of intermediate water control schemes, such as the one at Matam, where costs are low, yields are high, water security is good, and the socioeconomic dislocation associated with the large-scale developments is not a problem. Increased effort could be directed at the development of swamp rice in the Casamance. There is also an important need to give more attention to rainfed cultivation in areas of sufficient and reliable rainfall, such as the Casamance and Eastern Senegal.

If domestic rice cannot be produced more cheaply than imports, the government needs to weigh the loss of national income resulting from expanding domestic rice production against the benefits it perceives from producing more of its food domestically. For example, regional development outside the Groundnut Basin has been furthered by the large rice investments in the Fleuve, but the use of capital-intensive costly schemes is an inefficient method of income redistribution. Domestic rice production was also intended to increase the security of food supplies, even though efforts to date have not met with much success.

Finally, the government might consider policies other than increased rice production to help achieve its objectives. To reduce the instability of food imports, Senegal could join with other countries to establish an import insurance scheme in order to protect itself against high world prices and large domestic shortfalls. The need to rely on domestic rice production might also be reduced by the improvement of local production and processing of maize and millet, which could be low-cost substitutes for rice.

Citations

1 Virginia Adloff, "French West Africa," Area Reference Series, United Nations Interim Commission on Food and Agriculture, 2 vol. Washington, D.C., 1944, preliminary.

2 ———, "French West Africa—Agriculture," unpublished notes from minutes of the Grand Council, 1947–54, microfiche, (n.p.) 1955.

3 *Agronomie Tropicale,* "Le Problème Rizicole dans les Territoires Africaines de l'Union Française," 4, Nos. 7–8 (July–August 1949).

4 Banque Centrale des Etats de l'Afrique de l'Ouest, *Indicateurs Economiques,* 241 (August–September). Paris, 1976.

5 Elliot Berg, *The Recent Economic Evolution of the Sahel.* University of Michigan Center for Research on Economic Development, Ann Arbor, June 1975.

6 George E. Brooks, "Peanuts and Colonialism, Consequences of Commercialization of Peanuts in West Africa, 1830–70," *Journal of African History,* 16, No. 1 (1975).

7 M. J. Chailley and D. Zolla, eds., *Congrès d'Agriculture Coloniale: Compte Rendu des Travaux,* vols. 1, 2, 4. A. Challelmel, Paris, 1918.

8 Pascal Bye and Yvon Le Moal, *Commercialisation et Diffusion des Produits Alimentaires Importés.* Institut de Science Economique Appliquée, Dakar, January 1966.

9 Comité Inter-états de Lutte Contre la Sécheresse Sahelienne (CILSS); *Marketing, Price Policy and Storage of Food Grains in the Sahel,* 2 vols. University of Michigan Center for Research on Economic Development, Ann Arbor, August 1977.

10 Michael Crowder, "West Africa and the 1914–18 War," *Bulletin de l'IFAN,* 30 (January 1968).

11 Marie-Thérèse Debien, "L'Association du Sénégal à la Communauté Economique Européenne et les Problèmes de l'Arachide," *Bulletin de l'IFAN,* 28, Nos. 3–4, 1966.

12 Jean Denis, "L'Alimentation Outre-Mer: Sénégal et Mauritanie," *Marchés Tropicaux du Monde,* November 24, 1956.

13 Josue Dione, "Le Déficit Céréalier du Sénégal: Situation et Perspectives." Institut Sénégalais de Recherches Agricoles, Dakar, December 1975.

14 Huguette Durand, "Essai sur la Conjoncture de l'Afrique Noire." Paris, 1957.

15 Ediafric, La Documentation Africaine, *L'Agriculture Africaine,* special edition of *Bulletin de l'Afrique Noire,* vols. 1, 2. Paris, 1973.

16 ———, *L'Agriculture Africaine, 1976.* Paris, 1977.

17 Eugene Guernier, "L'Afrique Occidentale Française," *L'Encyclopedie Coloniale et Maritime,* vols. 1, 2. Paris, 1949.

18 A. G. Hopkins, *An Economic History of West Africa.* Columbia University Press, New York, 1973.

19 Elizabeth Hopkins, "Farmers' Responses to an Agricultural Extension Scheme: A Senegalese Example." Economics Club, Lusaka, 1972, mimeo.

20 Institut Colonial de Marseille, *Annuaire Economique Coloniale, 1930.* Marseille, 1930.

21 ———, *Le Commerce et la Production des Colonies Françaises.* Marseille, 1926.

22 Institut Universitaire de Technologie, *Etude: Budget Consommation,* 2 vols. Université de Dakar, Dakar, June 1976.

23 International Bank for Reconstruction and Development, *Senegal: Tradition, Diversification, and Economic Development.* Johns Hopkins University Press, Washington, D.C., 1974.

24 International Fertilizer Development Center, *West Africa Fertilizer Study,*

vol. 2, Senegal. Prepared for the U.S. Agency for International Development, Florence, Alabama, 1977.

25 International Monetary Fund, *Surveys of African Economies,* vol. 3. Washington, D.C., 1970.

26 *Journal de la Marine Marchande,* "Indochina: Ses Trafics et Ses Ports," n.d.

27 Seth La-Anyane, "Development Strategy of Agriculture in West Africa." Ph.D. dissertation, Food Research Institute, Stanford University, Stanford, June 1974.

28 P. Marchat, "L'Organization du Crédit au Sénégal." Institut de Science Economique Appliquée, Dakar, June 1962, mimeo.

29 Yves Mersadier, "La Crise de l'Arachide Sénégalaise," *Bulletin de l'IFAN,* Serie B, Nos. 3–4 (1966).

30 Yvon Mersadier, "Structures des Budgets Familiaux à Thiès," *Bulletin de l'IFAN,* 17, Serie B, Nos. 3–4 (1955).

31 G. Nesterenko, "Rapport pour la Mission d'Aménagement du Sénégal–Richard–Toll au 1er juillet 1953." Dakar, 1953, mimeo.

32 C. W. Newberry, "The Formation of the Government General of French West Africa," *Journal of African History,* 1, No. 1 (1960).

33 Joseph Nguekeng, *L'Economie du Riz en Afrique de l'Ouest: Situation en République du Sénégal.* United Nations, Groupe de Conseillers en Développement de l'Afrique de l'Ouest, Association pour le Développement de la Riziculture en Afrique de l'Ouest, Monrovia, December 1975.

34 Donal B. Cruise O'Brien, *The Mourides of Senegal.* Clarendon Press and Oxford University Press, London, 1971.

35 Michel Renaud and Jacques Brochier, "Etude Monographique sur la Diffusion des Unités de Culture Attele dans l'Arrondissement de Thiènaba." Institut de Sciences Economiques Appliquées, Dakar, 1964.

36 J. P. Rigoulot, "Demande de Financement à l'USAID pour PIDAC." Ziguinchor, 1977, mimeo.

37 J. Seguela, "L'Etat Actuel de la Culture de l'Arachide au Sénégal," *Bulletin Mensuel de l'Agronomie Coloniale,* 116 (August 1927).

38 Sénégal, Government of, Ministère du Développement Rurale et de l'Hydraulique, *Actions Planifiées de la Production Céréalière,* Dakar, 1977.

39 ———, Direction Générale de la Production Agricole, Direction des Etudes Méthodes et Plan, Ve Plan Quadriennal (1977–81), Commission 1A, Proposition d'Actions. Dakar, July 1976.

40 ———, Société d'Aménagement et d'Exploitation des Terres du Delta, "Bilan de l'Operation Hydroagricole du Delta et de la Basse Vallée." Saint-Louis, 1974.

41 ———, Société d'Aménagement et d'Exploitation des Terres du Delta, "Demand de Financement. Cuvette de Demet auprès de la B.O.A.D." Saint-Louis, March 1977.

42 ———, Ministère des Finances et Affaires Economiques, Direction de la Statistique, Bureau de Recensement, *Résultats Provisoires du Recensement Général de la Population d'Avril 1976.* Dakar, 1977.

43 ———, Direction de la Statistique, *Comptes Economiques du Sénégal.* Dakar, 1962 and various other years.

44 ———, "Essai d'Evaluation de la Production de l'Agriculture: Productions Vivrières." Dakar, 1975, mimeo.

45 ———, Ministère du Plan et de la Coopération, IVe Plan Quadriennal de

Développement Economique et Social, 1973–77. Les Nouvelles Editions Africaines, Dakar, July 1973.

46 ———, Service de Statistique et Mécanographie, *Bulletin Statistique et Economique Mensuel de l'A.O.F.* Dakar, various issues.

47 ———, *Le Commerce Extérieur de l'AOF en 1957: Résumé Rétrospectif.* Dakar, 1958.

48 Edmond Sere de Rivières, *Sénégal-Dakar.* Editions Maritimes et Coloniales, Paris, 1953.

49 Société d'Aide Technique et de Coopération, *Développement de la Riziculture au Sénégal, Rapport Général.* Paris, 1968.

50 Société Nationale des Etudes de Développement (SONED), *Etude sur la Commercialisation et la Stockage, des Céréales au Sénégal*, 2 vols. Dakar, 1977.

51 G. Spitz, *Terres Lointaines: l'Afrique Occidentale Française.* Société d'Editions Geographiques, Maritimes, et Coloniales, Paris, 1947.

52 J. Dirck Stryker, "Food Security, Self-Sufficiency, and Economic Growth in the Sahelian Countries of West Africa." WARDA/Stanford Study of the Political Economy of Rice in West Africa, Food Research Institute, Stanford, February 1978, preliminary.

53 Virginia Thompson and Richard Adloff, *French West Africa.* Stanford University Press, Stanford, 1957.

54 A. Hasan Tuluy, "Rice Production in Senegal." WARDA/Stanford Study of the Political Economy of Rice in West Africa, Food Research Institute, Stanford, July 1979, Chapter 8.

55 United Nations, Food and Agriculture Organization, "Food Balance Sheets, 1964–66." Rome, 1971.

56 ———, *Aspects Economiques du Problème des Cultures Vivrières.* Rome, 1970.

57 ———, "Projet Hydroagricole du Bassin du Fleuve Sénégal: Etude Economique de la Riziculture dans la Région du Fleuve Sénégal." Saint-Louis, 1969.

58 U.S. Department of Agriculture, Economics Research Service, "Agriculture Policies in Africa and West Asia." Foreign Agricultural Economic Report No. 49, Washington, D.C., December 1968.

59 U.S. Department of Agriculture and U.S. Agency for International Development, "Senegal," *Rice in West Africa.* Washington, D.C., 1968.

60 West Africa Rice Development Association (WARDA), *Report on an Inventory Survey on Rice Post-Production Technology in Senegal.* Monrovia, August 1976.

61 ———, "Prospects of Intraregional Trade of Rice in West Africa." Monrovia, November 1977.

8. Costs and Incentives in Rice Production in Senegal

A. Hasan Tuluy

Since Senegal achieved independence in 1960, the government has sought to diversify agricultural production away from peanuts into other export crops and import-substituting cereals. Increased rice production has been one of the principal elements of this strategy. The main objective of greater rice production is long-term self-sufficiency in cereals to enhance national food security. This goal gained in importance following the 1968–73 drought and the high world rice prices in 1974. As part of the policy to develop the country's rice sector, the government, through public land development agencies (LDAs), has invested in water control and protected domestic rice in the face of cheaper imports. In addition, a highly subsidized production package has been made available to a small group of rice farmers.

The government has expanded irrigated and mechanized production in the Senegal River Basin (Fleuve Region in Map 7.1) where rice cultivation was unknown before 1947. More recently, the government has also devoted increased attention to the principal rice-growing region of the country, the Casamance, where the thrust of policy has been to improve traditional cultivation methods. One of the central policy questions is whether, or to what extent, to invest in capital-intensive, secure, high-yielding, yet costly, irrigated polders in the Fleuve as an alternative—or a complement—to improving the already established labor-intensive, less secure, lower-yielding, but also lower-cost, cultivation practices in the Casamance.

The purpose of this chapter is to estimate the private and social profitabilities of several rice production techniques in different regions of the country in order to assess the impact of potential changes in factor constraints and government policies on the relative efficiency with which scarce resources are allocated. The resource cost methodology used in the study is particularly well suited to Senegal owing to the importance of existing rice imports and to the country's strategy of substituting local production for these imports.

The first section of the chapter describes the major rice production techniques found in Senegal. A summary discussion of rice policies fol-

TABLE 8.1. *Key Characteristics of Rice Production Techniques, 1976*

Production technique	Area[a] (ha)	Paddy yield (mt/ha)	Type of water control	Improved seeds	Fertilizer	Pesticides
Pump-irrigated, double-cropping manual technique, Matam village polders, Senegal River Valley[b]	400	4.75	pump irrigation	yes	chemical	if required
Pump-irrigated, double-cropping mechanized technique, Nianga, Senegal River Valley[c,d]	700	3.80	pump irrigation	yes	chemical	yes
Pump-irrigated, single crop, mechanized technique, Boundoum Project, Senegal River Delta[e]	9,000	2.5	pump	yes	chemical	yes
Traditional swamp and mangrove rice, Casamance	61,700	1.08	partially controlled flooding	no	organic	no
Improved swamp rice, manual technique, Casamance	4,200	2.25	limited bunding	yes	chemical	no
Improved swamp rice, manual technique with partial water control, Casamance	500	3.6	controlled drainage	yes	chemical	no
Improved rainfed technique, "gray soils," with animal traction, Upper Casamance	13,500	2.07	none	yes	chemical	yes

NOTE: In all techniques sickles were used in harvesting; however, in traditional swamp and mangrove, hand knives were also employed.
[a] Refers to the type of rice production technique and not solely to the model used in the estimation of cost data.
[b] Double-cropping with maize and vegetables.
[c] Experiments with a dry season rice crop, but mostly double-cropping with maize and canning tomatoes.
[d] Experiments with combine-harvesting.
[e] Most combine-harvesting abandoned.

lows. The third part presents estimates of private and social returns, costs, and profitabilities for each technique. The implications of these findings for national objectives and policies are then discussed in a concluding section.

Production Techniques

Senegal is a country of about 200,000 square kilometers situated on the most westerly point of Africa. Its climate is varied, ranging from the arid Sahelian zone in the north to the subtropical zone in the south. It is characterized by two distinct seasons—an 8- to 9-month dry period from October to June, followed by a 3- to 4-month rainy season. Annual rainfall varies from about 350 millimeters in the north to over 1,500 millimeters in the south and southeast. However, rainfall exhibits large interannual variation, and reliability of precipitation decreases from the south to the north and from the coast inland. Three rivers flow through Senegal, though their irregularity restricts full use of their water resources. With the exception of the clay hydromorphic soils found along the river basins, most of Senegal's soils are ferruginous, permeable, and poor in organic matter. Approximately 12 to 14 percent of the land area is under cultivation, and another 25 percent is potentially cultivable.

Table 8.1 contains characteristics of the principal rice production techniques in Senegal. Table 8.2 indicates the quantity and types of inputs employed by these techniques and some major cost components of production. The seven systems summarized in these two tables are described in detail in the sections that follow.

About 80,000 hectares (ha) of the total of 90,000 ha planted to rice in Senegal are located in the Casamance, where traditional techniques occupy three-fourths of the area. Despite the fact that the Casamance is the site of traditional rice production in Senegal, the main focus of governmental efforts has been to develop irrigated polders along the Senegal River. Currently, about 10,000 ha are under irrigated cultivation, including 9,000 ha in the Delta.

Traditional Swamp and Mangrove Rice, Lower Casamance

Swamp rice is traditionally grown throughout the Lower Casamance. It is extremely labor-intensive, especially in the mangrove swamps concentrated in the Lower Casamance. Earthen dikes are built around the rice plots. Soil preparation involves leaching out salts over successive rainy seasons and building ridges onto which seedlings are transplanted. Soil fertilization is accomplished either by putting rice straw, weeds, and other vegetable wastes back into the soil during the ridge building or by

TABLE 8.2. *Quantities and Costs Per Rice Crop Per Hectare of Major Inputs*

Production technique	Farm labor (man-days)	Fertilizer[a]			Seeds (kg)	Annualized land development costs (000 CFA fr)	Extension service costs (000 CFA fr)
		N	P	K			
Pump-irrigated, double-cropping manual technique, Matam village polders, Senegal River Valley	270	122	96	0	40	9.0[b]	13.0[b]
Pump-irrigated, double-cropping mechanized technique, Nianga, Senegal River Valley	135	81	92	20	120	49.0[b]	6.0[b]
Pump-irrigated, single crop, mechanized technique, Boundoum Project, Senegal River Delta	92	69	72	0	120	75.5	4.8
Traditional swamp and mangrove rice, Casamance	208	—	—	—	40	1.0	0
Improved swamp rice, manual technique, Casamance	229	61	36	54	40	1.2	5.3
Improved swamp rice, manual technique with partial water control, Casamance	266	87.5	45	67.5	40	14.0	11.0
Improved rainfed technique, "gray soils," with animal traction, Upper Casamance	111	57	27	40.5	80	3.0	5.6

[a] Recommended annual doses, in kg.
[b] Land development pump and extension costs have been prorated between the two crops according to the respective water requirements.

applying household wastes. In some areas, cattle are allowed to graze in the rice fields after the harvest. Throughout the Casamance, traditional swamp rice is planted in seedbed nurseries and is transplanted two weeks to a month after the rains have begun. Transplanting reduces the need for weeding, and in saline areas field soils can be leached longer while the seedlings remain in the nurseries. Harvest occurs from late November through January, depending on the rains, planting dates, and cycle duration, and is accomplished with either a small hand knife or a sickle.

Traditional rice employs few intermediate inputs. Farmers use numerous varieties of *Oryza sativa* and a number of local varieties of *Oryza glabberima*. During seedbed preparation, five to ten varieties, which are adapted to different soil and growing conditions, are selected and planted (15, p. 6). Varieties that ripen at different times are used to space out the harvest.

When fields are ready for transplanting, a strict order of planting is followed—rainfed fields first, rapidly flooded areas next, and then swamps fed by tributaries. Varietal selection and sequential transplanting reduce risk in an environment of uncertain water availability.

Capital inputs into traditional rice production are limited. A few hand tools are used, including a hand hoe (or specialized shovel) for land preparation, a machete for clearing brush and grass, and a small hand knife for harvesting. The only capital investments in this system are labor-intensive bunding, construction of perimeteral dikes in saline areas, and tree removal in mangrove swamps. The bunds and dikes require frequent repairs and maintenance.

Labor is a major constraint on production, and in some areas the problem is compounded by heavy out-migration (14). Although most of this migration is seasonal and conflicts only with certain tasks in swamp rice cultivation, some of it is becoming permanent. In addition, the relatively high profitability of peanuts has led to an increased emphasis on their production, which has diverted some labor from rice. This is less true in the Upper and Middle Casamance, where there is a stricter division of labor, with women cultivating rice and men growing upland crops, than in the Lower Casamance, where both men and women grow rice.

Yields from traditional swamp rice are rarely measured and vary considerably with differences in soils, climate, hydrology, and labor inputs. The total area under this technique is also difficult to determine; approximately 61,700 ha have been estimated to be devoted to traditional swamp and upland rice (see Table 8.1). Problems of soil and water salinity are common in the Lower Casamance mangrove swamps. When rainfall is insufficient to leach soils adequately, fields must be abandoned for several years.

Improved Swamp, Manual Technique, Casamance

An improved manual swamp technique is being introduced throughout the Casamance, particularly in the middle and upper regions. The improved technique is similar to the traditional techniques, but also includes use of improved seeds, fertilizers, sprayer-applied pesticides, and sickle harvesting. Better farm practices, such as timely seeding, weeding, and proper plant spacing, are important aspects of this technique. Although yields are double those of the traditional swamp techniques, sickle harvesting cuts harvesting times in half.

The major constraints are labor and credit. The financial requirements of the new inputs constitute an impediment to rapid adoption by the average farmer, especially since most of the rice farmers are women, who do not have access to government credit. New credit arrangements are currently being worked out.

Improved Swamp, Manual Technique with Partial Water Control, Middle Casamance

This technique requires some mechanical alteration of the land so that rice can be grown in areas that are waterlogged during the rainy season or where the slope results in too rapid a rate of runoff. In waterlogged lands, natural drainage lines are mechanically deepened and improved, new drainage channels and sluices are added to evacuate excess water, and contour field bunds are constructed to retain the requisite amount of water for cultivation. In the upstream sloped areas near streams, a weir and a system of distribution channels are constructed and fields are leveled to retain and distribute irrigation water evenly.

Production inputs are similar to those used in the techniques described above, although slightly greater use of fertilizers is recommended (see Table 8.2). Without proper credit, it is unlikely that farmers will be able to undertake the investments necessary for the implementation of the water control system. Moreover, these irrigation and drainage works require a restructuring of land tenure rules. Initially, at least, these types of investments have to be undertaken by various LDAs.

Improved Rainfed Technique, "Gray Soils" with Animal Traction, Upper Casamance

This technique is practiced outside of the swamp areas on transitional "gray soils," and oxen are used for land preparation, seeding, weeding, and transport of paddy from the field. Harvesting is done with a sickle. Yields are favorable on gray soils because plants continue to be nourished by a high water table after the rains have stopped. The area is planted with varieties that have longer cycles than do those that generally survive in true upland conditions. Relatively high fertilizer rates have a beneficial effect on yields, but also necessitate greater labor times devoted to weeding.

Major additional inputs are a pair of trained oxen, a plow, a cultivator, a seeder, and a cart for transport. Although these inputs considerably raise costs to the farmer, they save labor—111 as compared with 229 man-days per ha in improved swamp rice cultivation—permitting more land per farmer to be brought under cultivation.

The major limitations on the expansion of this gray soils system are the need to clear the fields properly, especially of tree stumps, so that animal-drawn equipment may be efficiently used; the requirement for more frequent and thorough weeding, a task that normally conflicts with labor demands of other crops; and problems of land tenure.

Pump-irrigated, Double-cropped Manual Technique: Matam Village Polders

Matam is an example of an irrigated village perimeter that constitutes one end of the spectrum of rice production techniques practiced by the Société d'Aménagement et d'Exploitation des Terres du Delta (SAED) along the Senegal River.[1] Before the kind of perimeter is set up, the village must ensure that there are about 20 ha of land available that are not subject to flooding, but are close enough to the river for pumping; that land tenure problems are settled, and that a village-level committee for perimeter management and operation is established. SAED helps choose an appropriate site and lays out the design of the major network. Clearing, stumping, and canal construction are then done collectively by the farmers, using traditional tools. Individual parcels in the polder are allocated by SAED lottery to polder members who have been chosen by the village. SAED occasionally limits the number who can join to assure that at least 0.20 ha are allotted per household. Each individual is responsible for the leveling and construction of bunds within his own plot. The polder pump is installed by SAED and operated by the villagers, and a SAED mechanic handles major maintenance.

Land preparation is accomplished in June with a traditional hand hoe or shovel and without the benefit of pre-irrigation rains to soften the soil.[2] A collective nursery is prepared in mid-June while the river level is still low. Farmers use a short-cycle, high-yielding variety—I Kong Pao—which is purchased from SAED and renewed every three years. Seedlings are transplanted three weeks later when sufficient field water becomes available.

---

[1] SAED is a public, semiautonomous land development agency responsible for agricultural development throughout the Senegal River Basin.

[2] At present the timing of the flood does not permit irrigation of major fields early in June. Pre-irrigation will be possible if the river is seasonally regulated by an upstream dam.

Fertilization rates are high.[3] Weeding is done manually or with a hoe (*daba*), though weeds are not generally a problem because the rice is transplanted, and the small size of individual plots allows thorough weeding.

Paddy is sickle-harvested by all family members from October through December and left to dry in the fields for about two days. It is then taken to a threshing floor on or near the polder and threshed over a barrel by women. Transport to on-farm storage sites varies considerably depending on the distance between the perimeter and the farmer's village. It may be done by renting, borrowing, or using one's own animal-drawn cart or by canoe or head-carrying.

In addition to hand tools, fertilizers, and improved seed, farmers use pump and extension services furnished by SAED. Pump repair is done by a SAED mechanic, and villagers pay for any spare parts required. Project farmers are expected to maintain an amortization fund out of which major repairs and pump replacement are paid.

The major constraint on expansion of this system is the availability of suitable land. In addition, if expansion increases the size of individual holdings, and there is no change in the technique used, labor could become a constraint. Finally, the effectiveness of the village committees in assuring pump maintenance and replacement has yet to be tested.

Pump-irrigated, Double-cropped, Mechanized Technique, Senegal River Valley: Nianga

The Nianga project is still at the pilot stage of 860 ha; its final design calls for 10,000 ha. Completion will depend on greater availability of water through the regulation of the Senegal River. Until an upstream storage facility is constructed, the off-take of river water for dry season irrigation will remain seriously constrained. Because Nianga does not experience saltwater incursion, farmers can practice double-cropping if sufficient water is available. They grow rice during the rainy season and tomatoes and maize during the dry season. Total investment costs are about 1,250,000 CFA francs per ha.

Individual parcels average about one ha. Mutual guarantee groups of 15 to 20 farmers are put in charge of water distribution and maintenance of the feeder canals. The pumping station and primary irrigation network are maintained by SAED, which also decides on the crops to be cultivated, the varieties to be grown, and the agricultural calendar.

Land preparation, seeding, and threshing are done mechanically. Land preparation requires deep-plowing every three to four years, and offset and cross-harrowing in intervening years. Seeding is done with a tractor-

---

[3] One reason for the high fertilization rates is that each farmer holding 0.2 to 0.25 ha is sold one 50 kg sack of urea and one of compound fertilizer to facilitate input distribution.

drawn seeder. A basal dressing of potassium chloride is applied at seeding, and during the crop cycle complex fertilizers and urea are manually applied. Weeding is done by hand in conjunction with chemical herbicides. Harvesting is manual using a sickle, and mobile Borga units of 800 kg/hour capacity are used for mechanical threshing.

Part of the cost of water delivery is covered by a water charge of 25,000 CFA francs per ha of rice crop. For all service charges and input costs, farmers pay the equivalent of about 1.5 to 2.0 tons of paddy per ha.

SAED has had difficulty in expanding at the planned rate for logistical and administrative reasons.[4] One of the key constraints on further expansion will be felt downsteam. Increased pumping during the dry season draws saltwater upstream in the Delta, influencing the dry season crop in downstream projects and making it more difficult for the Delta perimeters to have suitable irrigation water available during the rainy season.

Pump-irrigated, Single-crop, Mechanized Technique,
Senegal River Delta: Boundoum

In the large Delta perimeter at Boundoum, polders are at different stages of water control. Primary polders are surrounded by an earthen perimeter dike with floodgates and depend on the river rising high and long enough for irrigation. There is no supplementary pumping and the fields are not leveled. Secondary polders are serviced by a pumping station and improved canal networks within the polder. In the absence of leveling, however, water depth varies greatly throughout the fields, and dwarf, high-yielding varieties cannot be used. With leveling in tertiary polders, this problem is overcome.[5] The cost of adding a pumping station, a more extensive canal network, and field leveling, however, is considerable. Primary and secondary improvements cost approximately 450,000 to 500,000 CFA francs/ha, and tertiary development adds another 500,000 CFA francs.[6] Yields vary between 500 and 1,200 kg/ha in primary systems, depending on water availability, and between 2,000 and 3,000 kg in tertiary systems (*21*, pp. 2–3, 8–9, 16–40).

In bad years, no polder is secure because of saltwater incursion upstream. About 75 percent of the Senegal River's total annual flow of 25 billion cubic meters (m³) occurs during the three months of the rainy season.

[4] SAED developed the basin at an average rate of 2,178 ha/year between 1973 and 1977, although cultivated areas do not always correspond to the number of ha developed (*18*, p. 46).

[5] At present, only one-half of SAED-managed areas are under full water control. Because of the extended drought of 1968–73, however, all existing polders are being transformed to full water control, and all new projects are begun as tertiary developments (*21*, p. 1).

[6] These are minimum estimates. Current tertiary construction costs have in some instances risen to 1.25 to 1.50 million CFA francs/ha (*23*, p. 28; *6*; *19*, p. 9; *16*, pp. 39–43, 66–67).

Thereafter, the flow tapers off and reaches 10 cubic meters per second (m^3/s) or less in April/May. This results in the drawing of saline ocean water 220 km upstream. Such incursions of saltwater render dry season cropping in the Delta impossible. A minimum flow of about 100 m^3/s is required to repulse the saline incursion in the high flow period. Owing to large interannual variations, the date at which sufficient river flow is attained to push back the ocean water and to permit pumping in the Delta varies significantly.[7] In years of heavy floods, important delays occur in the planting of the rice crop in the Delta polders. In consecutive drought periods, salinization and acidification of soils become additional problems.[8]

The average holding per farmer is about 2.5 ha in the Delta, considerably larger than that of Nianga. Project control and farmer organization are the same as those in Nianga. Crop decisions are made by SAED, which also provides inputs, machinery, pumping, and extension services.

Seeding is done mechanically with line seeders. Fairly high seeding rates are required because of the low germination rates in the saline soils of the Delta. Land preparation and threshing are done with motorized equipment. As in Nianga, deep-plowing is undertaken every three years, with offset and cross-harrowing done in intervening years. At harvesttime, farmers pay about 1.7 tons paddy per ha for all seasonal inputs and services.[9]

The major problems confronting this sytem are water insecurity, the presence of salt, and the high cost of polder development. Farmers face the additional financial risk of being liable for machinery service charges and cash inputs without adequate production security.

SODAGRI, Upper Casamance

The Société de Développement Agricole et Industriel du Sénégal (SODAGRI) project being proposed for the Upper Casamance is similar in structure to Nianga. Plans have been made for two rice and other crops per year using large-scale, mechanized production with controlled irrigation

[7] Retreat of the saline incursion can vary from July 24 to August 8 for flood probabilities of 50 to 95 percent flow frequency. Furthermore, large-scale developments upstream may result in drawing the saline waters further upstream in the dry season, further delaying the date at which the saline incursion is repulsed in the rainy season (28, p. 28). For a discussion of the effects of planting delays on yields in the Delta, see 28, p. 42. Beye (1) also discusses the effects of salinity on yield in Boundoum.

[8] Earlier the Delta was thought to be an ideal area for rice cultivation owing to the lack of relief, the "complementarity of rainfall and flooding," the heavy clay soils, the lack of population and associated tenure problems, and market access (see 28, p. 24). The same report also pointed out potentially serious soil toxicity and acidity problems of certain Delta soils (28, p. 46 and 46 bis). Sène et al. observe that 25,300 of the 34,200 ha of the Delta are saline (20, p. 437).

[9] In 1975 farmers paid for seed, fertilizer, weed killer, machinery services, thresher services, and a flat per hectare water charge of 25,000 CFA francs, totaling about 70,000 francs/ha of rice crop.

provided by a retention dam. Estimated costs are close to those of Nianga, but because of the long distance from supply centers to the project site, higher costs can be anticipated. At the time of writing, the SODAGRI project had not solved some basic technical problems, and hence no cost information was available.

Collection, Milling, and Distribution

Almost all traditionally grown paddy is either dried in the field and threshed, or carried from the field, stored on the stalk, and threshed as needed (see Table 8.3). Traditional threshing is done with a stick over an oil drum, or in a wooden mortar with a pestle. The latter method is also used to hull nearly all traditional rice. One kilogram of paddy can be hulled in about 15 minutes, and the outturn is approximately 70 percent rice with about one-half brokens.

In a few areas paddy is hulled with small, private, diesel-powered machines, but this is usually only done during the rainy season when agricultural labor demands limit the time available for hand pounding. There are 30 or 40 of these small hullers in Senegal. They are officially discouraged because the national marketing agency, Office National de la Coopération et d'Assistance au Développement (ONCAD), has a statutory monopoly on rice marketing and milling.[10] The average rated capacity of these mills is about 0.2 tons/hours of rice,[11] or close to 500 tons of rice per year,[12] but actual capacity utilization is probably much less. Outturn from these units, which are primarily 20 horsepower, steel cylinder machines, appears to range between 50 and 70 percent, with 25 to 40 percent brokens. Milling charges on these small mills recently increased from 3 CFA francs/kg of rice to about 10 CFA francs owing to the increased cost of fuel and lubricants. This brings the cost of the technique fairly close to that of hand pounding. If women did not have to go so far to get to a milling unit, these small hullers would be used more widely as an alternative to hand pounding.

Threshed paddy for official sale to LDAs is brought by the farmers to collection centers, where it is weighed and purchased. At the time of weighing, the paddy equivalent of the cost of purchased inputs is deducted from the farmer's crop. In the Fleuve this often amounts to 40 percent of the harvest. The balance of the paddy is then paid for directly if purchased by the Société pour le Développement des Fibres Textiles (SODEFITEX) or settled several months afterward if purchased by

[10] There are an estimated 13 huller units in the Fleuve, 15 in the Casamance, and another five or ten in the Sine Saloum region of the country.

[11] These observations are based on the work by Spencer et al. (29).

[12] Theoretical capacity was calculated assuming eight hours of operation per day for 250 working days per year.

TABLE 8.3. *Key Characteristics of Rice-Milling Techniques*

Milling technique	Projected full capacity (mt paddy/ year)[a]	Percent brokens[b]	Milling ratio	Rice milled in 1976 (mt)	Unit cost of milled rice (000 CFA fr/mt)	By-products
Hand pounding	—	40%–60%	0.65–0.70	54,000	13	yes
Small mills	20,000	25%–40%	0.66	650–2,000	3	yes
Large mills	50,000	40%–50%[d] 85%–90%[e]	0.65	4,840	28	yes

[a] Projected full capacity for small mills assumes 40 such units multiplied by 250 days at 8 hours a day with a throughput of 0.2 mt paddy/hr. For large mills, full capacity is estimated on the rated per hour capacity of the three mills operating in Senegal in 1975–76; 2mt/hr plus 2 mt/hr plus 6 mt/hr equals 10 mt/hr times full operation at 5,000 hrs equals 60,000 mt.

[b] All rice is raw; no parboiling is done in Senegal. Percent brokens from large mills are in terms of percent of white product output.

[c] By-products from hand pounding are used as domestic animal feed; some of the by-products from mills are sold as bran, in the case of large mills through ONCAD.

[d] Séfa.

[e] Fleuve.

SAED or Projet Rural de Sédhiou (PRS). After purchase, paddy is evacuated by ONCAD or the LDA to the nearest mill, often with some delay.[13] Some paddy is retained and treated by the LDAs for seed.

Large milling units of between 2 and 7 tons/hour rated capacity are used for officially purchased paddy. There are four such large mills currently operating in Senegal. One mill at Ross Béthio (6 tons/hour) was constructed in 1971, and another at Richard Toll (7 tons/hour) has only recently been repaired; both are run by SAED. Their combined theoretical capacity is 13 tons/hour, or about 65,000 tons a year. At present, however, only 6,000 to 10,000 tons of paddy are processed annually by these two mills. The outturn is around 66 percent, with 80 percent brokens. There is also a private mill in Saint-Louis with good equipment, but it is not allowed to operate because of the ONCAD/SAED monopoly. The other two mills are in the south. One at Séfa, near Sédhiou, is run by a private company for ONCAD, and the other is at Kedougou in Eastern Senegal. Each has a rated capacity of 2 tons/hour, or about 20,000 tons per year combined. The milling outturn for these two mills is between 64 and 65 percent, with about 45 percent brokens.

The cost of milling in these large units is about 16 to 17 CFA francs/kg of paddy milled, largely because of low capacity utilization. At the three mills that are currently operating, only about 20 percent of capacity is

[13] These delays in the Fleuve result in important humidity losses because of the large diurnal temperature variation. Paddy frequently arrives at the Ross Béthio mill with no more than 7–8 percent relative humidity and large cleavages. In the absence of parboiling, this results in the high percentage (about 80–90 percent) of broken rice obtained from the SAED mills (35, pp. 54–70).

used. Of the 26 CFA francs/kg cost of milling rice, ONCAD pays 8.5 CFA francs/kg, and 17.5 CFA francs are absorbed by a subsidy.

Rice from the mills along the Senegal River is evacuated to ONCAD offices at Saint-Louis and is sold, along with imported rice, to quota-holding merchants in the region. Rice from the Kedougou mill is handled there in the same manner, whereas rice from Sédhiou is shipped to Ziguinchor for official sales throughout the Casamance. Only small quantities of domestically produced rice are sold in Dakar, the major import substitution market.

Government Policies Affecting Rice Production

A two-tiered system of government incentives exists in Senegal. At the general level, the central government affects all agricultural production through trade policy instruments, price setting, provision of infrastructure, and institution building. At the sectoral level, the public land development agencies have a direct impact on rice production in particular geographic areas through their extension of a subsidized technology package, credit, and marketing services.[14]

Generally, goods destined for the agricultural sector, such as fertilizers, pesticides, equipment, and machinery, are exempt from import duties. Although an additional duty of 5 percent is levied on most imported items, reductions or exemptions are commonly granted. An excise tax of 2.1 percent is also paid on agricultural machinery and irrigation pumps. Complex fertilizers, which are produced locally, are subject to a 13.5 percent sales tax. Some inputs—insecticides, for instance—are subject to all of these duties and taxes. Since most major investments in the rice sector are externally financed, they are exempt from taxes. However, several indirect taxes—fuel taxes, vehicle registration fees, annual road taxes, social security and other personnel taxes—are levied on projects in the rice sector.

Guaranteed producer prices for paddy were introduced in 1963–64. Prices are established by an interministerial advisory committee to the Prime Minister, the Comité des Grands Produits Agricoles (CGOT), and are announced each November, several months after planting (3, vol. 2, pp. 81–85 and 95–96). The paddy price is uniform across regions and varies only according to the red rice content, for which there is a dis-

[14] SAED, the LDA responsible for the development of the Senegal River Valley, was founded in 1965 and has incorporated all previous development agencies of the region. SODEFITEX, primarily a cotton agency with operations in the eastern part of the country, has also diversified into upland rice cultivation. A regional LDA did not exist in the Casamance until 1977–78, when the Société pour la Mise en Valeur de la Casamance (SOMIVAC) was founded to supervise the region's projects.

count. In November 1974, paddy producer prices were set on a par with peanuts at 41.5 CFA francs/kg, slightly below the price of cotton.

Paddy producer prices are set above the prices for paddy equivalent of comparable quality imported rice, and the consumer retail price is also generally set higher than the price of imports. In addition to this protection for local producers, subsidies are paid to the LDAs on local rice delivered at the millgate and to the marketing agency to cover the difference between its cost and the regulated retail price.[15]

ONCAD holds a monopoly on paddy purchases, thus legally eliminating private traders. ONCAD, or the LDAs on its account, purchase paddy at harvest. Payments are usually deferred until after the farmers' input and service debts have been deducted. Official purchases of paddy have not exceeded 8 to 10 percent of the estimated annual paddy production because of insufficient price incentives and the lateness and unwillingness of the marketing agency to enter the market.[16] Official producer prices thus apply only to those limited purchases.

The Senegalese government has emphasized policies that seek to increase the aggregate supply of rice through technical improvements. The most effective of these has been the formation of LDAs. These agencies develop the land, supply water, and exercise a substantial degree of control over the choice of crops. Furthermore, LDAs have direct access to technical expertise and to foreign funding at favorable terms for their capital investments. The Senegalese government is frequently required to provide funds only for certain recurrent expenditures, typically about 20 percent of total project costs.

The cost of developing a full water control, tertiary SAED polder in the Fleuve is about 60,000 CFA francs per ha per year when annualized over a 25-year life at 2.5 percent real interest rate.[17] The annual cost of providing partial water control in the Casamance swamps is 12,000 CFA francs/ha. These land development costs are fully absorbed by the LDA projects. In theory, maintenance as well as operating costs of the irrigation network are charged to project farmers through a flat per crop-hectare water charge. This charge, however, usually covers only the operating cost of providing water and falls short of assuring proper network maintenance. Along with infrastructure, LDAs provide free management and extension services to project farmers.

[15] The price of locally produced rice from the SAED mill has recently been above the official retail price, which has been above the c.i.f. import price (4).

[16] ONCAD's marketing difficulties, its cash flow problems, and the restrictive price structure are discussed in 3, pp. 96–116.

[17] The Matam village polders, although technically tertiary developments, are less expensive, since their development and operation are manual. Total costs are estimated at 300,000 to 350,000 CFA francs/ha.

As Table 8.4 indicates, inputs to rice farmers are subsidized. The formation of LDAs has facilitated the dissemination of technology packages using improved inputs, and, to encourage farmer acceptance, subsidies are placed on most key inputs, including improved seeds, fertilizer, some herbicides and pesticides, machinery services, extension, and management. In addition, a general program of agricultural subsidies, the Program Agricole, was begun in 1964 with major foreign financing.[18]

At present, the subsidy bill for both LDAs and the Program Agricole is largely met from national sources through current receipts of special accounts. The most important of these is the Caisse de Péréquation et de Stabilisation des Prix (CPSP), a stabilization fund that centralizes the net earnings from peanut and cotton sales and tax receipts on imported rice.[19] While the CPSP surplus has primarily been used to finance general budgetary deficits or part of the investment budget, a share of the earnings is earmarked for producer—and sometimes consumer—subsidies. With short-term supplier's credit extended by the Banque Nationale du Développement Sénégalais (BNDS), ONCAD purchases agricultural equipment and inputs at cost and delivers them to LDAs and cooperatives at subsidized prices. The price difference is then paid to ONCAD by the Stabilization Fund.[20] As with product prices, official input prices are uniform throughout the country.

The main recurrent input subsidy is on locally manufactured compound fertilizer (NPK) and imported urea. At its peak in 1974, the farm subsidy on compound fertilizer was 75 percent of its ex-factory price. In 1976 it was reduced to about 60 percent when ONCAD purchased NPK at 66.7 CFA francs per kg ex-factory but delivered it at 20 CFA francs. Imported urea, valued at 83.0 CFA francs c.i.f., was supplied in 1976 to farmers as seasonal credit-in-kind for 25 CFA francs.[21] The large government subsidy and nitrogen responsiveness of improved rice varieties have led to an impressive growth in the demand for fertilizers (*12*, p. 29). From about 1,600 tons in 1961–62, fertilizer consumption for rice climbed to 13,000 tons in 1967–68 and, after a sharp decline during the drought, reached about 20,000 tons in 1974–75.

Subsidies are also paid on selected high-yielding seeds. Research costs of developing suitable strains are not passed on to farmers, who only pay

[18] For a description, see 5, pp. 185–274, and 4.

[19] For a description of the Special Accounts prior to the establishment of the CPSP, see International Monetary Fund (*13*, pp. 554–95).

[20] See 3, vol. 2, pp. 33–36. Craven and Tuluy (*4*) describe the difficulties encountered by ONCAD and the BNDS in paying for agricultural subsidies during the 1973–76 period.

[21] The farmer price of fertilizers was further increased in 1977 to 25 CFA francs/kg for compound and 35 CFA francs/kg for urea, reducing the subsidy to about 45 percent.

TABLE 8.4. *Farm Subsidies and Charges*
(CFA francs/mt milled rice)

Production technique	Intermediate inputs			Subsidies								Fixed charges
	Seeds	Fertilizer	Pesticides and weed killer	Mechanical land preparation	Land development	Water	Extension	Medium-term credit	Other	Total		
Matam village polders	62	7,603	188	0	0	0[a]	3,091	36[b]	5,430[c]	16,411		0
Nianga, Senegal River Valley	617	6,989	0	−829[d]	19,547[e]	0	2,342	0	8,720[f]	37,386		9,968[g]
Boundoum project, Senegal Delta	921	8,494	1,430	−1,261[d]	58,798[e]	0	2,920	124[b]	13,204[f]	84,630		15,152[g]
Traditional swamp and mangrove rice, Casamance	0	0	0	0	0	0	0	0	0	0		0
Improved swamp rice, Casamance	−92	11,265	278	0	120[h]	0	3,798	22	0	15,391		0
Improved partial water control, swamp rice, Casamance	−58	9,477	174	0	5,558[i]	0	4,749	48	4,288[c]	24,236		0
Improved rainfed animal traction, Upper Casamance	−201	10,358	303	0	0	0	4,129	634[j]	0	15,223		0

NOTE: Per hectare subsidies were divided by the net yields and then divided by the milling ratio to obtain net subsidies per mt of milled rice. For the Fleuve techniques, the milling ratio of the SAED mill (0.66) was used. In the Casamance the milling ratio used was that of Séfa (0.65). Negative entries denote taxes on the farm.
[a] The cost of pump replacement was charged to the farmer even though the system of farm amortization of the pump has not yet been put into effect. The original pump subsidy was 1,745 FCFA/mt of milled rice.
[b] On the purchase of an ox cart.   [c] Operation and maintenance of polder vehicles, mechanics, drivers, and other technical personnel.
[d] Deep-ploughing and cross-harrowing.   [e] Pumping station and power generator costs were included in land development and not in water delivery costs.
[f] Operation, maintenance, irrigation network and equipment depreciation, personnel other than extension services, adjusted for farm taxes on certain machinery services.
[g] 25,000 FCFA water charge per hectare of paddy.   [h] Rock phosphate was included as a land development cost.   [i] Costs of drainage and retention canals.
[j] Subsidies on the purchase of animal traction equipment.

the marginal cost of seed multiplication. Moreover, cost estimates from the Nianga seed farm suggest that there is an added direct subsidy of about 15 to 20 percent on the 70 CFA francs per kilo the farmer pays for selected seeds.

Other subsidized services available on seasonal credit are the machinery services and water provided by large projects. Machinery services are made widely available to project farmers in the Delta and Lower Valley of the Senegal River, where heavy clay soils are not suited for manual land preparation and the growing season is constrained by a short flood cycle.[22] Most services charges are set approximately to meet costs. Certain tasks, such as deep-plowing, are subsidized about 10 percent, whereas others, for example drill seeding, are priced so that the farmer is actually taxed. As mentioned previously, the flat water charge—25,000 CFA francs per ha of rice in 1976—covers only the operating cost and not the maintenance and depreciation expenditures of the irrigation network.

Threshers, seeders, carts, and other equipment are made available to producer cooperatives with a state subsidy and on credit. The subsidy mechanism is similar to that for seasonal inputs. Agricultural equipment is purchased by ONCAD from the local manufacturer, the Société Industrielle Sénégalaise de Construction Mécaniques et Matériels Agricoles (SISCOMA), at cost and is sold to the producers at subsidized prices. Repayment takes place in three to five annual installments at an estimated implicit interest rate of 8 percent.[23] ONCAD's losses are reimbursed by the Stabilization Fund. The agricultural equipment program has had only a small impact on rice producers, whose purchases have been limited to carts, draft animals, and, less frequently, plows and seeders.

The government's rural credit program is essentially restricted to these schemes for purchasing seasonal inputs and agricultural equipment. Official credit is therefore rationed by channeling it exclusively through LDAs and producer cooperatives.[24] Cooperative members are collectively responsible for the repayment of the credit. Although this practice has virtually eliminated defaults, it has also skewed the provision of credit toward farmers with sound creditworthiness.

[22] The brevity of the flood peak does not permit pre-irrigation of the soils, which would allow for manual or animal traction land preparation. Thus there is a premium on time-saving techniques of early land preparation and shorter-cycle varieties.

[23] Since guarantee payments, made by the cooperatives to ONCAD, on which no interest is earned, constitute equity in the BNDS, the actual interest rate is probably higher. In 1976 these funds were about 46 percent of total disbursed credit, so that the true nominal rate was closer to 17 percent. See de Jonge et al. (*14*).

[24] Only 3 percent of agricultural credit has been allocated directly to individual farmers, mainly for vegetable and poultry production. No official credit has been given to rice producers not participating in government projects.

Shadow Prices

The factor markets in Senegal function well, causing market prices to reflect the scarcity values of factors.[25] The agricultural daily wage in the Fleuve (250 CFA francs) is less than that in the Casamance (300 CFA francs). Because this differential persists in the face of vigorous migration, the gap probably reflects actual relocation costs. Minimum wage legislation, especially for skilled labor, has had little impact, even in the urban industrial sector. There is no reason to suspect that the labor market has major imperfections, and the market wage is an adequate measure of the opportunity cost of labor.

The effective wage rate for traditional production techniques in the Casamance might, however, be less than 300 CFA francs. Ties between the traditional and commercial sectors in this area of underdeveloped infrastructure are weak, and the supply of family labor to the market could be quite inelastic. Hence, valuation of predominantly female traditional on-farm labor at the marginal wage rate of the commercial agricultural sector overstates the reservation price of family labor in traditional techniques. Instead, it is assumed that female labor should be valued at 250 CFA francs per day.

In view of the wide availability of land and the extensive nature of Senegalese rice cultivation, the shadow price for land is assumed to be zero. The Fleuve experiences occasional shortages of water rather than land, and hence the social opportunity cost of water might be positive there.[26] This constraint will not be binding for some time, except in years of drought, however, and by then there is likely to be some seasonal regulation of the river (25, p. 10). In the Middle Casamance, long fallow cycles and low labor intensity of cultivation suggest that land pressures will not develop soon (2, p. 23).

Capital markets in Senegal are segmented. For priority sectors and projects, capital is made available on preferential terms. LDAs borrow from the national development bank (BNDS) at 7.5 and 8.0 percent, or 2.5 to 3.0 percent real rates of interest, and capital from foreign donors may be made available at lower rates.[27] A small number of project farmers have access to concessionary government credit channeled through producer cooperatives.[28] Otherwise, rural interest rates are well above the government rates and typically vary according to the borrower's asset position.

[25] For a more thorough discussion, see Appendix B.

[26] For a brief discussion of the shadow price of water, see Appendix B.

[27] The difference between the nominal and real interest rates indicates the creditors' expectations concerning rates of inflation (33).

[28] Although nominal rates are generally around 8.0 percent, for reasons discussed in note 23, the rate at which cooperative farmers actually receive credit may be closer to 17 percent.

Because of higher transaction costs and default risks, these funds are generally available at between 20 and 35 percent nominal interest rates per annum. Capital market segmentation is perpetuated through strict rationing, absorptive capacity limitations, and project specificity of funding.

Private and Social Profitability

Combinations of the different production, collection, milling, and distribution systems are described in Table 8.5 as activities. The seven main production techniques are variously combined with informal and official collection methods, hand pounding, custom hulling or public milling, and distribution of rice through private and official channels. The most widely practiced combinations and those presenting greatest potential for expansion have been selected for analysis.

Private and social profitability estimates for the production of Matam village polders and the Nianga project in the Middle Valley were calculated for hand-pounded, on-farm consumption and for delivery to the Saint-Louis and Dakar markets, where rice is distributed by public agencies. Production from the Delta perimeter goes entirely through official channels mainly to the Saint-Louis and Dakar markets. Estimates for traditional swamp production of the Casamance were calculated only for home consumption, since rice is produced primarily for domestic household needs. The improved swamp and gray soils, animal traction production techniques were each evaluated for household consumption and for sales in Ziguinchor and Dakar through national agencies. In addition, cost estimates were made for rice collected from improved, manually cultivated swamps and custom-milled for sale on the illicit parallel market in Ziguinchor.

Private Profitability

Various indicators of private and social profitability are shown in Table 8.6. The most widely practiced technique in the country, traditional swamp rice in the Casamance, exhibits slightly positive private profitability when female unskilled labor is valued at 250 CFA francs per day and paddy is valued at an on-farm import substitution price of 58 CFA francs/kg. The switch to more remunerative crops and out-migration from traditional rice cultivation areas attest to this low level of private profitability (14).

The introduction of improved seeds, chemical fertilizers, and better agricultural practices into swamp rice cultivation areas has led to significantly higher yields compared with traditional methods. Although labor requirements and intermediate cash expenses have also risen, average yields have doubled to 2,250 kg/ha. Nevertheless, despite the fact that

TABLE 8.5. *Combined Production and Post-Harvest Activities*

Activity designator	Production technique	Collection	Milling technique	Distribution center
1.1	Matam village polders	None	Hand pounding	Home consumption
1.2		Public agency	Large public mill	Official channels, Saint-Louis
1.3		Public agency	Large public mill	Official channels, Dakar
2.1	Nianga project	None	Hand pounding	Home consumption
2.2		Public agency	Large public mill	Official channels, Saint-Louis
2.3		Public agency	Large public mill	Official channels, Dakar
3.1	Delta/Boundoum project	Public agency	Large public mill	Official channels, Saint-Louis
3.2		Public agency	Large public mill	Official channels, Dakar
4.1	Traditional swamps	None	Hand pounding	Home consumption
5.1	Improved swamp rice	None	Hand pounding	Home consumption
5.2		Informal channel	Small hullers	Informal channels, Ziguinchor
5.3		Public agency	Large public mill	Official channels, Ziguinchor
5.4		Public agency	Large public mill	Official channels, Dakar
6.1	Water control swamps	None	Hand pounding	Home consumption
6.2		Public agency	Large public mill	Official channels, Ziguinchor
6.3		Public agency	Large public mill	Official channels, Dakar
7.1	Rainfed animal traction	None	Hand pounding	Home consumption
7.2		Public agency	Large public mill	Official channels, Ziguinchor
7.3		Public agency	Large public mill	Official channels, Dakar

TABLE 8.6. *Indicators of Private and Social Profitability*
(CFA francs/mt of milled rice)

Activity designator[a]	(1) Net private profitabilty[b] (NPP)	(2) Net social profitability (NSP)	(3) Resource cost ratio (RCR)	(4) Effective protection coefficient (EPC)	(5) NSP – NPP	(6) C.i.f. – market price	(7) Total taxes (+) or subsidies(−)
1.1	21.25[c]	24.29	0.658	0.886	3.04	13.93	−10.89
1.2	22.47	−13.35	1.304	0.985	−35.82	−33.60	−2.22
1.3	22.47	−17.49	1.430	0.971	−39.96	−38.01	−1.95
2.1	16.55[c]	6.36	0.881	1.121	−10.19	12.83	−23.02
2.2	17.83	−28.71	2.037	1.503	−46.54	−32.05	−14.49
2.3	17.83	−32.85	2.340	1.547	−50.68	−36.47	−14.21
3.1	4.16	−62.02	18.939	10.566	−66.18	−31.34	−34.84
3.2	4.16	−66.15	231.439	114.340	−70.31	−35.74	−34.57
4.1	−0.42	−4.93	1.050	0.836	−4.51	−4.54	−0.03
5.1	16.46[c]	13.40	0.824	0.905	−3.06	10.38	−13.44
5.2	18.02	13.42	0.804	0.939	−4.60	3.20	−7.80
5.3	18.02	−3.59	1.055	0.897	−21.61	−14.00	−7.61
5.4	18.02	−12.84	1.224	0.858	−30.86	−23.67	−7.19
6.1	33.02[c]	16.67	0.772	0.952	−16.35	4.55	−20.90
6.2	34.57	2.03	0.969	0.917	−32.54	−17.49	−15.05
6.3	34.57	−7.21	1.127	0.934	−41.78	−23.70	−18.08
7.1	32.65[c]	26.55	0.641	0.930	−6.10	9.03	−15.13
7.2	34.21	11.92	0.818	0.893	−22.29	−16.44	−5.85
7.3	34.21	2.67	0.954	0.905	−31.54	−22.66	−8.88

[a] For a more detailed description of activity designators, see Table 8.5.

[b] Where post-harvest activities are in fact controlled by public agencies and all post-harvest costs are assumed to be met through state subsidies or transfers, PP shows only residual on-farm profitability.

[c] In contrast to (b), in the case where the farmer processes and markets—or consumes on-farm—his production, the PP measure reflects the profitability of the entire activity from production through marketing.

the costs of extension services are not included in the calculation of net private profitability (NPP), partially improved swamp cultivation remains only marginally profitable as a cash crop. Relative to other cash crops, such as peanuts and cotton, paddy prices do not offer large incentives. Profitability calculations per man-day in the Middle Casamance, where these improved swamp techniques are being introduced, indicate that peanuts are more remunerative than rice (3, vol. 1, pp. 129–31). Given the present price structure, therefore, paddy is often not the first choice for a cash crop in areas where switching between crops is possible (3, vol. 1, p. 115).

The construction of bunds, retention structures, and drainage canals on improved swamps results in greater water control and permits fuller exploitation of the fertilizer-responsive rice technology. Fertilizer application is further increased in partially improved swamps, farm labor inputs rise by 15 to 20 percent, and yields are increased 60 percent above those attained in the unimproved swamps. Costs of the dense extension network and the water control investments are fully subsidized and therefore are not charged to project farmers. As a result of these subsidies and the high yields, NPP is favorable in improved swamps with partial water control.

The key to the high NPP rates of the rainfed, gray soils technique is the reduction in the labor requirements realized by the introduction of animal traction. Draft animals and small farm equipment, like most intermediate inputs, receive important subsidies. Consequently, the farmer's cash expenses remain at the level of improved swamp techniques with water control. The activity is highly profitable because the drop in yields is more than compensated by a reduction in labor input.

As a result of impressive yields—4.5 to 5 tons/ha achieved through the use of transplanting, the application of chemical fertilizers, careful and intensive labor, and pump irrigation—private profitability in Matam village polders is high. Inputs consist mainly of unskilled family labor and subsidized seasonal inputs for which the farmer is required to make an outlay of about 15,000 CFA francs, either in cash or in credit repayable at harvest. Although the capital costs of polder development, about 25,000 CFA francs, and pump replacement are charged to the farmers, the high extension costs are not.[29] Even with the inclusion of these costs, however, the technique remains privately profitable. This high NPP has clearly contributed to the rapid acceptance of this labor-intensive technique and to the strong demand for new polders in an area where rice cultivation was unknown before 1973–74.

---

[29] The extension density appears to be higher in Matam than elsewhere in the country because of the dispersed nature of the village polders and the small average size of holdings per cooperative member, and hence the large number of farmers for whom each extension agent is responsible.

Downstream at Nianga, labor times are reduced with higher degrees of mechanization. Yields are 3,800 kg/ha, 80 percent of those at Matam. Although farmer charges for mechanical services, water use, and intermediate inputs are steep (about 70,000 to 80,000 CFA francs/ha), the activity is privately profitable because of the high yields, reduced labor inputs, and the opportunities for double-cropping, which permit the spreading of certain fixed costs between two crops. Production in the Middle Valley is secure, water is readily available at current levels of development, and to date salinity has not presented problems.

By comparison, NPP in the Delta is borderline.[30] With the yields obtained in the Delta, a single rice crop is only marginally profitable. Despite having the lowest number of man-days among all production techniques, rice farming in the Delta combines high farmer charges with average yields of only about 2,500 kg/ha to give a low NPP. Therefore, farmers in the Delta rely on a variety of secondary economic activities, including where possible a remunerative second crop, livestock, seasonal migration, or remittances from migrant family members. Rice, however, remains the principal food crop.[31] Assured availability of usable, salt-free water during the cropping season is also a critical problem in the Delta. Greater water security and more constant seasonal availability through river regulation should increase NPP by raising yields and permitting cultivation of a dry season crop.

Social Profitability

As the results presented in Table 8.7 indicate, all techniques of rice production in Senegal are socially unprofitable, with the exception of gray soils, animal traction cultivation and production for on-farm and sometimes regional consumption. The substantial divergence between private and social profitability (see column 5 of Table 8.6) is the result of a high degree of trade protection and of net input subsidies.

Unlike the Casamance, where rainfall is sufficient to permit rainfed cultivation, field water availability in the Fleuve can only be assured through costly investments in land development and water control. In spite of these investments, however, river water salinity in the dry season prevents the cultivation of a second crop in the Delta. In addition, heavy clay soils and the timing of fresh water availability result in a need for mechanized field operations. To ensure adequate private profitability, none of the capital costs on land, irrigation, or equipment are passed on to the farmers, and

---

[30] For a comprehensive socioeconomic discussion of NPP in the Delta, see Waldstein (*34*).

[31] Rice continues to be cultivated despite the low NPP because of the different paddy price that applies to on-farm consumption. The reference price then is not the official price but the paddy equivalent of the price of milled rice at the nearest consumption center plus transportation and handling costs to the farm.

TABLE 8.7. *NSP and RCR Results of Major Rice Production Activities at Various Consumption Centers*
(CFA francs/kg milled rice)

Technique	Home		Saint-Louis[a]		Ziguinchor[a]		Dakar[a]	
	RCR	NSP	RCR	NSP	RCR	NSP	RCR	NSP
Pump-irrigated, double-cropping manual technique, Matam village polders	0.658	24.29	1.304	−13.35	n.a.	n.a.	1.430	−17.49
Pump-irrigated, double-cropping mechanized technique, Nianga, Senegal River Valley	0.881	6.36	2.037	−28.71	n.a.	n.a.	2.340	−32.85
Pump-irrigated, single crop, mechanized technique, Boundoum project, Senegal River Delta	2.377	−33.01	18.939	−62.02	n.a.	n.a.	231.439	−66.15
Traditional swamp and mangrove rice, Casamance	1.055	−4.93	n.a.	n.a.	n.a.	n.a.	n.a.	n.a.
Improved swamp rice, manual technique, Casamance	0.824	13.40	n.a.	n.a.	1.055	−3.59	1.224	−12.84
Improved swamp rice, manual technique with partial water control, Casamance	0.772	16.67	n.a.	n.a.	0.969	2.03	1.127	−7.21
Improved rainfed technique, "gray soils" with animal traction, Upper Casamance	0.641	26.55	n.a.	n.a.	0.818	11.92	0.954	2.67

[a]Marketing and distribution by public agencies.

production in the Delta is highly unprofitable in social terms. The Middle Valley is not subject to annual saltwater incursion and can thus benefit from double-cropping. At Nianga, where double-cropping with a highly remunerative industrial tomato crop is feasible, resource cost ratios (RCRs) are around 2.0. Further upstream around Matam, soils are lighter and population densities higher so that two labor-intensive crops can be grown. In Matam, few capital costs are subsidized, all operations are manual, and RCRs are 1.3 to 1.4 on the major urban markets.

Subsidies are provided on most intermediate inputs and services. Total subsidies are 30–35 CFA francs per kg of milled rice in the Delta and 14–23 CFA francs in Nianga. Net subsidies are much lower in Matam, where farmers are collectively responsible for all land development and maintenance of the irrigation system, and production is less socially inefficient.

In the Casamance, subsidies per unit of output, mainly on seasonal inputs and extension services, are less than in the Fleuve. An exception is improved swamps, where LDAs provide large subsidies on water control investments. For traditional swamp cultivation with little access to improved inputs and extension services, subsidies are effectively zero. Hence, rice can be produced more efficiently in the Casamance, with its greater natural availability of water, than in the Fleuve.

The investment, operation, and maintenance subsidies that are required to ensure farmer acceptance of improved rice techniques have important budgetary consequences. In the past, LDA budgetary deficits have frequently been met through large transfers from an already strained national budget. Elimination of any major subsidy, however, especially from mechanized government projects, would render the activity privately unprofitable.

NPP estimates are based on the protected paddy price, whereas net social profitability (NSP) is based on the lower c.i.f. price of comparable quality imports, which have been available in most years at about $250–300 per metric ton c.i.f. The divergence between NPP and NSP is further influenced by the relatively long distances between the rice-producing regions and Dakar, the major port of importation. Because of high transport costs, the price of imported rice delivered to on-farm consumption centers often exceeds local production costs. At urban consumption points such as Saint-Louis, Ziguinchor, and especially Dakar, on the other hand, transport costs work in the other direction, increasing the cost of domestically produced rice delivered to the cities and lowering the price of imports with which local rice competes. Since the cost of transporting domestically produced rice to these cities is heavily subsidized, the divergence between NSP and NPP increases.

A summary of NSP and RCR indicators for the major production techniques at various consumption points is contained in Table 8.7. Rice from the Fleuve is highly unprofitable at off-farm consumption centers because of costly, mechanized production techniques and the high percentage of brokens (80 percent). Casamance rice is of better quality and effectively competes with more expensive imports; its RCRs are closely distributed around one.[32] Rice produced in the Casamance could substitute for imports in Dakar at $325–385 (1975 U.S. dollars) per metric ton. In contrast, to be competitive in major urban markets, Nianga would require a delivered c.i.f. price of about $440 for 80 percent brokens, whereas Delta rice could compete only at a world price of $570. This indicates that the Casamance has a distinct efficiency advantage over the Fleuve in supplying the Dakar market.[33]

Sensitivity Analysis

The sensitivity of the NSP results to changes in factor costs, yields, and milling ratios is shown in Table 8.8. The social profitability of all techniques is most sensitive to changes in yields and in milling ratios. For the traditional and improved swamp techniques of the Casamance and the Matam village polders, the next most significant influence on social profitability rates is the variation in the cost of unskilled labor. This effect is most pronounced for on-farm consumption activities, where unskilled labor is by far the largest component of costs.[34] The impact of changes in these costs diminishes as paddy is delivered to off-farm consumption points owing to increases in transport, processing, and distribution costs, all of which use a higher proportion of skilled labor and capital. With techniques using greater degrees of mechanization, the impact of skilled labor, including drivers, operators, extension agents, and other technicians increases and exceeds the influence of changes in the cost of unskilled farm labor.

An unanticipated result is the relative insensitivity of all activities to changes in the price of capital. This outcome was expected for the traditional and labor-intensive techniques. However, the low capital elasticity

[32] For all techniques, rice processed on the informal market by small hullers has the highest rates of private and social profitability. This observation is consistent with findings in other West African countries (31).

[33] The import substitution prices in the Dakar market at which the various production techniques are profitable also indicate that labor-intensive techniques are the most efficient competitors:

Technique	U.S. $/mt (1975)	Technique	U.S. $/mt (1975)
Rainfed, gray soils	324	Improved swamp	386
Improved swamps		Nianga	437
with water control	363	Senegal River Delta	570
Matam	376		

[34] The classification of family farm labor as unskilled labor is somewhat arbitrary. Clearly, certain tasks such as transplanting and weeding involve skills.

TABLE 8.8. *Elasticities of Net Social Profitability with Respect to Yields, Milling Ratios, and the Social Costs of Primary Inputs*

Activity designator[a]	Yields	Milling ratio	Social costs		
			Unskilled labor	Skilled labor	Capital
1.1	2.137	2.137	−1.470	−0.348	−0.102
1.2	3.830	4.157	−2.068	−1.691	−0.527
1.3	2.924	3.448	−1.579	−1.334	−0.414
2.1	10.947	10.946	−4.106	−2.577	−0.688
2.2	2.389	2.630	−0.634	−1.029	−0.302
2.3	2.088	2.299	−0.554	−0.992	−0.270
3.1	1.666	1.749	−0.301	−0.587	−0.168
3.2	1.562	1.640	−0.282	−0.562	−0.160
4.1	16.399	16.399	−17.578	−0.087	−1.466
5.1	4.687	4.687	−3.755	−0.493	−0.439
5.2	4.671	4.970	−2.867	−0.758	−0.487
5.3	17.457	19.471	−12.106	−4.403	−2.634
5.4	4.884	5.448	−3.387	−1.320	−0.761
6.1	3.427	3.427	−2.367	−0.465	−0.559
6.2	28.111	31.672	−16.060	−8.352	−6.347
6.3	7.921	8.925	−4.525	−2.511	−1.832
7.1	1.783	1.783	−1.233	−0.338	−0.219
7.2	3.973	4.581	−2.175	−1.526	−0.787
7.3	17.732	20.442	−9.705	−7.236	−3.629

[a] For a more detailed description of activity designators, see Table 8.5.

coefficients of both mechanized techniques in the Fleuve are at first surprising. These low elasticities occur because of the relatively high cost of intermediate inputs and assumptions concerning the longevity of certain capital items and low rates of discount, which result in low annualized capital values.

Table 8.9 shows changes in NSP in relation to changes in the world market price of rice over a range from $250 to $600 per metric ton. No techniques are socially profitable at the low figure and all are at the high one.

These results underscore the importance of yield data and of labor times and wage information. Few consistent time series studies of yields are available in Senegal. Much of the available information is either for experimental plot yields or estimates of production over harvested area which exclude investments made on seeded but abandoned land. Both measures tend to overstate economic yields. SAED has recently begun annual crop-cuttings, although it is too early to obtain time-series which permit an assessment of yield stability. For the majority of the paddy produced in the country from individual traditional farms, yield information

TABLE 8.9. *Net Social Profitability in Relation to the World Market Price of Rice*
(CFA francs/mt milled rice)

Activity designator[a]	Percent brokens	Net social profitability for a world price of							
		$250/mt	$300/mt	$350/mt	$400/mt	$450/mt	$500/mt	$550/mt	$600/mt
1.1	40%	-2,363	10,262	22,887	35,512	48,137	60,762	73,387	86,012
1.2	80	-28,899	-16,274	-3,649	8,976	21,601	34,226	46,851	59,476
1.3	80	-30,938	-18,313	-5,688	6,937	19,562	32,187	44,812	57,437
2.1	40	-20,464	-7,839	4,786	17,411	30,036	42,661	55,286	67,911
2.2	80	-44,422	-31,797	-19,172	-6,547	6,078	18,703	31,328	43,953
2.3	80	-46,462	-33,837	-21,212	-8,587	4,038	16,663	29,288	41,914
3.1	80	-77,969	-65,344	-52,719	-40,094	-27,469	-14,844	-2,219	10,406
3.2	80	-80,008	-67,383	-54,758	-42,133	-29,508	-16,883	-4,258	8,367
4.1	40	-31,397	-18,772	-6,147	6,478	19,103	31,728	44,353	56,978
5.1	40	-13,203	-578	12,047	24,672	37,297	49,922	62,547	75,172
5.2	40	-11,351	1,274	13,899	26,524	39,149	51,774	64,399	77,024
5.3	45	-30,291	-17,666	-5,041	7,584	20,209	32,834	45,459	58,084
5.4	45	-33,332	-20,707	-8,082	4,543	17,168	29,793	42,418	55,043
6.1	40	-7,585	5,040	17,665	30,290	42,915	55,540	68,165	80,790
6.2	45	-24,673	-12,048	577	13,202	25,827	38,452	51,077	63,702
6.3	45	-27,714	-15,089	-2,464	10,161	22,786	3,541	48,036	60,661
7.1	40	2,306	14,931	27,556	40,181	52,806	65,431	78,056	90,681
7.2	45	-14,782	-2,157	10,468	23,093	35,718	48,343	60,968	73,593
7.3	45	-17,822	-5,197	7,428	20,052	32,678	45,303	57,928	70,553

[a]For a more complete description of activity designators, see Table 8.5.

is available only from highly inadequate estimates made by the regional agricultural services.

In the absence of survey data on labor-intensive techniques, unskilled labor data are more reliable from mechanized, pump-irrigated projects in which labor tasks are delimited by pumping periods and mechanized operations and are directed by the extension staff. For labor-intensive techniques, on the other hand, labor times exhibit larger apparent spatial and interannual variation. Confidence, of course, is greater where survey data are available, as in Matam, and less reliable for the traditional swamp techniques. Input data for skilled labor were collected from personnel records and other studies and are more accurate and reliable than figures on unskilled labor.

Milling ratios, another parameter to which NSPs are highly sensitive, have been well studied in large public mills, and the evidence suggests that there is little room for much improvement in efficiency.[35] For small hullers, on the other hand, milling information is sketchy. Cost estimates indicate that small hulling is the most efficient milling technique because of reductions in paddy transport costs and fuller capacity utilization, which lead to lower average fixed costs.

Concluding Remarks

Since the period of unusually high rice prices in 1973–75, Senegal has been concerned with fluctuations in the price of rice on the international market. The experience of that era has increased the government's commitment to self-sufficiency in rice, defined as a reduction in imports brought about through increases in local production. On average, three-fourths of Senegal's total annual rice consumption of 245,000 tons is met by imports. In years of unusually bad domestic harvests (1973–74) imports have reached 200,000 tons, whereas in good years (1975) they have declined to around 100,000 tons. Between 1961 and 1975 imports averaged 155,000 tons, requiring the expenditure of about 10 percent of annual export earnings (3, pp. 8–9).

Imported rice is destined almost exclusively for Dakar (Cap Vert) and the Groundnut Basin.[36] But these consumption centers lie at considera-

---

[35] Increased efficiency in public mills is likely to be associated with increased throughput. However, gains in efficiency will be modest because fixed costs are only a minor component of total costs in large mills.

[36] Data by region on consumption, production, and deficit or surplus (in tons) are from *4*, p. 10a; *26*, p. 100:

Region	Consumption	Production	Deficit (−) or surplus (+)	Region	Consumption	Production	Deficit (−) or surplus (+)
Cap Vert	96,429	0	−96,429	Eastern			
Groundnut				Senegal	5,342	4,272	−1,070
Basin	55,850	4,435	−51,415	Casamance	49,333	56,134	+6,801
Fleuve	20,573	11,193	−9,380	TOTAL	227,527	75,034	−151,492

ble distances from the main producing regions in the Casamance and the Fleuve. Hence, domestic transportation costs act as a barrier to the competitive delivery of local rice, especially in view of the high cost of many of the techniques used in producing that rice.

Efforts to expand rice production and therefore to increase self-sufficiency can be analyzed in terms of their contributions to three fundamental economic objectives—increased national income, more even income distribution, and greater food security. If the efficient allocation of resources and hence generation of national income is the principal criterion, techniques that minimize net social costs should be promoted. With current prices and production conditions, labor-intensive techniques in the Casamance and in Matam and animal traction methods in the Middle and Upper Casamance are the most efficient techniques for local import substitution. However, only the animal traction technique of the Casamance is competitive with imports in Dakar. All other rice-producing techniques are inefficient suppliers of the Dakar market, as shown by the figures in the following tabulation:

Technique	Location	Water source	Level of security	NSP in Dakar
Matam	Upper Fleuve	Pumping	High	−17.5
Nianga	Middle Fleuve	Pumping	High	−32.9
Delta	Fleuve Delta	Pumping	Low-med.[37]	−66.2
Improved swamp with water control	Middle Casamance	Bunding	Medium	−7.2
Rainfed animal traction	Middle and Upper Casamance	Gray soils	Med.-low	+2.7
Improved swamp manual	Casamance	Rainfed	Low	−12.8

Senegal's primary objective related to rice policy, however, is to enhance the security of food availability by reducing fluctuations in domestic supplies. In the Fleuve and the Casamance, food security can be increased through investments in small-scale irrigation. By increasing water security, higher and less variable yields can be attained and more intensive technologies introduced. But the least costly, small-scale techniques often do not have high sales per farmer and thus do not make locally produced rice available to other consumers in years of drought when millet production falls short. Marketings can most easily be increased only at a substantial efficiency cost by expanding production from double-cropped, large-scale irrigated projects, like Nianga, in the Senegal River Basin.

[37] Delta pumping, even on tertiary developments, is not very secure, since saline water incursion can lead to total abandonment of the fields or to delays in planting, which result in important reductions in yields and increases in costs.

Increasing rice production also has implications for regional income distribution. The drought severely affected the incomes of the poorer, peripheral regions of Senegal, and the government has since devoted attention to increasing regional incomes and to regional development. Investments in the Fleuve would raise net farmer incomes and increase production security. Although both Matam and Nianga exhibit large private profitability, indicating that farmers stand to gain from rice production, social profitability in Matam is much higher than at Nianga, suggesting that the distributional gains there have a lower cost in terms of national income. Furthermore, production at Matam requires less LDA intervention and assistance and, owing to its scale and labor intensity, results in a wider distribution of benefits.

Animal traction and improved swamp production techniques in the Casamance have the highest private and social profitabilities in that region. However, in the Casamance, other crops (corn, peanuts fruits) are both privately and socially more profitable than rice. To the extent that other crops have higher social profitabilities, rice policy would be an inefficient means of effecting regional development.

With currently known technologies and factor prices, Senegal can best advance its objectives of rice policy by promoting low-cost irrigated projects, following the Matam model, which increase and stabilize farmer incomes without incurring high social costs. This approach can be complemented by expanding output in the Casamance with efficient techniques using animal traction. But before any dams regulating flows of the Senegal River are built, Senegal should seek to develop loc-cost techniques using larger irrigated polders that are well adapted to its needs and resources.

Citations

1 Gora Beye, "Une Année d'Etude de l'Evolution de la Salinité dans la Cuvette de Boundoum Ouest." Centre de Recherches Agronomique, Bambey, Senegal, October 1967.

2 J. M. Boussard and J. Bourliaud, *Possible Consequence of Some Agricultural Policies for the District of Sédhiou (Central Casamance, Republic of Senegal); An Application of a Linear Programming Model.* Institut National de la Recherche Agronomique, Paris 1974.

3 Comité Inter-états de Lutte Contre la Sécheresse Sahelienne, (CILSS), Working Group of Marketing, Price Policy, and Storage, *Marketing, Price Policy and Storage of Food Grains in the Sahel,* 2 vols. University of Michigan Center for Research on Economic Development, Ann Arbor, August 1977.

4 Kathryn Craven and A. Hasan Tuluy, "Rice Policy in Senegal." Stanford/WARDA Study of the Political Economy of Rice in West Africa, Food Research Institute, Stanford University, Stanford, July 1979; Chapter 7.

5 Valy-Charles Diarassouba, *L'Evolution des Structures Agricoles du Sénégal: Destructuration et Restructuration de l'Economie Rurale.* Editions Cujas, Paris, 1968.

6 A. B. Diop, *Société Toucouleur et Migration*. Initiations et Etudes, vol. 18. Institut Français d'Afrique Noire–Université de Dakar, Dakar, 1965.

7 Robert W. Herdt and Thomas H. Wickham, "Exploring the Gap Between Potential and Actual Rice Yield in the Philippines," *Food Research Institute Studies*, 14, No. 2 (1975).

8 Hydroplan Ingenieur Gesellschaft, *Aménagement Hydroagricole du Périmètre de Nianga/Sénégal: Premier Tranche de l'Aménagement Intérieur, Rapport Final sur l'Execution des Travaux*. Saint-Louis, Senegal, 1975.

9 Institut Universitaire de Technologie, *Etude: Budget Consommation*, 2 vols. Université de Dakar, Dakar, June 1976.

10 International Bank for Reconstruction and Development, *Senegal: Tradition, Diversification, and Economic Development*. Johns Hopkins University Press, Washington, D.C., 1974.

11 ———, *Migration and Employment in Senegal, An Introductory Report*. Washington, D.C., June 1976.

12 International Fertilizer Development Center, *West Africa Fertilizer Study*, vol. 2, Senegal. Technical Bulletin IFDC-T-4, Florence, Alabama, April 1977.

13 International Monetary Fund, *Surveys of African Economies*, vol. 3. Washington, D.C., 1970.

14 Klaas de Jonge, Jos van der Klei, Henk Meilink, and Jan Roeland Storm, *Les Migration en Basse Casamance*. Rapport Provisoire, Afrika-Studiecentrum, Leiden, 1976.

15 Olga F. Linares, "Intensive Agriculture and Diffuse Authority Among the Diola of Senegal." Center for Advanced Study in the Behavioral Studies, Stanford, 1979.

16 Leon A. Mears, "The Domestic Resource Cost of Rice Production in the United States," *Food Research Institute Studies*, 15, No. 2 (1976).

17 J. P. Minvielle, *Migrations et Economies Villageoises dans la Vallée du Fleuve Sénégal*. Office de la Recherche Scientifique et Technique Outre-mer, Dakar, May 1976.

18 Organisation pour la Mise en Valeur du Fleuve Sénégal, *Aménagements Hydroagricoles dans le Bassin du Sénégal: Rythmes de Développement et Modulation des Crues*. Dakar, April 1977.

19 John M. Page, Jr., and J. Dirck Stryker, "Estimating Costs and Incentives." Stanford/WARDA Study of the Political Economy of Rice in West Africa, Food Research Institute, Stanford University, Stanford, July 1979; Appendix A.

20 Djibril Sène, P. Deleuse, and J. Birie-Habas, "Le Riz au Sénégal: Production et Recherche; Situation Actuelle et Perspectives," *L'Agronomie Tropicale*, 26, No. 4 (1971).

21 Sénégal, Government of, Ministère du Développement Rurale et de l'Hydraulique, Société d'Aménagement et d'Exploitation des Terres du Delta, "Bilan de l'Opération Hydroagricole du Delta et de la Basse Vallée." Saint-Louis, 1974.

22 ———, "Demande de Financement, Cuvette de Demet, auprès de la B.O.A.D." Saint-Louis, March 1977.

23 ———, "Projet d'Intensification de la Riziculture dans la Delta, Programme 1974/75. Demande de Financement au Fonds d'Aide et de Coopération de la Republique Française." Saint-Louis, December 1974.

24 ———, Société d'Aménagement et d'Exploitation des Terres du Delta, Bureau d'Etudes et de Programmation, *Campagne Rizicole, 76–77, Sondage de Rendements*. Saint-Louis, March 1977.

25 ———, Société d'Aménagement et d'Exploitation des Terres du Delta, Bureau d'Etudes et de Programmation, "Superficies de la Moyenne Vallée Potentiellement Irrigables en Contre-saison à Partir des Eaux du Fleuve Sénégal avant la Réalisation du Barrage de Manantali." Saint-Louis, December 1977.

26 ———, Ministère des Finances et Affaires Economiques, Direction de la Statistique, *Le Sénégal en Chiffres*. Société Africaine d'Edition, Edition 1976–77, Dakar, 1977.

27 ———, Ministère des Finances et Affaires Economiques, Direction de la Statistique, Bureau de Recensement, *Résultats Provisoires du Recensement Général de la Population d'Avril 1976*. Dakar, 1977.

28 Société d'Aide Technique et de la Coopération, *Développement de la Riziculture au Sénégal*. Rapport Général. Paris, 1968.

29 Dunstan S. C. Spencer, I. I. May-Parker, and F. S. Rose, "Employment, Efficiency and Income in the Rice Processing Industry of Sierra Leone." African Rural Economy Paper No. 15, University of Sierra Leone–Michigan State University, East Lansing, 1976.

30 Stanford/WARDA Study of the Political Economy of Rice in West Africa, "Prospect of Intraregional Trade in Rice in West Africa." WARDA/77/STC7/9, West Africa Rice Development Association, Monrovia, September 1977.

31 ———, "The Political Economy of Rice in West Africa: A Preliminary Summary of the Results from the WARDA/Stanford Study." Food Research Institute, Stanford University, Stanford, August 1978.

32 J. Dirck Stryker, "Comparative Advantage and Public Policy." Stanford/WARDA Study of the Political Economy of Rice in West Africa, Food Research Institute, Stanford University, Stanford, July 1979; Chapter 12.

33 J. Dirck Stryker, John M. Page, Jr., and Charles P. Humphreys, "Shadow Price Estimation." Stanford/WARDA Study of the Political Economy of Rice in West Africa, Food Research Institute, Stanford University, July 1979; Appendix B.

34 Abraham S. Waldstein, " 'Peasantization' of Nomads and 'Nomadization' of Peasants; Responses to State Intervention in an Irrigated Agricultural Development Scheme in the Senegal River Delta." n.p., 1978, mimeo.

35 West Africa Rice Development Association, "Rice Project Managers Meeting." Seminar Proceedings 5, Monrovia, January 1976.

36 ———, "Presentation Succinte de la Société d'Aménagement et d'Exploitation des Terres du Delta (SAED) en République du Sénégal." Monrovia, June 1976.

PART FIVE

Mali

9. Rice Policy in Mali

John McIntire

The production of rice and other cereals stagnated after Mali, formerly French Soudan, became independent in 1960. Early policy relied on an expansion of collectivist agricultural production in the face of inadequate producer incentives and consequently neglected the new nation's two unique advantages in rice production. The first of these advantages was the Office du Niger, the largest irrigation scheme in West Africa, initiated by the French in the 1930's to produce cotton for export and rice for local consumption. The Office suffered from organizational and agronomic problems that the French and the first government were unable to solve. The second advantage was the potential for large-scale, low-cost rice land development in the flood plain of the Niger River.

The new government, which came to power in 1968, changed Mali's approach to expanding rice production and, after a temporary setback due to several years of drought, its policies enabled the country to regain self-sufficiency and even to export rice. An evaluation of Malian rice policy and of the country's options for the future is the subject of this study, which is organized into four sections: a description of the economic geography of Mali, a historical analysis of the rice policies of the governments since independence, an evaluation of current problems and policy options, and a prognosis of future policy developments.

Economic Geography

Data on the population of Mali, distributed into its six administrative zones, are shown in Table 9.1. Although there was no complete census until 1976, Mali's population is estimated to be growing at an annual rate of 2.5 percent, based on comparisons with partial surveys done in 1960 and 1967. Population density is low generally, and the vast Saharan section of the country is almost uninhabited. The population depends on the wetter sub-Saharan zones (the northern limit of which is defined roughly by a line extending between the cities of Kayes and Mopti), and that dependence appears to have grown since the late 1960's.[1] There is a good

[1] The Sahelian zone is defined as that area receiving between 250 and 500 millimeters of rain annually. When speaking of agricultural seasons, the expression 1974–75, for example, will be written 1975—that is, the year in which harvests became available for consumption.

TABLE 9.1. *Regional Distribution of Population, 1976*
(000)

Region	Total	Rural	Urban[a]	Density (per km²)
Kayes	871	797	74	7.3
Bamako	1,320	875	445	14.7
Sikasso	1,172	1,097	75	15.3
Ségou	985	897	88	17.5
Mopti	1,236	1,182	54	13.9
Gao	724	673	51	0.9
TOTAL	6,308	5,521	787	5.1

SOURCES: *19* and density of regional areas from *18*.
[a]Only those populations shown in the census as living in "communes" (the central, incorporated districts of the largest towns), of which there are 13; the urban estimate is probably low for that reason.

deal of dry season (January to May) migration out of Mali toward the coastal countries, especially the Ivory Coast, and some more permanent loss of population has occurred in the regions near Timbuktu and Kayes.

The urban proportion of the population grew from approximately 9 percent in 1967 to nearly 13 percent in 1976; because there is some reason to believe that the rural population was underrepresented in the 1967 survey, it is possible that the urban share has, in fact, increased more rapidly than these numbers imply.[2] Most of the urban migration has been to Bamako, the capital city, which is perhaps seven times as large as the next largest city, and it appears that there have been important movements toward the city of Mopti as well.

Mali, one of the poorest countries in the world, has a per capita income (in 1975 prices) of $85. Real gross domestic product has grown at roughly 3.5 percent annually since independence in 1960, and per capita income has grown at slightly more than 1 percent. The rural population has incomes of between $50 and $90, with an average of perhaps $60, and it depends largely on the subsistence rainfed cultivation of millet or sorghum. The wealthiest rural inhabitants are in the cotton and peanut areas, although livestock is a significant part of rural wealth, especially in the more arid northern zones.

National cropped areas and a recent regional division are shown in Tables 9.2 and 9.3. Areas, yields, and outputs of the major crops are shown in Table 9.4. Hectarage and production figures are hard to interpret as a consistent series because of the drought, but the patterns of output and of

[2]A 1967 population survey was done in Mali as part of the annual *Enquête Agricole* and was used for demographic projections in the 1974–78 Five-Year Plan. That survey has been used here to estimate changes in population according to ecological zones and rural or urban location.

TABLE 9.2. *National Distribution of Cropped Areas, 1968–74*
(000 ha)

Crop	1968	1969	1970	1971	1972	1973	1974
Millet/sorghum	882	580	546	1,008	574	736	668
Rice	198	162	133	172	169	167	185
Cotton	76	91	76	75	79	86	48
Peanuts	126	101	96	148	151	147	127
Corn	31	25	33	77	35	27	44
TOTAL[a]	1,788	1,805	1,673	1,696	1,748	1,775	1,553
Hectares/person[b]	0.35	0.34	0.31	0.30	0.31	0.30	0.26
Hectares/member rural population	0.38	0.38	0.34	0.34	0.34	0.34	0.29

SOURCE: *18.* Surveys after 1974 are not yet tabulated.
[a] Includes intercropped areas, which are mostly mixes of millet and cowpeas.
[b] Populations are estimated by assuming a constant annual rate of growth of 2.5 percent, using the 1976 census as the base year and extrapolating backward. Rural populations are estimated by assuming a constant annual rate of growth of 2.0 percent, using the 1976 census estimate of rural population as the base and extrapolating backward.

TABLE 9.3. *Regional Cropping Patterns, 1974*
(000 ha)

Crop	Kayes	Bamako	Sikasso	Ségou	Mopti	Gao	Total
Millet and sorghum	138	72	87	155	200	15	667
Rice	1	9	9	74	39	13	145
Cotton	—	11	29	6	2	—	48
Peanuts	82	7	15	17	6	—	127
Corn	21	4	10	3	5	—	23
Other[a]							523
TOTAL	242	103	150	255	252	28	1,553

SOURCE: Figures are from Mali, Government of, Ministère de Développement Rural, Institut d'Economie Rurale, *Rapport de l'Enquête Agricole*, Bamako, 1973–74.
[a] Includes cowpeas, fonio, and cassava in pure stands, plus millet or sorghum and cowpeas in mixed culture.

regional specialization are evident. Millet and sorghum[3] cover the largest areas, and the principal cash crops—cotton, peanuts, and rice in the Office du Niger—occupied only about 25 percent of the total average cropped area between 1968 and 1975. Yields of millet are low, between 600 and 800 kilograms (kg) per hectare (ha), and vary directly with rainfall. Yields of cotton, between 1,000 and 1,200 kg, and of peanuts, roughly 800 to 1,000 kg, have increased lately as a result of the introduction of new techniques in development projects.

The regional distribution of crops follows the distribution of rainfall,

[3] The word millet will be used for the words millet and sorghum as a matter of convenience, except when it is necessary to make a careful distinction between these two different crops.

TABLE 9.4. *Area Planted, Yield, and Production of Major Crops, 1961–76*

Year	Rice Area (ha)	Rice Yield (mt/ha)	Rice Production (mt)	Peanuts Area (ha)	Peanuts Yield (mt/ha)	Peanuts Production (mt)	Millet and sorghum Area (ha)	Millet and sorghum Yield (mt/ha)	Millet and sorghum Production (mt)	Cotton Area (ha)	Cotton Yield (mt/ha)	Cotton Production (mt)
1961	170	1.088	185	n.a.	n.a.	122	n.a.	n.a.	850	n.a.	n.a.	n.a.
1962	182	1.016	185	n.a.	n.a.	138	n.a.	n.a.	820	n.a.	n.a.	n.a.
1963	200	0.950	190	n.a.	n.a.	167	n.a.	n.a.	870	n.a.	n.a.	n.a.
1964	123	1.341	165	n.a.	n.a.	182	n.a.	n.a.	865	n.a.	n.a.	n.a.
1965	159	0.997	158	144	1.028	148	859	0.789	678	89	0.371	33
1966	169	0.959	162	122	1.254	153	830	0.869	721	76	0.237	18
1967	178	0.890	159	129	1.232	159	910	0.810	737	62	0.452	28
1968	198	0.857	169	140	0.850	119	1,035	0.802	830	76	0.487	37
1969	162	0.836	135	129	0.744	96	932	0.598	557	91	0.604	55
1970	133	1.174	157	118	1.152	136	745	0.809	603	76	0.605	46
1971	172	0.977	168	162	0.963	156	725	0.986	715	75	0.747	56
1972	169	1.348	158	174	0.874	152	1,258	0.568	715	79	0.899	71
1973	167	0.527	88	160	0.843	135	900	0.692	623	86	0.849	73
1974	185	0.724	134	n.a.	n.a.	132	n.a.	n.a.	660	69	0.797	55
1975	198	0.914	181	n.a.	n.a.	188	n.a.	n.a.	850	68	1.044	71
1976	224	1.116	250	n.a.	n.a.	205	n.a.	n.a.	865	87	1.149	100

SOURCES: For area, yield, and production in rice, 28 and 29. For areas in peanuts, millet and sorghum, and cotton, Center for Research on Economic Development, University of Michigan, *Mali Agriculture: Sector Assessment*, 1976; crop areas in that study appear to be based largely upon results from *Enquêtes Agricoles*. For production in peanuts, millet and sorghum, and cotton, 4. Yields in peanuts, millet and sorghum, and cotton are calculated from aggregate areas and production figures, and should only be taken as indicating orders of magnitude.

MAP 9.1. Mali, Showing the Principal Rice-Producing Region.

soils, and floodwater availability. Peanuts are grown on the eroded soils of northwestern Mali, which are generally suited to very extensive agriculture and to stock grazing (see Map 9.1). The productivity of those soils varies directly with rainfall, and there is little river or groundwater available to supplement rainfall. Cotton is grown on ferruginous soils to the south and east of Bamako toward Sikasso, which straddle the Niger River near Ségou and cover the Sahelian zone above the city of Kayes. In addition to cotton, these soils support more intensive cultivation of millet than do the eroded soils of the peanut zones, and they also support small amounts of rainfed rice in southern Mali. Nearly all of the rice in Mali is grown on the hydromorphic soils of the Niger and Bani river basins, which support the greatest densities of rural human and animal populations in the country. Little cotton or peanuts are grown there, and the opportunities for cash incomes are restricted to fishing, herding, and rice cultivation.

Annual rainfall at the six regional capitals during a 30-year period is as follows (in millimeters): Gao, 260; Mopti, 520; Kayes, 720; Ségou, 720; Bamako, 1,050; and Sikasso, 1,300. Cotton production is centered in the comparatively moist Sikasso region, peanut production in the somewhat drier Kayes region, and rice production in the arid delta around Mopti. The relation between rainfall and population density is obscured somewhat because the wetter Sikasso region is the site of endemic human and animal diseases (onchocerciasis and trypanosomiasis), which limit crop production in river basins, and the floods of the Niger and Bani support many people in a wide zone that receives less than half the rainfall of the Sikasso region.

Rice Production

Table 9.5 describes the types of rice cultivation in Mali. The first is limited to the semipermanent swamps around some of the main tributaries of the Niger, Bani, and Senegal rivers and occupies perhaps 15,000 ha. There is no water control in this system, and yields vary from 1 to 2 metric tons (mt)/ha without the use of improved seeds or fertilizers. Attempts have been made to introduce animal traction, improved Asian *Oryza sativa* varieties, and water control through simple diversion dams in the Opération Riz Pluvial et Bas-Fonds, Sikasso, run by the Compagnie Malienne pour le Développement des Fibres Textiles (CMDT), and in the Opération Petits Périmètres in Kayes. Although studies have asserted that the land potential of this type of cultivation approaches 200,000 ha in southern Mali, expansion of this area is now limited by the onchocerciasis noted earlier (*11*) and by the poorly developed infrastructure of the region.

Rainfed rice in southern Mali is the second type. It is closely associated with the swamp technique in that area and has also benefited from the work of the Sikasso rice project. Paddy yields using the swamp technique are 1 to 1.5 mt/ha, and they are expected to reach 2.5 mt with the improved system. The rainfed and swamp systems are essentially subsistence cultures; the aim of policy is to provide greater water security and higher, less variable, incomes to the participating farmers rather than to develop large marketed supplies for urban consumption.

The third and historically most important type of rice cultivation is the uncontrolled flooded culture of the Niger-Bani Delta north and west of Mopti (*7, 13*). That system has been the principal object of recent investment in Malian rice production and has been, along with the Office du Niger, the source of much of the recent growth of rice output. The traditional flooded system uses African *Oryza glaberrima* seed varieties, manual cultivation, and no chemicals or machines. Yields of paddy are 500 to 700 kg/ha and paddy is hand-pounded for home consumption. The tradi-

TABLE 9.5. *Key Characteristics of Rice Production Techniques*

Production technique	Area, 1976 (ha)[a]	Gross paddy yields, 1976 (mt/ha)[b]	Paddy production, 1976 (mt)	Water control	Means of land preparation	Improved seeds	Fertilizer	Pesticides
Gravity irrigation, Office du Niger	39,922	2.25	89,425	diversion dam	oxen	yes	yes	no
Controlled flooded, Ségou	34,355	1.58	54,281	partially controlled flooded	oxen	yes	no	no
Traditional swamp and rainfed	11,000	1.20	13,200	none	manual	no	no	no
Improved swamp and flooded, Sikasso	4,000	1.80	7,200	small diversion dam	oxen	yes	no	no
Traditional flooded, Delta	110,000	0.50	55,000	unimproved flooded	oxen & manual	no	no	no
Controlled flooded, Mopti	16,074	1.15	18,485	partially controlled flooded	oxen	yes	no	no

SOURCES: For gravity irrigation, Office du Niger, Bureau of Economic Affairs (Ségou). For controlled flooding, Opération Riz Ségou, Direction Générale (Ségou). For traditional swamp, Opération Riz Pluvial et Bas-Fonds, Direction Générale (Sikasso). For traditional flooded and controlled flooded, Opération Riz Mopti, Direction Générale (Sévaré); this includes some swamp rice production in Kayes region.

NOTE: One crop per year for all techniques, and manual harvesting everywhere.

[a] Seeded areas.
[b] Per seeded hectare.

tional system is being replaced in Opérations Riz Ségou and Mopti by an improved technique which provides limited water control in diked polders,[4] *Oryza sativa* seed varieties, animal traction, some mechanical threshing, and small amounts of fertilizer. Output from the improved areas is roughly 1.5 mt/ha, some of which is milled industrially for urban consumption. This improved system offers the advantages of low capital and maintenance costs and the disadvantage of poor water security. The potential of the flooded technique, traditional and improved, is estimated to be as large as one million ha, and its expansion to date has probably been at the same rate as population growth. Replacement of the traditional system by the improved system is constrained by the supply of foreign financing and by the risk associated with the development of poorly secured polders.[5]

The fourth type of rice cultivation is the irrigated technique at the Office du Niger. The basis of the system is a diversion weir at Markala, roughly 250 kilometers (km) downstream of Bamako and an 8-km main canal with two primary canals. Although the Markala barrage cannot store water, it does provide full water control to all the rice fields.[6]

The Office is a very extensive system with average holdings of 9.5 ha (the modal holding is about 7 ha) and some holdings of as much as 80 ha. Office farmers are more specialized in rice production than are rice farmers in the adjacent Opération Riz Ségou who have dry crop holdings equal on average to their rice fields. Yields in the Office are the highest in Mali: the average in three of the four sectors is nearly 2.5 mt/ha.[7] The Office discontinued mechanized state farms in the early 1970's and has since encouraged the use of animal traction and farmer management. The Office's management has also stopped trying to utilize fully its existing irrigation capacity; it has reduced the area sown to 38,000 ha (from a maximum of 40,500 ha in 1974–75), thus stalling the trend of

[4]The control is against too rapid entry or too early recession of floodwater. If the flood does not arrive, or if it arrives late, there is no way to get water to the fields. The systems can be adapted to pumping, but it would probably be necessary then to dig secondary canals and drains and to level the fields.

[5]Both rice projects, Opération Riz Ségou and Opération Riz Mopti, are financed by external donors. Though the polders in both projects are designed to be filled completely in nine out of ten years, filling rates vary widely between polders and some appear to be badly designed. The size of the first Mopti project (financed by the International Development Association) was reduced from 31,000 to 26,000 ha in order to improve the filling rates of some of its polders.

[6]See the history of the Office in John de Wilde (32, pp. 245–300). Cotton cultivation was discontinued in the Office as the result of agronomic problems, but it may be started again in the Office in an area that is not now planted to rice. Water is pumped onto the 3,000 ha of sugarcane in the Office.

[7]Yields in the Kolongo sector of the Office have usually been poorer than yields in other sectors because of drainage problems.

growth of cultivated rice area that took place after the termination of cotton cultivation in 1970.[8]

Collection, Milling, and Marketing

There are two channels, public and private, of rice milling and marketing in Mali. The public channel is the legal monopoly of the Office des Produits Agricoles du Mali (OPAM) and the government rice projects. Paddy is collected from farmers by OPAM or by the rice projects, such as the Office du Niger, at official prices which do not vary by region, system of production, or season, and is transported to mills at official cost-price schedules. These schedules include charges for losses, sacks, collection, transport, and handling.[9] Paddy produced in the Office du Niger is milled there and is the property of the Office until it leaves the mills, whereupon it is sold to OPAM at an official ex-mill price. Paddy from the other rice projects or from independent farmers is delivered by the projects to the OPAM mills, where it becomes the property of OPAM.

Milling capacity is shown in Table 9.6. Hand pounding dominates in the traditional rice areas of the Delta, as well as in remote producing areas along the loop of the Niger and in swamp rice areas in Kayes and Sikasso. Industrial mills are most important in the Office du Niger, where they have operated since 1948 and where capacity has usually been adequate to process all of the officially marketed paddy. Several of OPAM's mills that have not been used for years are now being rehabilitated because milling capacity has recently been inadequate to meet the growth of production from the Office and the flooded rice projects. Milling ratios in the industrial milling sector are good, although the quality of output is poor, with high percentages of brokens and impurities.

OPAM is responsible for the transport of rice to consumption centers. Until the spring of 1977, OPAM did not have sufficient truck capacity to move cereals and hired private truckers to supply urban consumers. Truckers were paid at official rates that, after their increase in 1974, seemed to cover costs and to provide a reasonable return to capital. More than three-fourths of OPAM's supplies of rice and other cereals are sent to Bamako, the largest city, and to the regions of Kayes and Gao, which are

[8] Areas cultivated to rice in the Office increased more or less continuously from 1945 to 1961. They stagnated in the 1960's, increased again in 1970 after cotton cultivation was stopped, and have been restricted lately as part of a policy to reduce the average size of holdings.

[9] Paddy prices vary by quality. The highest price is paid for white paddy, the lowest for red, and there is an intermediate price for mixed red and white. All of the paddy produced in the rice projects, with the exception of a small amount in Mopti, is white. Project officials complain that the cost-price schedules (*barèmes*) do not cover the costs of paddy collection and transport.

TABLE 9.6. *Rice Milling Capacity in Mali*

Milling technique	Annual capacity[a] (mt/paddy)	1976 output (mt/rice)	Ownership	Milling ratio	Percent broken
Industrial mills[b]					
Kourouba	6,000	0	OPAM	n.a.	n.a.
Tamani	11,000[c]	n.a.	private	n.a.	n.a.
Diafarabé	13,800	0	OPAM	n.a.	n.a.
Sévaré	15,000[c]	8,933	OPAM	0.63	50%–70%
Molodo	18,000	12,500	Office du Niger	0.65	60
Kolongotomo	12,000	4,325	Office du Niger	0.64	70
N'Diebougou[d]	18,000	0	Office du Niger	n.a.	n.a.
Kourouma	18,000	12,400	Office du Niger	0.67	60
Small hullers					
100–150[e]	38,000–56,000	n.a.	private	0.45–0.70	60–70
Hand pounding[f]	n.a.	50,000–70,000	rural women	0.70	80

SOURCES: Office du Niger mill data are from Office du Niger, Bureau of Economic Affairs (Ségou); OPAM and Tamani mill data are from OPAM Directorate (Bamako) and from Opération Riz Mopti (Sévaré).

[a]There are plans to construct a 20,000 mt (paddy) capacity mill at Dioro as part of Opération Riz Ségou and a 12,000 mt (paddy) capacity mill as part of Opération Riz Pluvial et Bas-Fonds, Sikasso, in addition to plans to double the capacity of the Sévaré mill by 1981.

[b]Hourly capacities of industrial mills have been converted to annual capacities on the assumption that mills work 5,000 hours per year (i.e. 20 hours/day and 250 days per year).

[c]These mills have parboiling capacity. Roughly 6 percent of the Sévaré mill's 1976 output of rice was parboiled.

[d]This mill was not installed until the end of the 1976–77 crop season.

[e]Diesel or electric machines with hourly capacities of 0.15 mt/paddy. Annual capacity has been estimated from the assumption that these machines work 2,500 hours/year. See WARDA, 30.

[f]This category includes traditional parboiled rice.

TABLE 9.7. *Distribution of Shares of Marketed Paddy, 1961–77*

Year	(1) Paddy output *(000 mt)[a]*	(2) Official marketings *(000 mt)*	(3) Office du Niger marketings as percent of (2)	(4) Official marketings as percent of (1)	(5) Residual supply *(mt)* (1) − (4)	(6) Residual rice *(mt)* (5) × 0.65
1961	148	45	83%	30%	103	67
1962	147	26	91	18	121	79
1963	159	33	68	21	126	82
1964	133	31	88	23	102	66
1965	125	27	83	22	98	64
1966	128	28	89	22	100	65
1967	123	37	71	30	86	56
1968	135	40	77	30	95	62
1969	108	32	79	30	76	49
1970	124	49	74	40	75	49
1971	134	42	91	31	92	60
1972	126	51	91	40	75	49
1973	61	50	94	82	11	7
1974	101	57	95	56	44	29
1975	141	83	78	59	58	38
1976	203	93	69	46	110	71
1977	n.a.	93[b]	70	n.a.	n.a.	n.a.

[a] Losses estimated at 10 percent of paddy output. Seed consumption estimated at 100 kg/ha multiplied by hectares planted in rice in the following crop year. Because hectarage and production data were unavailable for 1976–77, it was necessary to calculate seed consumption for 1975–76 as a function of hectarage planted in that year.

[b] Estimated on the basis of data provided by the Office du Niger, Bureau of Economic Affairs (Ségou); Opérations Riz Ségou and Mopti.

usually in deficit. OPAM incurs losses in supplying those regions because it is unable to vary its retail prices with the costs of transport, and profits from supplying more accessible regions do not cover those losses.

Although private cereals trade is banned in Mali, it is important in the collection and milling of paddy and in the transport and marketing of milled rice. Farmers bring paddy and rice to village markets, where it is bought by traders who arrange for processing and/or transport to markets. There is little direct entry by traders into villages except on weekly market days. Paddy traders are usually not hullers and the owners of hullers usually do not sell rice. Most private hullers are small (150 kg/hour of paddy) diesel or electric machines, and there is an increasing tendency to locate them near producing areas, especially the improved polders of the Ségou project.

Although the private trade is legally repressed, there is in practice some official reliance upon it, especially upon the hullers. There are no extensive studies of the private trade and therefore no reliable estimates of marketed quantities, prices, or margins.[10] Table 9.7 presents estimates of paddy production and of official marketings to estimate the amounts available for farm consumption or private marketings in the years immediately after independence.

[10] Studies of cereals marketing in Mali include 2, 4, and 6.

Consumption and International Trade

Rice consumption is low in Mali, on average less than 20 kg per capita, and rice's share in caloric intake is less than 15 percent.[11] Average consumption has been constant since independence except during the drought, when it increased somewhat owing to large imports. Consumption is highest in the producing areas (more than 100 kg per capita in the Office du Niger and in the traditional rice zones of the Delta) and in Bamako and other urban markets. Rice is, on the other hand, nearly unknown in the more remote rural areas of Kayes and Gao as well as in the drier zones near Mopti, where millet is the principal crop.

Malian rice imports are controlled by OPAM, and exports are managed both by OPAM and by the Office du Niger. Mali exported rice in the colonial era, usually from the Office du Niger to Senegal and, after independence until 1965, to the Ivory Coast. There was no recorded rice trade from 1966 to 1968, although rice imports, which might otherwise have occurred, were prevented by foreign exchange shortages. Imports began in 1969 and attained their peak of 70,000 metric tons in 1974. Small amounts of rice and paddy were exported by OPAM and by the Office du Niger to the Ivory Coast, Upper Volta, and Togo in 1976–77.[12] Trade in both directions is in 25–35 percent brokens, and Malian exports have historically been viewed as being of poor quality.

Rice Policy Between 1960 and 1968

The principal objective of Malian economic policy between 1960 and 1968 was to achieve a socialist transformation of the economy and society. The basis of the transformation was a system of planning, instituted with the assistance of French Marxist advisers. Major policies included nationalization of industries, establishment of state industrial and commercial firms, and large public investments in transportation. The new government set up producer and consumer cooperatives, banned both national and international private commerce, and established a group of state monopolies. Orthodox French political influence and commercial presence declined greatly in the period and, with the exception of that in Guinea, Mali's socialist program was the most radical and autarkic in French West Africa.[13]

---

[11] Even this low share may be an overstatement because rice output statistics are collected with comparative thoroughness, whereas the production of some foods—cassava and yams, for example—is not well surveyed. Table 9.8, below, presents a partial series of rice's share in consumption relative to those of millet and sorghum. Studies of food demand in Mali are 13, 15, and 25.

[12] Detailed data on Mali's international rice trade position are available upon request.

[13] Accounts of Malian history in this period are in Jones (10), Foltz (5), and Amin (1).

The main objective of Malian agricultural policy was to supply cheap cereals to urban populations in order to hold down wages in the new state enterprises. Another important objective was to increase exports of cotton and peanuts to finance imports of capital goods. The government intended to set up collective fields in every village, a scheme based on the premise that collective agriculture was the original and most productive African form. Other instruments were a system of fixed producer prices for cereals and export crops, fixed consumer prices for cereals, and the establishment of state agencies—OPAM with the legal monopoly on cereals trade, and the Société Malienne d'Importation et d'Exportation (SOMIEX), with the legal monopoly on the marketing of export crops. Policies were constrained by lack of financial and organizational capacity of the marketing agencies.[14] Rice received a large share of agricultural investments in the first National Plan (1961–65) and has consistently been the central instrument of cereals policy throughout the country's brief history.

Table 9.8 shows official producer prices of millet, paddy, peanuts, with cotton and gives official retail prices in parentheses. Price policy was driven by the need to keep prices low in order to derive the maximum amounts of export tax revenue from cotton and peanuts and to provide cheap domestic cereals. The official producer price of paddy was lowered after independence to maintain Mali's competitiveness in the formerly protected export markets of the franc zone; the price of millet was lowered also.[15] The effects of producer prices on farm incentives were not considered. Rather, it was hoped that collectivization and the expected reduction in transport costs and commercial margins would increase crop supplies in spite of the lower prices.

Official purchases of cereals between 1961 and 1968 are shown in Table 9.9. In the first years after independence, official marketings of millet and rice, plus net foreign trade in millet, wheat, and rice, were less than 10 kg per capita of the total population and 100 kg per capita of the urban population.[16] Nearly all of the officially marketed rice came from the Office du Niger, where control of access to the irrigation system enabled the Office's administration to collect most of the paddy grown there. Before the price increases of 1967 and 1968 OPAM was able to buy at most 6 percent of the total output of millet at a time when production, both absolutely

[14] The French firm, Compagnie Française pour le Développement des Fibres et Textiles (CFDT), operated the cotton production and marketing agency under an agreement with the new government. CFDT was exceptional because it had good financial and organizational capacity for marketing.

[15] Mali exported rice to Senegal before independence. See Jones (*10*, pp. 300–301) for a discussion of price policy after independence.

[16] However, Mali's record of official purchases of cereals is better than that of the other Sahelian countries. See CILSS (*4*, vol. 1).

TABLE 9.8. *Official Producer and Retail Prices for Selected Crops, 1959–77*
(Malian fr/kg)

Year	Official producer price for millet/ sorghum[a]	Official retail price for millet/ sorghum	Official producer price for paddy[b]	Official retail price for rice[c]	Official producer price for cotton[d]	Official producer price for peanuts[e]
1959	16	n.a.	12.5	n.a.	34	14.8
1960	11.5	n.a.	12	n.a.	34	15.6
1961	10	16.5	9	30	34	14
1962	10	16.5	9	30	34	14
1963	10	16.5	11	37	34	14
1964	10	16.5	11.5	42	34	14
1965	11	17.5	12.5	45	34	14
1966	11	17.5	12.5	45	34	14
1967[f]	16	25	16	57	34	16
1968	16	25	18	56	40	24
1969	16	25	18	56	40	24
1970	18	25	25	76.5	45	30
1971	18	35	25	76.5	50	30
1972	20	35	25	80	50	30
1973	20	35	25	88	50	30
1974	32	51	25	88	50	30
1975	32	51	40	111.5	50	40
1976	32	51	40	111.5	75	40
1977	32	51	40	111.5	75	45

SOURCES: *17*, except for 1959 and 1960, which are from *7*, p. 300.
NOTE: All years are the second year of the crop season; i.e., 1967–68 is written 1968.
[a] Mali has never established separate prices for millet and sorghum. Before 1967 there were separate prices by surplus or deficit region; the prices shown here are surplus region prices. To give an idea of the difference, in 1960 the "surplus" price was 16 and the "deficit" price 11.5.
[b] Paddy quality is white paddy.   [c] Rice quality is 40 percent brokens.
[d] First quality seed cotton.   [e] Peanuts in the shell.
[f] The Malian franc was devalued by 50 percent in 1967.

and per capita, was high. The disproportionate share of rice in official marketings was therefore the result only of the special control of the Office's production because millet, then as now, provided a much larger share of total cereals output.

Estimates of the urban availability of cereals are presented in Table 9.10. Total production, adjusted for seed, field losses, and official marketings, was divided by estimated rural population to estimate rural per capita availability. Cereals imports were added to official marketings and divided by urban population to estimate urban per capita availability on the restrictive assumption that imports and official supplies were sent only to urban areas. The harvests of 1961–64 were excellent and, although there are no surveys of privately marketed supplies, it is evident that the estimated gap between rural and urban per capita cereals availability, as shown in Table 9.11, should have encouraged trade. That impression is reinforced by OPAM's poor performance, as suggested by its financial

TABLE 9.9. *Official Marketings of Cereals and Cereal Products, 1961–68*
(000 metric tons)

Year	Domestic trade in:			International trade in:[d]		
	Millet/ sorghum	Rice (OPAM)[a]	Rice (Office)[b]	Millet/ sorghum	Rice	Flour[c]
1961	n.a.	n.a.	24.4	—	−12.6	4.3
1962	20	2	15.3	—	−15.0	6.3
1963	29	7	15.0	2.4	−5.2	9.7
1964	16	2.5	17.5	—	−1.1	3.9
1965	17	3	14.6	—	−3.7	20.4
1966	26	2	16.1	6.8	—	12.3
1967	56	7	16.8	10.0	—	10.0
1968	60	6	19.8	−8.0	—	10.0

SOURCES: Data for rice (Office) are from Office du Niger, Bureau of Economic Affairs (Ségou). Data for flour imports are from Food and Agriculture Organization, *Trade Yearbooks*, Rome, various years. All other data are from *10*, p. 367, and *4*, vol. 2.
    [a] Includes hand-pounded and parboiled rice purchased by OPAM (which began operation in 1962) as well as paddy purchased and converted to rice equivalent by using a milling ratio of 0.65.
    [b] Includes paddy purchased at the Office du Niger and converted to rice equivalent using a milling ratio of 0.65.
    [c] Includes wheat converted to flour at a milling ratio of 0.72.
    [d] Positive values are imports; negative ones are exports.

losses and by the growing gap between official and market retail prices.[17]

The government could not maintain cereal prices, either in the cooperative markets, which it controlled directly through OPAM, or in the private markets, which it tried to influence through its cooperative buying and selling policies. This is apparent in Table 9.11, which shows the evolution of cooperative and market prices of rice and millet, cooperative and market food price indexes, and the general price level. Although the table is necessarily incomplete, it does indicate that the government was increasingly less able to defend its retail price of rice and that only the increased marketings of 1967–68, due to the producer price increases, enabled it to defend its millet price. Consumers who could buy in the cooperatives bought at less than the market prices of cereals.

Nominal protection coefficients, defined as the ratio of domestic to c.i.f. prices, have been calculated for the entire period after independence. The results, presented in Table 9.12, show the bias of price policy toward consumers who bought in the cooperatives between 1960 and 1968. These consumers were better off with respect to the world price of rice than were consumers who had to buy in the private market.[18] When the coefficients are calculated with the c.i.f. price at Abidjan in the denomina-

[17] OPAM's losses are noted in Jones (*10*, pp. 239, 250).

[18] The bias observed in the consumer price policy of the first government was intended, but it was partly ineffective because of OPAM's incomplete control of cereals marketings. Consumers were organized into cooperatives, for which there was a membership fee per family, and each family was allowed a quota of purchases in the cooperatives.

TABLE 9.10. *Per Capita Cereals Availability, Selected Years 1961–75*
(Kilograms)

Crop	1961	1963	1965	1967	1969	1971	1973	1975
Millet and sorghum:								
Rural[a]	187	179	140	169	101	122	104	133
Urban[b]	0	62	41	140	47	40	105	65
Total	187	169	130	166	96	114	104	124
Percent imports in total	0%	0%	0%	1%	3%	2%	0%	0%
Rice:								
Rural	17	20	15	12	10	12	2	7
Urban	49	44	33	50	77	71	119	99
Total	19	21	15	16	17	18	15	18
Percent imports in total	0%	0%	0%	0%	22%	15%	54%	18%
Total:								
Rural	204	199	155	181	111	134	105	140
Urban	49	106	74	190	124	111	224	164
Total	206	190	145	182	113	132	119	142
Percent imports in total	0%	0%	0%	1%	6%	4%	15%	2%

SOURCES: Calculated from data in 12 and 19.

[a]Rural cereals availability is defined as the sum of production minus seed, losses, and official marketings (paddy marketings are converted to rice at a milling ratio of 0.65), divided by rural population.

[b]Urban cereals availability is defined as the sum of official marketings plus imports (or minus exports) divided by urban population.

TABLE 9.11. *Price Indexes, 1961–76*

	Food price indexes[a]		Millet price indexes		Rice price indexes[b]		
Year	Cooper-atives	Bamako market	Official retail	Bamako market	Official retail	Bamako market	GDP deflator[c]
1961	n.a.	n.a.	100	n.a.	81	n.a.	88
1962	n.a.	n.a.	100	n.a.	81	n.a.	91
1963	100	100	100	100	100	100	100
1964	n.a.	n.a.	100	128	114	118	110
1965	117	162	106	261	122	202	112
1966	122	179	106	311	122	236	117
1967	148	191	152	255	154	241	128
1968	171	191	152	233	151	227	161
1969	161	191	152	267	151	200	167
1970	169	193	152	233	207	222	188
1971	188	234	212	270	207	252	206
1972	197	251	212	400	216	284	218
1973	214	298	212	672	238	332	226
1974	256	331	309	461	238	361	241
1975	316	350	309	405	301	341	320
1976	351	279	309	383	301	332	n.a.

SOURCE: *17.*
[a] Mali has no true consumer price index.
[b] Rice quality is 40 percent brokens.
[c] Calculated from the World Bank, *World Tables,* Johns Hopkins University Press, Baltimore, 1976, and *19.* The Malian franc was devalued by 50 percent in 1967.

tor, they also show the protective effect of transport costs. A general reduction in road transport costs of 25 percent between Abidjan and Bamako in that period would have reduced the Bamako c.i.f. price, thus bringing nominal protection coefficients (calculated at official retail prices) close to unity.

OPAM, like many state enterprises of the era, lost money (*10*). Although detailed information on its cost-price schedules is no longer available, OPAM's losses increased with its sales volumes. The government could not afford to continue to subsidize consumers, nor did it want to raise consumer prices more rapidly because of the effect of such an increase on the incomes of urban government workers.[19]

The government's lack of an effective policy to increase cereal production, especially of millet, also contributed to OPAM's failure to supply off-farm markets. Agricultural investment in the first Plan, as described in Table 9.13, was limited mainly to cotton and sugarcane in the Office du Niger and flooded rice production. The Office du Niger accounted for

[19] The minimum wage and the government's wage paid to its own employees did not increase throughout the eight years of the first regime; retail prices of cereals did increase in that time. See CILSS (*4,* vol. 1, p. 136). See Jones (*10*) for the government's budget deficits.

TABLE 9.12. *Nominal Protection Coefficients, 1961–76*

Year	(1)[a]	(2)[b]	(3)[c]	(4)[d]
1961	n.a.	n.a.	n.a.	n.a.
1962	—	n.a.	—	n.a.
1963	—	n.a.	—	0.78
1964	—	n.a.	—	0.89
1965	—	n.a.	—	1.62
1966	—	0.74	—	1.60
1967	—	0.48/0.55	—	1.00
1968	—	0.53	—	0.94
1969	0.60	0.55	0.88	0.81
1970	0.96	1.02	1.12	1.19
1971	0.97	1.11	1.28	1.46
1972	0.97	0.92	1.40	1.33
1973	0.64	0.57	1.07	0.97
1974	0.35	0.34	0.58	0.57
1975	n.a.	0.70	n.a.	0.87
1976	n.a.	1.00	n.a.	1.21

[a] Ratio of wholesale cooperative prices of 40 percent broken rice to the c.i.f. Bamako price of similar grade rice. The source of cooperative price data is *17*. The source of import price data is *28*. Transport costs from Abidjan to Bamako were assumed to be 20 Malian fr/kg of rice between 1961 and 1968 and to increase by 2 fr/kg per year after 1968.

[b] Ratio of the wholesale cooperative price of 40 percent broken rice to the c.i.f. Bamako price of 25–30 percent Thai broken rice, for which the source of data is Charles P. Humphreys and Patricia L. Rader, "Rice Policy in the Ivory Coast," preliminary draft, June 1977. The source of cooperative price data is *17*.

[c] This is the ratio of the Bamako market price of 40 percent broken rice to the c.i.f. Bamako price of similar grade rice, for which the source of data is that cited above in note *a*. Market price data are taken from *17*.

[d] This is the ratio of the Bamako market price of 40 percent broken rice to the c.i.f. Bamako price of 25–30 percent Thai broken rice, for which the source of data is that cited above in note *b*. Market price data are from *17*.

one-half of total planned expenditures in agriculture,[20] and the rehabilitation of flooded rice polders, built during the colonial era, received nearly another fourth. Paddy collected in the Office du Niger declined between 1961 and 1968, and the polders program had no effect on production or on marketings. Poor performances were a result of the absence of adequate price incentives generally, and agronomic problems in the Office specifically. Planned expenditures on the rainfed cereals were confined to the provision of equipment and to the training of extension workers; expenditures were small and output of rainfed cereals showed no systematic tendency to increase throughout the era.[21]

[20] Planned expenditures were a good deal larger than actual, and the shortfall of spending contributed to the failure of the investment program of the first Plan. Of the 14.3 billion Malian francs (at par with the CFA franc until 1967) scheduled to be spent on agricultural equipment, the flooded rice polders, and the Office du Niger, only 8.08 billion were actually spent; 5 billion mf of the shortfall were in the Office. Much of the actual investment in the Office was in cotton and sugarcane, not rice. See Jones (*10*, pp. 352, 356).

[21] By 1968 only 13,535 ha of millet or sorghum were farmed collectively, with an average yield of 260 kg/ha, and 16,560 ha of rice were farmed collectively with a yield of 136 kg of paddy per ha. This compares with other farms of 867,000 ha in millet or sorghum, with

TABLE 9.13. *Investments in First National Plan of Mali, 1961–65*

Category	Percent	Billions of Malian francs
Industry	6.3%	4.0
Transport	48.3	30.5
Other (health, housing, administration)	13.1	8.4
Power and communications	6.7	4.3
Agriculture	25.6	16.4
Office du Niger	15.0	9.6
Polders program	3.8	2.4
Agricultural equipment	3.6	2.3
Other	3.2	2.1
TOTAL	100.0%	64.0

SOURCE: *10*, pp. 198, 204, 352.

The second objective of Malian agricultural policy under the original government was to increase production and exports of cotton and peanuts. The cotton program, run by the French textile firm CFDT under an agreement with the Malian government, was successful, and official exports of seed cotton increased throughout the period. The peanut program, which consisted mainly of making SOMIEX the official purchaser and exporter of the crop and the weak collectivization efforts, failed, and official marketings decreased from 97,000 mt in 1958 to 29,300 mt in 1968, although total production did not decrease by as much as official marketings. Cotton and peanut production and official marketings are shown in Table 9.14.[22]

Although the rank order of official producer prices did not change during the first government, there was a reduction in the producer prices of paddy and millet, and a lesser one in peanuts, after independence. Those reductions, motivated by the need to keep cereals prices low and to maximize export tax revenue, hampered agricultural development in Mali. In the absence of a land constraint, raising cereals prices did not run the risk of reducing the supply of export crops, with consequent adverse effects on tax revenues. Although seasonal labor constraints did exist, relative producer prices actually moved against the most successful crop—cotton.

---

yields of 840 kg, and 135,000 ha in rice, with yields of 727 kg (*8*, Table 11). Mali's first president, Modibo Keita, announced, "we ask each village to act so that its collective field develops in such a way as to attain in five years, at the end of the Plan [1966], an area of 1 hectare per family" (*10*, p. 227).

[22] The Malian franc was inconvertible from the time of its creation in 1962 until its 50 percent devaluation in May 1967. It is possible that Malian peanuts were being smuggled to Senegal and bartered there; that type of smuggling, caused by the inconvertibility, might have occurred even if peanut prices had not been higher in Senegal.

TABLE 9.14. *Cotton Marketings and Peanut Production and Marketings, 1959–68*
(000 metric tons)

Year	Cotton marketings[a]	Peanuts[b]	
		Production	Marketings
1959	7	n.a.	36
1960	7	n.a.	51
1961	11	122	83
1962	13	138	66
1963	20	167	71
1964	25	182	75
1965	29	148	49
1966	18	153	27
1967	28	159	40
1968	33	119	29

SOURCES: Most sources of production data give figures that correspond closely, sometimes exactly, to marketing figures. Since virtually all production of cotton in Mali in this era was controlled by the Office du Niger or by CFDT, marketing figures are probably good estimates of total production anyway. Source of cotton marketing is *10*, p. 367. Source of peanut production and marketing is *4*, vol. 2.
[a] First-grade seed cotton.
[b] Unshelled.

One therefore cannot conclude that the collapse of peanut marketings and the stagnation of rice marketings were due to any extraordinary price incentives given to cotton production.[23]

Failure to achieve the objectives of Malian agricultural policy between 1960 and 1968 did not lead to a change in policy during this period. There were indications that President Keita's government intended to tighten the controls it had established in agriculture and to retain its fixed producer/consumer prices and its legal ban on private trade in grains and in the export crops. What led to the change in some policies after 1968 was the change in government.

Rice Policy After 1968

The failures of socialist Mali's agricultural policy—the inability to collect and sell enough cereals to defend official prices, the government's incapacity to bear the deficits of OPAM and of other state enterprises, the ineffectiveness of the collective farms, and the near collapse of peanut exports—were the important and immediate influences on the agricultural policy of the military government which took power in November 1968.

---

[23] This argument does not consider the effects of input prices on supply or of the superior commercial organization of CFDT; this second factor seems to have been decisive in CFDT's success. See de Wilde (*32*, pp. 301–36) for an account of CFDT's work in the early 1960's.

In 1969 the new regime allowed a limited private trade in cereals, raised the producer prices of all crops, and abolished the collective farms. The immediate consequences were to reduce official purchases of millet (official purchases of paddy did not change much because of the overwhelming influence of the Office du Niger) and to double the official marketings of peanuts. The 50 percent devaluation of the Malian franc in May 1967, a condition of its convertibility being guaranteed by the Central Bank of France, influenced the government to raise its producer prices and enabled it to buy more peanuts.[24]

New circumstances—the change in the external grain trade balance due to poor harvests—were also important in forming the policy of the new government. The change in the trade balance, combined with the long-held belief that the rice-growing potential of the Niger-Bani Delta was underexploited, led to a renewed emphasis on rice production, with two important changes from the policy of the previous regime. The first change was the organization of improved production in integrated rural development projects, for which the model, Operation Arachide, had been established in 1967 by the old government. The second was the abandonment of the effort to increase production and marketing by collectivization and a return to the traditional system of land tenure.

The harvests of 1968–69 were the first of a series of subnormal crops, which culminated in the very low output caused by the severe drought of 1972–74 (3). During that period the government gave much greater, even dominant, importance to the objective of securing food production at the level of national self-sufficiency against drought. That objective had been given little attention by the first government, in part because it had been thought that the central problems of economic policy were political, and in part because rainfall and crop production in the early 1960's were good. In view of the importance given this objective, increases in rice production received even greater attention because rice was believed to be more secure against both subnormal rainfall and drought.

The principal objective of Malian cereals policy since 1968 has been to regain self-sufficiency and to improve the security of food supplies. Other objectives have been to increase rural incomes and to improve the nutrition of the population. The main goal of the first government—to control the production and marketing of cereals in order to maintain low prices—has been relegated to second rank. The central instrument of policy has been technical improvement of and investment in rice production in the Office du Niger and in the two main rice projects at Ségou and Mopti. Price policy has been used to stimulate production within the system of

[24] For the reason why the devaluation probably enabled the government to buy more peanuts, see footnote 21 above.

fixed producer/consumer prices and the official monopoly of national and international cereals trade established within the first government. Policy has been constrained by the budgetary restrictions of the government, flood failures, and poor rainfall.

The immediate origins of the principal new policy objective are shown above in Table 9.10. Domestic per capita availability of all cereals, both on and off farm, declined after 1969; concessional and commercial imports of millet, rice, and wheat grew in volume and value, and rice imports alone grew to 15 percent of the total value of imports by 1974.[25] The official explanation for the growth of imports was poor rainfall and the destructive activities of private traders, who had been allowed to operate between 1969 and 1971.

Planners in the second government emphasized rice production for several reasons.[26] First, investments in flooded and irrigated rice were thought to be more secure than investments in rainfed cereals. Planners also believed that the Office du Niger could be made more productive, especially after cotton production had been abandoned there in 1970. Finally, rice was thought to be more nutritious than millet or sorghum, and planners thought that the substitution of rice for either of those cereals would improve nutrition even if per capita consumption of cereals did not increase.[27]

The choice of policies showed the influence of the important constraints. Improvement and expansion of rice cultivation were concentrated in the flooded rice polders at Ségou and Mopti, where low-land development costs ($500–1,000 per ha), cheap technologies (for example, higher-yielding seed varieties), and restricted access to polder land allowed increases in output and official marketings with little strain on the national budget. Economic analysis of the polder projects at the time of requests for foreign financing was based on the assumption that the polders would be filled in nine out of ten years and that flooded rice cultivation was therefore more secure than rainfed agriculture. The emphasis of policy on the Office du Niger was based on the belief that its capacity was underutilized and that it was a cheap way of expanding drought-secure agriculture.

[25] The c.i.f. value of Malian imports in 1974 was 86.08 billion Malian francs; the c.i.f. value of rice imports (food aid excluded) was 13.26 billion (9, 28).

[26] Views of Malian planners from the National Direction of the Plan and from the Institute of Rural Economy in Bamako are contained in the *Bilan Cerealier* (12), a study used to make projections of cereals supply and demand for the 1974–78 Five Year Plan.

[27] These assumptions are discussed in the *Bilan Cerealier* (12), in the text of the Five Year Plan (16), and in an unpublished document (14) circulated within the Institute of Rural Economy on Malian cereals policy. Rice is not more nutritious than millet, at least in terms of grams of protein and of fats per kg of grain (4, vol. 1, pp. 208–13). At 1978 product prices, calories of millet were much cheaper than calories of rice.

TABLE 9.15. *Production and Marketings at Office du Niger and Flooded Rice Projects, 1970–77*

Category	1970	1971	1972	1973
Office du Niger:				
Production *(000 mt paddy)*	—	—	61.6	69.0
Yields[a] *(kg/ha)*	—	—	1,760	1,944
Marketings *(000 mt paddy)*	36.1	38.1	46.3	46.9
Opération Riz Ségou:				
Production *(000 mt paddy)*	9.8	—	18.6	4.2
Yields[a] *(kg/ha)*	1,240	—	—	1,342
Marketings *(000 mt paddy)*	—	—	—	0.8
Opération Riz Mopti:				
Production *(000 mt paddy)*	—	—	7.8	1.5
Yields[a] *(kg/ha)*	—	—	—	800
Marketings *(000 mt paddy)*	—	—	—	—

Category	1974	1975	1976	1977
Office du Niger:				
Production *(000 mt paddy)*	78.0	82.0	86.2	94.4
Yields[a] *(kg/ha)*	1,948	2,000	2,240	2,385
Marketings *(000 mt paddy)*	54.2	62.9	65.5	62.1
Opération Riz Ségou:				
Production *(000 mt paddy)*	8.1	45.0	55.7	38.6
Yields[a] *(kg/ha)*	1,556	1,752	1,750	1,373
Marketings *(000 mt paddy)*	3.1	14.7	20.9	n.a.
Opération Riz Mopti:				
Production *(000 mt paddy)*	3.7	19.4	18.4	25.4
Yields[a] *(kg/ha)*	948	1,618	1,451	1,705
Marketings *(000 mt paddy)*	—	4.6	8.0	8.9

SOURCES: *21, 22,* and *23.*
[a] Yields are estimated from crop cuttings at these projects.

The growth of output, yields, and official marketings from the Office and from the rice projects at Ségou and Mopti is shown in Table 9.15. The great improvement of Office yields and marketings was due to the 122 percent increase in the producer price between 1969 and 1974 and to the reorganization of production, which changed part of the system from a partially mechanized state farm to one using animal traction and private farm management.[28] The extensive control, by Sahelian standards, of paddy marketings by the Malian government after 1970 was the result of progress in the Office and, after 1975, of progress in the Ségou and Mopti projects.

Following the poor harvests at the Ségou and Mopti rice projects in 1973 and 1974, due to poor floods, production and marketings almost

[28] The real price of paddy (i.e. the nominal price deflated by the GDP deflator) increased by 64 percent in this period.

reached their planned levels in 1975–77. This success was due to some of the same factors that encouraged production in the Office, principally the increase in the real price of paddy, but was also attributable to technical changes and to subsidies on irrigation works and extension services. The yield increases from the use of improved seeds and better water control in the polders are an important example of the government's ability to exploit the resources of the Niger-Bani river system and, in so doing, to influence the marketed supply of paddy, which the first government failed to do.

The current government has also established regional rice projects to increase rural incomes and to maintain regional food self-sufficiency. The small dams in Sikasso and the pumping systems in Gao and Kayes are not intended to produce rice for export to the cities (as are the Mopti and Ségou projects), but to increase production in regions that tend to import even in normal rainfall years and to reduce the prospect of famine in subnormal and drought years. The installation of such projects is also intended to reduce the costs of supplying poorly accessible regions, two of which, Kayes and Gao, now receive important shares of OPAM's marketings.

Essential to the recent emphasis of policy on stimulating production and on reducing imports has been the attempt to raise real producer and consumer prices. The nominal protection coefficients presented above in Table 9.13 show that cooperative prices, after the increases of 1969, were close to Bamako c.i.f. prices and remained there except during the years of the drought, when world prices were high and large imports had to be subsidized. Market prices were always above cooperative prices and world prices except during the drought. Consumers without access to cooperative supplies were therefore taxed with respect to the c.i.f. prices, whereas consumers with access were subsidized. The cooperative price of rice has risen faster than the rate of inflation and faster than the market price of rice, whereas the market price has not grown so fast as the rate of inflation since 1974.

The marketing policies of the two governments have been similar. Private cereals trade was made legal after 1969, but was again banned in 1971, subject to the qualification that it is to become legal after OPAM has constituted a stock of 40,000 mt of cereal.[29] The immediate effect of liberalization in 1969 was to reduce OPAM's purchases of millet, which have never regained their level of 1968. OPAM has been able to buy much more paddy since 1969, but that success has been a counterpart to the production policies of the government. It is clear that the commodity

[29] See the CILSS (4) Mali country study for a record of OPAM's recent stockholdings. Stocks were greater than this amount at the end of the 1971–72 and 1973–74 seasons, but it is not known if private trade was then made legal.

composition of OPAM's purchases (much more rice than millet, sorghum, or maize) reflects more the system of physical control in projects than it does relative prices or total outputs. OPAM has again lost money since 1968, especially during the drought, when it absorbed the costs of supplying food aid to much of Mali.

In the 1974–78 Plan it was assumed that per capita consumption of rice would increase and that of millet and sorghum would decrease without any change in their relative prices. First, planners argued that higher paddy prices and input subsidies had to be maintained. Second, it was felt that the government could not offer consumer subsidies because it could not afford them, nor could it, as with production subsidies, rely on foreign aid to pay for them. Third, the possibility that surpluses could be exported regularly was rejected, although small exports in favorable years— and the emphasis of policy was on favorable crop years when the weather was good and not on years when world prices were favorable—were considered. Planners argued that rejecting an export policy allowed the loosening of the constraint that had forced the government to lower its paddy price in 1960 (*12, 16*).

Self-sufficiency was attained in 1976 and 1977, although imports were necessary in 1978 as a result of poor rainfall and the export of stocks of paddy and rice from the 1976–77 harvest. The problem of policy, as viewed by the government, is to maintain self-sufficiency subject to supply changes caused by weather and to fluctuations in world prices.

An Evaluation of Policies

Of the three classes of policy described in the Introduction, investment has been the most important in Mali, with tax/subsidy and price/trade being of lesser importance. Investments have been generally associated with direct government interventions.

The major investments have been made in construction and rehabilitation of irrigation works, whose scale and technical complexity necessitate public intervention of some form. Although Delta farmers had long grown flooded rice, they had no capacity to make yield-increasing investments, such as dike and canal construction or mechanical deep-plowing to kill wild rice, and productivity remained low until the end of the 1960's.

Secondary investments, such as those in mills and warehouses, followed from the primary investments in water control. Secondary government investments were associated with the decision to accept some paddy from producers in irrigated areas in order to amortize (in part) the capital costs of irrigation. They were also consistent with the government's desire to control cereal marketings.

Investments in irrigation, the central policy because of the country's natural and historical advantages in rice production, have been generally successful. They would not, however, have been so without complementary government interventions, especially the provision of inputs (for example improved seeds) and of technical advice to farmers. The failures of earlier investments (for example the construction of polders during the colonial period and the attempts to rehabilitate them during the 1960's) were due to the inability to provide such services. Inputs and extension were also essential to exploiting the Office du Niger's underused capacity. The new Office program after 1970 involved reorganization of production, such as the transfer of cotton lands to rice and of state farm rice land (*régie*) to settler management, which contributed to the growth of production in the 1970's.

The success of these public investments and interventions owes much to physical factors. The Office du Niger, with its sunk costs of irrigation, and the Niger-Bani Delta allow simple and inexpensive water control, to which entry can be restricted. Public agencies controlling the irrigation works are able to control some of the supply of paddy from them. Physical control of paddy, as opposed to control through the price system, allows the government to operate its mills at roughly full capacity and thereby to avoid underutilization of fixed capacity[30] without having to pay the market price to buy paddy. On the other hand, physical factors, especially the poor national transport system, work against the input supply and crop marketing agencies because of the high costs involved in supplying some regions.

Constraints on expansion of investments in production are financial and climatic. The government's ability to expand irrigated or empoldered areas depends almost entirely on foreign financing. Attracting that financing depends, in part, on relative rates of return, and rice is often less remunerative than cotton or groundnuts. Although the Malian government has argued that the costs of not having secure food supplies during droughts are greater than the costs of obtaining them by growing rice, some investments, especially those on the polders projects, are probably not secure enough to compensate for their lower rates of return than those on investments in export crops.

The objectives of price/trade policies have been to supply urban centers cheaply, to protect producers and consumers from price fluctuations, and to maintain fixed prices of cereals throughout the country. Policies have included the ban of private trade in cereals (except during the liberalization between 1969 and 1971), and the establishment of a national market-

---

[30] Analysis of Malian milling costs, especially those of its large industrial mills, shows that they are substantially lower than costs of similar mills in other West African countries and that much of the Malian advantage is due to lower average fixed costs.

ing agency, fixed producer prices, and cost-price schedules for collection, milling, and marketing. International rice trade was regulated by export restrictions before 1966 and by import quotas after 1968.

Constraints on price policy are mainly political and financial. Holding fixed consumer prices in all areas results in OPAM losses on cereals supplied to the Gao and Kayes regions, for which marketing costs are higher than for the Bamako market. OPAM has also carried over deficits resulting from its shipments during the 1972–74 drought and from the subsidies it paid to consumers of commercially imported rice during the period of high world prices, which coincided with the drought. The government, which establishes the prices at which OPAM must buy and sell, does not want to raise consumer prices in Bamako because of the effect on the incomes of its workers; it does not want to raise prices in the outlying areas to cover costs of OPAM's supplying them, for similar reasons.

Price policies have had mixed results. In the main urban market at Bamako they have led to a price structure in which market retail prices are greater than official retail prices. This implies that there is excess demand at the official price, which OPAM cannot meet, forcing consumers to buy in the private markets. On the other hand, the bias of the price structure in favor of consumers in outlying areas has probably enabled OPAM to control distant markets by making private trade unprofitable there.[31] That official retail prices have always been lower than market prices suggests that OPAM has established a floor price and thereby helped to insulate producers from downward seasonal and interannual price changes.

At the outset of the 1974–75 season, the official paddy price and the corresponding retail rice price were increased. This change was responsible for some of the expansion of production in the Office and in the flooded rice projects and the resulting growth of official marketings. Other forces behind that growth were the investments in the projects and other government interventions. The growth of production and of official marketings has not entirely eliminated the margin between the official and market retail prices, and therefore the price policy continues to effect a transfer from producers to consumers.[32]

Trade is secondary to price in this class of policies. Rice's principal substitutes, millet and sorghum, are almost entirely nontraded goods. Rice trade policy is limited by the fact that high transport costs to ports cause a wide separation between Malian f.o.b. and c.i.f. prices, thus allowing the domestic price of rice to change within a wide band without changes in

[31] OPAM's losses in these activities (or, alternatively, the subsidies paid to consumers in outlying areas of the country) are financed out of credits from the Central Bank of Mali, on which OPAM is one of the largest drawers.

[32] Effects of changing the producer price of paddy are analyzed in "Rice Production in Mali" (20).

exports or imports.[33] When domestic prices are within that band, trade policy has no role. Trade policy was most important when Mali began to import rice after 1968. The costs of the import quotas, instituted partly because of foreign exchange shortages, fell on market consumers who paid prices above the c.i.f. price, not on consumers of OPAM rice, who usually paid less than the c.i.f. price because of the limited availability of subsidized imports.

The government has used tax and subsidy policies infrequently because of budgetary limitations. Rice and other cereals are exempt from indirect consumer taxes, except those embodied in the inputs used to produce and market them. There is a small net subsidy throughout the rice sector, which benefits mainly farmers in the development projects. Subsidies fall largely on the capital costs of irrigation works and on the variable costs of extension services. Subsidies on farm equipment sold officially were removed before the 1977–78 crop year, and the government has announced its intention to remove all remaining subsidies, including those on fertilizers. The net subsidy on capital costs of irrigation arises from the insufficiency of in-kind water charges paid by participating farmers.[34] This particular subsidy is an effective policy, from the government's viewpoint, because it is paid partly out of foreign aid to the rice development projects; it allows greater incentives to farmers without higher paddy prices and thus without higher retail rice prices.

The policies described here have clearly helped the government to reduce imports and to increase official marketings of rice. They have been only partly successful in increasing rural incomes because of the small proportion of the rural population involved in rice production. Price policy favors those who can buy in the official stores and discriminates against those who buy in the private markets, where prices are higher and more variable. The main impediment to advancing governmental objectives is still weather fluctuations, which cause output to fall, leading to higher domestic prices and necessitating imports. A policy of achieving full production security through irrigation schemes, in order to eliminate imports in even the worst years and to allow almost complete control of marketings, has not been adopted because of its high investment costs, the serious budget restrictions in Mali, and the limited supply of concessional

[33] The difference between the Dakar c.i.f. price and the Bamako c.i.f. price is roughly 25,000 to 30,000 mf/mt. This implies that the band within which domestic prices can change is 50,000 to 60,000 mf/mt, i.e. twice the margin between the c.i.f. Dakar and c.i.f. Bamako prices, or 45 to 55 percent of official retail rice prices in Bamako.

[34] The incidence of taxes throughout the rice sector is discussed more completely in (20), where it is shown that farmers benefit from subsidies on irrigation works and extension services but that the official paddy price is far below import substitution levels. This means on the whole that the effective rate of protection on officially marketed output is significantly negative.

aid. An important associated policy to reduce imports and to achieve greater food security is the establishment of government grain reserves. This has only recently begun to be implemented and it is too early to evaluate its effects.

Prognosis

Since 1968 Malian rice and cereals policy has been built upon three main arguments. The first and most important is that rice production in irrigated and flooded perimeters is more secure against both subnormal rainfall and floods and severe drought than is the rainfed cultivation of millet and sorghum. That assumption is valid in the Office du Niger, but is less so in the controlled submersion polders at Ségou and Mopti, from which 35 percent of Mali's projected 1980 paddy output is expected (31). If the polders are less secure than planned, there will be greater fluctuations in paddy production, and the average supply will be less to the extent that increased risks impair intensification in those projects. Consequently, the government will confront a choice between price/trade policies (especially a reliance on imports in bad years and on stocking supplies in good years) and policies to make production more secure. But secure production techniques (for example, pumping schemes like the one at San, which operates as part of the Ségou project), if they cost more than current techniques, would impair Mali's export competitiveness[35] and its efforts to restrain domestic prices. The Malian government has stated that it is willing to bear the costs of more secure systems of water supply because any other policy "would bring, in case of drought, greater economic, social and political costs,"[36] but its capacity to do so depends upon concessional foreign financing.

The second argument is the incorrect perception that rice is more nutritious than millet and hence that rice consumption should be increased. Policy makers have argued also that if rice were available in increasing quantities and at lower prices, consumers would readily increase their purchases of it and that the stagnation of consumption after independence was the result of the failure of production to expand. The limited availability of foreign aid to fund, for example, large extensions of the Ségou and Mopti projects has not permitted the empirical validation of this line of reasoning, but there is no question that it influences the Malian government and its cereals investment decisions.

[35] Malian rice exports can be competitive in Ivorian markets at a world price (c.i.f. Abidjan) of $300/mt for 25 to 35 percent broken rice, but Malian rice is not competitive in Dakar at a world price of $250/mt for 80 percent broken rice (20).

[36] See (14). Also see Stryker (26) for a general discussion of food security policies in the Sahel.

The third assumption is the supposed necessity of complete government control of grain marketing. Attempts to achieve this control have never been successful and they remain unlikely. Continuing attempts at extensive public intervention result in fairly large financial costs (for example, OPAM's deficits) with almost no demonstrable benefits and in some hidden economic costs resulting from disincentives given to greater cereals output. This public policy has had less influence on rice than on other cereals because of the government's greater ability to control marketings from the irrigation projects. But it is probable a more flexible marketing policy for rice projects would reduce some of the costs of current policy and encourage more rapid adoption of the intensified techniques that are now the object of much extension effort.[37]

The retention of these assumptions implies that rice will continue to be the principal instrument of Malian cereals policy. This conclusion is reinforced by the slow development of technologies for the production and transformation of rainfed cereals.

Because the majority of Malian farmers do not grow rice and will not do so in the conceivable future, the objectives of increasing rural incomes will not be well achieved. The rice projects have increased rural incomes, but they affect fewer than 250,000 people in a rural population of 5,500,000. The emphasis of other Malian development projects is largely on production and marketing of cash crops, not on development of rainfed cereals.

If rice remains the basis of cereals policy, the high costs of some types of projects (for example, new large dam irrigation works or pumping systems) and the continuing experience of poor harvests, of which the 1977–78 crop year is the latest example, will impel policy to turn increasingly to the Office du Niger. For the Office to continue to increase productivity, policy makers must choose among intensification of production (incurring some capital costs not necessitated by a more extensive policy), extension of cultivated areas to utilize more fully the existing irrigation network, or some mix of the two policies (27, 31). The Office has the advantages that its main capital costs are sunk and that aid donors are now willing to pay for the rehabilitation projects. Foreign aid, which is necessary in view of the budgetary constraint, is related both to water security in the Office and to the recent advances there. An emphasis on intensification in the Office has advantages over other policies (for example more polder construction in the Delta or greater development of rainfed cereals in southern Mali). The main advantage is that it allows constitution of larger supplies, either for export (if world

[37]See (4) and (26) for criticism of Malian marketing policies, enumeration of some of the inefficiencies of the present system (in Mali and in other Sahelian countries), and proposals for reform.

prices are favorable) or for storage to maintain national self-sufficiency in deficit years, which could also contribute to greater price stability.[38]

Rice production involves less than 30 percent of the rural population, and consumption of rice provides only 15 percent of the national supply of calories. Nevertheless, the great dependence on rice of Malian cereals policy has persisted. This orientation has derived largely from the hydrologic qualities of the Malian segment of the Niger River and from the legacy of French policy in the Office du Niger, which provide Mali with special advantages in rice production. Although those physical bases of policy changed little from the first regime to its successor, the instruments and results of policy have differed greatly. The adoption of new policies by the current government enabled it eventually to regain self-sufficiency and even to export rice. But the government is now faced with the problem of limiting the costs of enhancing production security as it decides how to expand rice production in Mali.

Citations

1 Samir Amin, *Trois Expériences Africaines de Développement: Le Mali, La Guinée, et le Ghana*. Presses Universitaires de France, Paris, 1965.

2 P. Ballan et al., *Etude des Structures des Prix et des Mecanismes de Commercialisation des Mils et du Sorgho au Mali*, 3 vol. IDET/CEGOS, Paris, May 1976.

3 John C. Caldwell, *The Sahelian Drought and Its Demographic Implications*. Overseas Liaison Committee of the American Council on Education, Washington, D.C., 1975.

4 Comité Inter-état de Lutte contre la Sécheresse Sahelienne (CILSS), Club du Sahel, *Marketing, Price Policy and Storage of Food Grains in the Sahel*, 2 vols. University of Michigan Center for Research on Economic Development, Ann Arbor, August 1977.

5 William J. Foltz, *From French West Africa to the Mali Federation*. Yale University Press, New Haven, 1965.

6 Food and Agriculture Organization of the United Nations, *Rapport au Gouvernement du Mali sur le Problème de Commercialisation des Céréales*. Bamako, 1973.

7 Jean Gallais, *Le Delta Intérieur du Niger: Etude de Géographie Régionale*, 2 vols. Institute Fondamental de l'Afrique Noire, Dakar, 1967.

8 International Bank for Reconstruction and Development, *Economic Report on Mali*. Washington, D.C., 1970.

9 International Monetary Fund, *International Financial Statistics*. Washington, D.C., April 1978.

10 William I. Jones, *Planning and Economic Policy: Socialist Mali and Her Neighbors*. Three Continents Press, Washington, D.C., 1976.

---

[38] The intensified Office du Niger technique (described in 28 and 20) has a net social profitability slightly less than the current technique, but increases net yields by more than one metric ton of paddy/ha (20).

11 Mali, Government of, Ministère de Développement Rurale, Institut d'Economie Rurale, *Etude de Factibilité du Projet Mali-Sud*. Bamako, 1971.

12 ————, *Bilan Céréalier*. Bamako, June 1972.

13 ————, *Enquête Agro-Socio-Economique (Mopti)*. Bamako, 1974.

14 ————, unpublished document. Bamako, January 1977.

15 ————, *Enquête Agro-Socio-Economique (Ségou)*. Bamako, 1975.

16 ————, Ministère du Plan, *Plan Quinquennal*. Bamako, 1974.

17 ————, *Bulletin Mensuel de Statistique*. Bamako, various years.

18 ————, *Rapport de l'Enquête Agricole*. Bamako, various years.

19 ————, *Recensement Général de la Population du Mali: Résultats Provisoires*. Bamako, 1976.

20 John McIntire, "Rice Production in Mali." Stanford/WARDA Study of the Political Economy of Rice in West Africa, Food Research Institute, Stanford University, Stanford, July 1979; Chapter 10.

21 Office du Niger, Bureau of Economic Affairs, Ségou, questionnaire data gathered by Charles P. Humphreys and J. Dirck Stryker. Stanford/WARDA Study of the Political Economy of Rice in West Africa, Food Research Institute, Stanford University, Stanford, January 1977.

22 Opération Riz Mopti, Direction Générale, *Rapports Annuels*. Sévaré, various years.

23 Opération Riz Ségou, Direction Générale, *Rapports Annuels*. Ségou, various years.

24 Opération Riz Pluvial et Bas-Fonds, Direction Générale, *Rapports Annuels*. Sikasso, various years.

25 Société des Etudes pour le Développement Economique et Social, *L'Approvisionnement des Villes en Afrique de l'Ouest: Bamako*. Paris, December 1972.

26 J. Dirck Stryker, "Food Security, Self-Sufficiency, and Economic Growth in the Sahelian Countries of West Africa." Stanford/WARDA Study of the Political Economy of Rice in West Africa, Food Research Institute, Stanford University, Stanford, February 1978.

27 West Africa Rice Development Association (WARDA), *Etude Prospective de l'Intensification de la Riziculture à l'Office du Niger*. Monrovia, 1977.

28 ————, *Rice Statistics Yearbook*. Monrovia, 1975.

29 ————, *Rice Statistics Yearbook: Up-date*. Monrovia, 1976.

30 ————, *Survey of Rice Post-Production Technology: Mali*. Monrovia, 1976.

31 WARDA and Stanford Food Research Institute, *Prospects for Intra-regional Trade in Rice in West Africa*. Monrovia, September 1977.

32 John de Wilde, ed., *Experiences with Agricultural Development in Tropical Africa: The Case Studies*. Johns Hopkins University Press, Baltimore, 1967.

10. Rice Production in Mali

John McIntire

The principal objectives of Malian rice policy have been to reduce imports, to stabilize urban prices and supplies, to increase and stabilize the incomes of rice farmers, and to achieve national food security. These goals have been achieved partly by expanding production at decreasing costs without trade restrictions or price supports. The success of policy has been based on the exploitation of water resources at the Office du Niger, where the irrigation system provides complete production security, and on the Niger and Bani rivers, whose geographic and hydrologic characteristics permit extensive low-cost polder development. Farmers have been given some indirect subsidies, particularly on irrigation, and costs of production have been below world prices.

The purposes of this chapter are to examine the economic efficiency of the expansion of Malian rice production and to study the effects of government policy in providing producer incentives. The chapter contains five sections: a description of the main production, milling, and marketing techniques; a discussion of the system of economic incentives affecting the rice sector; a comparison of the private and social costs of rice production; an analysis of the sensitivities of those benefits to changes in important parameters such as labor costs, yields, and the world price of rice; and an evaluation of the effects of various policies on government objectives.

Description of Techniques

Production

Table 10.1 shows quantities and costs per hectare (ha) of inputs used in these techniques. The important characteristics of the principal rice production techniques practiced in 1976 are summarized in Table 9.5 in the previous chapter. The oldest production technique in Mali is the flooded system (TFM) in the Delta of the Niger and Bani rivers around the city of Mopti and along the flood plains of the Niger toward Ségou and of the Bani toward San. Traditional flooded rice is grown in holdings of about 1.5

TABLE 10.1. *Quantities and Cost Per Hectare of Major Inputs*

Production technique	Farm labor (man-days)	Fertilizer (kg) N	Fertilizer (kg) P_2O_2	Fertilizer (kg) K_2O	Seeds (kg)	Land development cost (mf)	Extension service cost (mf)
Gravity irrigation (ONC)	90	15	0	0	100	sunk	10,000
Gravity irrigation (ONI)	120	64	46	0	80	220,000	20,000
Controlled flooded, Ségou (CFS)	70	0	0	0	100	400,000	6,000
Controlled flooded, Ségou (CFSI)	95	32	23	0	80	400,000	10,000[b]
Traditional swamp and rainfed (TS)	120	0	0	0	100	0	0
Improved swamp and flooded (IPS)	115	0	0	0	100	300,000–1,000,000[a]	3,500
Traditional flooded, Delta (TFM)	60	0	0	0	120	0	0
Controlled flooded, Mopti (CFM)	80	0	0	0	100	500,000	5,000
Controlled flooded, Mopti (CFMI)	100	32	23	0	80	500,000	10,000[b]

SOURCES: ONC and ONI are from 6 and 13; CFS and CFSI are from 8; TS and IPS are from 9; and TFM, CFM, and CFMI are from 7, 1, and 2.

NOTE: Six of these techniques—ONC, CFS, TS, IPS, TFM, CFM, and CFM—are described in Table 9.5. ONI is a variation of ONC with an assumed use of more fertilizer resulting in yields of 3.5 mt/ha; CFSI is a variation of CFS with an assumed use of fertilizer resulting in yields of 2.5 mt/ha; and CFMI is a variation of CFM with an assumed use of fertilizer resulting in yields of 2.5 mt/ha.

[a]The maximum of this range is the estimated cost of an irrigation scheme in southern Mali that is not yet in production.

[b]The assumption was made that the levels of administrative overhead and extension density in the Ségou and Mopti projects were chosen with the goal of achieving yields of 2.5 to 3.0 mt/ha of paddy (i.e., the yields of the CFSI and CFMI techniques) and that some of the overhead and extension costs represent transient excess capacity. It was therefore decided to exclude some of those costs from the analysis of the CFS and CFM techniques.

ha over a total area of 80,000 to 100,000 ha.[1] Farmers have larger holdings of rice than of rainfed cereals (millet and sorghum) and combine agriculture with fishing, herding, and trade. There is no mechanization, although animal traction has been common since the 1920's and is now more important than manual cultivation.[2] The only water control is the construction of earth dikes to prevent too rapid entry of water onto fields of immature plants. Average gross paddy yields are 500 to 700 kilograms (kg) per ha; maximum yields are 1,000 to 1,200 kg without the use of inorganic fertilizers.

The crop cycle begins in late May with the first usable rains.[3] The flood starts to rise in the last two weeks of June, reaches most fields between August 15 and September 15, achieves its peak in October, falls rapidly from mid-November to January, and subsides slowly after January. Flood timing and height are highly variable and are not strongly correlated with regional rainfall because they are determined by rainfall throughout the basins of the Niger and Bani rivers upstream of the Delta.

The crop calendar is adapted to the supply of water. Soil preparation, whether accomplished manually or with animal traction, is possible only when there has been enough rain to break the hardpan that forms on the soil between February and mid-May. Hand seeding of *Oryza glaberrima* follows plowing, and the seeds are turned over with a hand hoe to prevent desiccation. Many seed varieties have been identified in the Delta, and the choice of variety depends upon the position of the farmer's field with respect to the river.[4] Farmers with lower fields choose longer-cycle, generally floating, varieties and plant earlier because their plants must germinate and grow rapidly enough to survive the earlier entry of water. Those having higher fields try not to plant too early because of the potential for desiccation in the interval between the first rains and the arrival of the flood. They also use shorter-cycle, generally standing, varieties to avoid desiccation of immature plants after the flood's recession.

Fields are weeded after seeding and before the flood's arrival, which drowns any remaining weeds that have not grown above the level of the water. There is little fieldwork between weeding and harvesting times. Harvesting is done with sickles from the end of November to the end of

[1] Rice parcels in eight areas of the Delta surveyed in 1973 and 1974 vary, on average, between 0.8 and 4.4 ha (2, 4). Accurate estimations of the total area devoted to traditional rice cultivation in the Delta are not available. The average holding of millet was about 1.3 ha per family, but the typical family did not have 1.5 ha of rice and 1.3 ha of millet; the mode seemed to be that families or zones would tend to be more specialized in one crop

[2] A description of the Delta's rice cultivators in the 1950's and early 1960's is contained in (1, vol. 1, pp. 199–228).

[3] Usable rain is defined as three millimeters in one day. Gallais observed that the first useful rain generally fell between May 15 and July 15 (1).

[4] Gallais writes that Pierre Viguier, an agronomist who worked in the French Sudan in the 1930's, identified 41 varieties of *Oryza glaberrima* (1, vol. 1, p. 99).

January. The paddy is piled in the fields to dry and threshed on the ground with flails.

Heavy fieldwork is done almost exclusively by men while women and children do ancillary tasks.[5] The introduction of animal traction reduces work per hectare at plowing time when off-farm labor is scarce because hired workers, who often come from millet-growing regions, are planting and weeding their own fields. Hired labor is more abundant at harvesttime because workers from millet regions have finished their harvests by late November and come to the rice regions in December to work. Use of hired labor has become more important since the introduction of animal traction enabled farmers to plant larger areas. Development of the traditional system is restrained by the variability of flooding, the severe infestation of fields with wild rice,[6] and the low yields of the local varieties.

The second important production technique, the controlled flooded polder system in Opérations Riz Ségou and Mopti, has been developed to solve the problems of the traditional uncontrolled flooded system. The new techniques (CFS and CFM) are designed primarily to shift traditional farmers into the improved polders. However, they have also introduced rice farming into some areas, especially near Ségou, that were formerly devoted largely to millet and sorghum cultivation, and to some absentee farmers, especially near Mopti, who have other primary activities such as trade.

The basis of the improved techniques is an unleveled polder consisting of an inlet gate, a common canal and drain, and an earth-protection dike encircling the cultivable area. Deep-plowing is done to kill wild rice. Empoldering allows control of the rate and timing of flooding and retention of water in the fields after filling; it thus prevents too early or too rapid filling and too rapid emptying. The system has no capacity to fill polders if the flood does not arrive and no capacity to empty them if the flood has not receded when the rice has matured. Polders are planted only in rice although some polder areas are left in pasture.[7]

Empoldering and improved control of flooding increase the yields of the improved *sativa* varieties. They permit better separation of varieties

[5] Women's work is usually limited to threshing, winnowing, and head-loading paddy home. Children help with plowing and crop protection.

[6] The occurrence of wild rice seems to be more serious in Opération Riz Mopti than elsewhere in and around the Delta (1). To combat wild rice, farmers sometimes plow after harvest, and mechanical deep-plowing is sometimes done. Chemical methods of control have been rejected as too expensive. At Mopti there has also been a program of weeding in deep water (*faucardage*), which necessitates taking land out of production for one season.

[7] In Opération Riz Mopti in 1975–76, 16,074 ha were planted and only 12,703 ha were harvested. Polders at that project were redesigned and reduced from a planned 31,000 to 26,000 ha (7). Reduction of areas within the protection dike causes an increase in the average fixed costs of irrigation works, but such increases are partly offset by the value of the pasture planted in areas taken out of rice.

according to water needs and resistance to flooding. They also allow better control of the crop calendar by regulating the timing of the flood. For both of those reasons, average gross paddy yields have grown to as much as 1,750 kg/ha on harvested areas, although they are smaller on seeded areas because of the incomplete filling of some polders.

Field techniques now differ little between the traditional and new systems, except for the introduction of harrowing and mechanical threshing. There is as yet no system of ox-drawn tool-bar weeding, and harvesting is still done with sickles. The more general use of animal traction in the new system has allowed an increase in the average size of rice holding, recently about 2.5 to 3.0 ha, and farmers generally have similar holdings in millet or sorghum.

Two new practices have been introduced to the Mopti and Ségou rice projects. The first, seeding in lines with ox-drawn seeders owned by the projects and rented to farmers, is designed to increase yields by permitting easier hoe weeding. The second, the application of small amounts of inorganic nitrogen and phosphate, is done after line seeding has been widely established in the projects. These practices, which now involve perhaps 10 percent of the farmers in the two projects, have been included in the definition of the intensified Mopti and Ségou techniques (CFMI and CFSI).

Amortization of flood control works, extension services, and project administration costs are partly supported by a land use fee levied on farmers in the two projects. The fee is a fixed amount of paddy per hectare and must be paid in kind. The projects also offer mechanical threshing and deep-plowing services to participating farmers, for which payment can be in kind or in cash.

The irrigated technique at the Office du Niger (ONC in Table 10.1) now supplies 35 to 40 percent of national paddy output and 60 to 70 percent of officially marketed paddy.[8] The Office is a semiautonomous public agency which rents land to farmers for a fixed in-kind fee per ha. After payment of the fee, farmers are supposed to surrender all their production to the project at the official price except an allowance for the consumption of family members. The Office sells seeds, oxen, and equipment to farmers on credit, provides extension services, maintains the irrigation network, and transports and mills paddy. Farmers do not own their lands, which can be taken from them for nonpayment of debts, and have little autonomy in production decisions, such as irrigation control.

[8] The Office du Niger was established by the French in the 1930's to grow cotton as an export crop and rice as a subsistence crop for settler farmers, many of whom were brought to Mali from what is now Upper Volta. The Office is in an arid area of low population density, and efforts to intensify production there have, until recently, failed, in part because of insufficient density of settlement. See (13) for a history of the Office.

Water control in the Office is maintained by a barrage at Markala, 250 km downstream from Bamako, which diverts water into a head canal of 8 km. This canal then bifurcates into two primary canals, the Canal de Macina, which runs northeast, and the Canal du Sahel, which runs north. The Canal de Macina feeds the production sector of Kolongo, and the Canal du Sahel feeds the remaining three—Molodo, Kourouma, and Niono. The four rice-producing sectors are divided into a total of 23 production units, all but one of which are managed by farmers.[9]

The Office is an extensive system supporting a comparatively small population of 47,000 on 40,000 ha of rice.[10] The average holding is 9.5 ha, the mode is about 7 ha, and a few farms are as large as 80 ha. Some farmers have millet and sorghum fields, about which little is known, although such holdings in the Office are probably less important than they are in the Ségou or Mopti project. Oxen draft power is universal and holdings of animals and equipment per ha by Office farmers are greater than those of Ségou or Mopti farmers, a fact that explains the larger Office holdings. The water control in the Office makes average gross paddy yields the largest in Mali, more than 2,250 kg/ha, and maximum yields sometimes approach 5,000 kg.[11]

The crop calendar begins in April with a shallow pre-irrigation to permit plowing before the first usable rains in late May. Plowing, broadcast seeding, and harrowing are done throughout May and June, and sometimes as late as the first third of August. A single hand weeding is done in July and August. Fertilizer is applied on roughly 30 percent of farms in doses of 50 to 100 kg of urea per ha. Fields are inundated in mid-August and drained after the first of November. Harvesting with sickles begins in December and continues until the end of January. The Office threshes 80 percent of the paddy mechanically with stationary threshers, using its own machine and crews of hired labor, and charges farmers a fixed amount of paddy per ton threshed. Farmers thresh about one-fifth of the harvest by hand and transport their share of the crop to the household by cart or by donkey.

Although the Office has recently been quite successful in raising yields, measures to intensify production are necessary if yields are to continue to grow. A program of intensification has been defined that includes mechanical field leveling, line seeding, and increased use of inorganic fertilizers. The resulting field technique is the intensified Office technique

[9]The remaining unit is used as a seed farm by the Office's administration. There are also some rice lands (*hors casier*) outside the officially developed area, which may produce 2,000 to 3,000 mt of paddy.

[10]Note, in comparison, that Opération Riz Ségou supports 115,000 to 125,000 people on a rice area of 35,000 ha (8).

[11]Yield estimates in the Office are made from crop cuttings and perhaps overstate true yields.

(ONI) noted in Table 10.1. Although this technique is not now practiced, estimates have been made of its costs and returns and they are discussed below (*14*).

The fourth production technique is the rainfed/swamp technique in southern Mali that is practiced on 15,000 to 20,000 ha spread over a network of small river basins in an area of 4.5 million ha. The only true rainfed technique is in the highest rainfall areas along the borders of the Ivory Coast and Guinea, where rice is sometimes grown in rotation with corn, sorghum, peanuts, and vegetables. Swamp variants of the technique use rainwater indirectly from the overflow of small rivers or from slope runoff, and water control is provided by simple hand-built earthworks. Rice is grown in swamps in rotation with other cereals and double-cropped with vegetables.

Cotton is the principal cash crop of the region, millet and sorghum the principal cereals, and rice is generally grown to complement these other crops. Average rice field size varies with the use of animal traction; it is 1.0 to 1.5 ha on farms that have oxen draft power and less than 0.5 ha on farms that do not. Large farms managed by men with manual cultivation, broadcast seeding, and sickle harvesting have been used in the analysis as the traditional swamp technique (TS). Women have smaller holdings which they cultivate manually with transplanting and finger-knife harvesting, but lack of data prevents the inclusion of that technique in the analysis. Average gross paddy yields are roughly 1,200 kg/ha with animal traction and slightly less without.

Opération Riz Pluvial et Bas Fonds (Sikasso) has begun a program to improve the traditional swamp/rainfed technique described above. The project provides water control to fields by constructing barrages across small rivers, which can be used to divert water to fields if rainfall is insufficient. Plows and other equipment are sold on credit, selected *Oryza sativa* seeds are distributed, and fields are deep-plowed when they are first cultivated. There are few other technical changes. Farmers do not use inorganic fertilizers or pesticides, seeds are broadcast, and paddy is threshed by hand. Average gross paddy yields have increased to 1,800 kg/ha with this technique (IPS in Table 10.1).[12]

Improved areas in the project have reached slightly more than 4,000 ha in 1975–76, with plans for expansion to 11,500 ha by 1980. Studies of the region have estimated the rice land availability to be 200,000 ha (*9*), but its development is restricted by low population density (owing to the presence in many areas of onchocerciasis, a disease commonly known as river blindness) and the very poor condition of roads, which makes input delivery and product marketing difficult.

---

[12] The project has raised average farm size to about 3 ha, thus contributing to the installation of rice as a staple crop in a region where it had been secondary.

TABLE 10.2. *Farm to Mill Transport*

Field technique	Mode and distance of transport	Designator
Office, current (ONC) and Office, intensified (ONI)	By 12-ton truck over 30 km of poor dirt roads or by barge over 30 km of canal	PCO
Ségou, current (CFS), and Ségou, intensified (CFSI)	By 12- ton truck over 45 km to small hullers in Ségou, or 80 km to industrial mill in Diafarabé, both on poor dirt and paved roads	PCS[a]
Mopti, current (CFM), and Mopti, intensified (CFMI)	By 12-ton truck over 35 km of poor paved road to industrial mill at Sévaré	PCM
Sikasso, improved swamp (IPS)	By 20-ton truck over 50 km of poor dirt and paved roads	PCR
Sikasso, traditional swamp and rainfed (TS), and Mopti, traditional flooded (TFM)	By cart or by head-lead, costs of which are included in field labor times and in capital value of farm equipment	PCF

SOURCES: PCO, 6; PCS, 8; PCM, 7; PCR, 9; and PCF, 1, 2, 9.
   NOTE: Distances are averages from field to mill, weighted by the percent of total paddy collected from each producing area.
   [a] An average distance of 60 km was used in the resource cost analysis.

Transport to Mill and to Household

Table 10.2 shows the systems of paddy collection and mill transport associated with each field technique. It has been assumed, except for field-to-household transport in the traditional Delta (TFM) and traditional swamp/rainfed (TS) techniques, that public agencies control the collection and mill transport of paddy. So little is known about the activities of private traders in paddy that their activities have not been included in the analysis.

Transport of threshed paddy from fields to households in the Delta and in Sikasso is done by head-load, donkey, or animal cart. Delta rice fields are planted farther from villages than are maize or millet fields, but no more than 90 minutes on foot. Sikasso rice fields are closer to villages, at most 30 minutes on foot. Net yields with these techniques are not much greater than 1,000 kg of paddy, so the output of one hectare could be carried in two cartloads or 30 to 40 head-loads. Labor costs of paddy transport to households are included in field labor times in these techniques.

Transport of threshed paddy to the warehouses and mills in the Ségou and Mopti projects and in the Office du Niger is in two stages. Paddy is first transported from fields to adjacent roads in tractor-pulled trailers. The Office then moves paddy to its mills over an average of 30 km, most of them on poor dirt roads. Paddy in the Ségou project is transported over similar roads an average of 45 km to small hullers in Ségou or 80 km to a mill in Diafarabé. Paddy in the Mopti project is hauled over an average distance of 35 km, mostly on a paved road to the Sévaré mill. The small quantity of paddy bought by the Sikasso rice project is collected and moved to the project's huller in Sikasso, an average of 50 km on very poor roads. Collection, sacking, and transport costs incurred by the projects are reimbursed at official cost schedules established by the Ministry of Finance.

Milling

Table 10.3 shows characteristics of the most important milling techniques. The Office des Produits Agricoles du Mali (OPAM) and the Office du Niger operate all large mills except one, at Tamani, which is not now in use. Apart from the Tamani mill, private milling in Mali is restricted to hand pounding and small electric or diesel steel cylinder hullers.

In the two traditional field production systems described (TFM and TS), women hand-pound paddy. The work is done either individually or in large groups after harvest for home consumption, and some women earn money by pounding for other families. Where electric or diesel hullers have been installed (for example in villages near the Ségou rice project), hand pounding has been rapidly displaced.

Small electric or diesel-powered hullers are not imported officially, but

TABLE 10.3. *Key Characteristics of Rice-Milling Techniques*

Milling technique	Projected full capacity[a] (000 mt paddy/yr)	Percent brokens[b]	Milling ratio	Rice milled in 1976 (000 mt)	Unit cost (mf/kg)	By-products[c]
Industrial[d]						
Molodo (IMO)	18	60%	0.65	12.5	16.4	yes
Kourouma (IMO)	18	60	0.67	12.4	16.4	yes
Kolongo (IMO)	12	70	0.64	4.3	16.4	yes
Sévaré (IMS)	15	40	0.57	8.4	13.3	yes
Small steel cylinder hullers (SM)	37.50	60–70	0.45–0.70	n.a.	8–15	some
Hand pounding (HP)	—	80–100	0.70	75–80	30	some

SOURCES: Industrial (except Sévaré), 6; industrial (Sévaré), 7; small steel-cylinder hullers, 11; hand pounding, 15; Diafarabé, Tamani, and Kourouba, 14 and 15.

[a] 5,000 hours of annual operation assumed (i.e. 250 days at 20 hours/day).

[b] Percentages of brokens are weighted averages (e.g. if the mill produces 10 tons of whole grains and 10 tons of 100 percent brokens, then the percentage of brokens shown would be 50). The Sévaré mill alone parboils rice.

[c] The industrial techniques have by-products that are used for animal feed and flour; small steel-cylinder hullers and hand-pounding techniques have by-products that are used for construction and animal feed, or have no by-products.

[d] The Molodo and Kourouma mills have apparently operated at greater than their rated capacities. Four other mills have either only just begun to operate or recently started up again after long periods of inactivity: N'Diebougou, which began operations in 1976–77, has a projected full capacity of 18,000 mt of paddy per year; Diafarabé (IMS) and Kourouba, both owned by OPAM, have projected capacities of 12,500 and 6,000 mt of paddy per year, respectively, and have just started up again after roughly ten years of inactivity; and Tamani, a privately owned mill whose owner must nonetheless buy paddy and sell rice at prices fixed by OPAM, has a projected capacity of 11,000 mt paddy per year.

they are widely used in rice-growing areas around Ségou, Mopti, and San. These machines do not have parboiling or polishing capacity and produce high percentages of broken grains. Outturns are 45 to 70 percent of rice, depending on the skill of the operator and the quality of the paddy. Hourly capacities are 0.15 metric tons (mt) of paddy, implying an annual capacity of 375 mt if they are operated 2,500 hours per year. Estimates of the total number of machines vary from 100 to 150, although many of them are not used, or are used infrequently owing to a shortage of spare parts and fuel. By-products are used to make plaster, or, if the husks have been removed by winnowing, to feed animals.

Industrial rice milling is done principally in the Office du Niger. The Office has four mills, three with hourly capacity of 3.6 mt of paddy, and the fourth with hourly capacity of 2.5 mt. No parboiling is done in these mills. Flour and bran are sold to Office farmers as feed; husks are used in one mill to drive a generator and are discarded in the others. The only other currently operating industrial mill in Mali is run by OPAM at Sévaré in association with Opération Riz Mopti. The Sévaré mill has parboiling capacity and produces mainly white rice. Husks are used to fuel a generator, and bran and flour are sold as feed.

Marketing and Consumption

Table 10.4 shows the principal markets and modes of supply in Mali and in the Ivory Coast and Senegal where Malian rice is or might be exported. OPAM has a legal monopoly on cereals marketing, but private traders are allowed to operate unofficially. (Analysis of the private sector has not been done here because so little is known about it.) OPAM accepts rice from the Office du Niger and Sévaré mills at rates established by the Ministry of Finance and transports it to urban markets. Until 1977 OPAM did not have sufficient truck capcity to transport its purchases of all cereals and had to hire private truckers to do so. OPAM has since increased its truck fleet and now seems to have sufficient capacity. Rice is transported to the 42 administrative subregions in Mali, where it is distributed to consumers through a system of stores.

The majority of OPAM's deliveries are to the Bamako market. The Gao and Kayes urban markets are of secondary importance, and OPAM loses money on its deliveries to these markets owing to insufficient handling and transport margins. This is in part due to the government's policy of maintaining a fixed consumer price of rice throughout the nation. Small amounts of rice are also delivered to the Ségou, Mopti, and Sikasso markets from the mills associated with the rice projects in those areas.

Export markets have also been identified in Abidjan, Bouaké, and Dakar. Recently, the Office du Niger exported rice to the Ivory Coast, and it has been assumed in the following analysis that future exports would also

TABLE 10.4. *Principal Markets and Modes of Supply for Malian Rice*

Market	Activity designator	Source of supply	Distance and modes of transport
Bamako	BK	Ségou (whether from Office du Niger or from Opération Riz Ségou)	240 km by 12-ton truck over new paved road
Bamako	BK	Sikasso (whether from traditional rainfed technique or from Opération Riz Sikasso)	380 km by 12-ton truck over paved road
Bamako	BK	Mopti (whether from traditional flooded technique or from Opération Riz Mopti)	620 km by 12-ton truck over 380 km of poor paved road and 240 km of new paved road
Kayes	KAY	Ségou (whether from Office du Niger or from Opération Riz Ségou)	240 km by 12-ton truck over new paved road and 495 km by rail from Bamako to Kayes
Abidjan	ABN	Ségou (whether from Office du Niger or Opération Riz Ségou)	1,145 km by 12-ton truck over mostly good paved road
Dakar	DKR	Ségou (whether from Office du Niger or from Opération Riz Ségou)	240 km by 12-ton truck over new paved road to Bamako and by rail to Dakar
Bouaké	BKE	Ségou (whether from Office du Niger or from Opération Riz Ségou)	773 km by 12-ton truck over paved road
Abidjan	ABN	Mopti (from Opération Riz Mopti)	1,305 km by 12-ton truck over mixed poor and good paved road
Dakar	DKRM	Mopti (from Opération Riz Mopti)	620 km to Bamako by 12-ton truck over 380 km of poor paved road and 240 km of new paved road; by rail from Bamako to Dakar
Bouaké	BKEM	Mopti (from Opération Riz Mopti)	933 km by 12-ton truck over mixed poor and good paved road
Gao	GAO	Mopti (from Opération Riz Mopti)	River transport from Mopti
Sikasso farm	SIF	Sikasso (from traditional rainfed technique)	Cart or head-load for on-farm consumption
Mopti farm	MPF	Mopti (from traditional flooded technique)	Cart or head-load for on-farm consumption

SOURCES: 3; except as follows: Sikasso farm, 9; Mopti farm, 2 and 7.

be sent by road. Rice sent to Dakar has been assumed to go from Ségou to Bamako by road and from Bamako to Dakar by rail.

Incentives and Shadow Prices

The system of incentives in the Malian rice sector has several general characteristics and several that are specific to the rice projects. The general characteristics are officially fixed product and input prices, plus a broad exemption of agriculture from tariffs and indirect taxes. Paddy and rice prices vary only by quality, never by location of market, production system, or season. They are established by ministerial decree, announced at the beginning of each crop season in June, and enforced by OPAM, which starts its crop-buying season in November. Rice imports are controlled by quotas and small tariffs.

All agricultural inputs are sold through a state agency, the Société de Crédit et d'Equipement Rural (SCAER), either directly to farmers or indirectly through the development projects. Direct subsidies on capital equipment such as plows, carts, and harrows existed in the past but were removed in the 1976–77 crop season; subsidies, however, were maintained on fertilizers and fungicides, although the government plans to remove them soon. There are also some small indirect subsidies on inputs resulting from the insufficiency of margins allowed for shipment of inputs from the SCAER depot in Bamako to development projects. SCAER's costs of operation are partly supported by taxes on cotton and peanut production.

The exemption of agriculture from tariffs and indirect taxes on direct inputs (for example fertilizers) or indirect inputs (for example gasoline used in project vehicles) is nearly complete. In principle, SCAER and the rice projects (including the Office du Niger) pay only a 3 percent tax on imports and no sales taxes at all, but, in practice, some additional duties are paid, especially those included in the prices of vehicles, petroleum products, and construction services bought from local suppliers. The rice projects and other state agencies also pay taxes on wages and salaries. There are no significant indirect taxes on traditional agriculture except a cattle head tax, which is poorly enforced.

The rice projects and the Office are semiautonomous public agencies, which provide participating farmers with two classes of goods and services unavailable to others. First, in return for a fixed in-kind fee per ha of land allocated to them, farmers receive construction and maintenance of irrigation works and extension and administrative services. Farmers are also required to meet a marketing quota per ha of project land; this requirement amounts to a tax because the official price of paddy has generally been below market prices.

TABLE 10.5. *Farm Subsidies and Water Charges*
(Malian francs/mt rice)

| | Class 1 | | | | Class 2 | | | | | | | | Class 1 and Class 2 total |
Technique	Water and land development costs	Extension and administrative overhead costs	Water charges	Net subsidy	Seeds	Fertilizer	Pesticides	Land preparation	Threshing	Credit	Other[a]	Total subsidy	
ONC	0	15,664	11,514	4,150	-115	790	0	0	-3,052	40	237	-2,100	2,050
ONI	5,470	14,528	11,105	8,893	513	3,163	0	0	-3,844	168	287	287	9,180
CFS	22,800	5,514	6,683	21,631	446	0	0	0	-277	0	-245	478	-21,153
CFSI	12,768	5,146	5,347	12,567	-200	2,285	0	0	-199	78	378	2,342	14,909
TS	0	0	0	0	-14	0	0	0	0	0	-167	-181	-181
IPS	16,064	2,803	0	18,867	255	0	0	0	0	12	356	623	19,490
TFM	0	0	0	0	-39	0	0	0	0	0	-167	-206	-206
CFM	28,501	4,595	4,774	28,322	-48	0	0	0	5,507	0	197	5,652	33,974
CFMI	15,951	11,161	6,416	20,706	-21	2,285	0	0	2,164	78	377	4,883	25,589

SOURCES: ONC and ONI, 6; CFS and CFSI, 8; TS and IPS, 9; and TFM, CFM, and CFMI, 7.

NOTE: Calculations are made with producer price of paddy equal to 1975–76 level of 40 mf/kg. Costs per ha have been converted to costs per mt/rice by dividing by the paddy yield and by the milling ratio commonly associated with the field technique.

[a]Tools, animal feed, animal traction equipment.

The second includes goods and services for which payment is made directly—machine services, inputs and their delivery, selected seeds, and credit. Farmers pay in kind for threshing, and in cash for inputs, selected seeds, and credit at the time of harvest. The projects sell inputs and provide credit for SCAER's account and deliver paddy to mills for OPAM's account. For those services the projects are reimbursed at rates established by the Ministry of Finance. Estimates of the real costs of each class of services, farmer payments, and net subsidies are shown in Table 10.5.

Marketing quotas and the requirement that some service payments be made in kind are intended to enable OPAM to control sales of paddy and rice and to ensure the financial viability of the projects. Official marketings have grown rapidly as a result of the growth of project output, but the financial status of most projects is poor (with the possible exception of the Office du Niger), and they are subsidized, usually with concessional aid. Subsidies finance the construction and maintenance of irrigation works and the costs of extension services and administrative overhead. The government hopes to reduce subsidies on recurrent costs.

Shadow Prices

The shadow prices used in the net social profitability analysis are shown in Table 10.6. (Their derivation and theoretical justification are discussed in Appendix B.) Of the adjustments necessary to derive shadow from market factor prices, only the removal of production taxes, specifically those on exports of cotton and peanuts, is likely to be of any quantitative significance. Such taxes will have important effects on shadow prices only if factors used in rice production are withdrawn from cotton or peanut production and only if they are in inelastic supply. Land is excluded because there appears to be little competition for land among rice, cotton, and peanuts (with the exception of areas in southern Mali, which are not now largely devoted to rice).

Unskilled labor is the only factor for which both of the above conditions might obtain. Evaluation of the sources of labor migration to rice producing areas suggests, however, that most laborers come from millet-producing areas (for example the Seno Plain and the Dogon Plateau) and not from the cotton and peanut areas. Unadjusted market prices were thus used as the basis of the shadow price estimates of unskilled labor. Surveys were made by the author in the informal labor markets at Sikasso, San, Mopti, Sofara, Ségou, and Bamako and in rice-producing areas west of Ségou to gather information on wages and the origins of workers. Wages across tasks (for example harvesting rice) were found to be roughly equal throughout the country, except in the Office du Niger and in the town of Sikasso, where they were higher. In Sikasso, this difference results from

TABLE 10.6. *Shadow Prices of Primary Factors by Region*

Primary factor	Nation	Office du Niger	Ségou	Mopti	Sikasso
Unskilled labor (*mf/man-day*):					
Men	500	700	400–500	400	600–700
Skilled labor (ratio of					
shadow to market wage):	1	1	1	1	1
Land (*mf/hectare*)	0	0	0	0	0
Capital (*percent*):					
Informal, rural	20				
Formal, public on					
irrigation works	2.5				
Formal, public on					
farm equipment	8				

SOURCES: For labor and land—surveys made by author. For capital—see Table 10.5.

the proximity of higher wage areas in the Ivory Coast; in the Office, it arises from the strong demand for hired labor in irrigated rice and sugarcane production.

Shadow prices of capital are based on the assumption that the capital market is segmented and that capital is available for certain purposes at concessionary rates of interest. Segmentation is maintained by credit rationing; farmers in the rice projects, for example, pay lower rates of interest than do other farmers because access to credit is restricted by project agencies.

The shadow price of land is assumed to be zero. Land rents are rare in Mali (in money or in kind) and seem to be paid only for site value, such as in the Office du Niger. Alternatives to rice production in the Office (for example wheat, long-staple cotton, or sugarcane) are discussed below in the sensitivity analysis.

Activity Combinations

Activity combinations used in calculations of social and private profitability for field collection, milling, and distribution techniques are presented in the tabulation on p. 346.

Activities were generally joined as they are in current practice, with two exceptions. The first is that the intensified field techniques (OMI, CFSI, and CFMI) were combined with post-harvest activities based upon Malian plans for expansion of milling capacity and judgments about the importance of these techniques in supplying the various markets listed in Table 10.4. The second is that export markets were identified and combined with production, collection, and milling techniques based upon an earlier study (16). These combinations have been used to estimate private

ONC/PCO/IMO/BK   CFSI/PCS/IMS/BK[13]   CFM/PCM/IMS/BK
ONC/PCO/IMO/KAY   CFSI/PCS/SM/BK
ONC/PCO/IMO/ABN   CFSI/PCS/IMS/ABN[13]   CFMI/PCM/IMS/BK
ONC/PCO/IMO/DKR   CFSI/PCS/IMS/DKR[13]   CFMI/PCM/IMS/GAO
ONC/PCO/IMO/BKE   CFSI/PCS/IMS/BKE[13]   CFMI/PCM/IMS/ABNM
                                                  CFMI/PCM/IMS/DKRM
ONI/PCO/IMO/BK   TS/PCF/HP/SIF   CFMI/PCM/IMS/BKEM
ONI/PCO/IMO/ABN   TS/PCR/IMS/BK
ONI/PCO/IMO/DKR
ONI/PCO/IMO/BKE   IPS/PCR/IMS/BK

CFS/PCS/IMS/BK[13]   TFM/PCF/HP/MOF
CFS/PCS/SM/BK   TFM/PCM/IMS/BK

and social costs and returns throughout the rice sector; varying the activities contained in any one combination (for example changing only the marketing activity) allows isolation of critical influences on those costs and returns.

Private and Social Profitability

Private profitability is defined only at the farm level. OPAM and other state agencies are assumed to cover their costs, including those resulting from taxes on the goods and services they purchase, through a combination of revenue and public subsidy to the three post-harvest activities. Farm private profitability is defined as gross revenue (i.e. net paddy yield times the paddy price) minus the sum of domestic factor costs and tradable input costs valued at market prices plus taxes.[14] The farmgate paddy price is assumed to be the 1975–76 official price of 40 Malian francs (mf)/kg.

Net social profitability is equal to the c.i.f. price of imported rice minus tradable input and domestic factor costs. The c.i.f. price of rice in Bamako is assumed to be 182.2 mf/kg (i.e. $364.4/mt). The prices of imported rice and of tradable inputs are converted to domestic currency units at the official exchange rate; domestic factor costs are valued at their social opportunity costs or shadow prices shown in the tabulation. The difference between private and net social profitability (per unit of rice output) equals the differences between the actual and the c.i.f. prices of rice in any given market, plus the sum of taxes or subsidies that affect the four activities in the rice sector, plus the sum of divergences between market and shadow factor prices.

[13] The Ségou field and collection techniques have been combined with the milling activity corresponding to the industrial mill at Diafarabé, which OPAM has recently rehabilitated.

[14] Seed costs and in-kind service fees are converted to money at the official paddy price of 40 mf/kg.

TABLE 10.7. *Net Private and Social Profitability, Lower Producer Price*
(Malian franc/mt of rice)

Field technique	(1) Net social profitability	(2) Resource cost ratio	(3) Effective rate of protection	(4) Private profitability[a]	(5) (1) − (4)
ONC	67,355	0.560	−0.275	−3,945	71,300
ONI	58,374	0.592	−0.358	−6,137	64,511
CFS	39,464	0.736	−0.354	−10,546	50,010
CFSI	60,364	0.591	−0.386	3,858	56,506
TS	36,998	0.720	−0.419	−32,080	69,078
IPS	47,252	0.648	−0.337	−3,681	50,933
TFM	1,829	0.988	−0.411	−58,864	60,693
CFM	16,681	0.886	−0.226	−12,504	29,185
CFMI	35,350	0.749	−0.282	−2,634	37,984

Field technique	(6) C.i.f. price minus market price	(7) Total net taxes	(8) Net farm taxes	(9) Net off-farm taxes
ONC	66,404	4,896	−2,751	7,647
ONI	66,404	−1,893	−9,540	7,647
CFS	62,975	−12,965	−21,923	8,958
CFSI	62,975	−6,469	15,427	8,958
TS	61,269	7,809	−1,585	9,394
IPS	61,269	−10,336	−19,730	9,394
TSM	53,319	7,374	−3,292	10,666
CFM	53,319	−24,134	−34,800	10,666
CFMI	53,319	−15,335	−26,001	10,666

[a]Calculated with official producer price of paddy of 40 mf/kg.

Net social profitability measures the natural comparative advantage in rice production of a country, as defined by its resource endowments, geographic position, and technical efficiency of production, with respect to a given set of world prices. Private profitability measures the incentives provided to economic agents in rice production by government policies. Therefore, analysis of social and private profitabilities and comparisons of divergences between the two can help to assess the relative importance of government policies, resource endowments, choice of production techniques, and world prices of inputs and outputs.

Private and social profitabilities of each technique, when combined with its lowest cost collection and milling technique for marketing to Bamako, are shown in Table 10.7. Techniques are grouped by degree of water control.

The Office du Niger

The current field technique (ONC) is less privately unprofitable than the intensified one (ONI) at the 1975–76 producer price, although the difference is small. This suggests that farmers will be more or less indifferent to the adoption of the intensified technique when the two are compared on a per hectare basis, but that the reduction in average holding associated with the introduction of the intensified technique will reduce total farm income and thus discourage use of the intensified technique.

The current field technique is also more socially profitable than the intensified one, and here the difference is more marked than the difference between private profitabilities. The intensified technique, which benefits more from farm level subsidies than the current one, is less socially profitable because the investment and variable costs (for example, field leveling and fertilizers) are not entirely offset by the higher paddy yields.

In both techniques the great difference between the social and private profitability is largely due to the difference between the c.i.f. price of rice and the official price, as columns (5), (6), and (7) of Table 10.7 show. Net taxes or subsidies in the rice sector are much less important.

Opérations Riz Ségou and Mopti

Three of the four techniques (CFS, CFM, CFMI) are privately unprofitable when official producer prices for paddy are used to evaluate output. The intensified techniques are more privately profitable than the current ones. That farmers have not yet adopted the more profitable technique can be attributed largely to the risk and unfamiliarity of fertilizer use in the intensified techniques. Important private cost differences exist between the two projects; the higher-cost Mopti farmers benefit from greater subsidies on irrigation work and extension services (Class 1 in Table 10.5).

On the other hand, all four techniques are socially profitable. The intensified Ségou technique is the most socially profitable, followed by the current Ségou and intensified Mopti techniques and, at a substantially lower level, the current Mopti technique. The social profitabilities of the Mopti techniques are reduced, in comparison to the Ségou techniques, by the extra margin of transport costs from Mopti to the Bamako market, as well as by their higher irrigation and labor costs.

The differences between social and private profitabilities in these four techniques are largely due to the margin between the c.i.f. and official prices of rice. Net subsidies explain a greater part of those differences in this group of techniques than they do in the two Office du Niger techniques, but their magnitude is still not very large, except in the two Mopti techniques (CFM and CFMI) where farm level subsidies are quite important.

Improved Rainfed/Swamp, Sikasso

The improved rainfed/swamp technique (IPS) is only slightly less privately profitable than the most profitable technique (CFSI). This is due in part to the fairly large farm subsidies it enjoys (there is, for example, no water charge in this project) and to the good yields achieved without fertilizer. This technique is also quite socially profitable, although less so than several of the others. The difference between the large positive social profitability of this technique and its private profitability is mainly due to the margin between the c.i.f. and official prices of rice, although the element of subsidies is important also, especially at the farm level.

Traditional Techniques

The two traditional techniques are both privately unprofitable at official prices. The rainfed/swamp technique (TS) is less unprofitable, although it is much less favorable than the improved technique (IPS) that is replacing it. The traditional Delta technique (TFM) is the least privately profitable of all nine and is greatly inferior in this respect to the four improved Delta techniques (CFS, CFSI, CFM, and CFMI) that are replacing it. These results explain the quite rapid adoption of the improved rainfed and Delta techniques in the last eight years.

The two traditional techniques are, in contrast, socially profitable, the Sikasso technique more so than the traditional Delta one. The large differences between social and private profitability in these two techniques are, again, due to the difference between the c.i.f. and official prices of rice; taxes and subsidies have little role in those differences, especially at the farm level, where producers are almost entirely unaffected by the fiscal system.

These comparisons show that all field techniques have strong comparative advantages for production of rice for the Bamako market, but that the current difference between the c.i.f. and official prices of rice makes farm private profitability often negative for rice delivered to Bamako. It is useful then to consider the incentive effects of increasing the farm price of paddy. This is done in Table 10.8, which presents comparisons of social and private profitability for a producer price of 50 mf. This change makes private profitability more generally positive. The two traditional techniques (TS and TFM) remain privately unprofitable, and the current Ségou and Mopti techniques (CFS and CFM) become only marginally so. Both Office techniques and the pair of intensified Ségou and Mopti techniques become positively profitable. Such a change would, therefore, encourage adoption of techniques that are more socially profitable in all four groups, with the exception of the Office du Niger, where the current technique is less privately but more socially profitable than the intensified one at the higher producer price.

Raising the producer price of paddy would also allow the government to reduce some subsidies now given to farmers. This would increase the value of the water charges paid by project farmers and reduce the major-

TABLE 10.8. *Net Private and Social Profitability, Higher Producer Price*
(Malian francs/mt rice)

Field technique	(1) Net social profitability	(2) Private profitability[a]	(3) (1) − (2)	(4) C.i.f. price minus market price	(5) Total net taxes	(6) Net farm taxes	(7) Net off-farm taxes
ONC	67,355	13,488	53,867	51,720	2,147	−5,500	7,647
ONI	58,374	14,416	43,958	51,720	−7,762	−15,409	7,647
CFS	39,464	769	38,695	47,800	−9,105	−18,063	8,958
CFSI	60,364	16,448	43,916	47,800	−3,884	−12,842	8,958
TS	36,998	−17,976	54,974	45,396	9,578	184	9,394
IPS	47,252	12,065	35,187	45,396	−10,209	−19,603	9,394
TFM	1,829	−46,500	48,329	37,446	10,883	217	10,666
CFM	16,681	1,341	15,340	37,446	−22,106	−32,772	10,666
CFMI	35,350	11,265	24,085	37,446	−13,361	−24,027	10,666

[a] Private profitability calculated with producer price of paddy equal to 50 mf/kg.

ity of current subsidies on irrigation works and extension services. This increase in the official price of paddy would essentially transfer the costs of subsidies from the government budget to consumers. Changing the paddy price from 40 to 50 mf/kg would increase wholesale rice prices about 13 percent.

The continued existence of the traditional techniques is attributable to several factors. One is that the relevant farmgate price of output is not the official paddy price but rather some average market price, which in Mali is generally higher than the official price and tends to increase returns to traditional production. A second is that the relevant farmgate price in subsistence production should include the imputed value of collection, processing, and marketing services included in the price of rice that the farmer would have to buy if he were not a producer. This would also tend to increase returns to traditional techniques. A third is that the improved techniques are not freely available; that is, the Delta polders and the improved lands in southern Mali cannot now accommodate all producers who desire improved lands. These problems have not been analyzed in detail here because there is so little reliable information on market prices of paddy and rice and because a uniform change in prices would not change the ranking of techniques.

One important source of the social profitability in Malian rice production is the low cost of water control and the increased yields resulting from the introduction of water control into traditional production systems. This is shown most clearly in the comparisons among the traditional Delta technique (TFM) and its replacements in the Ségou and Mopti projects (CFS, CFSI, CFM, and CMFI). On the other hand, the social profitability of the traditional swamp technique in Sikasso (TS) is comparable to those of all but one of the improved Delta techniques, implying that

that region has a natural advantage in rice production because of its superior rainfall.

A second source of the social profitability of these techniques has been the introduction of improved seed varieties. This is most noticeable in the Sikasso rice project, where the improved technique (IPS) shows significantly greater social profitability than the traditional technique after the introduction of limited water control and of improved seeds. In the improved Delta techniques, the replacement of local *glaberrimas* with improved *sativas* explains perhaps half of the improvement in yields over the traditional technique.

The use of inorganic fertilizers explains little of the difference in social profitabilities among the current field techniques (ONC, CFS, CFM, and IPS).[15] Fertilizers are used only in small doses in the current Office du Niger technique, the most socially profitable one, where water control seems to be the most important factor in raising yields. Fertilizers do, however, increase yields in the improved techniques (ONI, CFSI, and CFMI). In spite of the fact that greater extension and overhead costs have been allocated to each hectare using the more intensive techniques, social costs fall per mt of rice, except in the Office du Niger. The small current use of fertilizers appears to be due more to risk and to farmers' unfamiliarity with them than it does to relative prices; the fertilizer-using techniques are generally more privately profitable than the current ones.[16]

Malian Rice Exports

Table 10.9 presents the net social profitabilities for several field techniques (ONC, ONI, CFSI, and CFMI) able to generate large marketings. Exports to Dakar are generally unprofitable because Senegalese consumers prefer cheaper broken grains, which sell at a discount of roughly 17 percent. Exports to Abidjan are profitable from all producing areas but Mopti, and exports to Bouaké are profitable from all areas. The differences between the net social profitabilities of these techniques for delivery to the Bamako market and to these neighboring foreign markets again demonstrate the importance of transport costs in the Malian comparative

[15] Current fertilizer use in the Office du Niger, for example, is roughly 30 kg of urea (46-0-0) per ha. Assuming the response of yields to nitrogen is 15 kg of paddy to 1 kg of nitrogen, the average yield increase in the current Office technique would be about 210 kg (i.e. $30 \times 0.46 \times 15 = 207$), or only 25 percent of the difference between yields there and yields in the flooded rice project (i.e. $2250 - 1400 = 850$; $210 \cong 850 \times 0.25$) (6).

[16] Experience with the San project is too short to serve as a guide to its long-run costs and returns. It is clear, nonetheless, that the project is more costly than its relevant alternatives in the flooded projects. Capital costs of irrigation in the San project are 2 million to 2.5 million mf/ha; capital costs in other polders of Opération Riz Ségou are less than 500,000 mf/ha (8).

TABLE 10.9. *Net Social Profitability of Malian Rice Exports*
(Malian francs/mt rice)

Technique[a]	Abidjan[b]	Dakar[c]	Bouaké[d]
ONC	12,310	−9,060	30,502
ONI	3,329	−18,041	21,521
CFSI	5,319	−16,051	23,511
CFMI	−12,294	−33,664	5,898

[a]The improved rainfed-swamp technique (IPS), although socially profitable for delivery to the Bamako market, has been excluded from the analysis of export potential because so little rice (less than 1,000 mt on average) is marketed through official channels from the Sikasso Rice Project. There is probably some rice sent to the northern Ivory Coast and to Upper Volta from the Sikasso region, but data on quantities, cost, and prices are unavailable for analysis of private marketing.
[b]C.i.f. price is 150,000 mf/mt ($300/mt), 25 to 35 percent broken rice.
[c]C.i.f. price is 125,000 mf/mt ($250/mt), 80 percent broken rice.
[d]C.i.f. price is equal to 159,800 mf/mt ($319.6/mt), 25 to 35 percent broken rice.

advantage. Exporting rice reverses the transport margin between the port c.i.f. price of rice and the c.i.f. price at Bamako, subtracting it from the Bamako price and adding it to the sum of domestic factor and tradable input costs, thus reducing net social profitability.

Sensitivity Analysis

The sensitivity of the results to changes in factor costs, paddy yields, milling outturns, and the world price of rice was analyzed to test the weak points in the data and to identify changes in the rice sector that might occur as a result of external changes in the Malian or world economy. Results of the sensitivity analysis are presented in Table 10.10. The elasticities are functions of the size of net social profitability and of the share of the factor in total initial costs.

Net social profitabilities are shown to be most elastic with respect to changes in the costs of unskilled labor. Unskilled labor enters the production process almost exclusively as direct field labor, and it is not, therefore, surprising to see that its effect is greatest in the traditional field techniques (TS and TFM), where farm costs are the largest component of total domestic factor costs. Unskilled labor elasticities are greater where production is for export or where it takes place at greater distances from the border (in Mopti, for example) because of the smaller net social profitabilities of those activities.

Net social profitabilities are less elastic with respect to changes in the cost of capital. The smallest elasticity is in activity combinations involving the current Office du Niger field technique (ONC), where the special assumption has been made that the costs of irrigation works are sunk. Elas-

TABLE 10.10. *Elasticities of Net Social Profitability with Respect to Yields and the Social Cost of Primary Inputs*

| Techniques | Yields | Milling outturns | Social costs | | |
			Unskilled labor	Skilled labor	Capital
ONC/PCO/IMO/BK	1.011	1.281	−0.623	−0.321	−0.338
ONC/PCO/IMO/KAY	1.797	2.275	−1.118	−0.595	−0.637
ONC/PCO/IMO/ABN	5.639	7.141	−3.484	−2.351	−2.312
ONC/PCO/IMO/DKR	7.262	9.196	−4.571	−2.599	−2.473
ONC/PCO/IMO/BKE	2.245	4.083	−1.385	−0.844	−0.854
ONI/PCO/IMO/BK	1.314	1.624	−0.632	−0.375	−0.476
ONI/PCO/IMO/ABN	23.828	29.448	−11.493	−8.906	−10.231
ONI/PCO/IMO/DKR	4.217	5.212	−2.077	−1.356	−1.690
ONI/PCO/IMO/BKE	3.577	4.421	−1.723	−1.207	−1.443
CFS/PCS/IMS/BK	3.527	4.140	−1.591	−0.659	−1.688
CFSI/PCS/IMS/BK	1.337	1.653	−0.630	−0.390	−0.462
CFSI/PCS/IMS/ABN	19.769	24.937	−9.354	−7.457	−8.110
CFSI/PCS/IMS/DKR	4.542	5.615	−2.194	−1.483	−1.737
CFSI/PCS/IMS/BKE	3.558	4.398	−1.681	−1.217	−1.370
TS/PCR/IMS/BK	2.600	3.163	−2.580	−0.370	−0.392
TS/PCF/HP/SIF	14.281	14.281	−1.775	−0.003	−0.072
IPS/PCR/IMS/BK	1.515	1.907	−0.963	−0.338	−0.589
TFM/PCM/IMS/BK	24.984	29.499	−20.807	−3.329	−6.253
TFM/PCF/HP/MOF	21.831	21.831	−2.909	−0.010	−0.432
CFM/PCM/IMS/BK	7.470	8.941	−3.589	−1.588	−3.625
CFMI/PCM/IMS/BK	2.558	3.175	−1.078	−0.741	−1.128
CFMI/PCM/IMS/ABN	14.890	18.479	−6.277	−4.696	−6.842
CFMI/PCM/IMS/DKR	2.217	7.751	−0.957	−0.714	−1.049
CFMI/PMC/IMS/BKE	7.537	9.354	−3.177	−2.609	−3.621
CFMI/PCM/IMS/GAO	1.578	1.959	−0.666	−0.394	−0.664

ticities of capital costs are slightly higher in the other combinations, but there are no systematic differences among them except for the groups involving the current Ségou and Mopti field technique (CFS and CFM); this is due to the comparatively large share of irrigation in total costs and low yields in those projects.

Elasticities of skilled labor costs are smaller than those for unskilled labor costs because the share of that factor in the total costs of all techniques is smaller. Skilled labor enters the production process almost entirely off-farm (for example in wages paid to mill workers or to drivers), and its importance increases, therefore, with growing farm yields of paddy owing to the inverse effect of increased yields on farm costs. Elasticities are slightly greater when production is for export, reflecting the added weight of skilled labor in transport costs.

The economic costs of land have been assumed throughout to be zero. Although land rents exist in some areas (for example Opération Riz Ségou), they are paid for specific types of empoldered rice land, i.e. as

TABLE 10.11. *Changes in Social Profitability of Office du Niger Techniques at Selected Land Rents*
(Malian francs/mt rice)

Net social profit when land rent[a] is equal to:	ONC	ONI
0	67,355	58,374
10,000	60,159	53,747
50,000	31,376	35,239
93,493	0	15,245
126,026	−23,250	0

[a] In mf/ha, converted to mf/mt rice by dividing by appropriate net paddy yields and milling outturns.

payments for capital embodied in land that can be used only in rice production. The only producing area where land may be said, perhaps, to have alternative value is in the Office du Niger, where the capital costs have been considered sunk and where alternatives include cotton, sugarcane, and wheat. As Table 10.11 shows, the rents generated in any of those activities would have to be equal to more than 90,000 mf/ha for net social profitability in the current Office technique to be negative. Although the intensified Office technique is less socially profitable than the current one, land rents must be greater than 125,000 mf/ha if that technique is to become socially unprofitable because of the higher paddy yields produced in the proposed intensified technique.

The elasticity of net social profitability with respect to paddy yields is large, reflecting the fact that the average cost of the most expensive of the four activities, farm production, varies inversely with yields. Social profitability can be improved to the extent that yields increase faster than the costs of production associated with the yield-increasing innovations. There is no general pattern of sensitivity to yields across the group of activities except that the net social profitabilities calculated for export markets are more sensitive to changes in yields than are those calculated for domestic markets, because their central values are smaller initially.

The elasticity of net social profitability with respect to changes in milling outturns is also large, reflecting the influence of outturns on average costs of field production, paddy collection, and milling. Outturn elasticities are greatest in the highest-cost activities and least in the lowest-cost ones. One peculiarity of this elasticity calculation, which is shared by the ones for yields, is that increases in social profitability from efficiency improvements (for example increases in outturns due to more careful milling) are greatest in export activities in which net social profitabilities are smaller than in domestic markets

Changes in net social profitability with respect to the world price of rice, shown in Table 10.12, must be interpreted cautiously. The natural protection provided rice production for domestic markets allows quite large changes in world rice prices without endangering the social profitability of import substitution; for example, the world price of rice would have to fall to roughly $230/mt (about three-fourths of its average value in this study) before the social profitability of the current Office due Niger technique would become negative. On the other hand, rice production for export is very sensitive to changes in world prices, and its social profitability would become negative even in the best market at Bouaké if the world price were to fall only 10 percent.

Among the factor prices, only changes in the costs of unskilled labor are likely to make much difference in total social costs and net benefits of rice production. This implies that errors in measurement in field labor times would have important consequences for estimations of net social profitability, as would errors in estimation of the shadow price of unskilled labor. Such errors are unlikely to affect the relative social profitability of rice compared with that of its important Malian alternatives (millet, cotton, and peanuts), however, because changes in the shadow price of labor would influence the entire agricultural sector, thus changing net social profitability more or less evenly across it. There is also no reason to believe that labor times in rice production are systematically understated.

Only the special case of the social profitability of rice production in the Office du Niger was analyzed with respect to the costs of land, as shown in Table 10.11. This analysis demonstrates that if land has no alternative economic value, more extensive techniques (i.e. the current technique), though lower-yielding, are more socially profitable.

The elasticities of net social profitability with respect to paddy yields and milling outturns are large, implying that errors in estimation of these parameters would have important effects on net social profitabilty estimates. A conservative approach is taken to estimation of yields in this analysis, however, so, if anything, net social profitabilty is underestimated. The problem of yield estimates is probably most important in the rainfed and flooded field techniques, where production varies because of rainfall and flood variability. Discount factors were applied to average yields in average rainfall and flood years to include some of the cost effects of yield variations in the analysis; to the extent, therefore, that production is less variable than it has been in those techniques, net social profitability will be improved. Milling outturns can be increased by more careful milling, which would improve net social profitability. Comparison of Malian outturns to outturns in similar milling techniques in other areas of the world suggests, however, that little further improvement can be expected because Malian mills have already approached their technical limits.

TABLE 10.12. *Net Social Profitability in Relation to the World Market Price of Rice*
(Dollars/mt)

Activity	Net social profitability for a world price (Thai 25–35 percent brokens) of								
	$200/mt	$250/mt	$300/mt	$350/mt	$400/mt	$450/mt	$500/mt	$550/mt	$600/mt
ONC/PCO/IMO/BK	−27	24	74	123	173	224	274	325	375
ONC/PCO/IMO/KAY	−44	6	56	106	156	206	256	306	356
ONC/PCO/IMO/ABN	−72	−22	25	78	128	178	228	278	328
ONC/PCO/IMO/DKR	−65	−18.12	32	85	135	185	235	285	335
ONC/PCO/IMO/BKE	−55	−5	44	95	145	195	245	295	345
ONI/PCO/IMO/BK	−44	6	56	106	156	206	256	306	356
ONI/PCO/IMO/ABN	−90	−40	7	60	110	160	210	260	310
ONI/PCO/IMO/DKR	−82	−32	16	68	118	168	218	268	318
ONI/PCO/IMO/BKE	−73	−23	26	77	127	177	227	277	327
CFS/PCS/IMS/BK	−99	−49	1	51	101	151	201	251	301
CFSI/PCS/IMS/BK	−42	8	58	108	158	208	258	308	358
CFSI/PCS/IMS/ABN	−88	−38	12	62	112	162	212	262	312
CFSI/PCS/IMS/DKR	−81	−31	19	69	119	169	219	269	319
CFSI/PCS/IMS/BKE	−72	−32	28	78	128	178	228	278	328
TS/PCR/IMS/BK	−86	−36	14	64	114	164	214	264	314
TS/PCF/HP/SIF	−30	20	70	120	170	220	270	320	370
IPS/PCR/IMS/BK	−54	−4	46	96	146	196	246	296	346
TFM/PCM/IMS/BK	−150	−100	−50	0	50	100	150	200	250
TFM/PCM/HP/MOF	−76	−26	24	74	124	174	224	274	324
CFM/PCM/IMS/BK	−130	−80	−30	20	70	120	170	220	270
CFMI/PCM/IMS/BK	−89	−39	11	61	111	161	211	261	311
CFMI/PCM/IMS/GAO	−74	−24	26	76	126	176	226	276	326
CFMI/PCM/IMS/ABN	−103	−53	−3	47	97	147	197	247	297
CFMI/PCM/IMS/DKR	−128	−78	−25	22	72	122	172	222	272
CFMI/PCM/IMS/BKE	−120	−70	−20	30	80	130	180	230	280

Conclusions

The Malian rice sector enjoys a strong comparative advantage owing to low labor and irrigation costs, fairly high paddy yields, and efficient milling. Although private profitability in the sector is much less than social profitability, efficient techniques have been adopted rapidly by farmers because they are more privately profitable than traditional rice production techniques or than competing crops. This supply response has enabled the Malian government to achieve most of the objectives of its rice policy.

The results of this analysis bear on two additional policy problems. One is the preference of the government to subsidize OPAM's losses rather than to pass OPAM's costs on to consumers in the form of higher cereals prices or to encourage more widespread private marketing. The demonstration above that private profitability is less than social argues for a possible increase in official producer and consumer prices to reduce the subsidy on consumers and the tax on producers, who buy and sell, respectively, at officially controlled prices.

The second is the problem of food security. The approach of Malian policy makers has been to improve food security by using the current Office du Niger technique, the flooded projects at Ségou and Mopti, and to a lesser extent the improved rainfed areas of southern Mali. In an average rainfall and flood year, this policy is successful, allowing satisfaction of that year's demand as well as the accumulation of stocks for the following year. But in a drought or subnormal rainfall and flood year, this policy fails because of poor harvests in the flooded projects and in southern Mali, not only of rice but also of rainfed cereals such as millet and sorghum.

There are several approaches to achieving greater food security. One is to invest not in more rice production but in export crops (for example cotton and peanuts) which may have greater social profitabilities (11, 10) and to hold the foreign exchange thereby earned or saved as reserves with which to import cereals in drought years. This would be the reliance upon the international cereals market which has apparently been rejected by the Malian government as involving unacceptable social, economic, and political costs (5, 10).

A second approach is to constitute reserves out of domestic production (and, perhaps, aid) in good or even average years and to hold stocks as physical reserves against production failures in drought years. The size of necessary reserves and associated management problems might make such a policy very costly at the national level, and it is doubtful that constitution of such stocks can do much except relieve emergencies until international aid arrives if deficits are very large (10). Mali, for example, received more than 430,000 mt in cereal aid in 1973 and 1974, in addition to more than 100,000 mt of commercial rice imports. Consequently, whereas the first two policies (constitution of foreign exchange or grain reserves) could obviously improve security, they would imply some reliance upon food aid and commercial markets.

A third strategy is to invest more in irrigated agriculture in the Office du Niger or in pumping projects along the major rivers, such as the one at San on the Bani. This policy would guarantee production as opposed to holding interannual stocks. The analysis of net social profitability here shows that, whereas the most important irrigated techniques (ONC and ONI) are also the most socially profitable, their comparative advantage depends partly upon the assumption that the capital costs of irrigation are sunk. To expand large-scale irrigated agriculture in other areas would require new infrastructure and would probably be prohibitively expensive.

A fourth approach is to expand the small-scale rainfed and irrigated agriculture in southern Mali, not only of rice but of other cereals. The social profitability of the principal improved rice production technique there (IPS) is high, and the physical area suited to that technique is large if the problems of human and animal disease can be solved. This approach would rely on rainfed agriculture to some degree and would therefore be less secure than large-scale irrigation, but it would allow the exploitation of areas in which rainfall variation is less and the natural security of production is greater.

A fifth approach, now being adopted by the Malian government, is to expand the empoldered areas in the Mopti and Ségou projects, while promoting intensification there and in the Office du Niger. Expansion of the polders is now limited by the technical requirement that they achieve a high probability of filling (in 95 of 100 years) in order to provide a greater amount of security. Empoldering smaller areas will also promote the intensification in the flooded systems by reducing risk associated with flood failure. Promotion of the intensified technique in the Office du Niger, although slightly less socially profitable than promotion of the current technique, would increase total output, whether for constitution of stocks for domestic use or export and is thus consistent with increased security and only slightly diminished economic efficiency.

There are several important consequences of this approach. One is that it helps to ensure official control of urban supplies by concentrating rice production in limited areas to which entry of producers is restricted and in which economic incentives encourage entry. A second is that it neglects the broad exploitation of mixed rainfed and irrigated agriculture in southern Mali, which might provide greater economic benefits and more production security[17] (at least if compared to the flooded projects at

[17] Correlation analysis was done for time series data on production of rice (using all techniques), millet and sorghum, cotton, and peanuts from 1960 to 1976. Outputs of rice, millet and sorghum, and peanuts were found to be positively correlated at roughly 0.6, with signifi-

Ségou and Mopti), but which is not likely to expand as rapidly or to produce as much marketed rice as are the techniques in the Delta or in the Office du Niger.

Another consequence of this approach to achieving greater security is that it restricts benefits of investments in rice production to relatively small sectors of the rural population. This is likely to be true to the degree that intensification in the flooded projects and the Office du Niger, without reduction of holdings, is promoted, and construction of new polders is neglected. The maldistributive effect among producers will be heightened to the extent that producer prices are raised in order to increase the private profitability of the intensified techniques. Finally, this approach to production security establishes the basis for Malian rice exports within West Africa, an advantage that none of the other approaches provides, and it thereby helps to advance other objectives of Malian agricultural policy, including increasing producer incomes and supplying urban markets regularly without recourse to imports.

Citations

1 Jean Gallais, *Le Delta Intérieur du Niger: Etude de Géographie Régionale*, 2 vols. Institut Fondamental de l'Afrique Noire, Dakar, 1967.

2 Mali, Government of, Ministère de Développment Rural, Institut d'Economie Rurale, *Enquête Agro-Socio-Economique (Mopti)*. Bamako, 1974.

3 ———, Office National de Transports, Bureau d'Etudes, data furnished in response to questionnaire submitted by Food Research Institute/West Africa Rice Development Association. Bamako, 1977.

4 ———, *Rapport de l'Enquête Agricole, 1973/74*. Bamako, 1976.

5 John McIntire, "Rice Policy in Mali." Stanford/WARDA Study of the Political Economy of Rice in West Africa, Food Research Institute, Stanford University, Stanford, July 1979; Chapter 9.

6 Office du Niger, Bureau of Economic Affairs, Ségou, data furnished in response to questionnaires submitted by Stanford/WARDA, 1977.

7 Opération Riz Mopti, Direction Générale, *Rapports Annuels*. Sévaré, various years.

8 Opération Riz Ségou, Direction Générale, *Rapports Annuels*. Ségou, various years.

cance greater than 95 percent. (Cotton was correlated with the other crops at about 0.3, at significance levels less than 70 percent; this is probably due to the rapid growth of cotton output in Mali during the 1960's and 1970's.) The coefficients of variation of rice and millet and sorghum output were, respectively, 20 percent and 14 percent, which suggests that rainfed cereals may, in fact, be less subject to variation than rice production. The significance of this result is open to some doubt because the trend of rice output since independence has been affected importantly by government policies (especially increases in producer prices and investments in improved polders) in a way that the trend of millet and sorghum output has not. It is obvious, however, that output in the Office du Niger (which produced more than one-third of Malian paddy output in 1976) is secure, and it is probable that production in most of the improved polders is more secure than exclusively rainfed cereals in all but the wettest areas of Mali.

9 Opération Riz Pluvial et Bas-Fonds, Direction Générale, *Rapports Annuels*. Sikasso, various years.

10 J. Dirck Stryker, "Food Security, Self-Sufficiency, and Economic Growth in the Sahelian Countries of West Africa." Stanford/WARDA Study of the Political Economy of Rice in West Africa, Food Research Institute, Stanford University, Stanford, February 1978.

11 ———, "West Africa Regional Project: Mali Agriculture." Paper submitted to the World Bank, Washington, D.C., June 1975.

12 J. Dirck Stryker, John M. Page, Jr., and Charles P. Humphreys, "Shadow Price Estimation." Stanford/WARDA Study of the Political Economy of Rice in West Africa, Food Research Institute, Stanford University, Stanford, July 1979; Appendix B.

13 John de Wilde, *Experiences with Agricultural Development in Tropical Africa*. Vol. 2, *Mali Republic: Irrigated Agriculture*. Johns Hopkins University Press, Baltimore, 1967.

14 West Africa Rice Development Association (WARDA), *Report on Intensification of Rice Cultivation in the Office du Niger*. Monrovia, June 1977.

15 ———, *Survey of Rice Post-Production Technology: Mali*. Monrovia, 1976.

16 WARDA and Stanford Food Research Institute, "Prospects for Intraregional Trade in Rice in West Africa," Monrovia, September 1977.

PART SIX

Comparative Analysis

11. A Comparative Analysis of Rice Policies in Five West African Countries

Scott R. Pearson, Charles P. Humphreys, and Eric A. Monke

A framework for policy analysis has been presented in the introductory paper and applied in the country studies. The intent of this comparative essay is to use the objectives, constraints, and policies approach to push the analysis further than can be done within the context of a single country. The principal advantage of planning and carrying out similar policy studies in a number of countries is the scope presented for obtaining comparative insights. A search for patterns within a group of countries also aids understanding of each government's choice of policy. This search begins with a summary of the main elements of policy in the Ivory Coast, Liberia, Mali, Senegal, and Sierra Leone to provide convenient points of reference for the comparative evaluation of policies that follows.

Background information on comparative levels of per capita income, population density, road networks, advanced schooling, and per capita rice consumption is presented in Table 11.1. The Ivory Coast has the highest income and best-developed infrastructure in the group, and Mali lags behind in all indicators of development. Reflecting its heavy reliance on mineral and plantation exports, Liberia shows a high level of income compared to its relatively poorly developed infrastructure. The reverse holds for Sierra Leone, with its relatively low income but better network of roads and levels of education. Senegal has a more balanced state of development and more consistent levels of income and infrastructure. None of the countries is densely populated, although Senegal and Sierra Leone appear to have the least room for agricultural expansion, and Mali has the greatest. Rice is the principal staple food in Liberia and Sierra Leone, a main supplementary staple in the Ivory Coast and Senegal, and a fairly minor foodstuff in Mali.

Issues

A useful insight that emerged from the Food Research Institute's earlier study of rice policy in Asia was an understanding of the complex rela-

TABLE 11.1. *Background Information on Five Countries*

Indicator	Ivory Coast	Liberia	Mali	Senegal	Sierra Leone
GNP per capita (US $, 1975)	540	410	90	360	200
Population density[a] (persons per km² of agricultural land, 1976)	42	34	19	62	54
Density of all-weather roads (km per 000 km² of land area)	44	23	13[c]	24	39
Advanced students per 000 persons[b]	19	6	9	14	17
Average rice consumption (kg per capita, 1965–76)	41	117	18	51	125

SOURCES: Chapters 3, 5, 7, 9; *Africa North and West*, map No. 153, published by Pneu Michelin, Paris, 1975; Food and Agriculture Organization, *Production Yearbook, 1977, 31* (Statistical series No. 15), Rome, 1978; Rolf Gusten, "Chapter on Transport in Senegal, Mali, Ivory Coast, Ghana," Letter No. 511, Regional Mission in Western Africa, International Bank for Reconstruction and Development, Abidjan, November 18, 1974; Charles P. Humphreys and Patricia L. Rader, "Background Data on the Ivorian Rice Economy," Stanford/WARDA West Africa Rice Project, Stanford, July 1979; United Nations, Statistical Office, *1977 Statistical Yearbook*, New York, 1978; World Bank, West Africa Regional Office, Agricultural Projects Department, "Appraisal of the Mopti II Rice Project—Mali," Report No. 1561c-MLI, Washington, D.C., November 1977; World Bank, *ATLAS*, Washington, D.C., 1977; and World Bank, *World Tables, 1976*, Johns Hopkins University Press, Baltimore, 1976.

[a] The following percentages of arable to total land areas are used: Ivory Coast, 50; Liberia, 47; Mali, 25; Senegal, 42; and Sierra Leone, 77. Values for Senegal and Sierra Leone are the midpoints of extreme low and high estimates.

[b] Dates vary but cover the period 1973–75.

[c] This value is calculated using one-third of the total area of Mali.

tionships among a country's comparative advantage in producing rice, pressures on its government in allocating scarce budgetary revenues, and the government's scope for implementing policies, especially trade policy (6, p. 282). If a country has a comparative advantage in rice production, its limited supplies of foreign exchange and scarce domestic resources (labor, land, capital, and water), when priced at their opportunity costs, can be used to produce rice profitably.[1] In this event, the government has a great deal of flexibility in its choice of policies affecting rice. The government can choose to do nothing, permitting its rice producers to compete efficiently with potential imports, or it can decide to tax rice producers to obtain government revenue (and to lower rice prices to consumers if it taxes exports of rice). In the first instance, the budget is unaffected, and in the second, rice contributes positively to revenues.

The ability of government to tax staple food production effectively has received substantial emphasis in development theory and in the historical experience of a number of Asian countries, of which Japan is the most successful example. Tax revenues provide potential investment capital for industrialization, but perhaps more important in the growth and industrial-

[1] The measurement of comparative advantage in rice is discussed in detail in 9.

ization process is the impact of food production taxes on wage rates. When a staple food is an effective wage good, comparative advantage in food production allows a country to maintain lower food prices and, other things being equal, lower wage rates than countries that find the cost of calories higher. Such an advantage is most pronounced at the initial stages of industrialization, when unskilled labor is the dominant resource used in manufacturing. Comparative advantage in food production thus presents governments with a policy choice—to exploit the existing advantage to a maximum and export food, or to maintain relatively low prices to consumers through taxation of food exports, thus influencing the growth and industrialization process and satisfying consumer distributional objectives.

In the opposite situation, the country does not have a comparative advantage in rice production because its costs of production exceed the costs of comparable imports. If the government wants to promote local production, it has little choice but to subsidize it. There is still a range of options available to transfer resources to producers. But all of them involve either higher prices, forcing consumers of rice to pay the costs of inefficient local production, or direct subsidies from the government treasury. Subsidies can be paid on inputs (e.g., fertilizer), on investment in production projects (e.g. land clearing and water control systems), or on output (with payments made to farmers, millers, or merchants).

Herein lies the bind for policy. Unless consumers can be forced to carry the entire burden, subsidization of local rice production means continued calls on the budget. To an important extent, foreign aid donors might be willing to provide assistance for investment in rice, but the government is then left with the possibility of drains on its recurrent budget. Such drains will be continuing if the government subsidizes intermediate inputs or renewal of capital equipment. Hence, even though the government might desire to expand production, use of trade policy is often constrained by consumer pressures, and the use of subsidies is limited by budgetary shortages. Such shortages, in turn, can be caused by pressures from other taxpayers or by strongly competing demands on government resources from outside the rice sector. The country might then opt to continue to import rice.

As discussed in detail in Stryker's companion paper (9), the costs, profitability, and comparative advantage of the various techniques of producing rice in five countries considered in this volume vary widely. In general, rice production is most profitable for home consumption in remote regions, because costs of transportation make delivery of imported rice relatively expensive. But in three of the five countries studied—Mali and Sierra Leone are the two exceptions—imports of rice at normal levels of world prices are cheaper than most locally produced rice delivered to the main consumption center (9).[2] Since these countries cannot efficiently substitute for most imports, they would be able to generate greater national income by using their resources in other, more productive activities and continuing to purchase rice from abroad. Yet their governments desire to reduce imports and become more self-sufficient in rice by increasing production.

Explaining this drive for self-sufficiency is crucial for understanding rice policy in West Africa. Four possible reasons are relevant. First, the governments might lack adequate information and not appreciate that import substitution for rice has been and is likely to continue to be costly. This information gap might be a reflection of a historical inertia through which attitudes, policies, and perceived circumstances have not changed much. Conversely, governments might have overreacted to transitory phenomena that briefly increased the comparative advantage of rice production, such as the surge in world rice prices in 1973–75. Information is costly, but empirical results of this study point to a high return to expenditures on rice analysis.

Second, policy makers might understand the current situation fully but hold different expectations about the future levels of key parameters— especially the world price of rice, the yields of improved techniques, and the relative costs of domestic resources. The expansion of local production to substitute for imports could be profitable—and the analysis of this study proved incorrect—if the world price of rice were to be considerably higher than that projected, if yields were much larger than anticipated, or if the alternative opportunities for domestic resources were not so lucrative as expected, causing factor prices to be lower than those used in the analysis. Sensitivity analysis has been carried out using more optimistic assumptions, however, and for the most part the level of optimism must be very high before any techniques in the Ivory Coast, Liberia, and Senegal become socially profitable ways of substituting for rice imports in the main cities (9).

The third and fourth explanations are related and thus can be conveniently discussed together. It is possible that governments understand that import substitution for rice is inefficient and believe it will continue to be so, but have other objectives that might be furthered by increased production and self-sufficiency. As argued in the Introduction, the economic aspects of the goal of self-sufficiency in rice can be analyzed in terms of three fundamental objectives—increased generation of income, changes in the distribution of income, and enhanced food security. There is certainly no reason why improved efficiency should not receive total or even primary

[2] $350 per metric ton (in 1975 prices) for 5 percent broken-quality rice, f.o.b. Bangkok, is taken as a reasonable long-run base for the world price of rice.

weight in a government's decision process. In the discussion below, an attempt is made to evaluate the extent to which alternative objectives are furthered by use of policies that enhance self-sufficiency.

Finally, government objectives in rice development are often complemented by those of foreign aid donors. Donors might provide concessional assistance to rice projects, including land development, provision of infrastructure, and investment in water control facilities. If this aid is in the form of grants or concessional loans, the costs in efficiency terms to the recipient country of expanding rice production could be very low unless other efficient projects are foregone when rice activities are aided. Usually, however, rice production projects impose costs on the local economy, including recurrent subsidies on inputs, misallocation of domestic resources, and welfare losses of consumers. These costs might be viewed as bearable, however, if distribution and security objectives—of both recipient and donor—are furthered.

The issues to be discussed in this chapter are now clear. First, which countries, if any, have a comparative advantage in rice production? Second, given that West African governments cannot influence the world price of rice, what techniques of production, if any, should governments promote?[3] Third, in view of the fact that governments have multiple objectives, how have various kinds of rice policies advanced each objective? Finally, in what ways has the availability of foreign aid for rice projects complemented government objectives and influenced the direction of rice policy? Comparative answers to these questions await summaries of the evolution of policy.

Comparison of Objectives, Constraints, and Policies

The methodological framework for policy analysis used in this study emphasizes interactions among a country's objectives, constraints, and policies.[4] This framework is summarized in the introduction to this book (7, pp. 4–5):

Governments are viewed as having several objectives that they try to achieve within a framework of constrained optimization. Constraints are limits on the availability or deployment of resources and on the flexibility of consumer preferences that prevent the full attainment of all objectives. Policies are the instruments used by governments to achieve objectives by influencing the allocation of resources and patterns of consumption. Constraints on resources thus limit the extent to which policies succeed and hence the degree to which objectives are attained. The method of implementing policies can also affect their success or failure. Policy analysis consists of identifying the relevant government objectives,

[3]The first two issues are examined in detail in Stryker (9).
[4]This approach is introduced in Timmer (12).

specifying the nature of resource or consumer constraints, delineating the policy options, and tracing the interactions.

Objectives

All WARDA-member countries have the attainment of self-sufficiency in rice as a central objective of policy, and self-sufficiency in rice can be viewed as part of the broader objective of self-sufficiency in staple foods.[5] It is useful, therefore, to explore whether increases in rice self-sufficiency through expansion of local production contribute positively or negatively to the three fundamental economic objectives—efficient generation of income, more equal distribution of income, and security of food supplies.[6] In particular, it is helpful to assess the relative effectiveness of various ways of increasing rice production in contributing to these objectives. In contrast to political economy analyses, which put political motivations at the fore, this approach initially looks for economic rationales for policy. If policies contribute negatively to all economic objectives, purely political motivations can sometimes explain a government's decisions.

Some insights into the weights that governments attach to objectives emerge from a comparison of the recent historical performance with respect to objectives of the five countries, as shown by the indicators in Table 11.2. Security of rice production is a tertiary goal in the three forest zone countries—Ivory Coast, Liberia, and Sierra Leone—because climatic variation does not cause wide swings in annual levels of rice production. Food availability is not a critical problem. Furthermore, food imports do not place a large demand on foreign exchange in these countries, giving them a wide margin in which instability of world rice prices can be tolerated. Finally, these countries have diverse and fairly stable opportunities to earn foreign exchange to pay for the additional cost of cereal imports that might be occasioned by unexpected shortfalls in domestic food output.

Conversely, Mali and Senegal seem to place primary emphasis on security because shortfalls in food crops are more frequent and severe in these Sahelian countries. High variation in food production—three times that found in forest zone countries—occurs in both countries. In addition, these countries have less flexibility in adjusting to unexpected reductions in local food production. For Senegal, this problem is exacerbated by fairly high instability in foreign exchange earnings and relatively large ce-

[5] Progress toward self-sufficiency is readily measured by observing increases or decreases in import shares of total rice consumption.

[6] Substitution in consumption between rice and various other foodstuffs can be an important issue of food policy. However, the focus of this study is on expanding production of rice because West African governments (with the possible exception of Senegal) desire to substitute for rice imports by increasing output, not by reducing consumption.

TABLE 11.2. *Objectives*

Indicator	Ivory Coast	Liberia	Mali	Senegal	Sierra Leone
Growth of GNP per capita, 1960–75 (*percent/year*)	3.5	1.8	0.9	−0.7	1.5
Ratios of different income groups[a]	0.37	0.19	n.a.	0.18	0.41
Food security:					
Variation in per capita food production[b]	5	3	19	21	6
Export instability (1968–74)[c]	9.3	3.5	5.0	12.5	9.5
Net cereal imports as a percent of earnings from merchandise exports (1960–61 to 1974–77)[d]	3.7	5.4	29.6	17.9	7.7
Rice self-sufficiency (1965–76)[e]	0.75	0.75	0.82	0.26	0.92

SOURCES: Chapters 3, 5, 7, 9; Charles P. Humphreys and Patricia L. Rader, "Background Data on the Ivorian Rice Economy," Stanford/WARDA West Africa Rice Project, Stanford, July 1979; Robert P. King and Derek Byerlee, "Income Distribution, Consumption Patterns and Consumption Linkages in Rural Sierra Leone," African Rural Economy Paper No. 16, Department of Agricultural Economics, Michigan State University, East Lansing, and Department of Agricultural Economics, Njala University College, Njala, Sierra Leone, 1977; Johns Hopkins University Press, Baltimore, 1976; World Bank, Regional Projects Department, Western African Regional Office, "Appraisal of a Second Sedhiou Project Senegal," Report No. 1094-SE, Washington, D.C., June 4, 1976; West Africa Rice Development Association, *Rice Statistics Yearbook*, Monrovia, 1975 (and subsequent updates); and United Nations, Department of International Economic and Social Affairs, Statistical Office, *Yearbook of International Trade Statistics, 1977*, vol. 1, "Trade by Country," New York, 1978.

[a]The figure for the Ivory Coast is the ratio of rural incomes in the savannah and forest zones, in 1974. The figures for Liberia, Senegal, and Sierra Leone are the ratios of rural and urban incomes in 1976, 1975, and 1974–75, respectively.

[b]These figures are coefficients of variation for estimated per capita food production, converted to grain equivalents. Years and crops for each country are: Ivory Coast, 1960–74, rice, maize, yams, plantains, and cassava; Liberia, 1965–76, rice; Mali, 1961–76, rice, maize, millet, and sorghum; Senegal, 1961–76, rice, millet, and sorghum; and Sierra Leone, 1970–76, rice. Except for Mali, no account is taken of seeds and losses. Because of revisions in statistical series, the years 1975–76 are not included for the Ivory Coast and 1960–69 are excluded for Sierra Leone.

[c]Export instability is based on five-year moving averages centered on the years covered. See explanation in the *World Tables, 1976*, p. 19.

[d]Data for the Ivory Coast cover the years 1960–77; for Liberia, 1960–75, excluding 1964; for Mali, 1961–76, excluding 1973; for Senegal, 1960–75; and for Sierra Leone, 1960–74.

[e]Self-sufficiency is defined as the ratio of net domestic production to total disappearance.

real imports. Consequently, increased rice production with reliable methods of water control is viewed by both countries as an important way to ameliorate the security of their food supplies.

Among the three southern countries, increasing incomes through an efficient allocation of resources is viewed as a much more important objective than enhancing food security, and the expansion of rice production is seen as a potential way of contributing to this goal. For the Ivory Coast, income growth is undoubtedly the main objective of economic policy in general and probably also of rice policy. In Liberia, recent agricultural development policy, including rice policy, has aimed at finding a long-term, gradual complement for growth based on exports of iron ore and rubber. Income generation through an expansion of agricultural and silvicultural activities lies at the center of this approach. In Sierra Leone, which has the highest per capita production and consumption of rice in the WARDA region, policy makers desire to achieve additional income out of more rice production primarily through the introduction of new techniques.[7] As Table 11.2 shows, these three countries have achieved growth rates exceeding those in the Sahelian countries, with the Ivory Coast by far the most successful. What the table does not show, and what is doubtful, is the contribution of expanded rice production to this growth.

In view of the wide disparity in income levels within countries, summarized in Table 11.2, each of the five countries has clearly stated goals to spread economic development more evenly by means of rice policy. In Liberia and Sierre Leone, the distributional concern is to generate higher rural incomes in general. The Ivory Coast has focused rice investment in its northern savannah, since that part of the country has not benefited from agricultural and silvicultural exports to the same extent as the forest zone. In Senegal, rice investment has been mainly concentrated in the Senegal River valley and, more recently, in the Casamance, the area of traditional rice production. Both areas are more remote and less developed than many other regions of the country. Finally, only Mali has emphasized low rice prices to consumers.

If this analysis is correct, the fundamental objectives of rice policy in each country can be ranked from primary (1) to tertiary (3) importance:

	Ivory Coast	Liberia	Mali	Senegal	Sierra Leone
Generation of income	1	1	3	3	1
Distribution of income	2	2	2	2	2
Security of food supplies	3	3	1	1	3

Whereas these rankings show differences between Sahelian and forest countries, the importance of such differences should not be exaggerated.

In summary, self-sufficiency is the major stated objective of rice policy in all five countries, and this goal can be viewed as essentially a means of enhancing economic growth, redistributing income, or improving security. Both the possibility of achieving self-sufficiency and its effects on the three fundamental objectives vary importantly among the five countries. In particular, the two Sahelian countries diverge widely from one another. Whereas both emphasize food security, Mali is an efficient rice producer and is nearly self-sufficient in rice in normal years. In contrast, Senegal lacks efficient production techniques and produces only one-

[7] In the terms of comparative statics, the interest of the three forest-zone countries centers on the search for the optimum point on the production possibilities frontier. The success of policies in achieving this goal will be reflected over time in the observed rate of growth.

quarter of its rice consumption, which, on a per capita basis, is nearly triple that of Mali. For the forest zone countries, the scope for import substitution is substantial, though not so large as in Senegal. Liberia and the Ivory Coast each produce about three-fourths of their rice needs, and Sierra Leone is more than 90 percent self-sufficient.

Constraints

Constraints to increasing rice production in West Africa are seldom absolute. It is usually possible to obtain the additional resources required to raise production, but the costs of attracting them can be substantial. Public policies can try to alleviate these cost constraints through the promotion of improved production techniques and the development of economic infrastructure.[8] The best way to assess the constraints facing countries in their efforts to increase rice production is to estimate both the costs required to overcome shortages of necessary resources and the capacity of the public sector to intervene.

Although constraints vary widely among the five countries, in all of them expansion of rice production is limited by the range of feasible production techniques, the costs of domestic factors of production, and the capacity to design and carry out effective public interventions. Table 11.3 contains information that can be used to assess the importance of different constraints on increased rice production. For a number of reasons, mostly associated with its level and rate of development, the Ivory Coast has the greatest degree of technical flexibility among the countries considered here in choosing methods of production. Although Mali has a comparative advantage in rice, its production is nevertheless constrained, as discussed below. For differing reasons, the other countries fall between the extremes of the Ivory Coast and Mali.

Rainfall is the most important constraint in traditional production. With the exception of areas around Sikasso, Mali cannot grow rainfed rice and requires irrigation to produce rice in other regions. Floodwaters in the interior Delta of the Niger River and in lowland basins along the Ivorian border have traditionally provided the necessary water to produce rice but with high uncertainty and no water control. In the other four countries, rainfed rice provides nearly all of traditional production, reflecting their relatively better endowment of rainfall.

---

[8]The relaxation of expected future constraints usually requires long lead-times, often as much as 20 to 30 years. Irrigation investment is an inherently long-term process, in terms both of constructing the infrastructure and of farmers' learning to manage water resources. These long-term effects could make rice production that is uncompetitive today more efficient in the future. Corden (1), among others, has argued, however, that future gains from learning seldom repay current losses from the inefficiencies caused by protection and subsidy policies. Whether future gains from improving management in irrigation projects will be sufficient to offset short-term costs is an important empirical question.

TABLE 11.3. *Information on Constraints*

Indicator	Ivory Coast	Liberia	Mali	Senegal	Sierra Leone
Rainfall in rice-producing areas (mm/yr)	1,300–1,500	2,000	620[a]	750–1,800[b]	2,500–3,000
Daily agricultural wages (US $/man-day, 1975–76)	1.40–1.80[c]	1.25	1.00	1.00–1.20	0.60–0.80
Direct farm labor cost per kg paddy (US $, 1975–76)[d]	0.118	0.247	0.065	0.147	0.118
Investment costs of water control[e]					
Complete (US $/ha, 1975)	3,983–5,978[f]	—	444[g]	400–4,311[h]	—
Partial (US $/ha, 1975)	1,390[i]	750[i]	600–900[j]	712[i]	249[i]
Annual costs of water control[e,k]					
Complete (US $/ha/crop, 1975)	373–348[f,l]	—	43[m]	103–340[h,n]	—
Partial (US $/ha/crop, 1975)	173[i,o]	124[i,p]	42[m,q]	80[i,r]	112[i,s]
Government consumption as percent of GDP (average, 1960–73)[t]	15.2	11.8	16.2	18.1	7.6
Government investment as percent of GDP (average, 1960–74)[u]	2.4	1.4	−2.15[v]	4.0	2.7
Debt service as percent of export earnings	8.3[w]	7.1[x]	22.0[y]	6.9[z]	16.1[aa]
Shipment to capital city (US $/mt)	93	71	76	70[bb]	34

SOURCES: Chapters 1–8 and 10; International Monetary Fund, *Balance of Payments Yearbook*, 29 (December 1978); Charles P. Humphreys and Patricia L. Rader, "Background Data on the Ivorian Rice Economy," Stanford/WARDA West African Rice Project, Stanford, July 1979; Charles P. Humphreys, "Data on Costs of Ivorian Rice Production," Stanford/WARDA West Africa Rice Project, Stanford, July 1979; Liberia, Government of, Ministry of Planning and Economic Affairs, *Quarterly Statistical Bulletin of Liberia* (June 1977), Monrovia; Dunstan S. C. Spencer, "The Economics of Rice Production in Sierra Leone—1, Upland Rice," Bulletin No. 1, Department of Agricultural Economics and Extension, Njala University College, University of Sierra Leone, Njala, 1979; World Bank, *World Tables, 1976*, Johns Hopkins University Press, Baltimore, 1976.

[a] This figure is an average for Mopti and Ségou.

[b] This first figure refers to the Senegal river valley and the second to the Casmance region.

[c] The first figure refers to the savannah zone and the second to the forest zone.

[d] These figures are the averages of labor costs in all techniques of paddy production in 1975, weighted by each technique's share in total output.

[e] These costs are net of most, if not all, taxes. As such, they represent social, not private costs.

[f] The first figure refers to mechanized schemes in the forest zone relying on pump irrigation, the second to gravity irrigation using dams in the savannah zone.

[g] This figure is the cost for improvements in the Office du Niger, consisting mainly of leveling and rehabilitation of canals. The cost of the basic infrastructure is considered sunk.

Water constraints in West Africa, coupled with the high water demands of the rice plant, make the objective of providing enhanced food security through increased production expensive to obtain. The cost of overcoming the water constraint varies enormously among countries and techniques. Complete control generally requires an investment of $4,000 or more per ha, whereas partial control costs as much as $1,000 per ha. The two notable exceptions, where complete control is not so expensive, are unlikely to be replicated on a large scale. The Office du Niger in Mali and the Matam polders in the Fleuve region of Senegal provide full water control at costs beneath those required for full control elsewhere. But in the former, enormous infrastructure costs, which were made in the 1930's, are now considered sunk. For the latter, the area in which low-cost projects can be carried out is restricted to land directly bordering the river.

Although estimates vary according to the type and lifetime of investment and the interest rate used, information in Table 11.3 gives some

---

(*Notes to Table 11.3 continued*)

[h] The first figure is for small-scale pumping in Matam, the second for large-scale, mechanized pump irrigation in the Delta.

[i] Partial water control here refers to the improvement of lowlands, usually in forest areas, by bunding and diversion weirs.

[j] The first figure is for improvement of lowlands in the Sikasso area, the second for controlled flooding improvements near Ségou and Mopti, excluding initial deep plowing.

[k] Unless otherwise noted, recurrent costs include both the annuity on the investment and operation and maintenance of the irrigation system.

[l] Annuities for dam irrigation are based on an average of 27 years for the system as a whole and an average annual interest rate of 5.8 percent. For pump irrigation, the expected average life is 15.4 years, with an average annual interest rate of 5 percent. These annuities for pump and dam irrigation both assume 1.85 crops per year and utilization of 80 percent of the total area.

[m] This figure covers only the annuity, based on a 25-year service life and a 2.5 percent annual interest rate. Operation and maintenance costs are not included.

[n] Annuities for Matam are based on an average service life of about nine years and an average annual interest rate of 1 percent. For the Delta, the average life is 24 years and the average annual interest rate is 2.75 percent. For Matam, 65 percent of the annuity is allocated to the rice crop.

[o] The annuity is based on a 15-year service life and an average annual interest rate of 7.7 percent. This figure assumes 1.3 crops per year and utilization of 90 percent of improved land.

[p] This cost covers the annuity and is based on a 20-year service life and an annual interest rate of 15 percent. Repairs are based on five man-days, or $6. Such costs are also included in direct labor charges.

[q] This cost represents only the annuity and is the average for controlled-flooding polders near Ségou and Mopti and for lowlands around Sikasso.

[r] The annuity is based on a service of life of 20 years and an average annual interest rate of 3 percent.

[s] The annuity is based on a ten-year service life and an average annual interest rate of 24 percent. Maintenance costs of 45 US $/ha are also included in farm labor costs.

[t] Government consumption is defined as recurrent expenditures on goods and services and includes all defense expenditures.

[u] Government investment excludes defense expenditures.

[v] Data for Mali cover only 1965–73.

[w] Data cover the years 1969–76, and debt service includes repayment and interest on government debt, loan repayments by government enterprises, and retirement of government securities.

[x] Data cover the years 1970–75, and debts service includes repayment and interest on government debt, IMF repurchases and reconstitution of assets. This ratio increases substantially (to 10.2 for 1967–73) when debt service is compared to total current account earnings.

[y] Data cover the years 1970–77, the debt service includes repayment and interest on long-term government debt, interest on overdrafts with the French treasury, charges paid to the IMF, and repurchases of IMF credit.

[z] Data cover the years 1968–75, and debt service includes repayment and interest on government debt, and repayment of trade credits issued to the government.

[aa] Data cover the years 1969–76, and debt service covers repayments of loans to the government, of issues by the central government, and of prefinancing by foreign contractors.

[bb] This figure is for shipment from either Matam or the Casamance. Shipment from the Delta costs about 51 US $/mt.

orders of magnitude of the high costs generally involved. For rice produced under full water control, the annual capital costs and charges for maintaining the irrigation system can be as high as $150 per mt of milled rice. The annual capital and maintenance costs vary more widely for rice produced under partial water control, but are estimated to be about $65 per mt.[9] However, with partial control, the security of production is often only marginally better than under traditional production, since the delivery of water remains largely dependent on natural rainfall and flooding. The one significant exception, which still has considerable potential for expansion, is the controlled flooded technique in Mali. The security of flooding is estimated to be 90 percent of that with complete control, whereas annual capital and maintenance costs are probably only about one-half those for the partially controlled, improved lowlands in the forest zone countries.

As suggested by the low population densities in these countries, wage rates are relatively high throughout West Africa and pose an important near-term economic constraint on the efficient expansion of rice production. Daily wage rates are clearly highest in the Ivory Coast, ranging from $1.40 to $1.80 per day for men, reflecting the success of the country in promoting agricultural exports and attaining a rapid rate of development. At the other extreme, wage rates in Sierra Leone are less than half those in the Ivory Coast. Such low rates stem from a lack of natural resources and agricultural capacity in that country. The wage rates for Mali, Senegal, and Liberia are in the range bounded by those in the other two countries.

The pattern of unit labor costs among countries requires consideration of worker productivity as well as wage rates. (By definition, unit labor costs are the product of the wage rate and the inverse of labor productivity.) Marginal unit labor costs can be approximated by the value of direct farm labor in each additional kg of rice produced domestically. Labor costs per kg of rice are clearly lowest in Mali, where inexpensive water control schemes, the extensive use of animal traction, and high rates of insolation all help raise the productivity of labor. For Mali, wage rates are relatively low and labor productivity is relatively high.

[9] These figures are based on the following assumptions:

	Partial water control	Full water control		Partial water control	Full water control
Annual capital and maintenance costs (US $/ha)	125	350	Milling outturn (percent)	65	65
Yield (mt paddy/ha)	3.0	3.5	Cost per mt milled rice	65	150

There is no clear trend in the future direction of irrigation costs. Upward cost pressure will result from using up the best locations for irrigation projects. But cost reductions can be expected as construction activity expands. In addition, increased regulation of water flow in the major rivers will likely lower costs, e.g. owing to the reduced size of the perimeter dikes required following better flood control.

On the other hand, the unit cost of labor in Sierra Leone is the same as in the Ivory Coast, where wage rates are twice as high. Senegal also seems to lose the benefits of its relatively low wage rates, and labor costs per kg of rice are the second highest in the five countries. In Sierra Leone, low wage rates appear to be more than offset by high labor input, and in Senegal they seem to be counteracted by low yields in Casamance. In the Ivory Coast, relatively high wages seem to be offset to a considerable extent by higher productivity—reflecting perhaps the favorable climatic conditions and greater use of other inputs. Liberia has the highest unit labor cost in rice production because of high wage rates coupled with very inefficient traditional production techniques. In short, differences in natural environments, including quality of land and supply of water, permit labor productivity in rice production to vary widely among the five countries. This differing productivity strongly influences the pattern of labor costs, since these costs depend on both productivity and wage rates.

The variations in worker productivity are not in themselves unusual. Research by Timmer and Falcon on nine Asian countries demonstrates the importance of complementary inputs in production—environmental conditions, irrigation investments, and high-yielding seed varieties (13). These factors accounted for a three-fold difference in yields among the nine Asian countries studied. The key point is that increases in complementary inputs reduce the relative importance of labor costs in total costs. But in the West African context, high wage rates, coupled with low worker productivity, cause very high unit labor costs.

Capital becomes constraining at the national level primarily when large-scale investment must be made in land clearing and water resource development. The necessary capital must come from either domestic savings or foreign borrowing and aid. Because the size of most of these investments demands that they be undertaken by a government agency, the capacity of the government to allocate tax revenues for investments and its ability to obtain foreign funds can importantly constrain the expansion of rice production.

Capital is also an important constraint at the farm level, as reflected by high real rates of interest that prevail in the informal rural capital markets. With the exception of cooperative projects, which only affect a small number of farmers, farm capital comes primarily from savings and short-term borrowing. Improvements in rural lending facilities are thus an important constraint on the dissemination of new techniques, with high levels of recurrent expenditures for improved seeds and fertilizer.

Of the five countries, Liberia would appear to have the greatest scope to increase both government investment and foreign borrowing. Neither the ratio of government investment to GDP nor the debt-service ratio is particularly high compared with the other countries. Moreover, favorable rainfall might reduce the need for large-scale investments relative to that in other countries, although the institutional constraint due to the scarcity of rural lending facilities is somewhat greater. At the other extreme is Mali, which has actually suffered negative government investment owing to the difficulties of the Sahelian drought. It also has a debt-service ratio that is three times as large as that for most of the other countries. Unlike Liberia, however, Mali has concessional aid available to it, and such capital is relatively inexpensive. Senegal also has access to considerable foreign aid, especially for projects in the Senegal River Basin, where water development is most expensive. Hence, Liberia, Mali, and Senegal probably have the least restrictive constraints on capital, although none has the flexibility to divert large sums into rice projects solely of its own choosing.

Existing high levels of government investment and a growing debt-service ratio in the Ivory Coast mean that additional investments come only at increasingly higher capital costs. In a country where natural conditions make water development especially expensive, the presence of such a constraint could hamper efforts to expand irrigated rice production on a large scale. Sierra Leone is the most severely constrained in terms of capital, with a very high debt-service ratio and a relatively high share of GDP already devoted to investment. Some concessional foreign aid is available but less than for the drier countries to the north.

In West Africa, land is widely available and hence has a very low opportunity cost. This situation can be expected to change in the future as population densities increase. But during the next 25 years or so, the period in which the longest investments in rice might be amortized, land is likely to remain inexpensive. Investments in land development to increase water control are considered under the constraint on capital. With respect to rice production, the surplus of land provides little in the way of economic advantage. Irrigated rice is relatively ill-suited to land-extensive production. Moreover, the potential for efficient utilization of inexpensive land (i.e. the substitution of capital and land for labor) in upland rice production, prominent in the forest zone countries, remains largely unknown.

Other constraints include the availability of revenue for recurrent financing of government programs, managerial talent, rural infrastructure, and the location of rice production relative to major consuming centers. Perhaps the most immediate concern of government leaders is the capacity of the budget to sustain the sizable recurrent expenses that accompany intervention in the rice sector. Such recurrent costs are particularly important when subsidies are paid on the output, and they can also be significant when high levels of modern inputs are heavily subsidized. Total current government expenditures are about one-sixth of GDP in the three francophone countries, wheras the share in the other two countries is much smaller. Owing to the severity of the budget constraints, alternative demands on funds could preclude additional expenditures on rice, unless foreign assistance is forthcoming.

With respect to domestic managerial talent, Table 11.1 provides some insights based on the importance of advanced education. The Ivory Coast currently has the highest level of advanced education, and it also hires large numbers of foreign technical experts. In contrast, Mali and Liberia have education levels for advanced students roughly one-half of those in the other countries. Moreover, Mali does not depend significantly on foreign talent. In between are Senegal and Sierra Leone. Of course, many other factors affect the capacity to intervene in the rice sector, among them the willingness to divert scarce talent into these areas. Mali and the Ivory Coast have probably had the best past experience and Liberia the least satisfactory. But in all countries managerial skill is scarce and policies demanding significant public intervention may be severely constrained.

With respect to rural infrastructure, as measured in Table 11.1 by the density of all-weather roads, the Ivory Coast is the least constrained and Mali is the most affected. However, since most of expanded production in Mali is likely to occur along the Niger River between Bamako and Mopti, where a reasonably adequate road system already exists, this constraint might be considered relatively unimportant for rice production. In Senegal, especially in the Casamance, the lack of good roads presents an important constraint that increases the costs of production and marketing. The cost of transporting local rice to the major consumption centers is greatest for Sengal, as a result of the long distance of major production from Dakar. It is least in Liberia and Sierra Leone, reflecting the small size of these two countries.

The predominance of producer-oriented rather than consumer-oriented constraints on West African rice policy contrasts significantly with the conduct of rice policy in much of Asia where consumers play a much more prominent role in the creation of objectives and constraints on policy formation (3, 4). Consumer-related issues of rice availability and price, particularly in urban areas, have been of critical concern to Asian policy makers. In part, differences between Asian and West Africa reflect an unavoidable bias in the method of policy analysis. Consumer-related constraints are often hidden, awaiting new policy actions or events to call them into existence. But, more important, the differences are due to the more severe income problems of some countries in Asia, particularly in cities, and the more central role of rice in Asian diets. Only in Liberia and Sierra Leone does rice play anywhere near as important a role in consumption patterns as in Asia, and in these two countries consumers seem to demonstrate a high degree of substitution between rice and wheat, plantains, cassava, yams, and other staples. This substitutability is important in understanding the relatively passive reaction of West African consumers to price policy.

Several generalizations can be offered regarding constraints on rice policies. First, because of the levels and variability of rainfall, the Sahelian countries have a greater need to control water supplies than do the forest countries. Second, all countries are constrained by labor costs in rice production, and none yet face land shortages. Third, all countries face difficult tradeoffs in allocating government revenues. Foreign aid can play an important role in easing the capital budget constraint of the Sahelian countries, whereas Sierra Leone is in the most difficult position with respect to capital. Finally, regarding administrative talent, rural infrastructure, and the location of production, the constraints facing countries vary. On balance, the Ivory Coast is probably the least constrained and Liberia is the most limited by these factors.

Policies

Whereas the number of constraints can be large, the range of options available to governments in West Africa for the purpose of implementing policy is quite narrow. In this study, policies are classified into three areas—trade and price policies, domestic tax or subsidy policies, and investment policies (7). These policies have an impact on the rice economy through their effects on output prices and on input and capital costs. The effectiveness of each policy is heavily influenced by the opportunity costs of domestic resources and by choices of production and milling techniques, which together influence social profitability of rice production. Information that can help measure the application and impact of government policies is contained in Table 11.4.

Trade policy is similar in all countries, except Mali. Contrary to the belief that West African governments subsidize rice consumers, domestic wholesale prices of rice between 1965 and 1976 have been about one-fourth to one-third higher than comparable c.i.f. import prices in the four protecting countries. Mali is the exception, where official domestic prices have been lower than c.i.f. import prices. Typically, the governments of Ivory Coast, Liberia, Senegal, and Sierra Leone restrict imports of rice with variable levies or quotas, thereby forcing consumers to pay higher than international prices, permitting producers to receive high prices, and generating government revenues from the rice imports. In Mali, the reverse holds, and consumers of officially marketed rice tend to be subsidized relative to c.i.f. import prices. When world rice prices increased rapidly in 1974, this pattern was temporarily interrupted because Senegal, Sierra Leone, and Mali subsidized imports to maintain more stable prices to consumers, whereas Ivory Coast and Liberia raised domestic prices in line with the hike in world prices, allowing their producers to benefit. Generally, though, the thrust of trade policy has been to transfer resources from consumers to producers or to the government treasury.

All countries set official producer prices for paddy and consumer prices for rice. The structure of official prices alters the pattern of prices created by trade policies only to the extent that the countries are successful in enforcing official prices, which requires a large involvement in the marketing and milling sectors. In this regard, the Ivory Coast and Mali have been most successful in purchasing paddy, handling from one-fourth to two-fifths of production in peak years. The remaining countries have never purchased more than a minor share of output, and marketing is dominated by private traders.

Since independence in 1960, the Ivory Coast has used a restrictive trade policy to raise consumer prices and thereby to buttress a floor price to producers. Producer prices—especially during the mid-1970's—have been further supported by farm subsidies channeled through collection and milling activities without equivalent increases in the consumer price. In 1975, roughly one-half of the official producer price consisted of government subsidy. Starting in 1978, however, government paddy purchases were severely curtailed. On balance, consumers still face domestic prices that are higher than comparable import prices for rice, but not so high as the official producer price would require if there were no government subsidy.

Price policy in Mali is more complicated. The government's established producer price applies to about one-half of rice marketings because a high proportion of commercial sales arises from large, geographically confined development projects whose farmers are required to pay fees in kind and to meet quotas for marketed paddy. The government through its state marketing agency then rations this rice by selling it at a price below the market-clearing level to selected consumers who belong to cooperatives. In effect, Malian price policy forces producers in government projects to subsidize consumers who have access to rationed rice in cooperative markets. The official price structure in Mali thus tends to tax farmers relative to the c.i.f. import price because the official consumer price is below that price. This policy raises the free market price of rice for both consumers and producers who are not part of government programs.

The other three countries also establish official prices for rice. But the impact of such policies has been small, since only a slight percentage of national production is marketed through government channels, especially in Liberia. The pattern varies, however, among the three countries. Both Senegal and Liberia tend to subsidize producers who use the improved techniques, although by amounts less than in the Ivory Coast. But tradi-

TABLE 11.4. *Information on Policies*

Indicator	Ivory Coast	Liberia	Mali	Senegal	Sierra Leone
Ratio of domestic official wholesale to c.i.f. price	1.38^a	1.29^b	0.61^c	1.25^d	1.36^e
Rate of subsidy or tax $(-)$ in official domestic producer price $(1975-76)^f$	0.56	0.11	0.05	0.25	-0.29
Government purchases of paddy $(000\ mt\ paddy)$	124^g	3^h	88^i	10^i	11^j
(Percent of total production)	(28)	(1)	(41)	(4)	(7)
Rate of subsidy on fertilizers^k (1975)	0.45^l	0.00	0.27^m	0.74	0.62^n
Additional output due to fertilizer $(000\ mt\ paddy,\ 1975)^o$	21	1	8	33	19
(Percent of total production)	(5)	(0)	(3)	(25)	(3)
Rate of subsidy on irrigated land development $(1975)^k$	0.76^p	0.00	1.00^q	1.00^r	0.41^s
Area under irrigation $(000\ ha,\ 1975-76)$	23	1	90	15	5.5
(Percent of total area in rice)	(6)	(1)	(40)	(16)	(1)
Rate of subsidy on motorized services $(1975-76)^k$	0.04^t	0.00	0.50^u	-0.16^v	0.77^w
Area plowed or threshed by motorized services $(000\ ha,\ 1975-76)$	8	1^x	90	10	17^y
(Percent of total area in rice)	(2)	(0)	$(57)^z$	(11)	(4)
Area planted in improved rice seeds $(000\ ha,\ 1975-76)$	27	2	94^{aa}	29	84
(Percent of total area in rice)	(7)	(1)	(42)	(32)	(19)

SOURCES: Chapters 1–10; Charles P. Humphreys and Patricia L. Rader, "Background Data on the Ivorian Rice Economy," Stanford/WARDA West Africa Rice Project, Stanford, July 1979; Charles P. Humphreys, "Data on Costs of Ivorian Rice Production," Stanford/WARDA West Africa Rice Project, Stanford, July 1979.

[a] Data cover the period 1960–77 and are for imports of 25–35 percent broken rice only. For 1960–72, the official wholesale price is estimated by subtracting 3 CFA francs from the official retail price.

[b] Data cover 1967–76 and are for all rice imports.

[c] Data cover 1969–74 and are for all rice imports. These are the only years in which Mali imported rice.

[d] Data cover 1965–76 and are for 100 percent broken rice imports only.

[e] Data cover the period 1960–1976, excluding 1964–65 and 1975. Imports include all types of rice.

[f] The rate of subsidy (or tax, for negative values) is defined as $S = [\ (P + M) - C]/C$, where S is the rate of subsidy, P is the official producer price, in rice equivalent, M is the sum of the private costs for collection of paddy, milling, and distribution of rice, using the most common techniques in the country, and C is the official wholesale price of rice. A value of zero implies that the structure of official prices neither taxes nor subsidizes. A positive value implies that official purchases of paddy and sales of rice produce a budgetary deficit, which is transferred to producers, whereas a negative value implies the generation of a budgetary surplus, which is taxed from producers.

[g] This value is the average for the two crop years 1974–75 and 1975–76. It may represent as much as 70 percent of all marketed rice.

[h] This value is for the crop year 1976–77 and may be equal to 8 percent of all marketed rice.

[i] This value is the average for the two crop years 1974–75 and 1975–76.

[j] This value is for the crop year 1974–75.

[k] All subsidies are based on social costs, net of all taxes.

[l] This rate is the unweighted average for rainfed and irrigated rice in the forest and savannah zones. The subsidy on fertilizer is estimated from the total subsidy on current inputs as the share of fertilizer costs in total costs. This estimate may overestimate the subsidy rate if other inputs are, in fact, more heavily subsidized.

[m] This is the rate of subsidy on fertilizer used in the Office du Niger.

[n] This is the average rate of subsidy on fertilizer used on bolilands (0.66) and on improved uplands and improved inland swamps (0.58).

[o] Additional paddy output due to fertilizer application has been estimated by assuming that 1 kg of N gives 14 kg of paddy on rainfed, upland rice and 25 kg of paddy on irrigated rice.

tional techniques, which dominate total production, are not effectively subsidized, because of either limitations in the funds to purchase paddy or the absence of purchased inputs in traditional farm production. In all three countries price policy depends critically on the capacity and will of the government to pay for subsidies on paddy that is purchased and milled by government agencies. Moreover, the incidence of taxation on consumers increased significantly after 1974 in these three countries because domestic prices were not allowed to fall in line with world prices during the 1974–78 period. Hence, producers have received increasing transfers from consumers.

Sierra Leone is the only country where the official producer price implies a tax on farmers relative to the official consumer price. Unlike Mali, however, the country lacks the large, geographically concentrated projects that make enforcement of unfavorable producer prices possible. As a result, the government is largely unable to compete in the private market for paddy, and the tax has very little effect on actual production.

The extent of subsidies on intermediate inputs into rice production differs greatly among the five countries. At one end of the spectrum, Liberia has no effective input subsidies save that on its extension service. As data in Table 11.4 show, virtually no Liberian production benefits from motorized services, fertilizers, or improved seeds. These results reflect the very low level of government intervention in the past, and rice projects being planned will surely contain higher rates of subsidy.

In the other countries, extension services are universally subsidized, and government policies also affect fertilizer, motorized services, and improved seeds. The Ivory Coast has concentrated on fertilizers, subsidizing roughly half their price, and to a lesser extent on improved

---

(*Notes to Table 11.4 continued*)

[p]This value represents the average rate of subsidy on investments in swamps, pump irrigation, and dams, weighted by the share in total costs of each type of land development between 1960 and 1976. Subsidy rates for swamps, pumps, and dams are 0.63, 1.00, and 0.89, respectively.

[q]This rate of subsidy applies to investments in the Niger River Basin. For swamps, the rate is 0.87.

[r]This rate of subsidy applies to investments in the Delta and at Nianga. Rates for investments at Matam and in swamps in the Casamance are 0.35 and 0.92, respectively.

[s]This rate applies to improved inland swamps only.

[t]This is the average rate of subsidy on large- and small-scale motorized services in rainfed and irrigated production weighted by the share in total area under each type of mechanization. Rates are zero, except for large-scale motorized services on irrigated production, where it is 0.15. Purchases of oxen and equipment are slightly subsidized at a rate of 0.12.

[u]This rate is the average for motorized threshing at Mopti and the Office du Niger. Deep plowing is subsidized at the rate of 1.00, but it is not done annually. Purchases of ox equipment are subsidized at the rate of 0.09.

[v]The purchase of ox equipment for rainfed production in the Casamance is subsidized at a rate of 0.28.

[w]This is the subsidy rate for land preparation of bolilands and riverain grasslands.

[x]This value is less than one, resulting in the zero percentage.

[y]This is the average of 1975 and 1976, which were 11,000 ha and 22,000 ha, respectively.

[z]This percentage is the share of paddy production threshed, not the percentage of land on which the paddy is produced (that percentage equals 0.42).

[aa]These seeds are renewed, in theory, every three years; hence the annual plantings in new improved seeds equal one-third the value shown.

seeds.[10] However, these subsidies have had relatively little impact, and only a small percentage of domestic rice production benefits from modern inputs.

Mali has provided little encouragement to fertilizer use through subsidization. On the other hand, mechanical services, especially motorized threshing, are subsidized by about 50 percent, and over half of domestic production is handled in this way. Improved seeds, introduced once every three years, have also been strongly promoted. Mali has, therefore, concentrated on inputs that best complement other elements in the production systems. Inexpensive water control does not require fertilizers in order to be profitable, and extensive tillage practices using privately owned oxen and equipment benefit from mechanized threshing. In addition, mechanized threshing allows the government to buy a large share of paddy marketings at the relatively low official price.

Senegal has relied most heavily on fertilizers and improved seeds, providing the largest subsidies (up to 75 percent on fertilizer) and increasing production by as much as one-fourth as a result. On the other hand, mechanical services are slightly taxed. Consequently, such services are used only in Delta, where soils are heavy and farmers participating in projects have little choice concerning their use.

Although Sierra Leone has the second highest rate of subsidy on fertilizers and the highest on mechanized services, the impact on national production has been quite small—owing in part to constraints on the government budget and the already high level of traditional production. As in most of the other countries, the use of improved seeds is the most widespread of all modern inputs.

All countries except Liberia have maintained important government investment policies in order to develop irrigated rice production, and Liberia is beginning developments in this area. Subsidies have been highest in Mali and Senegal, probably because of the large scale of the projects required to control water along the Niger and Senegal rivers. With small-scale irrigation schemes, farmers participate in the investment, and subsidy rates on land development costs are generally lower. They range from about two-fifths to two-thirds in Sierra Leone and the Ivory Coast, respectively, compared with 100 percent in Mali and Senegal. For Senegal, the major exception to this pattern is Matam, where low costs, small scale, and relatively high population densities make it possible to obtain participation with subsidies amounting to only about one-third of total costs.

A common theme underlies rice investment policy in all five coun-

[10] In the Ivory Coast, a package of inputs is provided through a fixed contract, the total value of which is subsidized. Hence, allocation of subsidies to specific inputs is arbitrary. The method used in this study is to prorate the total subsidy to the different inputs that make up the package according to their respective shares in the total value of the contract.

tries—the overwhelming importance of foreign aid donors in designing, financing, and implementing rice development projects. The Ivory Coast, the only country in the group that has the resources to fund major rice production projects without concessional foreign assistance, has not done so to an important extent—preferring instead to use its own available funds for other, more profitable investments. All the governments, however, take some equity participation in donor-assisted projects. Liberia is at the high end of the range with about 50 percent, and Mali and Senegal are at the low end with 15 to 20 percent.

The amount of direct government participation in rice production, processing, and distribution is greatest in Mali, decreasing through Senegal, Ivory Coast, and Sierra Leone, and least in Liberia. In all countries state farms for rice are almost nonexistent, and direct government participation in paddy production is restricted to land and water development, production of improved seeds, research and extension, provision of credit, and input delivery. The degree of government involvement increases through the rice production chain. In all countries, small private hullers coexist with large government-owned mills, but the former are much more important in the anglophone and the latter in the francophone countries. This same pattern is also true for rice marketing. All five countries have state marketing agencies, but they often move a large proportion of paddy or rice in Mali and the Ivory Coast, whereas the private trade is predominant in Liberia, Senegal, and Sierra Leone. This participation provides a potentially important source of pressure on policies by creating influential interest groups—such as state development agencies and marketing boards—which usually have easy access to decision makers.

Evaluation of Policies

Policies should be evaluated in terms of their effectiveness in achieving one or more government objectives in the face of resource constraints that limit both the choice and implementation of those policies. Effective policies successfully advance objectives at minimum cost.[11] The effectiveness of a policy depends, first, on its ability to make a positive contribution toward advancing an objective, such as increasing national income, dis-

---

[11] Imperfections in factor or product markets, caused by segmentation of markets, externalities, and natural monopolies, among others, create divergences between private and social evaluations of resources and products (1). Government intervention can generate additional income efficiently by offsetting these divergences, wholly or in part. In the absence of such imperfections, however, policies affecting production will result in reductions of income through an inefficient use of resources, and those affecting consumption will involve losses in consumer welfare. For example, a government might choose to promote an inefficient method of producing or milling rice in order to advance distribution or security objectives or for noneconomic reasons. But unless significant market imperfections are simultaneously offset, the policy will engender costs because of productive inefficiency or consumer losses. This is the nature of the trade-offs among multiple objectives, discussed earlier.

tributing income more evenly, or improving the security of food supplies. This aspect of effectiveness can be readily measured by changes in appropriate indicators used to define the objective. As will be clear from the discussion that follows, some policies do not advance certain objectives, irrespective of the level of economic costs. Second, the effectiveness of a policy depends on the costs associated with it in obtaining a given improvement in an objective. The methodology used in this study to assess the social profitability of rice production techniques can be applied to measure the loss (or gain) in economic efficiency and potential national income engendered by policies that cause a divergence between social and private profitability. Additional costs can be associated with the political effects caused by the transfers required to enact policies and with the administration of the policy interventions. The evaluation of policies, therefore, consists of two steps—determining whether desired objectives are furthered and measuring the associated cost (or gain) of resource reallocation.

The fundamental objectives of efficient income generation, income redistribution, and food security can be furthered by either increases in the level or changes in the structure of rice production. But self-sufficiency through import substitution demands increased national output if consumption levels are to be maintained. To achieve these ends, rice policies provide either universal or specific incentives. Universal incentives are available to all farmers and include tariffs, fertilizer subsidies, and paddy price supports. If productive inputs are highly mobile, universal incentives are relatively easy to administer and cause the least distortion in efficiency. However, these policies can bring about large transfers among economic groups, such as all rice consumers and all rice producers, that are both unintended and unwanted. On the other hand, if resources are, or can be, tied specifically to the production of rice—such as systematic insecticides, mechanical threshing, and, to a lesser extent, irrigated land—specific incentives might offer the most efficient and most easily administered type of policy with fewer unintended transfers.[12]

Nevertheless, all government policies are likely to bring about at least some unintended transfers among various groups in the country. In general, rice producers stand to gain from policies aimed at increasing production. Since consumers are the only losers from import restrictions, a strong consumer bias in a country (creating, for example, pressure to hold down urban prices) would be required to dissuade governments from us-

[12] Specific incentives thus require that segmented factor markets exist, which government policies can exploit to achieve objectives effectively. In less developed countries, such segmentation is common and arises from diverse causes. These causes include the immobility of assets and productive resources, the time required to learn about new techniques, and the large scale of many investments in land development. Segmentation permits the government to ration its incentives among selected groups, with minimal leakage to other groups.

ing this policy instrument to increase production. The government budget—as well as the taxpayers outside the rice sector—are likely to be the strongest forces in favor of trade policy and against output and input subsidies. Between these last two policies, producers can be expected to favor an output subsidy, because each producer will then be free to allocate inputs in production optimally.

The government treasury's position is, however, indeterminate. It depends on the relative costs of administering input and output subsidy programs and on the impact of the alternative policies on rice production. Input subsidies can be ineffective relative to output subsidies if the inputs (e.g. fertilizer) are used in the production of alternative crops. But if input subsidies can be tied solely to the marginal costs associated with additional production, such as through the development of irrigated perimeters, input subsidy programs are usually preferred to universal output subsidies.

The two countries with a comparative advantage in rice, Mali and Sierra Leone, are not required to subsidize inefficient local production. For Mali, security of food production production appears to be the primary goal, and income generation and its regional distribution are of somewhat lesser importance. The country has been able to expand socially profitable rice techniques that improve the security of rice production, notably in the Office du Niger and to a lesser degree in projects at Ségou and Mopti. This expansion of competitive rice production to improve food security clearly generates additional national income. In addition, much of this extra income accrues to farmers and other rural residents, although urban consumers gain from Malian price policy.[13] In the future, Mali is likely to face a difficult decision in its rice investment policy between further intensification of existing projects with improved packages based mainly on fertilizer and better water control, which would raise recurrent costs, and extensification through the construction of additional polders in the Niger-Bani Basin. Although the second approach involves a somewhat lower degree of security, it is likely to be the most profitable, given the availability of concessional foreign aid for polder construction and the existence of additional land that could be developed.

The principal objective of rice policy in Sierra Leone, the other country in this group that has a comparative advantage in rice production, appears to be increasing incomes and staple food supplies in its rural areas. In order to achieve this objective, the country is investing—using capital supplied by foreign aid donors—in several rice projects to introduce and spread im-

---

[13] Malian price policy, which keeps retail prices below c.i.f. import prices, has the effect of transferring income from producers to consumers, thereby redistributing income largely from rural to urban residents. This policy option is made possible by the absence (in normal production years) of the need to protect local production with higher consumer prices or to provide government subsidies.

proved techniques of production in both upland and swamp regions. Land development is subsidized for improved swamps, and modern inputs, such as fertilizer, seeds, and mechanical services, are also subsidized. In addition, farmers receive significant protection from the world price of rice. If these improved techniques are socially profitable, as they appear to be, they can contribute to efficient income growth. This income growth could involve rice only indirectly if new technologies are substituted for their traditional counterparts, thus freeing domestic resources for other cash-crop opportunities. If the new techniques increase production rather than simply substituting for traditional cultivation, they contribute to self-sufficiency in rice without requiring a trade-off with the growth and distribution objectives.

It is curious that Sierra Leone uses strong incentives to promote expansion of a commodity in which the country enjoys a comparative advantage. The issue becomes especially important because the government budget has been constrained historically and has been unable to provide all the subsidized inputs demanded. The explanation may be twofold. Since redistribution of income is also an objective that is strongly held, trade protection and subsidized production projects may serve as a means of transferring income to poorer regions. For example, in two of the poorer regions of the country, the north plains and bolilands, improved rice production techniques have raised the net return per unit of labor input by three to five times that earned in traditional rice production.[14]

Second, rice in Sierra Leone may not be competitive with other crops that can be produced, even though it is competitive with imports of rice. In that event, the government would have to adopt policies that discriminate in favor of rice in order to expand domestic production. Only improved, highly subsidized rice is competitive in regions that produce three of the more important cash crops—oil palm, coffee, and cocoa.[15]

[14] A study of Sierra Leone farm systems reports the following private returns per unit of labor input for 1974–75, in Le per man-hour, net of capital charges and operating expenses (8, p. 60). In the riverain grasslands, which is a relatively rich area, the increase in income resulting from improved production is less than in the two other, poorer areas.

Region	Traditional rice	Improved rice
Northern plains	0.085	0.25
Bolilands	0.053	0.28
Riverain grasslands	0.105	0.17

[15] Private returns (in Le per hour for 1974–75) net of capital charges and operating expenses were as follows (8, p. 60). (The figure for traditional rice in the Moa Basin includes returns on minor other crops.)

Region	Traditional rice	Improved rice	Oil palm	Coffee and cocoa
Northern plains	0.08	·0.25	0.17	—
Riverain grasslands	0.10	0.17	0.36	—
Moa Basin	0.10	—	—	0.14

Sierra Leone could therefore be undertaking rice projects and policies that, while competitive internationally with rice, are not the most efficient use of resources when compared with other domestic production opportunities. Because export taxes on oil palm, coffee, and cocoa generally depress domestic prices of these crops whereas import restrictions raise the domestic price of rice, the existing gap in private returns between traditional rice production and cultivation of export crops is smaller than would exist in the absence of these trade policies.

The choice of policy is much more difficult in the other three countries, which do not have a comparative advantage in exporting rice or in competing with imports in the urban consumption centers. If they desire to promote local rice production to replace imports in the cities, governments in these countries must protect or subsidize producers, which entails losses in national income. Consequently, their freedom of policy choice is circumscribed because they face difficult trade-offs.

The two objectives of rice policy in the Ivory Coast are to increase incomes generally and to ensure that the northern part of the country, in particular, benefits from this growth. Unfortunately, neither of these goals has been furthered by recent policy. In light of the unprofitability of rice production in Ivory Coast, any policy to expand output is bound to be costly. Recent Ivorian trade and price policy has resulted in welfare losses to consumers, government subsidies to producers, and a decrease in GNP that has been estimated at 3 billion francs CFA annually (5).

Moreover, distribution objectives have not been sufficiently well served by rice policies to offset these highly negative income effects. Rice investment policies have, by design, clearly favored the north, which is consistent with the objective of income redistribution. But the irrigated techniques in the north are less efficient than improved rainfed production in the forest zone, because of the greater cost of water control in more arid areas, and no improved technique is as efficient as traditional production in the northern savannah zone. Therefore, investment subsidies have primarily served to offset higher costs rather than to redistribute income toward northern farmers. Moreover, despite high investment subsidies, costly trade and output price policies have also been required to make improved, irrigated rice production in the north privately profitable. The greatest proportion of transfers resulting from these policies, however, has gone to farmers in the forest zone, where most rice is grown. Hence, Ivorian rice policy does not advance either the income generation or the regional redistribution objectives effectively.

Excellent opportunities to produce other crops efficiently exacerbate the problem of making rice policy effective. Greater social profitability of other crops, such as coffee, cocoa, cotton, copra, and palm products, results in high opportunity costs for national resources devoted to rice production. Moreover, strong incentives are necessary to bring forth increased rice production, and incentives of such size increase the magnitude of unintended transfers and the costs of administration. In the Ivory Coast, large budgetary deficits coupled with the unwillingness of consumers to pay high prices for rice have thwarted the implementation of government rice policies aimed at increasing the share of output from modern techniques and at transferring production resources and income to the north.

A more effective rice policy for the Ivory Coast would involve reduction of protection and elimination of the milling subsidy paid to government mills to support domestic producer prices. Beginning in 1977, the government has followed this strategy. Meanwhile, the government can continue the search for a new technology, probably based on divisible labor-saving techniques for rainfed rice production, that can relax the most immediate resource constraint of expensive labor. Both income growth and more equitable distribution of incomes, however, are better promoted by producing other crops that can be grown efficiently, especially in the poorer north. Because the security of food supplies has historically been a relatively unimportant issue in the Ivory Coast, there is little pressure to achieve self-sufficiency in rice production at high costs for this purpose.

Liberia has objectives for rice policy similar to those of the Ivory Coast—a primary emphasis on income generation, a secondary desire to have the increases in income occur in rural areas, and little concern with food security. Despite the place of rice as the principal staple food in Liberia, until very recently government intervention has been limited to trade policy. Investment policies in rice have only lately begun, no important subsidies on inputs exist, and government expenditures on rural infrastructure have not been large. Trade policy has consisted of taxing rice imports to collect government revenue and to protect local production. Because Liberian rice cannot be delivered efficiently to Monrovia, which is the main market for rice imports, government efforts to increase rice output run counter to the objective of generating income. Government policy does transfer resources from urban consumers to rural producers, but only at a significant loss in national income. Furthermore, the limited volume of marketings suggests that actual urban-rural income transfers are of a small magnitude. This situation will only be rectified by cost-reducing improvements in techniques of production and distribution. Such improvements might best be promoted by investment subsidies and research. As in the Ivory Coast, maximization of rural incomes requires attention to other crops, such as coffee and cocoa, that utilize available resources more efficiently.

Evaluation of rice policy in Senegal is more complicated. This Sahelian country is mainly concerned with improving food security, although the government also wants to change the regional distribution of income and to increase national income. Senegal does not have a comparative advantage in rice generally, and the most secure techniques are often the least efficient. The objective of increasing national income is thus contravened by policies that expand secure rice production. Moreover, the evidence is not convincing that increased production of irrigated rice will necessarily reduce the long-run instability of food supplies until numerous technical problems, such as management, maintenance, and salinity, are resolved.

Because areas where rice is produced coincide with those designated to benefit from improved income distribution, expansion of rice production by building irrigated polders can be an effective means of achieving this goal as well as of improving the security of local food supplies relative to traditional production. But as in Ivory Coast, Liberia, and Sierra Leone, rice production policies do not usually maximize the income growth potential of these areas.

A second aspect of the food security issue involves the willingness to rely on imports to offset shortfalls in domestic production (10). Food security must consider the reliability and costs of improved production relative not only to traditional production but also to the variability of prices and availabilities of rice on the world market. Food security is thus not fully realized until domestic production is increased to a level where imports are usually not necessary. This dynamic problem depends on the variability of domestic production, the variability of world prices, and the subsidies needed to sustain irrigated domestic production. Unfortunately, analytical techniques to relate these trade-offs within a framework of maximum economic efficiency are not available. Some general remarks based on the results of this study are possible, however.

If variability in c.i.f. prices is the concern of policy, the government has three broad categories of policy response open to it. First, it can substitute other staples for rice during periods of high rice prices. Second, the government can establish a financial buffer fund to cover the expected change in c.i.f. prices. Only the difference between the actual c.i.f. price and the expected long-run average c.i.f. price needs to be covered by this fund. Third, the government can subsidize irrigated production. Subsidization is necessary because, on average, irrigated techniques are socially unprofitable in delivering rice to Dakar, the main center of import substitution.

The social profitability results for Senegal indicate that subsidies of $70–265/mt are needed to support production, given a long-run c.i.f. Dakar price of $250/mt. This level of subsidy thus amounts to an average percentage subsidy of 28–106 percent of c.i.f. prices. If the government desires protection against fluctuations in c.i.f. prices equal to 100 percent of the average price (the maximum historical variation), a buffer fund is clearly more efficient than the subsidization of domestic production as a means of providing food security because real rates of interest on government loans to the Senegalese government are only 2.5 to 8 percent. Furthermore, these calculations assume that domestic production is 100 percent reliable and that no substitution in consumption occurs. Only if the government has strong inclinations that rice will be unavailable at any price on the world market does the alternative of domestic production appear economically rational. In summary, given the high cost of rice production imposed by severe resource constraints in Senegal, a trade-off arises between losses of income that must be incurred in order to obtain increases in food security relative to traditional production and positive regional distributional effects.

A number of general observations emerge from these evaluations of individual country policies. First, for countries with a comparative advantage in rice, Mali and Sierra Leone, rice policy can be used to further all of their objectives simultaneously. Second, countries such as the Ivory Coast and Liberia, which desire to redistribute income to certain rural areas but produce rice inefficiently if it is used to replace imports in urban consumption centers, are likely to reach their objectives more effectively by focusing on more profitable crops. Finally, in Senegal, where improved food security is the primary objective, the effectiveness of policy depends on the choice among alternative techniques and regions with different costs and degrees of security. Expansion of rice production under existing techniques can only increase food security at a high cost in terms of forgone national income and recurring subsidies, and hence it is a less desirable policy than establishment of a buffer fund.

The role of foreign aid donors is very important in influencing the costs of rice development that are borne by West African governments. Sometimes donors provide concessional assistance to countries that can produce rice efficiently. Donors might also share a recipient country's goal of improving food security or aiding the rural poor, and for this reason they might justify giving aid for a project that cannot compete without protection or subsidy. In the five countries discussed here, aid has been a predominant force behind rice investment. Given budgetary constraints, it appears unlikely that any of them would choose to make large investments in rice projects in the absence of foreign aid. If this observation holds true, donors will continue to help shape rice policy in West Africa through their roles as contributors to rice investment projects and as spokesmen for various trade, price, and subsidy policies.

The results of the social profitability analysis confirm that some kinds of production techniques can compete efficiently in all countries with imported rice for consumption on-farm or in markets in the producing area. Transportation costs of delivering rice imports to distant rural areas provide natural protection to much local production. Accordingly, a potentially fruitful approach for countries that are unable to substitute efficiently for imports in main urban consumption centers and that desire to increase food security or to improve income distribution is to concentrate on production that can be carried out efficiently for local and regional markets. Such projects would have to be carefully designed, probably combining features of better water security with relatively small scale and a modest degree of capital intensity. In the longer term, reductions in marketing costs through improvements in the transportation system would reduce both the natural protection of rice produced by these projects and the cost of delivering rice to main consumption centers.

Summary and Conclusion

All WARDA member countries desire to achieve self-sufficiency in rice production. The economic significance of increasing local production of rice to substitute for rice imports can be examined with reference to the national objectives of income generation, redistribution of income, and food security.

Detailed economic analysis has been undertaken in five countries. Two, Mali and Sierra Leone, have a comparative advantage in producing rice to substitute for imports and, with some techniques, for export to neighboring countries. Central issues for these two countries involve the choice of technique for continued expansion of production and the selection of policies that will provide necessary incentives to farmers as well as maximize the contribution to other objectives.

In Mali, policy makers face a choice between intensification of production, based on the use of fertilizer, improved water control, and mechanical weeding practices in existing projects, and extensification, principally by creating more polders with controlled flooding in the Niger-Bani Basin. A main feature of this choice is weighing the enhanced security of production, rising costs, and high recurrent public expenditure requirements associated with intensification, against less secure, more socially profitable production in new polder schemes.

The decision in Sierra Leone is between promoting rice or encouraging other crops in which it enjoys an even stronger comparative advantage. If it chooses to continue its efforts to accelerate rice production, the government needs to select policies that encourage farmers to undertake more rice cultivation. In the face of more lucrative alternatives elsewhere in the economy, such policies require large transfers to producers. Fortunately for Sierra Leone, this result is consistent with its objective of improved income distribution, although rice subsidies have a severe impact on a budget that is already tightly constrained. Since food security is not a strongly held objective, the choice between full and less complete water control is wholly an efficiency issue.

Ivory Coast, Liberia, and Senegal are unable to produce rice efficiently with existing techniques for delivery to either export markets or main domestic consumption centers. Some techniques of production in these countries can compete with imports of rice in rural areas of production and thus do not require protection from import competition. But substantial portions of local rice production cannot survive in the absence of restrictive trade policy, which results in income losses from the inefficient use of resources and in welfare losses from the higher price of rice.

Several factors might help to explain why the Ivorian, Liberian, and Senegalese governments desire to promote unprofitable rice production. The first is an information gap, a lack of understanding that rice does not have a comparative advantage. This explanation is not particularly convincing, given the existence of rice imports and the observable high costs of much rice production. It is true, however, that policy makers often base decisions on distorted private, rather than social, prices.

An additional explanation is based on governmental expectations that rice will become competitive in the future because of dynamic learning effects that accompany intensification, rising world prices for rice, or worsening prospects for other domestic activities, usually exports, that would cause the costs of local land, labor, or capital to decrease. Sensitivity analysis based on reasonable changes in these parameters does not indicate that future competitiveness is in sight for these countries.

One central issue for this group of governments, therefore, is to examine the sources of their inefficiency in rice production and the likelihood that greater efficiency might be achieved in the foreseeable future. Generally, advanced techniques have not improved productive efficiency in these countries because they simply substitute more expensive intermediate inputs for small reductions in relatively expensive domestic resources. Either the technology does not exist that can overcome existing constraints competitively or the choice of technique in the past has been inconsistent with prevailing and expected factor prices.

A third possible reason why these countries are devoting scarce resources to rice when they could generate more income in alternative uses is that they believe expanded rice production contributes to other objectives. Governments may not be fully aware of the trade-offs inherent in making choices among policies to advance conflicting objectives.

The government of Senegal, for example, can be viewed as holding food security as a primary objective. If greater security of food supplies can be obtained by increasing rice production, despite Senegal's comparative disadvantage in rice, the government should weight its security and income objectives and decide how much to forgo of one in order to enhance the other. But it is quite possible that self-sufficiency in rice or food may not be the most effective way to secure food supplies, which would mean that Senegal's rice policies have caused a loss in income with little or no offsetting gain in security.

The Ivorian government has attempted to transfer income to the northern savannah area by promoting rice production in that region. Even though the northern zone has benefited from a larger share of heavily subsidized investments than the richer south, these investments must still be coupled with trade protection and price subsidies to make the improved rice techniques attractive to farmers. Because trade control and price subsidies apply to all domestic rice production, the south has, on balance, benefited more than the north from rice policy simply because most Ivorian rice is produced in the southern forest zone. The key issue for the Ivory Coast is to review whether emphasis on rice is desirable in view of more profitable alternatives available in Ivorian agriculture to achieve the same objectives.

The government of Liberia has tried to increase incomes in rural areas by encouraging rice development. This goal, which does not have a particular regional focus, might be met better, with a gain instead of a reduction in potential national income, if the government promoted expansion of agricultural commodities, such as coffee and cocoa, which can be grown efficiently.

The study has several implications for WARDA's goal of reaching regional self-sufficiency of rice in West Africa. First, most rice produced with existing techniques is socially profitable if the output substitutes for imports on-farm or in markets near the site of production. It is thus desirable to expand production for many regional markets with current and improved techniques. The replacement of traditional methods with more efficient improved techniques can also release domestic resources for use in other productive activities, including cash cropping in many areas.

Second, outside of Mali and Sierra Leone, rice production to replace imports in urban consumption centers is socially unprofitable with existing techniques. Furthermore, the advanced techniques, especially those using full water control, are usually less efficient than traditional rainfed production. Hence, research into and development of more appropriate technologies are required before future rice production will become socially profitable. Critical areas for research include development of chemical and mechanical techniques to substitute for labor, more efficient use of irrigation water, additional investment in infrastructure, and cost-reducing changes in processing and distribution. This technical research should be complemented by continuously updated analysis of policy changes needed to accompany the introduction of new techniques and of the effectiveness of policies in furthering objectives as constraints gradually change.

The development and dissemination of new technologies is no small order. But if the historical experience of Asian rice policy is any guide, the agenda outlined above is of critical importance. In most Asian countries, both price policy and research were critical preconditions for the success of production programs. Malaysia, the Philippines, Taiwan, and Indonesia, for example, achieved rapid production gains as a result of the dissemination of seed-fertilizer packages, once appropriate price incentives were established (3, 4). Nor do such revolutions occur overnight. The experiences of Taiwan and Malaysia, where 20 to 30 years were required for the development of effective varieties and irrigation facilities, are relevant to the current West African situation (2). Creation of the International Rice Research Institute and other research institutions has reduced but not eliminated this time lag.

Yet to note that prices matter overlooks some fundamental differences between the economic environments of West Africa and Asia. The Green Revolution that took place in Asia during the 1960's represented a technological package very well suited to Asian factor endowments and institutional settings. Labor was relatively low-cost or seasonally unemployed, thus allowing profitable increases in double-cropping and land-use intensity. Irrigation infrastructure had been in place for decades, if not centuries, reflecting substantial farmer experience with water control. As this study has shown, these conditions differ greatly from those in contemporary West Africa. Most Asian technologies are not transferable without substantial sacrifices in economic efficiency, and hence the successful development of rice production in West Africa will likely prove to be a highly indigenous process.

Citations

1 W. M. Corden, *Trade Policy and Economic Welfare*. Clarendon Press, Oxford, 1974.

2 Dana G. Dalrymple, *Development and Spread of High-Yielding Varieties of Wheat and Rice in the Less Developed Countries*, 6th ed. Foreign Development Division, Economic Research Service, U.S. Department of Agriculture, Washington, D.C., 1978.

3 *Food Research Institute Studies*, 14, No. 3 (1975).

4 *Food Research Institute Studies*, 14, No. 4 (1975).

5 Charles P. Humphreys and Patricia L. Rader, "Rice Policy in the Ivory Coast." Stanford/WARDA Study of the Political Economy of Rice in West Africa, Food Research Institute, Stanford University, Stanford, July 1979; Chapter 1.

6 Eric A. Monke, Scott R. Pearson, and Narongchai Akrasanee, "Comparative Advantage, Government Policies, and International Trade in Rice," *Food Research Institute Studies*, 15, No. 2 (1976).

7 Scott Pearson, J. Dirck Stryker, and Charles P. Humphreys, "An Approach for Analyzing Rice Policy in West Africa." Stanford/WARDA Study of the Political

Economy of Rice in West Africa, Food Research Institute, Stanford University, Stanford, July 1979; Introduction.

8 Dunstan S. C. Spencer and D. Byerlee, "Small Farms in West Africa: A Descriptive Analysis of Employment, Incomes, and Productivity in Sierra Leone." African Rural Economy Working Paper No. 19, Department of Agricultural Economics, Michigan State University, East Lansing, and Department of Agricultural Economics, Njala University College, Njala, Sierra Leone, February 1977.

9 J. Dirck Stryker, "Comparative Advantage and Public Policy." Stanford/WARDA Study of the Political Economy of Rice in West Africa, Food Research Institute, Stanford University, Stanford, July 1979; Chapter 12.

10 ———, "Food Security, Self-Sufficiency, and Economic Growth in the Sahelian Countries of West Africa." U.S. Agency for International Development, Washington, D.C., February 1978.

11 J. Dirck Stryker, John M. Page, Jr., and Charles P. Humphreys, "Shadow Price Estimation." Stanford/WARDA Study of the Political Economy of Rice in West Africa, Food Research Institute, Stanford University, Stanford, July 1979; Appendix B.

12 C. Peter Timmer, "The Political Economy of Rice in Asia: A Methodological Introduction," *Food Research Institute Studies*, 14, No. 3 (1975).

13 C. Peter Timmer and Walter P. Falcon, "The Impact of Price on Rice Trade in Asia," in G. S. Tolley, ed., *Trade, Agriculture and Development*. Ballinger Press, Cambridge, Mass., 1975.

12. Comparative Advantage and Public Policy in West African Rice

J. Dirck Stryker

Pearson, Stryker, and Humphreys (9), in the Introduction, suggest that the fundamental policy objectives concerning rice in West Africa are generating more income, distributing income in a more equitable fashion, and reducing the risk associated with production and consumption of essential foods. Self-sufficiency in these foods, another important goal of West African governments, is seen as a proximate objective which contributes in one way or another to each of the three fundamental objectives.

If self-sufficiency in rice is defined as a situation in which a country does not import the cereal, the effect on income of achieving this goal will depend on whether or not the country has a comparative advantage in the production of rice. If it does have such an advantage, income will be greater; if it does not, income will be less. At the core of the concept of comparative advantage lies the notion of social profitability, a measure of the economic efficiency with which a good is produced from the point of view of the nation. As shown in Appendix A, this measure differs from private profitability, the net incentive to individual producers, primarily because of government taxes, subsidies, and policies affecting prices and trade.

Combinations of all these policies may be used to promote self-sufficiency by encouraging the expansion of domestic production. Estimates of the contribution that such an expansion makes to each of the three fundamental national objectives are presented here for a number of different rice producing-processing-marketing activities in West Africa. These estimates are taken from individual country studies of the Ivory Coast (4), Liberia (7), Mali (6), Senegal (15), and Sierra Leone (10). Whereas the contribution of each activity to national income is measured quantitatively, the effects of the activities on the distribution of income and food security are discussed largely in qualitative terms. Comparisons are also made with other agricultural activities in West Africa and with those involving rice production in the United States and several Asian countries.

The next section reviews some of the geographical and institutional factors underlying comparative advantage of rice in West Africa. This is followed by a brief survey of the available production techniques. Estimates of private profitability are then presented and analyzed, and the effect on each activity of government taxes, subsidies, and price and trade policies is discussed. That section also analyzes the social profitability of producing rice in each of the West African countries in comparison with the cultivation of other crops in West Africa and rice in the United States and Asia. The impact of each rice-producing activity on the distribution of income and the level of food security is then explored, along with requirements of these techniques for scarce public revenue. Finally, some general conclusions are presented concerning the relative efficiency of the activities in achieving desired objectives.

Geography and History

Several generalizations can be made about West Africa that influence its comparative advantage in rice production. First, because it is a region of low population density, the value of land is slight relative to that of labor, and the costs of transport, marketing, and the provision of government services in rural areas are relatively high (12). Second, in comparison with many other areas of the world, most countries in West Africa have a poorly developed infrastructure. Irrigation is in its infancy, water flows are not regulated on any of the major rivers, and poor transport facilities impede access to many areas. Finally, there is often a severe shortage of government budgetary resources and of people with the training required to undertake major development projects.

In other respects, suggested in Table 12.1, there are substantial differences among countries and even between regions within countries. The first two indicators in this table describe key geographical features of these regions—mean annual rainfall and distance to the nearest major port. There is a fairly sharp distinction between regions with 1,300 millimeters or more of rainfall, which can produce rice using rainfed techniques, and those which receive 700 millimeters or less but have substantial water resources for irrigation or flooding. In addition, important differences in water conditions exist within these rainfall zones. The Senegal River, for example, suffers from a number of disadvantages in comparison with the Niger-Bani system: smaller water flow, greater intra-annual variation and uncertainty of flooding, and salt incursion from the sea.

Rice-growing regions also vary substantially in their distance to the nearest seaport. The interior regions are provided with natural protection against rice imports because of high inland transport costs. But these transport charges also raise the cost of using inputs supplied from abroad

TABLE 12.1. *Key Characteristics of Several Rice-Producing Regions of West Africa*

Region	Mean annual rainfall (mm)	Average distance to major seaport[a] (km)	Rural population density (persons/km²)	Unskilled rural wage rate (US $/day)	Rural per capita income (US $/person)	Degree of urbaniza-tion[b] (percent)	Density of all-weather roads (km/km²)
Ivory Coast:							
Forest	1,550	406	19	1.80	150	37%	
Savannah	1,300	667	9	1.40	53	21	0.0435
Liberia	2,000	225	7	1.25	168	23	0.0189
Mali:							
Mali-Sud	1,300	853	14	1.20	50	6	0.0144[c]
Mopti	520	1,317	13	1.00	40	4	
Segou/Office du Niger	700	1,145	16	1.00/1.25	50/85	9	
Senegal:							
Fleuve, delta	320	338	16	1.00	56	49	0.0170
Fleuve, valley	420	513	7	1.00	75	18	0.0170
Casamance	1,400	395	22	1.20	100	16	0.0660
Sierra Leone	2,500/3,000	175	23	0.60/0.80	70	25	0.0238

SOURCES: 4, 6, 7, 10, 15.

[a]Distance is defined from a major town or city near the geographical center of each region.

[b]Urban population is defined generally as those living in towns of 10,000 people or more.

[c]Total kilometers of all-weather roads divided by the one-third of Mali's total land area that is in the zone where agriculture is possible.

and make it difficult for the interior countries to export to other West African countries, especially to the coast where the major markets are found.

The third indicator in Table 12.1, rural population density, is a key variable determining the types of production techniques that are appropriate for West African conditions.[1] As already noted, population density is generally low throughout the region. There is variation, however, and the density in Liberia is only about one-third that of the Casamance in southern Senegal or the southern Ivory Coast. In addition, there are important concentrations of population, not shown in the aggregate data, in such areas as the Senegal River Valley and the Lower Casamance.

The low ratio of labor to land influences the next two indicators, the wage rate of rural unskilled labor and rural per capita income. The wage rate is often considerably greater than in many Asian countries, where population densities are much higher. The world's most important rice exporter, Thailand, for example, had a rural wage rate equal to about $.60 at official exchange rates prevailing in the mid-1970's, or only one-third that of the forest zone of the Ivory Coast and lower than that of any West African country studied here except Sierra Leone.[2] Aside from the southern Ivory Coast, which has experienced considerable rural development and where relatively high wages and per capita income are attracting large numbers of migrants from other regions, the Office du Niger in Mali, a fully irrigated scheme for rice and sugar production, has been able to raise wages and incomes above the lower levels found in the surrounding regions of Mopti and Ségou.[3]

The last two variables of Table 12.1, degree of urbanization and density of all-weather roads, are indicators of the level of commercialization and state of infrastructure development that exist in each region. These affect the cost of transporting and marketing rice, delivering inputs, and providing administrative and extension services. Urbanization and road development are generally much more advanced in the coastal than in the interior regions. This is a reflection of the sequential nature of development, which started during the colonial era along the coasts and only recently moved to an important extent into the interior (11).

Finally, there are special factors that influence comparative advantage but that are not easily summarized in tabular form. One is the continued existence of capital the investment of which was made in the past and therefore may be considered sunk.[4] The most important example is the

[1] For a discussion of the relationship between population density and agricultural technique, see Boserup (1).

[2] The problems associated with making comparisons of this type at official exchange rates have been discussed elsewhere (3).

[3] See Appendix B for a discussion of the West African labor market.

[4] Capital investments that were made in the past and the costs of which are not going to be incurred again over the relevant planning horizon are considered sunk, that is, their costs are not included in the calculation of private or social profitability.

Office du Niger: its diversion dam and principal canals were constructed during the 1930's. The regions also differ with respect to the availability of data required for development. More is known about the hydrology of the Senegal River, for example, than about that of the Niger. Another special factor is past agricultural research. The results of this research, most of which has been conducted outside of West Africa, give rice an important advantage over food crops such as millet and sorghum, on which research is only beginning. Standing rice has also been the subject of considerably more research than have the floating and upland varieties traditionally grown in West Africa. There is a question, however, of whether rice techniques developed in Asia are appropriate for West African conditions, especially in view of the differences in population density on the two continents.

Rice Production Techniques

The techniques used to produce rice in West Africa, which are described in detail in the individual country chapters, range from traditional upland cultivation with no modern inputs and long periods of fallow to intensive mechanized cultivation under total water control. The techniques vary substantially with respect to yields, costs, dependence on outside inputs, and labor-land ratios. Several of these characteristics are shown in Table 12.2.

It is clear from this table that the techniques listed are far from homogeneous. Labor times, especially, vary enormously between countries, and this variation is not very well correlated with differences in yields, which tend to be fairly similar for each technique.[5] Part of the variation in labor times may be due to the substitution of capital for labor, especially in the form of mechanization. But different methods used for estimating labor times may also have produced results that are not entirely comparable. Labor inputs in the Ivory Coast, Mali, and Senegal, for example, have been calculated from a number of sources, including information provided by farmers and extension workers, which suggest how much time should normally be required for each agricultural task. Estimates for Liberia and Sierra Leone, on the other hand, are based primarily on multiple-interview surveys of farmers, which indicate the number of days actually devoted to each task but say relatively little about the amount of time or effort expended in performing that task on a given day. It is likely that the former approach tends to underestimate and the latter to overestimate actual labor inputs.

Another source of variation in labor times relates to the treatment of

[5] The simple correlation coefficient between labor and yields across both techniques and countries in Table 12.2 is 0.34.

labor used for land development. If freshly cleared land is cultivated for several consecutive years, the time involved in clearing is treated in Table 12.2 as a land development cost. In several instances, however, time spent developing the land is included with other labor inputs as a current operating cost, either because the land is only farmed for one year or because, as for Sierra Leone, the data do not allow a distinction to be made between these two types of labor input. Although this results in some error in these variables in Table 12.2, whether these land development costs are treated as a capital or current input probably does not affect very much the overall calculation of private and social profitability.

Although the data in Table 12.2 are not strictly comparable, it is useful, nonetheless, to draw a few conclusions from the table. One is that yields are positively correlated with degree of water control.[6] Not only does better water control by itself improve yields, but also the fixed cost associated with land development encourages greater use of yield-increasing variable inputs such as fertilizer.[7] The yield response to fertilizer is particularly great, moreover, because of the improved seed varieties used and because existing levels of fertilizer utilization are generally very low.[8]

Labor inputs, on the other hand, are not very closely linked with degree of water control.[9] Traditional and improved manual rainfed techniques, including upland and swamp cultivation, use up to 400 man-days per hectare or more, whereas manual irrigated cultivation, such as that found in the Senegal River Valley, uses less than 300 man-days. Animal traction techniques yielding about 2 mt/ha employ 88 to 111 man-days per hectare in rainfed cultivation, 70 to 100 man-days in controlled flooding, and 90 man-days under irrigated conditions. Fully mechanized cultivation, on the other hand, appears to use at least as much labor with rainfed as with irrigated techniques, though this conclusion is based on only one example of rainfed farming. Some of the reasons why labor, in contrast to other variable inputs, is used relatively less in irrigated agriculture might be that less land clearing is required, good water control lessens the need

[6] The simple correlation coefficient between land development cost and yields, using data from Table 12.2, is 0.48. Although highly correlated with degree of water control, land development cost is not a perfect proxy for this variable, since this cost also depends on natural conditions and input prices, which differ substantially between countries.

[7] The simple correlation coefficient between nitrogen fertilizer and land development cost is 0.32; that between fertilizer and yields is 0.78. One instance in which land development has not led to relatively high rates of fertilizer use is the Office du Niger ("Animal traction irrigated single crop" in Table 12.2), which with its capital costs already sunk is able to operate using a very land-extensive technique resulting in yields of only 2.25 mt of paddy per hectare.

[8] Humphreys and Pearson (5) estimate that response rates of 10–15 kg of paddy per kg of nutrient should be attainable in irrigated cultivation under average West African conditions.

[9] The simple correlation coefficient between land development cost and labor input per crop is −0.16.

TABLE 12.2. *Characteristics of Rice Production Techniques*

Production technique	Paddy yield (mt/ha/crop)	Land development cost (U.S. $/ha)	Farm labor (man-days/ha/crop)	Fertilizer (kg/ha/crop)			Extension service costs (U.S. $/ha/crop)
				N	P₂O₅	K₂O	
				N	P_2O_5	K_2O	
Traditional manual upland:							
Ivory Coast	0.89–1.30	28–54	85–113	0	0	0	0
Liberia	1.05	0	214	0	0	0	0
Sierra Leone	0.81–1.17	0	205–238	0	0	0	0
Improved manual upland:							
Ivory Coast	1.50–2.20	28–72	97–117	50	27	27	31
Liberia	1.57	0	231	42	42	42	25
Sierra Leone	1.46–1.87	0	225–258	50	50	0	8
Animal traction upland:							
Ivory Coast	1.80	51	88	50	27	27	31
Senegal[a]	2.07	60	111	57	20	40	22
Mechanized upland:							
Ivory Coast	2.00	520	30	50	27	27	31
Traditional manual swamp:							
Liberia	1.55	50	243	0	0	0	0
Mali	1.20	0	120	0	0	0	0
Senegal	1.08	17	208	0	0	0	0
Sierra Leone	2.20–2.83	34–39	274–356	0–8	0–8	0–8	0
Improved manual swamp:							
Ivory Coast	3.50	1,460	240–247	50	27	27	63
Liberia	3.50	750	331	42	42	42	49
Senegal	3.60	818	266	88	45	68	44
Sierra Leone	2.78–3.03	173	336–390	53	53	0	17
Improved manual mangrove:							
Sierra Leone	1.74–2.80	0	445	0	0	0	0
Animal traction swamp:							
Mali	1.80	600–2,000	100	0	0	0	7
Partially mechanized swamp:							
Ivory Coast	4.00	1,680	181	50	27	27	63
Liberia	3.50	1,504	235	42	42	42	135

TABLE 12.2—continued

Production technique	Paddy yield (mt/ha/crop)	Land development cost (U.S. $/ha)	Farm labor (man-days/ ha/crop)	Fertilizer (kg/ha/crop)			Extension service costs (U.S. $/ ha/crop)
				N	P₂O₅	K₂O	
Improved manual uncontrolled flooding:							
Sierra Leone	0.96	0	112	9	9	0	0
Animal traction uncontrolled flooding:							
Mali	0.60	0	60	0	0	0	0
Mechanized uncontrolled flooding:							
Sierra Leone	1.13–1.82	0	68–91	0–13	0–13	0	0
Animal traction controlled flooding:							
Mali	1.40	800–1,000	70–80	0	0	0	10–12
Improved animal traction controlled flooding:							
Mali	2.50	800–1,000	95–100	32	23	0	20
Animal traction irrigated single crop:							
Mali	2.25	sunk	90	15	0	0	20
Improved animal traction irrigated single crop:							
Mali	3.50	440	120	64	46	0	40
Mechanized irrigated single crop:							
Senegal	2.50	4,794	92	69	72	0	10
Manual irrigated multiple crop:							
Ivory Coast	4.00	6,812	247	50	27	27	63
Senegal	4.75	315	270	122	96	0	52
Mechanized irrigated multiple crop:							
Ivory Coast	2.75	4,972	34	50	27	27	63
Senegal	3.80	3,116	135	81	92	20	24

SOURCES: 4, 6, 7, 10, 15.
ᵃYields are higher than is usual for upland rice cultivation because of the peculiar "gray soils" on which the rice is grown. These permit the plants to be nourished by a high water table after the rains have stopped.

for weeding, and, where natural conditions permit, rainfed swamp techniques are often similar to those involving irrigation.

There is a fairly strong correlation between yields and extension costs, but this may be because these costs are closely associated with the delivery of modern inputs not otherwise included in Table 12.2, rather than because of the usefulness of extension advice per se.[10] Furthermore, the relation between these costs and the services actually provided by the extension agents is complicated by such factors as the locational concentration of farmers, the extent to which project overhead expenses are charged to the extension service, and the administrative efficiency of that service. Nevertheless, the relatively high correlation between yields and extension costs suggests the usefulness of this variable as a proxy.

To test the validity of these conclusions with several different influences on yields operating at once, a single-equation, least-squares regression was run, using the 33 country-specific techniques shown in Table 12.2. With paddy yield as the dependent variable, the following regression coefficients were obtained (with standard errors in parentheses):

Land development cost	Farm labor	Nitrogen fertilizer	Extension service cost
.151	3.212	17.427	11.708
(.057)	(.845)	(3.184)	(3.591)

All coefficients are significant at the 95 percent confidence level, as is the coefficient of determination of 0.813.

One of the most interesting conclusions from this analysis is that each of the last three input variables exerts a separate influence on yields beyond that of land development costs. Nevertheless, water control clearly remains a necessary, if not a sufficient, condition for obtaining high yields in many of these areas because of insufficient rainfall and other natural conditions.[11]

To test for the possibility that yields might vary systematically among countries, dummy variables were introduced into the regression equation, with the following results:

Constant	Ivory Coast	Liberia	Senegal	Sierra Leone
878.032	−455.811	−867.400	−526.138	−643.439
	(247.620)	(305.571)	(315.141)	(290.249)

[10] The simple correlation coefficient between yields and extension costs is 0.72.

[11] The statistical results reported here are only a first step toward estimating a production function for rice cultivation in West Africa. Many of the econometric problems associated with estimation have not been considered, only a linear relationship has been fitted, and several important variables have been either excluded (e.g. level of mechanization) or somewhat misspecified (e.g. degree of water control). Nevertheless, the conclusions of the analysis, even at this stage, are suggestive of interesting possibilities for further research.

Land devel- opment cost	Farm labor	Nitrogen fertilizer	Extension service cost
.138	4.697	17.748	12.782
(.055)	(1.012)	(3.954)	(4.146)

The inclusion of the dummy variables has relatively little influence on the regression coefficients of the inputs, except for that of labor, which is almost 50 percent greater than without the dummies. All these coefficients remain significant, as does the coefficient of determination, which equals 0.842.

The most interesting results are revealed by the coefficients of the dummy variables, which allow for shifts in the constant term of the regression equation. This term is a maximum for Mali, for which the dummy variable equals zero. For each of the other countries, however, the coefficient of the dummy variable is negative, indicating that, other things being equal, yields are lower than in Mali. The downward shift of the constant term is greatest and highly significant for Liberia. It is less pronounced, but still significant, for Sierra Leone. The coefficients for the Ivory Coast and Senegal, on the other hand, are sizable but not statistically significant at the 95 percent level of confidence. Overall, country-specific yield variations may be as great as 867 kg/ha.[12]

These results suggest that Mali has a significant natural advantage compared with the other countries in producing rice. One reason may be the good flood conditions of the Niger River. Another is the sunk capital investment in the irrigation system of the Office du Niger. At the other extreme, Liberia and Sierra Leone appear to have natural disadvantages in rice production. This may be because, as Spencer (*10*, p. 15) observes, "cultivation in heavy rain forest areas is more labor demanding than in more open savannah regions and thinner rain forests such as in the Ivory Coast."

Milling and Marketing Techniques

A range of techniques also exists for milling and marketing. Chief characteristics of three different milling techniques—industrial, small hullers, and hand pounding—are given in Table 12.3. The three techniques vary substantially with respect to scale. The largest industrial mills are capable of milling 30,000 tons of paddy per year, whereas a single person can hand

---

[12] These coefficients help to explain why the regression coefficient of farm labor is increased when the dummy variables are introduced into the regression equation. Liberia and Sierra Leone generally have high labor inputs in relation to the other countries. Without the dummy variables, the coefficient of labor is biased downward because of the effect of these omitted country specified dummy variables. Once the variables are introduced, the bias is eliminated.

TABLE 12.3. *Characteristics of Rice-Milling Techniques*

Milling technique	Projected full capacity[a] (mt paddy/year/unit)	Capacity utilization in 1976	Unit cost (U.S. $/mt milled rice)	Milling ratio	Quality of output
Industrial:					
Ivory Coast	15,000–20,000	0.64	56	0.66	25–35% broken
Liberia	10,000	0.06	119	0.67	25–35% broken
Mali	6,000–18,000	0.94	27–33	0.57–0.67	40–70% broken
Senegal	10,000–30,000	0.15	104	0.65	40–90% broken
Sierra Leone	3,750–15,000	0.25	67[b]	0.64	10% broken
Small huller:					
Ivory Coast	500	0.10	20	0.63	fresh, some parboiled
Liberia	400	0.38	52	0.66	24–45% broken
Mali	375	0.44–0.69	16–30	0.45–0.70	60–70% broken
Senegal	500	0.05–0.15	28	0.66	25–40% broken
Sierra Leone	433	0.50	14	0.67	20–40% broken
Hand pounding:					
Ivory Coast	6.25	n.a.	133	0.65–0.69	stones, some parboiled
Liberia	6.90	n.a.	78	0.60	40–50% broken
Mali	4.50	n.a.	60	0.70	80–100% broken
Senegal	6.00	n.a.	21	0.65–0.70	40–60% broken
Sierra Leone	5.70	n.a.	31	0.67	20–40% broken

SOURCES: *4, 6, 7, 10, 15.*

[a]Assumes 5,000 hours (250 days at 20 hours/day) of operation per year for industrial mills, 2,500 hours (250 days of 10 hours/day) per year for small hullers, and 1,500 hours (250 days at 6 hours/day) for hand pounding.

[b]Adjusted from the cost estimate of Spencer (see Table 12.1) to reflect actual rates of capacity utilization.

pound only about five or six tons during the same period. The former technique is also very capital-intensive, employing on a single shift only ten to 15 workers in an entire mill, whereas one woman pounding rice by hand uses only a crude mortar and pestle. Between these two extremes are the small-scale mills, with which two persons at a time can annually hull 400 to 500 tons of paddy.

Processing costs differ substantially between techniques and countries. The cost of milling a ton of rice in the large industrial mills varies inversely with rates of capacity utilization and at current rates is considerably higher than in the small hullers for all countries. Hand pounding is cheaper than large-scale milling except in the Ivory Coast and Mali, where rates of capacity utilization in the industrial mills are fairly high.[13] The cheapest milling technique, however, is small-scale hulling. Only in Senegal, where rates of capacity utilization are very low because of a highly fragmented market for rice which is locally produced and consumed, are small-scale mills more expensive than hand pounding. On the other hand, paddy and milled rice must be transported to and from the small hullers, which for on-farm consumption decreases their advantage over hand pounding.

Milling ratios do not seem to differ markedly between techniques or countries in any consistent way. Hand pounding yields a higher percentage of broken rice, and quality is decreased by the presence of foreign matter. Small-scale hulling also increases the percentage of brokens compared with large-scale milling, except in the drier regions, where breakage rates in the large mills are quite high.

The collection of paddy and distribution of rice generally take place within a dual marketing system. On the one hand, public marketing agencies purchase paddy from the farmer at an officially prescribed producer price, deliver it to publicly or privately owned industrial mills, and provide for its distribution and sale to the consumer at an official retail price. Large quantities of paddy and milled rice, however, are also typically traded in a private marketing network, where prices are established principally by supply and demand and where processing is done either by hand or in small-scale mills.

Marketing costs are influenced by the density of population and the relative adequacy of the existing road network. They are also affected by the location of consumption in relation to production. There are numerous different possibilities, including delivery from interior producing regions to coastal markets, distribution to markets within the producing regions, and on-farm consumption. In general, the further apart are the points of pro-

---

[13] Hand pounding in the Ivory Coast is also relatively expensive because of the high wage rates prevailing in that country.

duction and consumption, both physically and vertically within the marketing chain, the higher is the total cost of collection and distribution and the lower is the border price used to calculate the social value of rice output.[14] The first stage in the marketing chain is often the most costly, however, because of high charges for short-distance transport, which are especially important if the paddy is collected from widely dispersed areas for delivery to a few centrally located industrial mills.[15] This is very important in areas, such as Liberia, where the population density is low and the road network is poorly developed.

Private Profits, Public Incentives, and Net Social Profitability

Appropriate production, milling, and marketing techniques have been combined to form the rice sector activities analyzed in detail in the individual country papers. Indicators of private profitability (PP), the effects of public incentives, and net social profitability (NSP) are shown in Table 12.4 for the major rice-producing activities, with consumption assumed in each case to take place in the capital city.

Private Profitability

Private profitability equals the value of output minus the value of all inputs, each measured in terms of the appropriate domestic market prices faced by farmers, millers, or traders. These prices are inclusive of government taxes and subsidies. The resulting PP indicator shows the incentive, for each activity, to alter the existing allocation of resources. If private profitability is positive, resources are encouraged to flow into the activity; if PP is negative, the direction of resource flow is likely to be away.

Private profitability, as shown in Table 12.4, is nearly always positive. The only exceptions are in Liberia and Mali, where PP, assuming delivery to the capital city, is negative for several activities that, instead, usually produce only for the farm. On-farm consumption raises the value of output and causes private profitability in most of these cases to be positive.[16] In addition, part of the harvest from these activities in Mali is sold on the free market at prices higher than the low official producer price used to calculate private profitability in Table 12.4. When both these adjustments are made, virtually every activity is privately profitable.[17]

[14] See the individual country papers and Appendix B for more detailed discussions of this point.

[15] On this issue, see Humphreys and Pearson (5).

[16] Monke (7) shows this to be true, for example, in Liberia.

[17] The only technique remaining unprofitable is traditional flooded cultivation, which yields of only 600 kg/ha, resulting partly from the drought conditions of recent years. This may be a case, discussed in Appendix B, in which the market wage overestimates the shadow price of family labor in a traditional technique.

The incentive to produce for market, however, varies enormously among countries and techniques. At one extreme, private profitability is so low that efforts to improve manual cultivation of rice in the uplands of Liberia are unlikely to succeed as long as this rice is distributed to Monrovia. In the Ivory Coast, on the other hand, price incentives and input subsidies enable farmers to earn profits of over $200/mt for several different production techniques.

Most of this variation is due to differences in profitability between countries rather than between techniques within the same country. An unweighted average of private profitability for each activity shown in Table 12.4 is $197/mt in the Ivory Coast, $111/mt in Sierra Leone, $46/mt in Senegal, $.3/mt in Liberia, and $−26/mt in Mali. Variation among techniques, on the other hand, is much less. For the Ivory Coast, Liberia, and Sierra Leone taken together, private profitability averages $76/mt in traditional manual upland, $109/mt in improved manual upland, and $130/mt in improved manual swamp cultivation. To take another example, private profitability in Ivory Coast upland, swamp, and irrigated cultivation varies only slightly with different levels of mechanization, and the direction of variation is not consistent. Finally, private profitability for each of the improved techniques in Mali differs remarkably little from zero.

Public Incentives

Private profitability of rice production in West Africa is influenced to an important extent by public incentives consisting of taxes, subsidies, and price and trade policies. In Table 12.4, these incentives are aggregated into two groups. The first is the net effect of price and trade policies causing the domestic price of rice to differ from its border price—either c.i.f. or f.o.b., depending upon whether rice is imported or exported. The second consists of net taxes and subsidies on intermediate and capital inputs. The sum of these two groups of incentives is equal to the difference between private and net social profitability.[18]

There are substantial differences between countries in the magnitude and type of public incentives offered to encourage rice production. In Mali, the low official price of paddy purchased by the government, which controls about 50 percent of the total tonnage marketed, tends to discourage substitution of domestic production for rice imports. Mali is practically self-sufficient in rice, however, and in good years even exports the grain. Therefore, the f.o.b. rather than the c.i.f. price may be more rele-

[18]This is true of each country except the Ivory Coast, for which PP and NSP also differ because of relatively small differences between private and social prices of land. Other possible reasons for private profitability varying from social profitability, such as the existence of externalities or monopoly power, could not be measured for any of the countries.

TABLE 12.4. *Private Profitability, Public Incentives, and Net Social Profitability*
(U.S. $/mt milled rice)

Production technique	Private profitability	Domestic price minus border price	Net subsidy	Net social profitability
Traditional manual upland:				
Ivory Coast forest	156	48	226	−117
Ivory Coast savannah	213	48	233	−70
Liberia	−96	144	−9	−231
Sierra Leone south	80	25	—	55
Sierra Leone north	26	25	—	1
Improved manual upland:				
Ivory Coast forest	189	48	262	−104
Ivory Coast savannah	213	48	288	−120
Liberia	−62	144	13	−219
Sierra Leone south	128	30	36	62
Sierra Leone north	75	33	46	−4
Animal traction upland:				
Ivory Coast savannah	235	48	286	−95
Senegal Casamance	106	78	36	−8
Mechanized upland:				
Ivory Coast savannah	230	48	328	−143
Traditional manual swamp:				
Liberia	−6	144	−9	−141
Mali	−64	−122	−16	74
Senegal Casamance	n.a.	n.a.	n.a.	n.a.
Sierra Leone south	137	30	—	107
Sierra Leone north	92	31	3	58
Improved manual swamp:				
Ivory Coast forest	136	48	291	−180
Ivory Coast savannah	174	48	305	−155
Liberia	42	144	12	−114
Senegal Casamance	79	65	72	−58
Sierra Leone south	158	44	49	65
Sierra Leone north	140	44	51	45
Improved manual mangrove:				
Sierra Leone south	117	23	—	94
Sierra Leone north	64	16	—	48
Animal traction swamp:				
Mali	−7	−123	21	95
Partially mechanized swamp:				
Ivory Coast forest	144	48	288	−146
Liberia	108	144	138	−174
Improved manual uncontrolled flooding:				
Sierra Leone boliland	147	33	6	108
Animal traction uncontrolled flooding:				
Mali	−118	−107	−15	4
Mechanized uncontrolled flooding:				
Sierra Leone boliland	165	24	117	24
Animal traction controlled flooding:				
Mali	−21	−126	26	79
Improved animal traction controlled flooding:				
Mali	8	−126	13	121

TABLE 12.4—*continued*

Production technique	Private profitability	Domestic price minus border price	Net subsidy	Net social profitability
Animal traction irrigated single crop:				
Mali	6	−133	4	135
Improved animal traction irrigated single crop:				
Mali	6	−133	22	117
Mechanized irrigated single crop:				
Senegal Fleuve	4	131	138	−265
Manual irrigated multiple crop:				
Ivory Coast savannah	202	48	360	−178
Senegal Fleuve	90	150	8	−68
Mechanized irrigated multiple crop:				
Ivory Coast forest	166	48	456	−334
Senegal Fleuve	64	139	57	−132

SOURCES: *4, 6, 7, 10, 15.* The sum of the last three columns equals private profitability, except for the Ivory Coast, where net social profitability differs from private profitability because of differences in social and private prices of land in addition to the effects of government incentives.

vant as a yardstick with which to compare the domestic price. If this is the case, the disincentive resulting from Mali's official price policy is greatly reduced and is offset for most improved production by net subsidies paid by the government on inputs.

The Ivory Coast, Senegal, and Sierra Leone all protect domestic production to a moderate degree through their trade policies and pricing systems. They differ substantially, however, in the extent to which they subsidize inputs. The Ivory Coast provides input subsidies averaging over $300/mt, mostly as a subsidy to the government-owned mills, which enables them to offer the farmer a high purchase price for his paddy. Input subsidies in Senegal and Sierra Leone, on the other hand, average only $62/mt and $27/mt, respectively. In all of these countries, net input subsidies increase with mechanization and higher degrees of water control. In addition, these subsidies are somewhat greater in the drier northern regions of each country than they are in southern areas, which receive more rainfall. This is partly because the subsidies are measured per ton of milled rice and yields in the southern areas are higher, but it is also due to the more elaborate water control structures required in some of the drier regions, such as northern Senegal.

Liberia differs from the other countries in that it relies primarily on import restrictions as a means of promoting local rice production. The domestic price in Monrovia in 1975–76 was $144/mt greater than the comparable c.i.f. price, and agricultural inputs received only very slight subsidies, except in partially mechanized cultivation, where the net subsidy was $138/mt. Inputs into traditional rice production are, on balance, slightly taxed in both Liberia and Mali.

Social Profitability

The magnitude of incentives offered to rice production is such that private profitability is an unreliable guide to the efficient allocation of resources. Net social profitability, measured in world prices or their equivalents, in fact, diverges widely from the private measure of net benefits. Furthermore, there is significant variation in NSP among countries and between techniques.

The NSP indicators in Table 12.4 suggest that only two countries—Mali and Sierra Leone—are able to substitute profitably local production of rice for imports consumed in the capital city. In all of the other countries, NSP is negative for each activity. Liberia appears to be especially disadvantaged, since there is no technique that can be used to produce rice without a loss of at least $114/mt. An unweighted average of NSP for each activity in Liberia is $−176/mt, compared with $−148/mt in the Ivory Coast and $−106/mt in Senegal. This is consistent with the regression analysis of yields, presented earlier, which gives as the coefficients of country-specific dummy variables −867 for Liberia, −456 for Ivory Coast, and −526 for Senegal. Although productivity, according to these results, might be slightly greater in the Ivory Coast than in Senegal, this advantage is more than offset by higher Ivorian wages.[19]

The opposite occurs in Sierra Leone, for which the coefficient of its dummy variable is −643, indicating a substantial productivity disadvantage. Yet Sierra Leone's net social profitability is positive in every activity but one—improved manual upland cultivation in the north—and even that activity is only marginally unprofitable. The principal explanation for this apparent anomaly seems to be that wage rates in Sierra Leone are very low in comparison with other West African countries.[20] In addition, as was noted earlier, techniques of production in Sierra Leone are very labor-intensive.[21] Since differences between countries in the shadow prices of the other inputs are relatively unimportant as a percentage of total production costs, low wage rates in Sierra Leone could compensate for its productivity disadvantage.

Mali clearly has the highest rates of net social profitability of any of the five countries. These range from $4/mt for ox-drawn cultivation under un-

[19] The shadow price of unskilled labor in the Ivory Coast varied in 1975–76 from 275 to 450 CFA francs per day, depending upon sex and region. In Senegal, it was 250 CFA francs in the Fleuve and 300 CFA francs in the Casamance for both males and females. For further details, see Appendix B.

[20] Wages of unskilled labor in Sierra Leone in 1975–76 were the equivalent of about 130–200 CFA francs per day, compared with a range of 200–450 CFA francs in the other countries.

[21] This may be at least partly because wages are low, inducing the use of labor relative to other inputs.

controlled flooding conditions to \$135/mt for single-crop, irrigated cultivation using animal traction at the Office du Niger. High rates of NSP in Mali appear to reflect the natural advantages mentioned earlier, such as the relatively predictable flooding of the Niger River and the sunk capital investment in the Office du Niger. In addition, wages in Mali are fairly low compared to other countries, especially the Ivory Coast.[22] Finally, the c.i.f. price of rice in Mali, which is used to value the benefits from production, is higher than that of the other countries because of the cost of transport from the coast to the frontier.

Within countries, there are also some generalizations that can be made concerning net social profitability of different techniques. In both the Ivory Coast and Sierra Leone, for example, social profitability is increased by improving manual rainfed cultivation in the south, but it is decreased when this is done in the north. NSP is also increased in the same way throughout Liberia, where the amount of rainfall is uniformly high. This appears to be because the yield response to the introduction of improved varieties and fertilizer is lower in zones of lesser rainfall than in those where rainfall is abundant.[23] It suggests that increased water control is necessary if the full potential of improved cultivation is to be realized in drier areas. The evidence concerning the effects on NSP of introducing improved practices into swamp cultivation is mixed, however, indicating that water there appears to be less critical. In Liberia, social profitability is increased, whereas in both northern and southern areas of Sierra Leone and the Ivory Coast NSP is reduced, with improved methods of cultivation. On the other hand, control of flooding in Mali raises NSP considerably and, in addition, makes possible further increases in social profitability through greater intensification of cultivation.[24]

There is some evidence that use of animal traction is more profitable than manual cultivation. This is true of improved upland cultivation in the Ivory Coast and of traditional swamp production in Mali. The advantage of partially mechanized swamp cultivation, involving use of power tillers, is less well established, however, since it is more profitable than manual cultivation in the southern Ivory Coast but less profitable in Liberia. In any case, full mechanization with tractors has a lower NSP than other less

[22] Wages of unskilled labor in Mali varied from 200 to 350 CFA francs in 1975–76.

[23] In Sierra Leone, the profitability of improved cultivation in the north is also decreased because of relatively high wages in the vicinity of Makeni, where this technique is being introduced.

[24] Although intensification of cultivation increases profitability in the controlled flooding perimeters of Mali, it does not do so in the fully controlled irrigation system of the Office du Niger. This is primarily because all investment costs are considered to be sunk for the current Office du Niger technique, but if further intensification is to raise yields from 2.25 to 3.5 mt/ha, greater investment to improve water control will be required. In the controlled flooding perimeters, on the other hand, no increase in water control is necessary to raise yields from their current level of about 1.4 mt/ha to close to 2.5 mt/ha.

mechanized techniques in every instance for which comparisons can be made—upland cultivation in the Ivory Coast savannah, uncontrolled flooding in Sierra Leone's bolilands, and irrigated multiple-crop production in Senegal. The intermediate stages of mechanization, therefore, appear to have the greatest chance for success.[25]

Finally, the data show clearly the problems associated with trying to produce rice in total water control irrigation systems where only one crop per year can be grown. In the delta region of the Senegal River, salt incursion from the sea during the dry season prevents pumping water from the river for a second crop, resulting in very high social costs of production. Further upstream, two crops can be grown, which spreads some of the high overhead and capital costs associated with this kind of irrigation.

Exports

The profitability estimates shown in Table 12.4 all assume that consumption of locally grown rice takes place in the largest urban center—in each case the capital city. Positive net social profitability in Mali and Sierra Leone, however, suggests that these countries might have a comparative advantage in exporting to other countries in West Africa. This would reduce net social profitability, however, because the f.o.b. value of rice is less than its c.i.f. value owing to the influence of transport costs. Revised NSP estimates for rice exported from Mali and Sierra Leone to several West African markets are given in Table 12.5 for a few selected techniques, which could be expanded to generate exportable supplies.

It appears from this table that there are opportunities for profitable exports of rice from both countries. The decline in net social profitability resulting from using an f.o.b. rather than a c.i.f. price to value rice output, however, is considerably greater for Mali than for Sierra Leone because of the much longer distances involved. This is evident from the gain in NSP that results from shipping rice to Bouaké, in the center of the Ivory Coast, rather than to Abidjan on the coast. It can also be seen by comparing the NSP of rice exported from Ségou with that produced near Mopti, closer to the center of the interior delta of the Niger River, the major area for potential growth of rice cultivation. Aside from long distances, Mali suffers an additional disadvantage in trying to supply the Dakar market because of competition from inexpensive broken rice imported from Asia.

The ranking of NSP by technique is the same as when rice is produced

[25] This evidence concerning the relative social profitability of different levels of mechanization is consistent with the findings of Humphreys and Pearson (5), which suggest that only the use of animal traction is more profitable than manual cultivation, and even this leads to only about a 10 percent reduction in social costs.

TABLE 12.5. *Net Social Profitability of Rice Exports*
(U.S. $/mt milled rice)

Production technique	Consumption point					
	Monrovia	Freetown	Abidjan	Bouaké	Dakar	Bamako
Traditional manual upland:						
Sierra Leone south	40	55	—	—	—	—
Improved manual upland:						
Sierra Leone south	47	62	—	—	—	—
Traditional manual swamp:						
Sierra Leone south	92	107	—	—	—	—
Improved manual swamp:						
Sierra Leone south	50	65	—	—	—	—
Improved animal traction						
controlled flooding:						
Mali (Ségou)	—	—	11	47	−32	121
Mali (Mopti)	—	—	−25	12	−83	71
Animal traction irrigated						
single crop:						
Mali (Office du Niger)	—	—	25	61	−18	135
Improved animal traction						
irrigated single crop:						
Mali (Office du Niger)	—	—	7	43	−36	117

SOURCES: *6, 10.*

for domestic consumption. The techniques vary, however, in the degree to which they can contribute to exports. In general, the improved techniques, with their higher yields, offer greater supplies of rice available for export. An increasing proportion of expanding exports will therefore come from improved cultivation. This implies some increase in the profitability of upland rice, but a considerable decline in that of swamp rice, produced for export in Sierra Leone. Similarly, the profitability of growing irrigated rice for export at the Office du Niger in Mali is likely to decline in the future as an increasing proportion of that rice comes from improved techniques of cultivation. On the other hand, rapid development of rural areas in the Ivory Coast is enlarging the interior Ivorian market for rice that Mali could supply. This will increase the f.o.b. price of Malian exports and improve their social profitability.

Local Consumption

In the Ivory Coast, Liberia, and Senegal, production of rice for consumption in the capital city is socially unprofitable for each activity. This does not imply, however, that rice that is consumed closer to the areas of cultivation is also necessarily unprofitable. Just as costs of transport and handling make it easier to meet competition from imports than to export profitably, so the cost of imported rice, and thus the shadow price of domestically produced rice, rises as the consumption point is shifted closer

TABLE 12.6. *Net Social Profitability of Rice for Local Consumption*
(U.S. $/mt milled rice)

Production technique	On-farm[a]	Local market[b]	Regional market[b]	Capital city[c]
		Consumption point		
Traditional manual upland:				
Ivory Coast forest	−8	−44	−87	−117
Ivory Coast savannah	34	18	−38	−70
Liberia	−128	n.a.	n.a.	−231
Improved manual upland:				
Ivory Coast forest	n.a.	−31	−75	−104
Ivory Coast savannah	n.a.	−34	−87	−120
Animal traction upland:				
Ivory Coast savannah	n.a.	−8	−62	−95
Senegal Casamance	87	n.a.	29	−8
Traditional manual swamp:				
Senegal Casamance	−39	n.a.	n.a.	n.a.
Improved manual swamp:				
Ivory Coast forest	n.a.	−110	−150	−180
Ivory Coast savannah	n.a.	−72	−122	−155
Liberia	−4	n.a.	n.a.	−114
Senegal Casamance	37	n.a.	−21	−58
Mechanized irrigated single crop:				
Senegal Fleuve	n.a.	n.a.	−249	−265
Manual irrigated multiple crop:				
Ivory Coast savannah	n.a.	−95	−146	−178
Senegal Fleuve	100	n.a.	−51	−68
Mechanized irrigated multiple crop:				
Ivory Coast forest	n.a.	−272	−305	−334
Senegal Fleuve	−25	n.a.	−116	−132

SOURCES: *4, 7, 15.*
  [a]Assumes rice is hand-pounded.   [b]Assumes rice is processed in small-scale hullers.
  [c]Assumes rice is processed in large-scale mills.

to the areas of production. In addition, of course, the cost of marketing local rice also declines.

The effect of shifting the consumption point closer to the farm is seen in Table 12.6. Net social profitability rises as consumption is transferred from the capital city to regional and local markets and finally to the farm. Transport costs are reduced, and various stages in the marketing chain may also be eliminated. In addition, the use of small-scale hullers to process rice for the local market results in further savings over large-scale milling. Furthermore, when consumption takes place on the farm, the elimination of collection and distribution costs more than offsets the higher cost of hand pounding over that of small-scale hullers.

As a result, a number of techniques that are not profitable when rice is consumed in the capital city become socially profitable when consumption takes place closer to the farm. This is especially true for Senegal because of the long distances between the producing regions and the

Dakar market, but it is also true for traditional manual upland cultivation in the Ivory Coast savannah and almost true of traditional manual upland cultivation in the Ivory Coast forest, upland cultivation involving animal traction in the Ivory Coast savannah, and improved manual swamp production in Liberia. On the other hand, mechanized cultivation is not profitable for consumption anywhere, and except for the techniques mentioned, which have NSPs close to zero, the Ivory Coast and Liberia do not appear to have a comparative advantage in rice production even for on-farm consumption.

Comparison with Asia and the United States

Since the production of rice in West Africa to supply its major markets is socially unprofitable for some countries and is only moderately profitable for others that are potential exporters within the region, the question arises as to which countries elsewhere in the world have sufficiently low social costs that they can profitably supply West Africa with rice.[26] A previous study of comparative advantage of rice production in Asia and the United States, using a methodology similar to that employed here, provides estimates of net social profitability for four countries, including two of the world's most important rice exporters—Thailand and the United States (8).

These estimates are shown in Table 12.7, together with some obtained for the West African countries. The results are straightforward. Thailand, the world's most important rice exporter, has by far the highest net social profitability. Production in the United States, which is the second largest exporter, is only marginally profitable. The other two Asian countries, the Philippines and Taiwan, as well as the Ivory Coast, Liberia, and Senegal, have NSPs that are negative as long as they are substituting for imports of rice going to their major markets.[27] Of the African countries, only Mali and Sierra Leone have positive NSP, and this would be reduced if these two nations were to export within the region. If they should try to export outside of West Africa, moreover, the f.o.b. price of rice would drop to the point where NSP would be negative. This problem should not arise, however, since the size of the West African market considerably exceeds the capacity of these two countries to supply it. In summary, world, as well as national, economic efficiency would be improved if West Africa would not try to be self-sufficient in rice but would rely more on foreign producers.

---

[26] Although it might be answered that these countries can be easily identified as those that actually export rice, this overlooks the possibility of subsidies to encourage exports of rice that is not being profitably produced.

[27] Local production for on-farm consumption might be socially profitable, but calculations employing this alternative assumption have not been made for the Asian countries.

TABLE 12.7. *Net Social Profitability of Rice in Asia, the United States, and West Africa*
(U.S. $/mt milled rice)

Production technique	Philippines	Taiwan	Thailand	United States	Ivory Coast
Improved upland	—	—	84/122	—	−143/−95
Improved swamp, mangrove, and flooding	—	—	165/196	—	−180/−155
Partially mechanized irrigated	−99	−161/−68	—	—	−146
Mechanized irrigated single crop	—	—	—	−30/20	—
Mechanized irrigated multiple crop	−105/−123	—	—	—	−334

Production technique	Liberia	Mali	Senegal	Sierra Leone
Improved upland	−219	—	−8	−4/62
Improved swamp, mangrove, and flooding	−114	79/121	−58	24/108
Partially mechanized irrigated	−174	—	—	—
Mechanized irrigated single crop	—	—	−265	—
Mechanized irrigated multiple crop	—	—	−132	—

SOURCES: Data are from Table 12.4 and 8. The latter estimates have been revised to reflect world prices of rice in 1975–76 rather than 1974, the year for which they were originally calculated. All estimates have been corrected for any differences that may exist between official and shadow rates of foreign exchange. Where estimates exist for several activities within each category of production technique, low and high values are given. Prices used to value rice output are f.o.b. for Thailand and the United States and c.i.f. for the other countries.

Comparison with Other West African Agricultural Activities

If resources in the agricultural sector of some of the West African countries were to be reallocated away from rice production, are there other rural activities that are more profitable? Estimates of net social profitability are available for a number of other crops in the Ivory Coast, Mali, and Senegal. These results cannot be compared with NSP in rice production, however, because the units are not comparable. Instead, it is necessary to make use of an indicator that is independent of units, such as the resource cost ratio (RCR) described in Appendix A. This ratio compares the social value of domestic resources used to produce a given quantity of output with the value added in world prices created in producing that output. If this ratio exceeds unity, the opportunity cost of the domestic resources, expressed in terms of world prices, is greater than value added in world prices, and net social profitability is negative. If the RCR is less than one, on the other hand, NSP is positive. Since both numerator and denominator of the RCR are expressed in the same units, the ratio itself is independent of these units, and comparisons can be made between activities producing different products. The lower the RCR of a given activity in relation to all other activities, the greater the comparative advantage that the country has in that activity.

An indicator of incentives comparable to the resource cost ratio is the effective protection coefficient (EPC), also discussed in Appendix A. This ratio compares value added in domestic market prices with value added in world prices. Since both are measured in the same currency, the EPC, too, is a ratio that is independent of units. If the EPC is greater than unity, there is an incentive for value added to be created locally; if the EPC is less than one, there is no such incentive. Unlike private profitability, which includes all taxes and subsidies, however, the EPC takes into account only those assessed on tradable outputs and inputs. It is only a partial indicator, therefore, of total net incentives affecting the allocation of domestic resources.

Estimates of the RCR and the EPC are given in Table 12.8 for each rice production technique plus a number of other agricultural activities. Valuation is c.i.f. or f.o.b., depending upon whether the product is customarily imported or exported.

The results for the rice activities parallel those discussed previously from Table 12.4. In addition, Table 12.8 permits comparison of rice with alternative rainfed crops. The results are striking. In the Ivory Coast, rice competes very poorly with coffee, cocoa, palm products, copra, and maize. Each of these other crops can be produced with at least one technique for which the resource cost ratio is less than unity. Hence, each is socially profitable, and in some cases very much so, whereas none of the RCRs for rice is less than one. Moreover, the effective protection coefficients are greater than unity for all of the rice activities but less than or equal to one for each of the other crops in the Ivory Coast—an indication of the protection that is required if rice is to be produced locally using inefficient techniques.

In Senegal, on the other hand, the competition between rice and other crops is closer. Peanuts seem to have a strong advantage and maize and cotton a slight edge, but rice production in some cases is more profitable than cultivation of millet, particularly in the Casamance region of the south. In the Fleuve area to the north, rice is more expensive to produce than is either rice or millet in the Casamance. The EPCs in Senegal tend to correspond fairly closely to the RCRs except for rice, which, with the EPC less than the RCR, in each case receives less protection in relation to need than the other crops. This is offset for a number of techniques by subsidies on nontradable inputs, such as irrigation, which increase private profitability. Maize, on the other hand, appears to be somewhat overprotected, since the EPC is generally greater than the RCR.

TABLE 12.8. *Resources Cost Ratios and Effective Protection Coefficients for Various Agricultural Activities*

Production technique, crop	Resource cost ratio	Effective protection coefficient
Traditional manual upland:		
Ivory Coast forest, rice	1.43	1.16
Ivory Coast forest, coffee	0.58	0.60
Ivory Coast forest, cocoa	0.46	0.84
Ivory Coast savannah, rice	1.26	1.17
Ivory Coast savannah, maize	0.88	1.00
Liberia, rice	1.78	1.46
Senegal Casamance, peanuts	0.80	0.76
Senegal Casamance, millet	1.30	1.01
Sierra Leone south, rice	0.87	1.02
Sierra Leone north, rice	1.09	1.02
Improved manual upland:		
Ivory Coast forest, rice	1.43	1.24
Ivory Coast forest, coffee	0.44	0.60
Ivory Coast forest, cocoa	0.42	0.84
Ivory Coast forest, palm products	0.43	0.91
Ivory Coast forest, copra	0.38	0.92
Ivory Coast savannah, rice	1.53	1.29
Ivory Coast savannah, cotton	1.03	0.49
Ivory Coast savannah, maize	0.84	0.98

Production technique, crop	Resource cost ratio	Effective protection coefficient
(Improved manual upland:)		
Liberia, rice	1.99	1.62
Sierra Leone south, rice	0.82	1.12
Sierra Leone north, rice	1.13	1.15
Animal traction upland:		
Ivory Coast savannah, rice	1.41	1.26
Ivory Coast savannah, cotton	0.84	0.52
Ivory Coast savannah, maize	0.81	0.99
Senegal Casamance, rice	1.04	0.90
Senegal Casamance, peanuts	0.48	0.78
Senegal Casamance, millet	1.27	1.25
Senegal Casamance, cotton	0.80	0.76
Senegal Casamance, maize	0.80	1.25
Mechanized upland:		
Ivory Coast savannah, rice	1.67	1.26
Traditional manual swamp:		
Liberia, rice	1.48	1.46
Mali, rice	0.72	0.58
Sierra Leone south, rice	0.69	1.02
Sierra Leone north, rice	0.90	1.03

TABLE 12.8—*continued*

Production technique, crop	Resource cost ratio	Effective protection coefficient	Production technique, crop	Resource cost ratio	Effective protection coefficient
Improved manual swamp:			Mechanized uncontrolled flooding:		
Ivory Coast forest, rice	1.75	1.22	Sierra Leone bolilands, rice	1.01	1.06
Ivory Coast savannah, rice	1.65	1.23	Animal traction controlled flooding:		
Liberia, rice	1.44	1.52	Mali, rice	0.74	0.65
Senegal Casamance, rice	1.26	0.93	Improved animal traction controlled flooding:		
Sierra Leone south, rice	0.82	1.08	Mali, rice	0.59	0.61
Sierra Leone north, rice	0.94	1.09	Animal traction irrigated single crop:		
Improved manual mangrove:			Mali, rice	0.56	0.72
Sierra Leone south, rice	0.74	1.02	Improved animal traction irrigated single crop:		
Sierra Leone north, rice	0.98	1.02	Mali, rice	0.59	0.64
Animal traction swamp, rice:			Mechanized irrigated single crop:		
Mali, rice	0.65	0.66	Senegal Fleuve, rice	232.22	114.34
Partially mechanized swamp:			Manual irrigated multiple crop:		
Ivory Coast forest, rice	1.61	1.22	Ivory Coast Savannah, rice	1.74	1.22
Liberia, rice	1.69	1.70	Senegal Fleuve, rice	1.41	0.97
Improved manual uncontrolled flooding:			Mechanized irrigated multiple crop:		
Sierra Leone bolilands, rice	0.72	1.04	Ivory Coast forest, rice	2.99	1.28
Animal traction uncontrolled flooding:			Senegal Fleuve, rice	2.35	1.55
Mali, rice	0.99	0.59			

SOURCES: *4, 6, 7, 10, 15, 13.*

Rice grown in Mali does not have to compete with other crops for land, since it is not an upland crop, but it does compete for labor and capital. The RCRs from Table 12.8 suggest, however, that all these crops can be grown profitably using a number of different techniques. The low level of producer prices in Mali, which reflects this comparative advantage in agriculture, is indicated by the low EPC for each activity.

Sensitivity Analysis

The two appendixes, concerned with methodology and shadow price estimation, discuss the specific assumptions underlying the previous estimates and the errors to which these estimates may be subject. The implications of these assumptions and errors for the empirical conclusions presented here are tested in the individual country studies, using sensitivity analysis. One of the most useful indicators calculated as part of this analysis is the proportion by which yields, the world price of rice, or the cost of each primary factor input would have to change before NSP would equal zero.[28]

The sensitivity analysis suggests that the empirical results indicating whether rice production is socially profitable are quite robust. In most instances, the results are very insensitive to changes in the cost of capital or skilled labor. They are more sensitive to variation in the cost of unskilled labor, especially for traditional and improved upland production. Nevertheless, costs of unskilled labor would have to be less than half the assumed values for any technique in the Ivory Coast and Liberia and for irrigated cultivation in northern Senegal before production for the major urban market would become socially profitable. At the other extreme, unskilled labor costs for irrigated or controlled flooding cultivation in Mali would have to be twice their assumed values before NSP would be reduced to zero. Lesser variation in the cost of unskilled labor would be required to adjust estimated values of NSP to zero in Sierra Leone and the Casamance region of Senegal.

The effects of assuming a positive rather than a zero shadow price of land have been analyzed only for Mali and the Ivory Coast. In Mali, rents of U.S. $180–250 per ha would be required to make production in the Office du Niger socially unprofitable. Irrigated land in the Ivory Coast, on the other hand, would have to be subsidized at U.S. $400–600 per ha before production would become socially profitable.

The net effects of changes in yields or the world price of rice are similar to those for unskilled labor costs. In the Ivory Coast, Liberia, and the Fleuve region of Senegal, yields or the world price would have to rise 25

[28]The costs of each primary factor input can vary because of errors in either the shadow price of that input or the technical coefficient relating it to output.

to 80 percent before production would become socially profitable. The value of these two variables for irrigated or controlled flooding in Mali, on the other hand, would have to fall by 30 percent or more to make production for consumption in Bamako unprofitable. The results for Sierra Leone and southern Senegal, as well as for traditional upland cultivation in Mali, are again more sensitive to errors in these variables.

Other Objectives and Constraints

The analysis thus far has been concerned primarily with indicators of economic efficiency as a national objective. The countries of West Africa, however, are not interested solely in generating more income. They are also concerned about distributing income in a more equitable manner and reducing the risk of shortfalls in the availability of food. In addition, many of the countries hold self-sufficiency in essential foods to be an important proximate objective which helps them to achieve their more fundamental goals. The contribution of rice production to each of these other objectives, however, often implies some loss of national income.

Some indicators that are related to national objectives other than increasing income are given in Table 12.9, with consumption again assumed to take place in the capital city. One is the number of man-days of unskilled labor per ton of milled rice, which is an indicator of employment opportunities generated and thus of the distributional effects of each activity.[29] A second indicator is the degree of security attached to production relative to rainfed cultivation in areas of lower rainfall. This is very roughly estimated on the basis of past experience with yields, variability of rainfall or flooding, and the type of water control system, if any, employed. The third indicator suggests the degree to which each technique is oriented toward market sales as opposed to on-farm consumption and, consequently, the extent to which it can contribute to national food self-sufficiency, especially in urban areas. This will depend on yields, size of land holding, family size, and the system that exists for paddy collection. The effectiveness of this system, at least when the public sector is involved, will be influenced by such factors as the degree to which fields are

[29] An excellent survey of the determinants of rural income distribution, including its link to employment, is contained in Cline (2). It is possible, of course, that agricultural employment has a less important influence on income distribution in Africa, where population density is relatively low, than in other areas of the world. A recent survey of irrigation projects in the Sahelian countries by Stryker, Gotsch, McIntire, and Roche (14), however, suggests that labor-intensive schemes generally have better distributional characteristics than those that are more capital-intensive. Other important dimensions of income distribution are its urban-rural and regional characteristics. To the extent that rice production contributes more than other alternatives to rural income in general and to rural income in poorer regions in particular, this aspect of the distributional objective is furthered. Among these poorer regions are the Ivory Coast savannah and both the Casamance and the Fleuve in Senegal.

TABLE 12.9. *Objectives and Constraints Indicators for Milled Rice Activities*

Production technique	Unskilled labor[a] (days/mt)	Degree of security[b]	Market orientation[b]	Budgetary loss[c] (US $/mt)
Traditional manual upland:				
Ivory Coast forest	134	medium	low	274
Ivory Coast savannah	145	low	low	281
Liberia		medium	low	144
Sierra Leone south	290	medium	low	25
Sierra Leone north	488	low	low	25
Improved manual upland:				
Ivory Coast forest	99	medium	medium	310
Ivory Coast savannah	166	low	medium	336
Liberia		medium	medium	157
Sierra Leone south	206	medium	medium	66
Sierra Leone north	298	low	medium	79
Animal traction upland:				
Ivory Coast savannah	98	low	medium	334
Senegal Casamance	102	low	medium	114
Mechanized upland:				
Ivory Coast savannah	48	low	high	376
Traditional manual swamp:				
Liberia		medium	low	135
Mali	191	medium	low	−138
Senegal Casamance	n.a.	medium	low	n.a.
Sierra Leone south	196	medium	low	30
Sierra Leone north	280	medium	low	34
Improved manual swamp:				
Ivory Coast forest	124	high	medium	339
Ivory Coast savannah	124	high	medium	353
Liberia		high	medium	156
Senegal Casamance	133	high	medium	137
Sierra Leone south	186	high	medium	93
Sierra Leone north	235	high	medium	95
Improved manual mangrove:				
Sierra Leone south	219	high	medium	23
Sierra Leone north	265	high	medium	16
Animal traction swamp:				
Mali	115	high	medium	−102
Partially mechanized swamp:				
Ivory Coast forest	96	high	medium	336
Liberia		high	medium	282
Improved manual uncontrolled flooding:				
Sierra Leone bolilands	215	low	low	39
Animal traction uncontrolled flooding:				
Mali	202	low	low	−122
Mechanized uncontrolled flooding:				
Sierra Leone bolilands	160	low	medium	141
Animal traction controlled flooding:				
Mali	98	medium	medium	−100

TABLE 12.9—*continued*

Production technique	Unskilled labor[a] (days/mt)	Degree of security[b]	Market orientation[b]	Budgetary loss[c] (US $/mt)
Improved animal traction controlled flooding:				
Mali	74	medium	medium	−113
Animal traction irrigated single crop:				
Mali	84	high	medium	−129
Improved animal traction irrigated single crop:				
Mali	74	high	high	−111
Mechanized irrigated single crop:				
Senegal Fleuve	76	high	medium	269
Manual irrigated multiple crop:				
Ivory Coast savannah	115	high	medium	408
Senegal Fleuve	108	high	medium	158
Mechanized irrigated multiple crop:				
Ivory Coast forest	48	high	high	504
Senegal Fleuve	74	high	high	196

[a] Calculated from the first and third columns of Table 12.2.
[b] Based on best judgments of the authors of the individual country studies, *4, 6, 7, 10, 15*.
[c] Equal to the sum of "domestic price minus border price" and "net subsidy" in Table 12.4.

physically concentrated, the level of mechanization, and the way in which payment for services or credit is linked with the sale of the crop. In addition, of course, factors not directly related to the production technique, such as paddy and other crop prices and the availability of consumer goods, will be important.

The last indicator in Table 12.9 shows the influence of each activity on the public budget. Although this is not a fundamental constraint influencing the pattern of comparative advantage in the long run, it can act powerfully in the short run to lessen the options of governments. The expansion of rice production as a substitute for imports has two principal budgetary effects the sum of which, as was shown in Table 12.4, equals the difference between private and social profitability. First, it increases public expenditures on input subsidies; second, it decreases revenues earned from levies on rice imports. These effects may be partially offset by an increase in tax receipts from expanded inputs, but on balance there is usually a budgetary loss associated with the expansion of rice production.

The data in Table 12.9 show a high degree of labor intensity of activities in Liberia and Sierra Leone relative to the other countries. Although this is consistent with profitable production in Sierra Leone because of the low wages in that country, it is not so in Liberia, where wages are higher and all production for the Monrovia market is socially unprofitable. High labor intensity helps to ensure abundant employment opportunities and an equitable distribution of rural income in both countries. But this is accomplished with a much higher loss of economic efficiency in Liberia than in Sierra Leone.

Requirements for unskilled labor also vary markedly between techniques. They are generally greatest for traditional methods of cultivation because of the limited use of other inputs and low yields involved. As cultivation is improved, labor requirements decrease to some extent, but they continue to remain high as long as advanced stages of mechanization are avoided. Even the introduction of animal traction and hand-operated power tillers has only a modest impact on labor requirements. Once tractors are introduced, however, the use of unskilled labor drops off markedly, and income distribution is adversely affected.

Security of production depends largely on two factors—amount and variability of rainfall and degree of water control. In southern West Africa, security of rainfed agriculture is fairly high, but as cultivation is undertaken farther north, fluctuations in yields become more pronounced. This is less important for swamp rice than for upland varieties, but even cultivation of lowlands is hazardous in the drier regions. As a result, rainfed agriculture becomes increasingly impractical, and either irrigation or river flooding must be employed. These methods vary markedly in the costs and risk involved and often there is a trade-off between profitability and security. For example, the controlled flooding perimeters in Mali are quite low-cost and profitable, but they involve a fair degre of risk since the flooding may be late or inadequate. Total water control systems, such as those in Senegal, on the other hand, are secure but expensive.

There are some schemes for which this trade-off does not have to be made. Because of the capital already sunk into the Office du Niger, Mali has the capability of producing rice both profitably and securely. In Senegal, even though the cost of constructing total water control systems using capital-intensive methods is high, there are also labor-intensive schemes employing manual techniques of construction and cultivation that not only are less unprofitable than the mechanized perimeters, but also have a high degree of production security.

The average quantity marketed from traditional rice production is low on average, even though total marketings may be fairly substantial in some countries, such as Liberia and Sierra Leone, because most West African rice is grown in this way. As improvements are made, raising yields and farm size, the quantity harvested by each family that is available for marketing generally increases. Sometimes, however, expanded production is accompanied by a increase in the amount of rice eaten relative to other foods, so that the gain in marketings is less than anticipated.[30] Governments wishing to reduce such switching can require payment of fees and debt service to be made in kind or demand that a certain amount of each harvest be sold to the public buying agency.[31] These practices are more successful for large-scale, mechanized irrigated schemes than they are for holdings that are dispersed and use few modern inputs. Consequently, governments trying to increase government-controlled supplies of locally produced rice for the cities have a special interest in promoting these types of projects, especially if they can be used to produce a high level of output per family.

Budgetary losses associated with the promotion of domestic rice production to replace imports in the Ivory Coast are very large. They are less, but still substantial, in Liberia and Senegal, though the form taken differs between these two countries. In Liberia, the major loss is of revenue forgone from taxes on rice imports, which does not show up explicitly in the budget as a subsidy. Senegal's budget, on the other hand, should register most of its losses as subsidies on inputs. In practice, however, many of these subsidies are financed through foreign aid. As long as this continues, Senegal will not feel the full budgetary implications of its policies to promote rice. Mali has a net budgetary gain from rice production used to replace imports that otherwise must be subsidized in order to be sold at the low domestic official retail price. In most years the country is self-sufficient, however, so that this apparent budgetary gain is largely spurious and would disappear if the border price were taken to be the f.o.b. rather than the c.i.f. price. Nevertheless, Mali provides only very modest input subsidies, so that there would be little claim on the public budget in any case. Finally, Sierra Leone occupies an intermediate position, with moderate budgetary losses associated with domestic rice production.

As expected, budgetary losses within each country are greater for techniques of production that are less profitable. Especially high losses are associated with mechanized cultivation in the Ivory Coast, Liberia, and Senegal and with irrigation in the Ivory Coast. Losses are lowest for all countries in traditional agriculture.

Summary and Conclusions

The aim of this chapter has been to examine the contributions made by rice-producing activities in five West African countries to some of their

[30] This has occurred, for example, in Senegal, where rice has tended to replace millet in the diet.

[31] Both of these requirements exist at the Office du Niger and the controlled flooding perimeters in Mali.

major national objectives—increasing income, distributing income in a more equitable way, and reducing the risk of food shortages. In addition, the analysis has shed light on the question of how movement toward self-sufficiency in rice helps to fulfill these objectives.

The countries vary markedly in the extent to which there are conflicts between these goals. Mali, for example, clearly has a strong comparative advantage in domestic production as a substitute for rice imports and also for export to other West African countries. Its advantages include the relatively predictable flooding of the Niger River, the sunk capital investment in the Office du Niger, fairly low wages, and the high c.i.f. border price that results from its interior location. This last factor is an advantage, however, only when Mali imports rice. In addition to generating more income while assuring self-sufficiency, expansion of rice production using improved techniques in Mali leads to greater security of cereal production and has few, if any, harmful distributional effects.

Sierra Leone has higher costs than Mali, resulting from a relatively low level of productivity, but production is still profitable because wages are very low. Furthermore, Sierra Leone can export rice more profitably than Mali because of lower transport costs to neighboring markets. Finally, as in Mali, the growth of rice cultivation in Sierra Leone is consistent with all of its other national objectives.

At the other end of the spectrum, the Ivory Coast and Liberia have a comparative disadvantage in producing rice for their national markets. This is true of every technique and region in which rice is grown. In Liberia, inefficiency stems from high costs of annual clearing, major pest and weed problems, and heavily leached soils. Negative net social profitability in the Ivory Coast, on the other hand, is principally the result of the relatively high wages in that country. This is due to competition from other crops such as coffee, cocoa, and cotton in which the Ivory Coast has a strong comparative advantage. The same high wage phenomenon is probably also true to a lesser extent in Liberia. Only if production is limited to on-farm use can these countries grow rice at the margin of profitability. There is thus a considerable conflict between the goal of self-sufficiency and the more fundamental objective of generating additional income.

Although rice production in the Ivory Coast and Liberia is unprofitable, the governments of these two countries have persisted in promoting this sector in order to redistribute income toward the north in the Ivory Coast and toward the rural sector in general in Liberia. Private profitability of production in the Ivory Coast savannah is quite high under the current incentive structure, so that the expansion of rice production should benefit farmers in that region, albeit with large public resource transfers. In Liberia, on the other hand, private profitability is negative in upland cultivation, so that this activity cannot be used to increase rural incomes beyond the need for rice consumed on the farm unless incentives are altered. Some progress might be made, however, with swamp rice. The chief problem in both countries, though, is that these redistribution goals can be much more effectively accomplished by promoting crops other than rice. The only way that expanded rice production can be justified, then, is through the increased security of production associated with irrigation and cultivation in swamps. This gain may be important in northern Ivory Coast, but it is hardly likely to be so in the southern part of that country or in Liberia.

Senegal occupies an intermediate position between Mali and Sierra Leone, on one hand, and the Ivory Coast and Liberia, on the other. With Dakar as the market, net social profitability is negative for every technique, but there are several activities—animal traction upland and improved manual swamp cultivation in the Casamance and manual irrigated production in the Fleuve—for which social losses are relatively modest. An important reason for negative NSP is the remoteness of the producing regions from Dakar. When the location of consumption is shifted toward the producer, several techniques are socially profitable. In addition, most of the rice-growing activities, especially the full water control systems of the Fleuve, offer some improvement in the security of food production. Finally, the regional distribution of income is improved by expanding rice production in Senegal, though the extent of that improvement depends upon whether labor-intensive or capital-intensive techniques are employed. Incomes in the Casamance, however, can be increased more by expanding peanut and maize production than by promoting rice.

The evidence suggests that outside of the areas of highest rainfall, improvements in production are only profitable if there is some degree of water control. Where natural conditions are appropriate, as in the swamps of southern Senegal or along the Niger River in Mali, this can be accomplished fairly easily. In other areas, such as the Senegal River Basin, the risks are such that expensive total water control systems are required. This implies a need for double-cropping—with the many agronomic, hydrological, and sociological problems that that entails.[32] Except for the Office du Niger, with its sunk capital costs, these systems of total water control have not yet proved to be profitable for other than local consumption. As a result, they must be justified on the basis of the additional security of food production they provide, the potential rice marketings they create, or their desirable distributional effects. In this respect, there may be a trade-off between the greater marketings generated by the mechanized schemes and the better distributional characteristics of manual production methods. Given the high costs and substantial subsidies

---

[32] Some of these are discussed in Stryker, Gotsch, McIntire, and Roche (*14*).

associated with these mechanized techniques, however, it may well be more efficient to induce greater marketing of rice from labor-intensive activities by offering farmers a higher paddy purchase price.

Intermediate techniques of rice cultivation involving animal traction may present the best opportunities for fulfilling all objectives. The empirical results suggest that they have reasonably high rates of social profitability, distribute their benefits fairly equally, place limited demands on budgetary resources, and generate regular marketings if paddy prices are adequate. Fully mechanized techniques, on the other hand, are very socially unprofitable, worsen the distribution of income, and are a continuing drain on the public budget. The only advantage these techniques have, which may partially explain their popularity, is that they generate fairly sizable marketings that can be shipped to the towns and cities.

In summary, the empirical results discussed here suggest that there should be greater specialization in West Africa rice production from the point of view of both location and technique. Outside of cultivation for on-farm or local consumption, which is profitable nearly everywhere and generally improves equality of income distribution and security of food production, rice destined for the major West African markets should be promoted primarily in Mali, Sierra Leone, and possibly Senegal. Highly mechanized techniques should be avoided, since they are ineffective in achieving most national objectives, and greater stress should be placed on use of animal traction and improved manual techniques. Increased water control is necessary in the drier regions to improve security of production and to take advantage of other technical innovations, but this should be limited, where conditions permit, to low-cost control systems. Where extensive total control structures are necessary, the gain in security must be weighed against the corresponding loss in national income that such structures entail.

Comparative advantage in West African rice production will not remain static. Supplies of labor coming from traditional agriculture are not infinitely elastic, and growing population density will increase the social opportunity cost of land. As a result, capital-intensive techniques are likely to increase in profitability relative to those that are intensive in the use of labor or land. Furthermore, many of the conditions that make irrigated production in the more arid regions unprofitable today—high construction and transport costs, lack of river regulation, low population density giving rise to weak local demand and scarcity of labor—will change in the future as the major river basins are developed. It is important, therefore, to perceive the social costs and benefits of rice production in a dynamic way that will evolve with the course of West African development.

Citations

1 Ester Boserup, *The Conditions of Agricultural Growth*. Aldine, Chicago, 1965.

2 William R. Cline, "Interrelationships between Agricultural Strategy and Rural Income Distribution," *Food Research Institute Studies*, 12; No. 2 (1973).

3 M. Gilbert and I. Kravis, *An International Comparison of National Products and the Purchasing Power of Currencies*. OEEC, Paris, 1954.

4 Charles P. Humphreys, "Rice Production in the Ivory Coast." Stanford/WARDA Study of the Political Economy of Rice in West Africa, Food Research Institute, Stanford University, Stanford, July 1979; Chapter 2.

5 Charles P. Humphreys and Scott R. Pearson, "Choice of Technique, Natural Protection, and Efficient Expansion of Rice Production in Sahelian Countries." Report prepared for the U.S. Agency for International Development, Food Research Institute, Stanford University, Stanford, June 1979.

6 John McIntire, "Rice Production in Mali." Stanford/WARDA Study of the Political Economy of Rice in West Africa, Food Research Institute, Stanford University, Stanford, July 1979; Chapter 10.

7 Eric A. Monke, "The Economics of Rice in Liberia." Stanford/WARDA Study of the Political Economy of Rice in West Africa, Food Research Institute, Stanford University, Stanford, July 1979; Chapter 4.

8 Eric A. Monke, Scott R. Pearson, and Narongchai Akrasanee, "Comparative Advantage, Government Policies, and International Trade in Rice," *Food Research Institute Studies*, 15, No. 2 (1976).

9 Scott R. Pearson, J. Dirck Stryker, and Charles P. Humphreys, "An Approach for Analyzing Rice Policy in West Africa." Stanford/WARDA Study of the Political Economy of Rice in West Africa, Food Research Institute, Stanford University, Stanford, July 1979; Introduction.

10 Dunstan S. C. Spencer, "Rice Production in Sierra Leone," Stanford/WARDA Study of the Political Economy of Rice in West Africa, Food Research Institute, Stanford University, Stanford, July 1979; Chapter 6.

11 J. Dirck Stryker, "Colonial Investment and Agricultural Development: The French Empire." Fletcher School of Law and Diplomacy, Tufts University, Medford, Mass., October 1975.

12 ———, "Optimum Population in Rural Areas: Empirical Evidence from the Franc Zone," *Quarterly Journal of Economics*, 91, No. 2 (May 1977).

13 ———, "Western Africa Regional Project: Ivory Coast, Chapter II, Economic Incentives and Costs in Agriculture." Fletcher School of Law and Diplomacy, Tufts University, Medford, Mass., April 1977.

14 J. Dirck Stryker, Carl H. Gotsch, John McIntire, and Frederick C. Roche, "Investments in Large Scale Infrastructure: Irrigation and River Management in the Sahel." Fletcher School of Law and Diplomacy, Tufts University, Medford, Mass., and Food Research Institute, Stanford, August 1979.

15 A. Hasan Tuluy, "Rice Production in Senegal." Stanford/WARDA Study of the Political Economy of Rice in West Africa, Food Research Institute, Stanford University, Stanford, July 1979; Chapter 8.

APPENDIXES

Methodology for Estimating Comparative Costs and Incentives

John M. Page, Jr., and J. Dirck Stryker

The authors of the country studies have used a common methodology in estimating comparative costs and incentives for various rice activities in each country. This methodology, which is developed in detail in this appendix, also involves the estimation of shadow, or accounting, prices of rice output, tradable inputs, and nontradable primary factors of production, discussed in Appendix B.

The next section presents the indicators of economic efficiency and comparative advantage used in the country studies and explores their theoretical significance. This is followed by a section that defines two measures of private incentives and shows how these are related to the efficiency indicators. There is then a section in which the principal assumptions of the empirical studies are outlined. Finally, the last section examines the comparability of the assumptions and results across countries and production activities. An annex to this appendix provides a sample calculation of each of the indicators.

Economic Efficiency and Comparative Advantage

The measures of economic efficiency and comparative advantage employed in these studies build upon the methodology of the Food Research Institute's work on the political economy of rice in Asia (15). At the core of the analysis are the twin concepts of private and social profitability and the concomitant distinction between market and social accounting (or shadow) prices.[1] Individuals or firms make private investment decisions and evaluate the success of past activities on the basis of observed or expected market prices. In the absence of distortions, market and accounting prices coincide and individual investment decisions at the margin result in social benefits equaling social costs for all activities.

[1] The terms "accounting price" and "shadow price" are used here interchangeably to refer to the social opportunity cost per unit of a scarce resource.

In practice few, if any, of the conditions for the coincidence of market and accounting prices are met in the countries of West Africa. Government taxes and restrictions on foreign trade distort product and factor prices as true measures of social opportunity costs. In addition, public interventions in labor and capital markets, such as minimum wage legislation and subsidized credit, result in market segmentation and introduce divergences between market and social prices. Investments made in response to market incentives without adjustment for these distortions may be economically inefficient and consequently reduce social welfare.[2]

Net Social Profitability

A technique for producing an additional unit of rice is efficient if the social value of its output is equal to or greater than the social opportunity cost of the commodities and factors of production employed in producing it. The measure of efficiency, thus defined, is the net social profitability (NSP) of the activity,

$$\text{NSP}_j = \sum_{i=1}^{n} a_{ij}p_i^* - \sum_{s=1}^{m} f_{sj}p_s^* + E_j \, , \tag{1}$$

where a_{ij} is the quantity of the ith output produced by, or intermediate input used by (inputs having negative signs), the jth activity, p_i^* is the accounting price of the ith output or intermediate input, f_{sj} is the quantity of the sth factor of production used by activity j, p_s^* is the accounting price of the sth factor of production, and E_j is a pecuniary measure of the net external benefits or costs imparted by the jth activity to the domestic economy.[3]

Activities are judged efficient if the level of net social profit, NSP_j, is nonnegative. If the project to be evaluated is in the public sector and all

[2] Not all government-induced distortions reduce social welfare, since government policy may be used to achieve national objectives other than economic efficiency. This is discussed in detail in the text of this volume but is ignored here because this appendix is concerned only with measures of economic efficiency. In addition to distortions induced by government policy, market prices may not equal social costs because of other market imperfections, such as economies of scale, externalities, or monopoly elements in the economy. These imperfections are difficult to measure, however, and are probably not so quantitatively important as the distortions introduced by government.

[3] The NSP criterion was first developed in a programming context by Chenery (5). The most general method of evaluating the efficiency of an investment project, on the other hand, is based upon the discounted value of cash flow (DCF) arising from the investment. The simple decision criterion associated with the DCF is to approve all compatible investment projects that exhibit present values greater than zero. The NSP criterion proposed in Equation (1) is formally equivalent to the DCF analysis if the former is expressed in gross rather than per unit terms and if all entries are dated and discounted to the intitial period by means of an appropriate social rate of discount. The empirical problems encountered when observations are confined to a single period are discussed below. For expositional convenience, time and the social rate of discount are suppressed in the presentation.

intermediate inputs and outputs are assumed to be tradable, the interpretation of the NSP criterion is straightforward. Assuming no joint production, the first term in Equation (1) gives the level of value added at world prices per unit of output generated by the activity,

$$\text{VAW}_j = a_{jj}p_j^* - \sum_{\substack{i=1 \\ i \neq j}}^{n} a_{ij}p_i^* , \tag{2}$$

where p_i^* and p_j^* are the "border" prices of the tradable intermediate inputs and outputs.[4] Value added at world prices is the net addition to national income evaluated at its social opportunity cost.

The second term in Equation (1) is the opportunity cost of primary factor inputs evaluated at accounting prices. When the opportunity cost of these inputs exceeds the net addition to national income and is not offset by significant positive externalities, net social profitability is negative, and the domestic factors of production could be employed more profitably in their best alternative uses.

Resource Cost Ratio

In evaluating past investments and the success of previous policies, it is useful to have a measure of the relative efficiency of activities producing different outputs. The net social profitability criterion is unsuited to this purpose, since the value of NSP varies with the measure of output of each activity. An alternative indicator which is independent of units of measurement can be derived from the NSP criterion. If domestic factor prices are expressed in terms of border prices, Equation (1) can be rewritten as the ratio of domestic factor costs at accounting prices to value added at world prices,

$$\text{RCR}_j = \frac{\sum_{s}^{m} f_{sj}p_s^* + E_j}{a_{jj}p_j^* - \sum_{\substack{i=1 \\ i \neq j}}^{n} a_{ij}p_i^*} \lessgtr 1 . \tag{3}$$

[4] The use of world prices as accounting prices for tradable goods and services is well established in the literature on social evaluation of investments in LDCs. References include Chenery (5), Bruno (2), Little and Mirrlees (12), Dasgupta, Marglin, and Sen (7), Findlay and Wellisz (8), and Srinivasan and Bhagwati (16). Border prices may be expressed either in units of local currency convertible at the official exchange rate or in a standard convertible currency, for example the U.S. dollar. The prices are assumed to be fixed because of the slight impact that individual West African countries have on most world markets.

The resulting resources cost ratio (RCR) represents, for the relevant activity, the rate of transformation between domestic resources and value added at world prices. If this rate of transformation exceeds unity, for example, the opportunity cost of domestic factors of production, expressed in terms of world prices, exceeds the net addition to national income, and the level of net social profit is negative.[5]

Minimizing the resource cost ratio in activities producing tradable goods is equivalent to maximizing value added at world prices per unit of domestic resources employed. Evaluating activities in terms of their RCRs thus provides a measure of relative economic efficiency. Activities with resource cost ratios less than one are efficient in the sense that the domestic factors they employ produce more value added at world prices than they would in the activities from which they are withdrawn. Alternatively, activities with RCRs greater than one are inefficient because they employ domestic factors whose opportunity cost is greater than the net income produced. To the extent that the government allocates resources among competing activities, it should select first activities with the lowest resource cost ratios.[6]

The usefulness of the resource cost ratio as an indicator of relative efficiency corresponds with neoclassical trade theory.[7] Domestic resources

---

[5]This result is demonstrated by setting NSP less than, equal to, or greater than zero and dividing by value added at world prices. The resource cost ratio is analogous to the domestic resource cost of foreign exchange (DRC),

$$\frac{\sum\limits_{s}^{m} f_{sj}p'_s + E_j}{a_{jj}p_j^* - \sum\limits_{\substack{i=1 \\ i \neq j}}^{n} a_{ij}p_i^*} \gtreqless v \,,$$

where p'_s is the shadow price of factor s expressed in local currency and v is the shadow exchange rate. Since the accounting prices of factors of production, whether expressed in foreign or domestic currency, are not independent of the assumptions made in estimating the shadow exchange rate, however, it is desirable to express the accounting prices of domestic factors directly in terms of foreign exchange, as is done with the RCR. An approximation to the shadow exchange rate could be made by computing an appropriately weighted average of the ratios p'_s/p_s^* for all domestic factors. The DRC has been widely used as an indicator of relative efficiency in distorted foreign trade regimes. References include Bruno (3), Krueger (10), Pearson (14), and Pearson, Akrasanee, and Nelson (15). The procedure of expressing individual factor accounting prices in terms of foreign currency was first advocated by Little and Mirrlees (12). Although the approach used here is similar in spirit, specific derivation of factor accounting prices differs substantially from the Little-Mirrlees method and is outlined in Appendix B.

[6]The use of the RCR, NSP, or any other partial-equilibrium criterion for resource allocation assumes that the introduction or expansion of an activity will have no effect on relative factor prices. The ranking of activities by resource cost ratios provides an adequate indication of relative efficiency, however, if the activities being evaluated are small relative to the rest of the economy and if differences in RCRs are relatively important.

[7]For derivations of the resource cost ratio in the context of this theory, see Krueger (11) and Srinivasan and Bhagwati (16).

in this theory are primary factors that are available in less than infinitely elastic supply. Their cost is shown in the numerator of the RCR. In the absence of distortions, the resource cost ratio for each activity in the economy will tend toward unity, and the resulting allocation of primary factors will maximize social welfare. Trade and factor market distortions bias domestic factor prices away from the optimum, giving rise to divergences of the resource cost ratio away from one and a misallocation of scarce domestic resources.

Considerable attention has been focused on the resource cost ratio (or its counterpart, the DRC) as a measure of comparative advantage, both among activities within a single country and between the same activities in different countries. Comparative advantage exists if the social opportunity cost of producing a commodity is less than its border price.[8] Hence activities with positive net social profitability or resource cost ratios less than unity are those in which the country exhibits a comparative advantage.[9] In addition, the smaller the RCR of an individual activity within an economy, the greater its relative efficiency. This kind of comparison is particularly useful in determining regions or techniques that offer the greatest scope for efficient expansion of production. Similarly, it is often interesting to contrast the relative efficiency of different countries in producing the same commodity. To the extent that factor accounting prices approximate values that would apply in the absence of distortions, countries with minimum resource cost ratios exhibit the greatest comparative advantage.

Incentives and Protection

In addition to measuring the economic efficiency of rice-producing activities, a major focus of research has been to measure the structure of

[8]This definition was first proposed by Chenery (5, pp. 19–25). It also appears in Pearson, Akrasanee, and Nelson (15, pp. 128–31).

[9]The relationship between social profitability and comparative advantage is more complex than has hitherto been recognized in the empirical literature. The factor accounting prices estimated in the computation of Equation (1) or (3) are second best in that they are estimated on the basis of an allocation of resources that corresponds to the existing pattern of distortions. Comparative advantage, on the other hand, should be measured using first-best accounting prices in order to indicate those activities that should be expanded if existing distortions were eliminated. Without these distortions, however, the pattern of resource allocation would be different, changing the accounting price of each primary factor. It is even possible for activities that are socially profitable under the existing structure of distortions to show negative social profits under the first-best allocation of resources. This problem, which has been discussed by Findlay and Wellisz (8) and by Srinivasan and Bhagwati (16), is an extension of the difficulty noted earlier relating to the assumption of constant factor prices in evaluating alternative activities. But whereas this assumption may be reasonably valid for second-best project evaluation, it may not hold sufficiently well to ascertain a country's first-best comparative advantage if there are large distortions in the economy.

incentives to producers within each country. At the core of this analysis are the concepts of net private profitability (NPP) and the effective protection coefficient (EPC). Net private profitability is given by

$$\text{NPP}_j = \sum_{i=1}^{n} a_{ij} p_i - \sum_{s=1}^{m} f_{sj} p_s , \qquad (4)$$

where p_i is the domestic market price of the ith output or intermediate input and p_s is the domestic market price of the sth factor of production. NPP is a signal to producers that indicates how they should allocate their resources to maximize profits. If NPP in a given activity is positive, they will gain by shifting resources into the activity; if NPP is negative, resources should be moved elsewhere.

Net private profitability differs from net social profitability because domestic market prices, rather than the social accounting prices, are used to value outputs, intermediate inputs, and primary factors of production, and net external benefits or costs (E_j) are excluded from NPP. Neglecting E_j, the difference between NPP and NSP can be written

$$\text{NPP}_j - \text{NSP}_j = a_{jj}(p_j - p_j^*) - \sum_{\substack{i=1 \\ i \neq j}}^{n} a_{ij}(p_i - p_i^*) - \sum_{s=1}^{m} f_{sj}(p_s - p_s^*) , \quad (5)$$

where p_j, p_i, and p_s are the domestic market prices of output, intermediate inputs, and primary factors, respectively, and p_j^*, p_i^*, and p_s^* are border prices expressed in domestic currency convertible at the official exchange rate. Differences between domestic and equivalent border prices may exist because of taxes and subsidies, quantitative restrictions on trade, import or export monopolies, various forms of price control, credit subsidies, and other market imperfections. The first term on the right-hand side of Equation (5) shows the influence of trade and price policies on the output price; the next two terms indicate the effects of taxes and subsidies, as well as other market imperfections, on the prices of intermediate and primary factor inputs.

Another indicator of incentives is the effective protection coefficient (EPC).[10] This measures the increase in domestic value added allowed by the structure of trade protection and domestic price control over the level of value added in the absence of such restrictions. If all commodities are traded, the EPC is given by

[10] This term was originated by Balassa (1).

$$\text{EPC}_j = \frac{a_{jj}p_j - \sum_{i=1}^{n} a_{ij}p_i}{a_{jj}p_j^* - \sum_{i=1}^{n} a_{ij}p_i^*} \; ; \tag{6}$$

that is, it is the ratio of value added in domestic prices to value added in world prices.

The effective protection coefficient and the resource cost ratio are closely related. Both have the same denominator and so the only difference is in their respective numerators. That of the EPC shows the scope that fixed world prices and trade and price incentives provide for some combination of domestic resource costs, rents, and profits, as well as government taxes and subsidies on each of these. The numerator of the RCR, on the other hand, contains only domestic resource costs adjusted for any net external benefits that may exist. The EPC thus indicates the potential for incurring domestic costs; the RCR measures the extent to which these costs are in fact incurred. The difference between the two depends on the structure of taxes and subsidies and on the supply functions for domestic primary factors.[11]

Principal Assumptions of the Empirical Studies

Indicators of net private and social profitability, effective rates of protection, and resource cost ratios are given in each of the country studies for a number of different activities consisting of various combinations of production, collection, processing, and distribution activities. Each activity is assumed to have fixed input-output coefficients for both intermediate inputs and primary factors. Since commodity and factor prices are also fixed, constant costs are assumed for each activity. This does not, however, preclude the existence of a rising supply curve for rice. At any given moment the level of each activity is constrained, and increases in supply may therefore come from activities with increasingly higher costs.

Technical coefficients and prices used in calculating the indicators apply primarily to the crop year 1975–76. Longer term averages are used when possible, however, where data for 1975–76 are not very representative. The most notable instance is the world rice price, which varied substantially over this period. The price of Thai 5 percent brokens, for ex-

[11] A distinctive advantage of the effective protection coefficient as a measure of incentives is that its incidence is always unambiguous because of the assumption that world prices are fixed. The impact of taxes and subsidies on specific domestic resources whose supply functions are not horizontal is less straightforward. For further discussion of this point see Corden (6).

ample, dropped from $385/mt to $246/mt, f.o.b. Bangkok.[12] A price of $350/mt, which is close to world price projections for the year 1985 (in 1975 U.S. dollars), is used as a basing price in the calculations, with adjustment for quality differences and transportation costs in each country.

Differences in production techniques are due primarily to variations in source of power, use of improved seed varieties and chemical inputs, degree of water control, and scale of operation. Techniques are also defined with respect to specific regions because of the importance of local ecological conditions in determining input-output coefficients and especially yields. Although indicators could be calculated for a relatively large number of these regionally specific techniques, only those that are either currently or potentially most important have been included in the country chapters.

Techniques for processing and marketing, on the other hand, are more limited. The only processing techniques considered here are hand pounding, small-scale hulling with rubber rollers or steel cylinders, and large-scale milling. Marketing techniques include official and private collection of paddy and distribution of milled rice. Various consumption points have been chosen ranging from the farm or village through regional and national urban centers to West African cities outside the producing country.[13]

Calculations of the indicators in each of the country studies are based on a number of common assumptions. The most important of these involve the treatment of nontraded goods, the role of capital and labor as factors of production, the handling of input taxes and subsidies, the problem of joint production, and adjustments made for various types of losses.

Nontraded Goods

The method chosen to incorporate nontraded inputs into the cost and incentive indicators accords with techniques currently advocated in the literature on social evaluation of investments. If it is assumed that production of a nontraded input occurs at constant costs to meet any growth in demand resulting from expansion of an activity in which that input is used, it is appropriate to value the input in terms of its marginal social cost of production.[14] The input-output structure of the supplying industry

[12] Data are from the Food and Agriculture Organization of the United Nations, Commodities and Trade Division, *Rice Trade Intelligence*, various issues.

[13] See Appendix B for a discussion of the appropriate border price of rice for each of these assumed centers of consumption.

[14] If nontraded inputs were not supplied under conditions of constant costs, changes in the level of demand for these inputs would engender both production and consumption effects, thus complicating the estimation of their accounting prices. The constant cost assumption concerning nontraded inputs is encountered in Bruno (2), Little and Mirrlees (12), and Dasgupta, Marglin, and Sen (7). It is valid as long as changes in the prices of primary factors are ignored, since physical input-output coefficients are assumed constant.

may then be used to break down costs into those associated with indirect tradable inputs and indirect primary factors, all measured in terms of border prices. The costs of these indirect inputs are treated in the same way as those of direct tradable inputs and primary factors. The numerator of the RCR given in Equation (3) now represents the opportunity cost of all domestic factors of production employed directly in activity *j* and indirectly in the production of inputs into this activity. The denominator gives the value of output less the cost of using direct and indirect tradable inputs.

A good or service is classified as nontraded if its domestic market price lies below the border price of imports but above that of exports, and, hence, the country neither imports nor exports the commodity. Certain categories such as construction, transport, and utilities are almost universally nontraded. There is another category, however, that may be labeled less than fully traded. These are goods or services that are tradable but that because of government policy or other distortions are not actually traded. Nontariff restrictions on imports, for example, are frequently imposed to protect domestic producers. Trade in this case may be eliminated even though the domestic price is far greater than the price of potentially competing imports.[15]

It has generally been assumed in the country studies that tradable commodities are fully traded. This implies that at the margin additional demand for protected inputs arising from expansion of rice production will be met through increased imports rather than from increases in domestic supply or deprivation of other consumers. Under this assumption the correct valuation of these inputs is their border price. The polar contrast to this approach is to evaluate all inputs subject to quantative restrictions as nontradable. Locally produced inputs would then be broken down into their marginal social costs of production in the manner described above.[16] High marginal social costs in the supplying industry would accordingly be reflected in the measured NSP or RCR efficiency indicators for rice. Because interest centers on the evaluation of rice production and processing, and the focus of the analysis is on the longer-term issues of comparative advantage and relative efficiency in that sector alone, the treatment of tradable goods as fully traded is appropriate.

Capital and Labor

Data employed in the country studies cover a single crop cycle, and efficiency measures are limited to a single period. The presence of capital inputs with economic lifetimes exceeding one year raises the problem of

[15] Joshi (9) discusses the distinction between fully traded and less than fully traded goods and its implications for the estimation of accounting prices.

[16] Bruno (4) makes this assumption.

specifying an annualized measure of capital services. The measure used here is the annual capital recovery factor, which yields a present value equal to the initial cost of the capital input when discounted over the economic lifetime of the asset at the relevant rate of interest. This annuity includes both interest and depreciation on the capital. When expressed in accounting prices, the annuity is entered directly as a cost item in the computation of net social profitability.

In practice, the major objection to the use of an annualized measure of capital services arises when differing levels of capacity utilization exist during the lifetime of the investment. Because interest here centers on long-run questions of relative efficiency, most efficiency comparisons are made at projected full, or at least constant, utilization of capacity.[17]

Capital also presents certain difficulties with respect to the specification of the resource cost ratio. The annual service charge associated with the stock of capital must be allocated to either domestic resource or tradable costs. If the social rate of discount is established by the cost of borrowing foreign funds, i.e. if capital is available in infinitely elastic supply from the international markets, the marginal source of investment funds is of foreign origin and the capital charge should be treated as any other tradable cost. No West African country, however, has unlimited access to foreign borrowing at a constant rate of interest. Part of the supply of capital, therefore, must come from alternative domestic uses, and capital charges should be treated as a domestic resource cost and assigned to the numerator of the resource cost ratio.[18] This has the added advantage of allowing the measure of value added in the denominator to correspond to its traditional definition, output less current intermediate inputs.[19]

Working capital is estimated on the basis of the average length of time during which labor, intermediate inputs, and output are tied up in pro-

[17] From the perspective of individual project analysis, this assumption ignores costs associated with agricultural infrastructure investments involving substantial periods of less than full capacity utilization. Many of these additional costs, however, are related more to the general process of development than to any particular technique of production and therefore are not relevant in making efficiency comparisons between techniques.

[18] This conclusion is consistent with Corden (6, pp. 171–74). Bruno (3) arrives at a similar formulation in the context of a general equilibrium programming model in which capital is included as a constraint.

[19] Considerable confusion has developed in the literature on the domestic resource cost (DRC) criterion as to the treatment of capital inputs. Suggested approaches have included treating depreciation on imported capital goods as imported intermediate inputs, whereas the financial return on the same items is considered to be a domestic resource cost, presumably because investable funds are in less than infinitely elastic supply. Such an approach fails to recognize that it is the total service cost of capital that represents the value of investable funds diverted from alternative employment. This service cost per unit of capital is given by $r = kp_K$, where r is the service cost, k is the capital recovery factor, and p_K is the price of the capital asset. Even if p_K is constant because of infinitely elastic supply of tradable capital goods, r is positively related to the size of total investment because the recovery factor k depends on the rate of interest at which funds can be borrowed.

duction, milling, and marketing. This is not done for inputs that are used throughout the year, such as extension and maintenance services, because of the difficulty of determining the timing of expenditures and because some capital items are already included in their costs. Thus the cost of working capital tends to be slightly underestimated.

Capital costs are considered sunk when they do not have to be reincurred while a given technique is being expanded to its potential during a period of a decade or so. As an example, construction of the diversion dam at Markala for the Office du Niger in Mali was undertaken in the 1930's for the purpose of irrigating an area much larger than that currently under cultivation. The costs of that dam are therefore ignored in calculating the profitability of rice production in the Office. On the other hand, if new investments would have to be made to realize the potential for expanding a particular technique during this period, the capital charge associated with those investments is included.

One area of ambiguity concerns the treatment of certain types of recurrent costs that could be thought of as capital investments. The most obvious examples are those associated with research and development and the initial establishment of extension services in a given area. Since it is difficult to charge these costs to any particular technique, they are assumed to be general developmental expenditures, and only the costs associated in the long run with a particular technique are charged to it.[20]

With respect to labor, the major problem is the lack of a standardized work unit. Labor inputs are estimated on the basis of the number of man-days required for different tasks, but it is clear that actual inputs vary in terms of the sex and age of the worker, the quality of the work, and the degree of skill and effort required. In some societies there is a clear distinction between tasks performed by men and by women. Elsewhere the lines are not so carefully drawn, but men may have an advantage over women when considerable physical effort is needed. Similarly, children are assigned jobs, such as bird-watching, which they can do as easily as adults, but they could have a hard time accomplishing other, more difficult tasks. If the sex/age category of a task is clearly defined and men, women, and children customarily receive different wages, these distinctions have been made in the calculations, and the cost of each category of labor has been estimated separately. More often, however, the distinctions are blurred and depend on individual family circumstances.[21] In these instances, the task has usually been defined in

[20] In Liberia extension costs were divided into capital as well as recurrent charges. This was possible because each project is clearly associated with a given technique of rice cultivation. The total matches fairly closely, however, extension costs in other countries.

[21] In some areas, for example, children who could otherwise guard against birds are in school and the task must be performed by adults at a relatively high cost.

terms of the number of man-days required and labor costs have been based on the wages of men.

Although most work is performed by the family, all farm labor is valued at the wage rate paid to casual, unskilled workers available from the local labor market. Yet a number of farm tasks require fairly high levels of skill. This is true, for example, of several operations associated with irrigated agriculture. Furthermore, important elements of farm management and supervision of hired labor are not directly linked with any particular labor operation. It is very difficult, however, to estimate the value of these skills because the tasks with which they are associated are not generally performed by hired labor and thus there are not data on relevant wage rates. Consequently, residual net farm income includes unmeasurable returns to skills, management, and entrepreneurship supplied by the farmer.

In many post-production operations, it is difficult to distinguish skilled from unskilled labor, even though the former clearly assumes greater relative importance. If operations such as loading and unloading plainly involve unskilled labor, this category is retained. Otherwise, labor costs are assumed to apply to workers with at least some degree of skill. This generalization introduces bias into the cost estimates, however, only if unskilled labor costs are adjusted for any difference between shadow and market wage rates. Since the error involved in not separating out all components of unskilled labor in post-production operations is small, the bias is not great.

Input Taxes and Subsidies

Input taxes and subsidies entering the calculations can be separated into two types. Some, for example, import tariffs on fertilizers, are assessed or paid prior to arrival of the inputs at the farmgate. In most cases, the incidence of these taxes and subsidies on tradables and on primary factors can be ascertained relatively easily. The farmer, however, may not pay the full cost of these inputs, inclusive of taxes and subsidies, at the farmgate. The difference between this cost and the price he actually pays is an additional subsidy, or more infrequently a tax, on the inputs. Since the cost at the farmgate, even of imported inputs, includes substantial nontradable elements, it is not usually possible to estimate the extent to which this last element of tax or subsidy falls on tradables or on primary factors. It is therefore assumed that its incidence is proportional to the relative importance of tradables and primary factors, inclusive of previous taxes and subsidies, in farmgate cost.

In general, the incidence of all these taxes and subsidies depends on how easily resources can be reallocated at the margin to avoid payment of taxes or to benefit from the subsidies. In some instances, however, the farmer may not be able to alter marginally his allocation of resources in response to any particular tax or subsidy. This is true, for example, when a flat water charge must be paid by the farmer if he is to participate in a given irrigation scheme. In this instance, the tax or subsidy is assumed to fall fully on the direct primary factors of production and not on the tradable component of cost.

There are several assumptions concerning the treatment of various forms of tax or subsidy on capital. These incentives are of two general types.

1. The price paid by farmers for capital equipment is taxed or subsidized, but there is no tax or subsidy on credit. In this instance, interest and depreciation are calculated with respect to these incentives in the same way as for any other input, and capital charges include taxes and subsidies on items of capital paid both prior to and at the farmgate.

2. Credit is offered to the farmer on subsidized terms. Usually this involves a limited period during which repayment must be made, and often this period is shorter than the service life of the capital asset. As a result, the credit subsidy is less than would appear from simply comparing the subsidized interest rate with the market or shadow price of capital.[22] To adjust for this, all credit payments made by the farmer are discounted at the market rate of interest back to the year in which the asset was purchased. The sum of these discounted payments is the present value of the asset measured in terms of its purchase price to the farmer. An annuity is then calculated at the market rate of interest based on the service life of the asset to show the annual cost of that asset to the farmer. The actual cost, on the other hand, is an annuity at the market rate of interest calculated on the basis of the cost of the capital asset delivered to the farmgate. The difference between this and the annuity based on the present value of the farmer's payments is the capital subsidy. It may include both an interest rate subsidy and a subsidy on the price paid by the farmer for the capital asset, though it is impossible to distinguish between the two when the asset is purchased only on credit.

Capital charges on taxes and subsidies are treated as if they were themselves taxes or subsidies. This is because a capital input is never measured explicitly as a stock but only as a flow of capital service charges. Consequently, a tax or subsidy on the stock of a capital input has as its flow counterpart the interest and depreciation charged on that tax or subsidy.

Joint Production

In a number of areas, rice is produced jointly with other outputs. Joint production occurs in several ways and in each case poses the problem of

[22]This assumes, of course, that the credit is tied specifically to the purchase of the asset and cannot be renewed except insofar as new assets are purchased. In addition, the interest rate used, whether real or nominal, must be consistent with the revaluation of capital assets associated with any assumed rates of inflation.

how to allocate fixed costs associated with the production of all outputs. Assumptions have been made in each instance that are, of necessity, somewhat arbitrary.

Rice produced using traditional rainfed techniques is frequently inter-cropped with other plants. Rice is usually dominant, however, generally comprising at least three-quarters of the land area. The approach here is to adjust for this by converting yields and labor inputs to pure stand equivalents. This ignores, however, some of the advantages that inter-cropping may have over pure stands.[23]

A second example of joint production occurs where capital equipment is used both for rice and for other crops. To charge the full value of that capital to rice would considerably overstate actual costs. Instead, costs per hectare of rice cultivated are estimated on the basis of the maximum area over which the equipment could be used if all land were devoted to rice. An appropriate adjustment is also made if animals or equipment are employed partly for nonagricultural purposes, such as carts used to transport people and other goods.

Irrigated infrastructure often supports the cultivation of two crops per year, one of which may not be rice. It is inappropriate to charge the full cost of the infrastructure to rice because this ignores the net benefit associated with having a second crop. Instead, the procedure used is to allocate these costs to rice in proportion to the estimated amount of water used to grow rice in comparison with the second crop. This can be justified on the grounds that the major purpose of the infrastructure is to supply irrigation water, though other uses such as flood protection also exist.

Another example of joint production is in milling. Although rice of various qualities is the major output, there are a number of by-products, such as bran and flour, which have a fairly important commercial value. To take this into account, the value at market prices of all by-products is subtracted from the cost of milling, with the breakdown into tradables, primary factors, and taxes/subsidies being proportional to the same breakdown for all milling costs.

Finally, there is the question of how to allocate overhead expenses when a single enterprise is involved in several activities. Earlier, it was noted that expenditures on research and development and the establishment of an extension service in a particular area are treated as investments in general development and are not charged to specific rice production techniques. A similar assumption is made with respect to overhead—only those expenditures that can clearly be identified with the cultivation, milling, and marketing of rice are included as costs. Generally, this can be

---

[23] See, for example, the work of David Norman in northern Nigeria (13).

construed as implying some degree of physical proximity of the items for which the overhead expenditures are made to the rice activity. If this relationship exists but several products or activities are involved, the expenditures are prorated in proportion to the relative importance of each. Extension services, for example, are allocated to rice production and marketing on the basis of the relative amount of time spent by extension agents working with each activity.

Loss Adjustment

Adjustment has been made for several different types of losses. Crop cuttings, for example, invariably tend to overestimate actual yields harvested. This may be because plants outside the sample area are harvested, more care than usual is taken with threshing, or other reasons, but overestimates of up to 20 percent of the harvest are not uncommon. In addition, there are losses associated with transport, milling, and storage. Data on these losses are very scanty. To avoid upward bias to estimates of paddy production, authors of the country studies have tried to determine as carefully as possible actual paddy harvested and have then decreased this quantity rather arbitrarily by 5 percent to account for losses between field and collection point. A similar adjustment, usually also about 5 percent, is made for losses due to drying, impurities, and damage in milling and marketing.

Another type of loss occurs even before harvest. If the quantities of rainfall and flooding are uncertain, part or all of the area planted to rice may never be harvested. Yields, however, generally refer to land harvested rather than to area planted. Accordingly, there is a tendency to understate costs in relation to benefits by ignoring seeds and other inputs involved in cultivating areas not harvested. Coefficients of these inputs have been adjusted upward to reflect the probability that some land that is prepared and seeded may not be harvested.

Comparability of Results

The individual country studies provide detailed descriptions and analyses of rice activities in five West African countries. Farming techniques range from very extensive traditional upland cultivation to highly mechanized irrigated production under systems of full water control. Processing includes both simple hand pounding and modern large-scale milling. For each activity, the relevant incentive system is described and estimates are given for the various indicators of private and social profitability. The sensitivity of the results to variations in yields, primary input costs, and the world market price of rice is tested. Finally, the activities

TABLE A.1. *Total Cost Summary of Production and Post-Harvest Techniques*

Technique	Unskilled labor	Skilled labor	Capital[a]	Land	Tradable inputs	Taxes and subsidies Tradables	Taxes and subsidies Primary factors	Total private cost	Total social cost	Total social cost milled equivalent (000 CFA fr/mt)	Total social cost milled equivalent ($U.S./mt)[b]
Farm production (000 CFA francs/ha), irrigation	68	29	39(26)	0	107	20	−88	175	230	118	472
Collection (CFA francs/kg paddy), public	1	1	1	0	3	2	1	9	6	9	37
Milling (CFA francs/kg milled rice), large-scale	2	3	3(2)	0	6	5	2	21	13	13	52
Distribution (CFA francs/kg milled rice), public (port city)	1	2	2	0	5	2	1	13	10	10	40

[a] Figures in parentheses are calculated in terms of accounting rather than market prices.
[b] Assumes an exchange rate of 250 CFA francs/U.S. dollar.

are assessed in relation to their contribution to different national objectives. Statistical appendixes for each country provide details concerning the data used and assumptions made in the empirical calculations.

Numerous sources of information were used in this empirical work, including farm management surveys in which farmers are visited two or three times a week during the cropping season, surveys with only one or at most a few interviews, direct observation, conversations with farmers, discussions with people with an abundance of firsthand experience, and documents and publications based on these and other primary sources. This multiplicity of source material obviously creates problems of comparability of results between different countries. There is evidence, for example, that labor times obtained from farmers on the basis of interviews conducted several times a week are often considerably higher than the opinions of experts concerning the amount of labor input required for a given technique.

To overcome this problem of comparability, frequent and extended discussions have taken place among the authors of the country studies to try to assure use of a common set of assumptions as a basis for estimation and to verify as much as possible the validity of the basic data. Although discrepancies that are unaccounted for undoubtedly still exist, a fairly high degree of confidence can be placed on comparisons of results between countries. In addition, the authors of the individual country studies have tried to evaluate the reliability of their results using sensitivity analysis, and the conclusions of these efforts are summarized in Stryker's comparative study, Chapter 12.

Annex

This annex provides an example of how the cost and incentive indicators described here and presented in the country studies are calculated. The reader should be able to follow the calculations all the way from the basic assumptions concerning technical coefficients and prices to the final empirical indicators. To the extent that he might choose to change any of the assumptions, he can calculate the effects of these changes on the final results.

Table A.1 gives cost breakdowns for a hypothetical rice activity that involves irrigated production, public agency collection, large-scale milling, and distribution by the public agency to the port city as the point of consumption. Total private cost and social cost can be obtained by adding, within the relevant columns, cost figures for all techniques included in the activity after adjusting these to a per-metric-ton of milled rice equivalent basis:

$$\text{Private cost} = \frac{175}{(3)(0.65)} + \frac{9}{(0.65)} + 21 + 13 = 137.6 \, ,$$

$$\text{Social cost} = \frac{230}{(3)(0.65)} + \frac{6}{(0.65)} + 13 + 10 = 150.2 \, .$$

The first two terms in the calculations assume a paddy yield of 3 mt/ha and a milling outturn of 0.65 mt milled rice per mt paddy.

The difference between private and social costs consists of two elements. First, taxes and subsidies on tradables and primary factors are included in private cost but not in social cost. Second, some of the primary factor inputs may have shadow prices that differ from their market prices. In the example, this is true of capital only.

Net private profitability (NPP) is calculated by subtracting the private cost per ton of milled rice from the market price of that rice at the consumption point.[24] This market price equals the c.i.f. price plus the tax (actual or imputed) on imported rice.[25] Assuming the c.i.f. price to be 100 CFAF/kg and the import tariff to equal 40 percent of the c.i.f. price, the domestic market price is 140 CFAF. Net private profitability is then given by NPP = 140 − 137.6 = 2.4. Net social profitability (NSP), on the other hand, is equal to the c.i.f. price of imported rice minus the social cost of delivering domestically produced rice to the consumption point, NSP = 100 − 150.2 = −50.2. In this case the value of NSP is negative, reflecting the relatively high degree of trade protection and farm subsidies that are necessary to offset the high production costs.

The effective protection coefficient (EPC) is equal to value added in domestic market prices divided by value added in world prices. The numerator equals the market price of milled rice minus the market value of all tradable inputs (per mt of milled rice), the latter being the c.i.f. value of tradable inputs plus taxes and subsidies on these inputs. The denominator equals the c.i.f. price of rice minus the c.i.f. value of tradable inputs (per mt of milled rice),

$$\text{EPC} = \frac{140 - \left(\dfrac{107 + 20}{(3)(0.65)} + \dfrac{3 + 2}{0.65} + (6 + 5) + (5 + 2) \right)}{100 - \left(\dfrac{107}{(3)(0.65)} + \dfrac{3}{0.65} + 6 + 5 \right)} = 1.67 \, .$$

[24] This is the market price to wholesalers in the consumption center.

[25] This assumes that rice is actually imported and abstracts from port handling charges. If rice is exported, instead, the market price equals the f.o.b. price minus any tax on exports.

This indicates that, whereas the tax on rice imports results in a domestic market price that is only 40 percent above the c.i.f. price, the combination of this tax and the somewhat lower taxes on tradable inputs allows value added in domestic market prices to exceed that measured in world prices by 67 percent.

Finally, the resource cost ratio (RCR) is calculated with the same denominator as the EPC but with a numerator equal to the social cost (per mt of milled rice) of all the primary factors of production—unskilled labor, skilled labor, capital, and land:

$$\text{RCR} = \frac{\dfrac{(68 + 29 + 26)}{(3)(0.65)} + \dfrac{(1 + 1 + 1)}{(0.65)} + (2 + 3 + 2) + (1 + 2 + 2)}{100 - \left(\dfrac{107}{(3)(0.65)} + \dfrac{3}{0.65} + 6 + 5\right)}$$

$$= 2.70.$$

The value of this ratio, 2.70, is greater than unity, so that the activity is socially unprofitable. This is consistent with the negative value of NSP. The RCR is also greater than the EPC, reflecting the relatively high subsidies paid on primary factors at the farm level, which allow costs to be sustained even beyond those permitted by the level of effective protection.

Citations

1 Bela Balassa, "Methodology of the West Africa Study." World Bank, Washington, D.C., n.d., mimeo.

2 Michael Bruno, "The Optimal Selection of Export-Promoting and Import-Substituting Projects," in *Planning the External Sector: Techniques, Problems, and Policies*. United Nations, New York, 1967.

3 ———"Development Policy and Dynamic Comparative Advantage," in Raymond Vernon, ed., *The Technology Factor in International Trade*. Columbia University Press, New York, 1970.

4 ———"Domestic Resource Costs and Effective Protection: Clarification and Synthesis," *Journal of Political Economy*, 80 (January/February, 1970).

5 Hollis B. Chenery, "Comparative Advantage and Development Policy," *American Economic Review*, 51 (March 1961).

6 W. M. Corden, *The Theory of Protection*. Oxford, Clarendon Press, 1971.

7 Partha Dasgupta, Stephen A. Marglin, and A. K. Sen, *Guidelines for Project Evaluation*. United Nations, New York, 1972.

8 Ronald Findlay and Stanislaw Wellisz, "Project *Evaluation*, Shadow Prices, and Trade Policy," *Journal of Political Economy*, 84 (June 1976).

9 Vijay Joshi, "The Rationale and Relevance of the Little-Mirrlees Criterion,"

Bulletin of the Oxford University Institute of Economics and Statistics, 34 (February 1972).

10 Anne O. Krueger, "Some Economic Costs of Exchange Control: The Turkish Case," *Journal of Political Economy*, 74 (October 1966).

11 ————,"Evaluating Restrictionist Trade Regimes: Theory and Measurement," *Journal of Political Economy*, 80 (January/February 1972).

12 I. M. D. Little and J. A. Mirrlees, *Project Appraisal and Planning for Developing Countries*. Heineman, London, and Basic Books, New York, 1974.

13 D. W. Norman, "The Rationalization of a Crop Mixture Strategy Adopted by Farmers under Indigenous Conditions: The Example of Northern Nigeria," *Journal of Development Studies*, 11 (October 1974).

14 Scott R. Pearson, "Net Social Profitability, Domestic Resource Costs, and Effective Rate of Protection," *Journal of Development Studies*, 12 (July 1976).

15 Scott R. Pearson, Narongchai Akrasanee, and Gerald C. Nelson, "Comparative Advantage in Rice Production: A Methodological Introduction," *Food Research Institute Studies*, 15 (1976).

16 T. N. Srinivasan and Jagdish N. Bhagwati, "Shadow Prices for Project Selection in the Presence of Distortions: Effective Rates of Protection and Domestic Resource Costs," *Journal of Political Economy*, 86 (February 1978).

Shadow Price Estimation

J. Dirck Stryker, John M. Page, Jr.,
and Charles P. Humphreys

The methodology described in Appendix A requires that outputs, intermediate inputs, and primary factors involved in the production, assembly, processing, and marketing of rice be valued at shadow or social accounting prices. These shadow prices reflect the value placed by society on the opportunities forgone by using scarce resources in the rice sector. As such, they serve as a guide to how resources can be allocated to maximize social welfare.

The next section develops some of the general concepts and outlines the major assumptions used in shadow price estimation. Following this explanation, there is a discussion of the specific procedures used to estimate the shadow prices of outputs, intermediate inputs, and various factor services. The estimates are presented and evaluated with respect to their limitations and biases. Major conclusions are summarized at the end of the paper.

Concepts and Assumptions

Shadow prices may differ from prices observed in the market because of noncompetitive behavior, externalities, and distortions introduced by government policy.[1] Policy-induced distortions are perhaps the most pervasive, and at the same time the most complex, cause of divergence between market and shadow prices in less developed countries.

There exist in the literature two basic approaches to the social valuation of resources. The first argues that shadow prices should be defined solely with respect to the objective of economic efficiency. Other objectives are recognized, but the use of policies to achieve those objectives is generally perceived as having an economic cost valued at these shadow prices.[2] An alternative approach is to build into shadow price estimates the weights

[1] See Bruno (8) for the derivation of shadow prices in an open economy, using a linear-programming framework of analysis.
[2] The theory of optimal policy to achieve noneconomic objectives is developed in Bhagwati and Srinivasan (3).

attached to various national goals.[3] Benefits accruing to different income groups, for example, may be assigned different values, taking into account the objective of improving income distribution. If this procedure is used, the optimal policy mix maximizes net social benefits expressed in terms of accounting prices that reflect these social weights.

The approach used in the West African rice project has been to define shadow prices solely in terms of economic efficiency. No adjustment is made for the contribution of policies to other objectives, though their impact on these objectives may be separately assessed.

If there are historical policy-imposed distortions in the economic system, the estimation of shadow prices will vary depending on whether these distortions are expected to be removed. In calculating "first-best" shadow prices, it is assumed that all government-imposed distortions will be eliminated except insofar as they contribute to the optimal allocation of resources. "Second-best" shadow prices, on the other hand, are estimated assuming that existing nonoptimal policies will remain in effect during the period of the analysis. This is equivalent to deriving the first-order conditions for welfare maximization with market distortions acting as constraints. This latter approach has been used in the West African rice project.

Given these assumptions, the general procedure used to estimate the shadow price of a given resource is to determine the decline in national income that occurs as the result of withdrawal of the resource from alternative uses. Under conditions where all goods are tradable, where government quantitative trade restrictions do not exist, and where the country has no monopoly power in international trade, this change in national income should be valued in world prices.[4] The justification for this valuation, rather than using domestic prices that reflect consumers' marginal preferences, is that consumer prices are invariant with respect to changes in the allocation of resources under these conditions. Welfare depends, then, only on the level of purchasing power over the world's goods and services and can be measured by national income denominated in world prices. Because of the possibility of trade, resources may be allocated so as to maximize production expressed in world prices without directly affecting the pattern of consumption except through alterations in income.

If some goods are nontraded and the number of primary factors equals the number of traded goods, Bhagwati and Srinivasan (4) have recently demonstrated that the shadow prices of the primary factors are uniquely

[3] This approach is discussed in detail in Little and Mirrlees (16), Squire and van der Tak (21), and Dasgupta, Marglin, and Sen (10).

[4] This approach to shadow price estimation is due primarily to Little and Mirrlees (16). Other more recent contributions have been made by Findlay and Wellisz (13), Srinivasan and Bhagwati (22), and Bhagwati and Wan (5).

determined by the world prices of the traded goods, and the world price equivalents of nontraded goods then depend only on the technology used in production. In the more general case where the number of primary factors does not equal the number of traded goods, however, shadow prices are not uniquely defined with respect to world prices but depend, in addition, on the pattern of policy distortions and market imperfections, together with the resulting allocation of resources (2). If government policy takes the form of a quantitative restriction on trade flows, any changes in the allocation of resources will, in general, alter the relative magnitude of the consumer price distortion resulting from this policy, and changes in social welfare will no longer depend uniquely on changes in output valued at world prices (4, p. 9).[5] Finally, where a country has some monopoly power in the international trade of certain goods, it is in general not sufficient to replace world prices with marginal revenue,[6] since this does not allow for changes in the domestic consumption of these goods which occur as resources are reallocated (4, p. 19).

In the literature on shadow price estimation, there is disagreement over some issues. In particular, the general equilibrium implications of partial equilibrium approaches to shadow pricing have only begun to be explored.[7] Nevertheless, most empirical estimation techniques are essentially partial equilibrium in nature. The methodology used here follows this tradition; it also explores some of the indirect effects induced by withdrawing resources from alternative uses and whether these indirect effects require adjustments in the shadow price estimates.

Although some attention has been given to identifying market imperfections resulting from noncompetitive behavior or externalities, the most important distortions appear to be those induced by government policy. The effects of these distortions are shown in the following generalized shadow price equation:[8]

[5] This is similar to the case of less than fully traded goods discussed in Appendix A. There it was assumed that if imports of tradable inputs are restricted by quota, this restriction would be relaxed to allow the additional demand for the protected input arising from the expansion of the rice sector to be satisfied from increased imports.

[6] As suggested, for example, by Little and Mirrlees (16, p. 161) and by Scott (20, p. 176).

[7] Some recent examples of general equilibrium analysis are Broadway (7), Dasgupta and Stiglitz (9), and Warr (24).

[8] This equation is an extension of one developed by Bertrand (2). The convention introduced by Little and Mirrlees of taking foreign currency as the unit of account and valuing nontraded goods in terms of foreign exchange is also adopted here. This is consistent with the method and assumptions of the resource cost ratio outlined in Appendix A. No adjustment is made for changes in the prices of goods subject to quantitative restrictions in lieu of taxes or subsidies, since controls of this type are relatively uncommon in the countries concerned. In addition, the potential effects of monopoly power in trade are ignored because of the small size of these countries. Finally, to simplify the analysis somewhat, all nontradable goods are to be used only for final consumption. The extension of this equation to include their use as inputs in production is straightforward.

$$p_h^* = \frac{1}{r}p_h + \sum_{i=1}^{m} p_i^w t_i^c \frac{dx_i}{dl_h} + \frac{1}{r}\sum_{k=1}^{n} p_k^d t_k \frac{dz_k}{dl_h} + \sum_{i=1}^{m} p_i^w t_i^d \frac{d\hat{x}_i}{dl_h} - \sum_{i=1}^{m} p_i^w t_i^d \frac{d\bar{x}_i}{dl_h} , \quad (1)$$

where p_h^* is the shadow price of the hth resource expressed in foreign currency, p_h is the market price of that resource, r is the official exchange rate, p_i^w is the world price of the ith tradable good measured in foreign currency, p_k^d is the producer price of the kth nontradable good, t_i^c is the tax rate that causes the price paid by consumers for tradables to differ from the border price (subsidy rates are expressed as negative tax rates), t_i^d is the tax rate that causes the price paid or received by producers for tradables to differ from the border price, t_k is the tax rate that causes the price paid by consumers for nontradables to differ from the price received by producers, x_i is final consumption of the ith tradable good, $\hat{x}_i$ is intermediate consumption of the ith tradable good, $\bar{x}_i$ is production of the ith tradable good, z_k is final consumption of the kth nontradable good, l_h is the fixed quantity of the hth resource available to the economy at the beginning of the relevant period, m is the number of tradable goods, and n is the number of nontradable goods.

Resources include primary factors, stocks of goods left over from the previous period, and foreign exchange. Market prices of resources that are fixed in supply are determined only by their scarcity values and not by the reservation prices attached to them by those who control their supply. Under this assumption, the value of leisure time and of savings, for example, can be ignored. Where quantities of resources are not predetermined, the scarcity value and reservation price should equal one another unless there are policy-induced distortions or other imperfections in the markets for these resources.

Given these assumptions, the estimation of shadow prices requires correction of the relevant market prices facing consumers and producers for distortions caused by government taxes and subsidies on flows of goods and nonfactor services. The first two terms to the right of the market price p_h in Equation (1) show that taxes on consumption goods raise the value of those goods to consumers. To the extent that withdrawal of a resource from alternative uses results in a change in consumption of tradables or of nontradables, the effect of distortions in consumer goods prices is taken into account through these terms.

The last two terms in Equation (1) correct for distortions faced by producers in the prices of both the inputs they purchase and the products they sell. When the utilization of tradable inputs and the production of tradable outputs are affected by withdrawal of a resource from alternative uses, a weighted average of the distortions in the prices of these inputs and outputs is subtracted from the market price of the resource. The weights are the amounts by which each of these inputs and outputs is increased or decreased.

In general, equilibrium, consumption and production of most goods would normally be expected to change as a result of a withdrawal of a unit of the resource from the rest of the economy. Estimating these changes in a world of many commodities, however, would be a formidable task. One alternative is to work with a limited number of aggregate commodity groups, such as imports, exports, and nontradables. The conditions for aggregation, however, are very stringent and the possibility of estimating parameters for these aggregates is quite limited.

A more practical approach is to investigate the activities from which a resource is likely to be withdrawn in order to make adjustments in the price of the resource based on the distortions existing in the major markets affected by the withdrawal. This procedure requires some knowledge of how relevant product and factor markets operate. To the extent that the government is directly involved in allocating resources, it also requires an understanding of how its decisions are likely to be made.[9]

If there are no direct government controls on consumption and if world prices, the exchange rate, and tax and subsidy rates are fixed, changes in the quantities of tradables and nontradables consumed may be related to changes in the availability of a resource via its price and income effects by the following equations:

$$\frac{dx_i}{dl_h} = \sum_{k'=1}^{n} \frac{\partial x_i}{\partial p_{k'}^c} \frac{dp_{k'}^c}{dl_h} + \frac{\partial x_i}{\partial y} \frac{dy}{dl_h}, \tag{2}$$

$$\frac{dz_k}{dl_h} = \sum_{k'=1}^{n} \frac{\partial z_k}{\partial p_{k'}^c} \frac{dp_{k'}^c}{dl_h} + \frac{\partial z_k}{\partial y} \frac{dy}{dl_h}. \tag{3}$$

Where k' represents any of n nontradable goods including the kth, $p_{k'}^c$ is the consumer price of the k'^{th} nontradable good, y is income, and the other symbols are the same as in Equation (1).

Although it may be impossible to estimate precisely the magnitude of these consumption effects, it is at least possible to determine the signs of the more important ones. Consider, for example, the withdrawal of one unit of labor from millet production. The first-order effect is likely to be a rise in the price of millet, and a consequent decline in consumers' real income ($dp_{k'}^c/dl_h < O$ and $dy/dl_h > O$). For goods such as rice (x_i) or cassava (z_k), which are substitutes for millet ($z_{k'}$), both their cross price and in-

---

[9]Two examples already mentioned are quantitative restrictions on trade and nonfully traded inputs, discussed in Appendix A.

come elasticities are positive ($\partial x_i/\partial p^c_{k'} > 0$ and $\partial x_i/\partial y > 0$ in Equation (2), and $\partial z_k/\partial p^c_{k'} > 0$ and $\partial z_k/\partial y > 0$ in Equation (3)). As a result, the two terms in the equations tend to offset one another and little change occurs in consumption. Hence, any distortions in the markets for substitutes can reasonably be ignored. If distortions in the market for millet exist ($t_{k'} \neq 0$), the shadow price of labor would still not have to be adjusted as long as the good's own price elasticity and its income elasticity are both negative ($\partial z_{k'}/\partial p^c_{k'} > 0$ and $\partial z_{k'}/\partial y > 0$), since the two terms in Equation (3) still tend to offset one another.[10] Under these conditions, the first and second terms in Equation (1) are close to zero and no adjustment to the market price of labor (p_h) is necessary because of changes in the consumption of millet or its substitutes.

On the other hand, if labor had been withdrawn primarily from fishing, and if fish ($z_{k'}$) and rice (x_i) are complements ($\partial x_i/\partial p^c_{k'} < 0$), both terms in Equation (2) would be positive. Therefore, the market wage rate in Equation (1) should be adjusted upward if rice is taxed and downward if rice is subsidized. The same result holds for Equation (3) in the case of non-traded complements.

The third term in Equation (1), showing the effects of changes in consumption of intermediate goods, is also influenced by changes in the production of both tradables and nontradables. With fixed input-output coefficients, the impact on intermediate consumption, $x_{i'}$, is straightforward, as shown in Equation (4):

$$\frac{d\hat{x}_i}{dl_h} = \sum_{i'=1}^{m} a_{ii'}\frac{d\bar{x}_{i'}}{dl_h} + \sum_{k=1}^{n} a_{ik}\frac{d\bar{z}_k}{dl_h}, \tag{4}$$

where i' represents any of the m tradables being produced including the ith, $a_{ii'}$ is the physical input-output coefficient for the i'^{th} tradable good, a_{ik} is the physical input-output coefficient for the kth nontradable good, and $\bar{z}_k$ is the production of the kth nontradable good. The rest of the notation is the same as for Equation (1). To illustrate, a shift of labor from millet to rice production causes rice output to rise and millet output to fall ($d\bar{x}_i/dl_h > 0$ and $d\bar{z}_k/dl_h < 0$). If rice is fertilized ($a_{ii'} > 0$) but millet is not ($a_{ik} = 0$), consumption of fertilizer will rise. If the market price of fertilizer is subsidized ($t^d_i < 0$), as it usually is in West Africa, the third term in Equation (1) would be negative. As a result, other things being equal, the shadow price for labor would be less than the market price.

---

[10] In this case, millet is assumed to be an inferior good. If the opposite were true, the right-hand side of Equation (3) would be positive and the market wage rate should be adjusted upward if millet consumption is taxed and downward if subsidized.

Similarly, the change in the quantity produced of each tradable good as the result of the reallocation of a scarce resource can be expressed by

$$\frac{d\overline{x}_i}{dl_h} + \frac{\partial \overline{x}_i}{\partial l_{hi}} \frac{dl_{hi}}{dl_h} . \tag{5}$$

This change thus equals the marginal product of the resource (l_h) in the production of the ith tradable good multiplied by the proportion of this resource withdrawn from that industry. This proportion may depend on the operation of resource markets, on government quantitative controls, and on factors specific to the activity toward which the resource is being diverted. The introduction of a new cash crop, for example, is likely to draw labor from both the local labor market and farmers' other activities. If an important proportion of a resource is withdrawn from sectors that are heavily taxed or subsidized, an adjustment to the market price of the resource should be made. Thus the value of the marginal product of labor withdrawn from the coffee sector in the Ivory Coast is greater than the market wage rate because exports of that sector are taxed.[11] In other words, the fourth term in Equation (1) is positive.

Shadow Price Estimates

In this section, shadow price estimates are presented for rice output, intermediate inputs, and primary factors. In addition, the section discusses the specific procedures involved in deriving those estimates, which involve two phases. The first is an estimation of the market price of each resource and an assessment of the extent to which this price is distorted by imperfections in the markets for primary factors. The second phase makes the adjustments, if required, indicated by Equation (1). Finally, the relationship between this approach and one requiring estimation of the shadow price of foreign exchange is examined.

Rice Output

The accounting price of rice output is the world price for the relevant quality at the assumed point of consumption. This consumption point is typically a port or capital city, but in some instances estimates were made for delivery to other markets. A basic reference price for Thai 5 percent brokens, f.o.b. Bangkok, was established at $350 per metric ton

---

[11] This result obtains even if the supply of labor is infinitely elastic, making the market wage invariant with respect to taxes or subsidies. The amount of the tax equals the difference between the value of the marginal product of labor and the market wage, and it is the value of the marginal product that is treated here as labor's shadow price.

(mt) in 1975 U.S. dollars, which is reasonably consistent with both past prices and future projections (*12*). This price is based on the long-term trend, which is more important for this analysis than short-term interannual variations. It is also fairly representative of 1975–76, the latest agricultural year for which complete data are available. To this export price, $50 per mt was added to cover insurance and freight.[12] The resulting c.i.f. reference price was then adjusted to account for observed quality differences between the rice actually imported and Thai 5 percent brokens.

The adjustment for quality differences varies by country. The discount for 25–35 percent broken rice is based on historical differences between its price and the reference price for Thai 5 percent broken rice. This discount averages about 30 percent.[13] The resulting c.i.f. price is reasonably consistent with the price of rice imported into the Ivory Coast and can also be used for Sierra Leone and Mali.[14] Liberia imports rice, mostly from the United States, which is of higher quality than Thai 5 percent brokens. Hence, a premium rather than a discount was applied to the price of Thai 5s. Retail price comparisons, however, indicate that imports are of better quality than domestically produced rice, so a $30/mt discount was then subtracted from the price of imported rice to arrive at an equivalent c.i.f. price for domestic rice. Senegal, on the other hand, imports two major qualities of rice—100 percent brokens for the mass market and 100 percent whole grains for a small, upper-income market. Since the rice produced in Senegal is of a quality intermediate between the two, the equivalent c.i.f. price of this local rice was estimated as the weighted average of the prices of the two qualities of imported rice, where the weights are the proportions of brokens and whole grains found in domestically produced rice.[15]

[12] The transport margin is a rough estimate based on shipping conference rates in effect between West Africa and various parts of the world in 1974. It does not allow for discounting below these negotiated rates.

[13] During the period 1955–74, the Thai export price for 25 to 35 percent brokens averaged 32 percent less than the price for 5 percent brokens. This discount has been somewhat smaller since 1974, ranging from 16 to 26 percent (*14*).

[14] Additional adjustments were required for Mali to account for the cost of transport between the port and Bamako. These are explained below.

[15] The equivalent c.i.f. price of domestically produced rice equals the observed c.i.f. price only when the two types of rice are of comparable quality. If domestic rice is of higher quality, its equivalent world price will exceed that of imports, and vice versa. This usually implies that locally produced rice is a nontradable, i.e. its domestic price lies between the f.o.b. and c.i.f. prices of rice of comparable quality traded on the world market. In the absence of government restrictions, the failure to import rice of comparable quality implies that this rice can be obtained at a lower price locally than on the world market. Therefore, the quality-adjusted world market price tends to overestimate the true social value of the domestically produced rice. The approach used in Senegal—to break down locally produced rice into its two tradable components—is valid only as long as there is an import market for each quality.

For imported rice that is transported from the coastal ports of Africa to the interior countries, the cost of transport and handling has been added to the c.i.f. price at the port to obtain a c.i.f. price at the interior border. On the other hand, for a country that is a potential exporter within the region, the cost of transport and handling between the frontier of the exporting country and the consumption center of the importing country has been subtracted from the price to wholesalers in that consumption center to yield a f.o.b. price at the frontier of the exporter. For example, the relevant shadow price of rice exported from Mali to Bouaké in the center of the Ivory Coast was calculated by first adding the cost of transport from Abidjan to Bouaké to the c.i.f. price, Abidjan, in order to obtain a price for imported rice in Bouaké. From this was subtracted the cost of shipping rice from the Malian frontier to Bouaké, which gives an f.o.b. price at the frontier. This price is the one that would make Malian rice competitive with Ivorian imports from other sources.

Within countries, various consumption points have been used to analyze the role played by transportation costs in providing natural protection against imports or in depressing the market for locally produced rice. The relevant shadow price is estimated by adding to the c.i.f. price the tradable component of the cost of transport and handling from the port or frontier to the internal consumption point. The value of primary domestic factors used in the internal distribution of these imports is, in turn, subtracted from the cost of distributing the domestic rice that replaces these imports, since these factors are saved by import substitution.

The results of these adjustments are shown in Table B.1, which gives the central values of the shadow price of rice of several different qualities in selected West African locations. The individual country studies examine the sensitivity of results to changes in these prices arising from variation in the world reference price.

Intermediate Inputs

The social accounting prices of intermediate inputs are equal to their border prices adjusted for the cost of internal transport and handling. If the input is tradable and likely to be imported, the cost of moving it from the frontier to where it is used has been added to its c.i.f. price. If the input is an exportable, the saving in transport and handling costs resulting from local use of the input has been subtracted from the f.o.b. price. The costs of internal transport and handling are treated like any other intermediate inputs, i.e. they are broken down into their indirect tradable and primary factor components, all valued at appropriate shadow prices. Customs duties and other indirect taxes and subsidies are excluded from the shadow price estimates because they are transfers among economic sectors, not real resource costs.

TABLE B.1. *Shadow Prices of Rice in Selected West African Locations*
(1975 U.S. dollars/mt)

Location		Quality	
	Whole grains	25–35 percent brokens	100 percent brokens
Ivory Coast:[a]			
Abidjan (c.i.f.)		300	
Abidjan (wholesale buying)		305	
Bouaké (wholesale buying)		329	
Forest farmers (consumer cost)[b]		385	
Savannah farmers (consumer cost)[b]		392	
Liberia:			
Monrovia (c.i.f.)	344	314	
Mali:[c]			
Bamako (wholesale buying)		364	
Ségou (wholesale buying)		369	
Mopti (wholesale buying)		378	
Mopti (consumer cost)[b]		398	
Gao (wholesale buying)		393	
Sikasso (consumer cost)[b]		361	
Senegal:[a]			
Dakar (c.i.f.)	400		250
Dakar (wholesale buying)	407		257
Saint Louis (wholesale buying)	445		295
Ziguinchor (wholesale buying)	452		302
Nianga (consumer cost)[b]		346	
Matam (consumer cost)[b]		346	
Lower Casamance (consumer cost)[b]		346	
Sierra Leone:[d]			
Freetown (c.i.f.)		300	
Freetown (wholesale buying)		309	

[a] The exchange rate is 250 CFA francs per U.S. dollar.
[b] Consumer cost equals the cost to farmers of the imported rice they must purchase at a market and transport to their homes if they were to replace rice that they produce and hand pound themselves.
[c] The exchange rate is 500 Malian francs per U.S. dollar.
[d] The exchange rate is 1.0 Leones per U.S. dollar.

Some intermediate inputs, such as composite fertilizers, selected seeds, and traditional tools, are not tradable but are manufactured locally. The social costs of these domestically produced intermediate inputs are calculated by disaggregating their production costs into indirect tradable and primary factor components, all of which are valued at appropriate shadow prices.

Unskilled Labor

The first step in estimating the shadow price of unskilled labor is to obtain data on market wage rates. The information from each of the countries included in the study reveals both the high degree of integration of the West African labor market and its complexity. Observed and estimated market wages vary systematically according to region. They are highest along the coast, especially in the Ivory Coast, and lowest in the more arid, interior areas. Within regions, the labor costs estimated on a per day basis are higher for males than females and higher for adults than children. Wage rates also depend on the labor task (for example, land preparation and transplanting cost more than other rice tasks), the method of payment (for example, piecemeal, per day, per month), and the inclusion of in-kind benefits such as meals. Other factors influencing wages include the length of the day, the extent of supervision required, and the magnitude and incidence of search costs.

On the other hand, there appears to be little, if any, seasonal variation in wage rates. Despite the impossibility of double-cropping in most of the region, opportunities for useful employment, as well as nonmarket activities and leisure, throughout the year are apparently sufficient to cause wages to remain relatively constant.[16]

A base wage was established in each region and then adjusted as necessary to account for some of the complexities involved in the determination of market wages. Where possible this estimated market wage was compared with the per unit labor cost found by dividing commercial contract rates by the number of days required to perform the relevant operation or with the returns per man-day earned by farmers engaged in subsistence food production. For example, the wage paid in irrigated rice in the forest zone of the Ivory Coast was estimated by adding to the nominal daily wage of 330 CFA francs the following items: 75 CFA francs for a meal provided by the employer, 25 CFA francs for travel costs paid by the employer, and 20 CFA francs for the implicit cost of supervision by the farmer. These costs total 450 CFA francs, an amount consistent with rates paid for contract labor. In the savannah zone to the north, nominal wages average about 275 CFA francs, and the only additional adjustment is for meals provided by the employer, yielding a total wage of about 350 CFA francs. A wage of this amount is similar to the return per man-day that can be earned in subsistence food production in that region.

The difference in wage rates between the forest and savannah zones of the Ivory Coast conforms to the difference in the relative favorability of ecological conditions in the two zones and to the pattern of north-to-south migration characteristic of this part of West Africa. Similar relationships exist, in fact, throughout the West African region, as suggested by Table B.2, which gives estimated average daily wage rates for adult males engaged in rice cultivation in several countries. The highest wages are paid in the forest zone of the Ivory Coast, which not only has ecological condi-

---

[16] Numerous interviews with farmers confirmed the wide range of ways in which they can occupy themselves during the dry season. Some of these include migration to the forest zone to harvest coffee and cocoa or to clear land, repairing of huts and tools, fishing either at home or elsewhere, and participation in social activities reserved for the off-season.

TABLE B.2. *Daily Agricultural Wage Rates for Adult Males in Selected West African Locations, 1975–76*
(U.S. dollars/man-day)

Location	Daily wage rate	Location	Daily wage rate
Ivory Coast:		Senegal:	
Forest	1.80	Fleuve	1.00
Savannah	1.40	Casamance	1.20
Liberia	1.25	Sierra Leone:	
Mali:		South	0.70–0.80
Ségou, Mopti	0.80–1.00	North	0.52–0.80
Office du Niger,			
Sikasso	1.20–1.40		

NOTE: Wages have been converted from local currency at the rates of 250 CFA francs = 1 U.S. dollar, 500 Malian francs = 1 U.S. dollar, 1 Liberian dollar = 1 U.S. dollar, and 1 Leone = 1 U.S. dollar.

tions favorable for cash crop agriculture but also has considerably exploited those conditions in recent years. The agricultural sectors in the other countries with coastal forest zones—Liberia and Sierra Leone—have grown less rapidly than in the Ivory Coast, which is reflected in their lower wages. In addition, there has been net emigration from these countries, especially from Sierra Leone.

The major migration that occurs, however, is from north to south. This pattern is indicated by the progressive increase in wage rates from the region around Mopti in Mali, one of the least favored ecological areas, through Sikasso in southern Mali and the savannah zone of the Ivory Coast, to the Ivorian forest zone. The only exceptions are Ségou, with its more favorable climate but higher population density than Mopti, and the Office du Niger, which has a large demand for outside labor because it is a large-scale irrigation scheme in a region with a low level of population. Senegal, though not closely linked to the other countries by migration flows, nonetheless also has a structure of wage rates related to ecological conditions and to the cost of traveling to Dakar and other labor markets.

Wage rates paid to men, women, and children often differ.[17] This distinction is useful, however, only if agricultural tasks are done specifically by one or another of these labor groups. In the Ivory Coast and Sierra Leone, women are generally paid about three-quarters of the male wage rate, except for tasks such as transplanting. The wage rate for children in Sierra Leone is about one-half that of men.

[17] The reason why women and children are paid less than men is presumably because they are less efficient, but the difference in efficiency varies enormously among tasks. Men, for example, clearly have an advantage in heavy clearing operations, but women are often better at weeding and some harvesting tasks. At the margin, however, the market wage is probably the best indicator of the value of foregone output resulting from the employment of each type of worker.

In the empirical calculations of this study, the only instance in which a distinction between age and sex categories has consistently been made is for bird watching, where the estimated shadow wage equals one-half of the male wage rate.[18] In addition, most operations—except transplanting—traditionally performed by women in the Ivory Coast have been valued at a rate about three-quarters of the value for men.

The next step is to investigate the extent to which market wage rates are distorted by imperfections in the labor market. One distortion is an official minimum wage that maintains wage rates above their market clearing levels. Official agricultural wages have been decreed in all the countries but do not generally appear to inflate wage rates actually paid by farmers. In the Ivory Coast, on the contrary, actual wages paid in the rural sector are greater than the officially established minimum. In other countries the reverse is often true, although the official wage is usually paid by public agencies.[19]

The analysis thus far indicates that the market wage rate approximates the shadow price of unskilled labor reasonably well if corrected for the kinds of product market distortions shown in Equation (1). Most of the evidence accumulated throughout the region suggests that both family and hired labor used for rice production are withdrawn from the cultivation of traditional food crops, such as cassava, yams, millet, and sorghum. Even labor initially taken from cash crops such as coffee and cocoa is likely to be replaced by workers coming out of the subsistence sector, who often travel long distances to find wage employment. Since these nontradable foods are neither taxed nor subsidized, no adjustment is necessary for distortions in their markets. On the other hand, a decline in their consumption resulting from labor being withdrawn from their production could lead to an increase in the consumption of rice—a food that is taxed in most countries. As noted earlier, however, the income and price effects move in opposite directions and thus tend to offset one another. Finally, the effect on the consumption of other goods and services of a decline in subsistence food production is not likely to be very important. The farmer simply replaces this food with part of the rice he produces in the new activity under consideration. Consequently, little or no adjustment seems necessary for distortions in the consumer markets.

[18] Because of the large uncertainty in the number of days required, in the number of hectares a single person can watch, and in the relative frequency of child and adult participation, this adjustment has little significance.

[19] A second possible imperfection in the labor market exists if wages paid to members of one ethnic group are lower than those paid to others equally qualified. This possibility was examined by Rader (19) in the Gagnoa region of the Ivory Coast, where different activities are pursued by local and immigrant ethnic groups. A series of interviews revealed that among workers paid by the day, there was no evidence of wage discrimination between ethnic groups.

There are, however, two situations in which the market wage may fail to approximate the shadow wage reliably. In both instances, the shadow rate may be overvalued. The first situation arises when family instead of hired labor is used, and the second occurs when hired labor has immigrated from abroad.

Since the market wage equals the supply price of the marginal man-day withdrawn from other employment, it indicates not only the value to the worker of his marginal product in that employment but also the transfer price necessary to induce him to move to a new job. When the change in output is brought about by shifting labor within the family farm, the cost of reallocating labor is less and requires a smaller return than the market wage to induce the shift. The problem may not be very great if agriculture is highly commercialized and the rural labor market well developed. It is more severe in traditional, smallholder agriculture which is incompletely linked to the cash economy. This makes comparisons between modern and traditional techniques difficult unless the results are relatively insensitive to changes in wage costs.

The second cause of possible overvaluation is illustrated by the Ivory Coast, where an important part of the labor force comes from other countries, particularly those to the north. Most of the wages paid to these workers, whether spent on consumption or sent home as remittances, is a claim on Ivorian goods and services and thus a real social cost. But part of the expenditures of foreign workers and their families are taxes on these products, which result in a transfer from the foreign workers to the local economy. The shadow wage for these workers is therefore less than the market wage. The empirical importance of this discrepancy, however, is not great. It is estimated that taxes comprise about 10 percent of the expenditures of foreign workers. Since these laborers amount to less than 25 percent of the total rural work force, the adjustment would be no more than 2.5 percent—well within the margin of error generally accepted in the calculations.[20]

The market wages shown in Table B.2 have thus been used as estimates of the shadow price of labor in each of the country studies. Comparative advantage in rice production depends, however, not only on current shadow prices but also on expectations concerning the future. Cash crop agriculture in West Africa has always drawn upon the traditional food crop sector for its labor supply. Although population in rural areas is still growing, the demand for labor may eventually begin to outstrip its supply and put upward pressure on real wage rates. This eventuality is particularly acute in the Ivory Coast, where past economic expansion has been based to

[20] A note has been prepared that documents these calculations and is available from the authors.

a considerable extent on immigration of foreign labor. Although there is no strong evidence that rural money wages in the Ivory Coast have risen faster than the cost of living, working conditions seem to have improved and non-pecuniary benefits have been increased. Search and travel costs may also have been shifted from workers to employers. Any upward pressure on wage rates in the Ivory Coast should have implications for relative wages over much of West Africa because of the extent to which labor markets in different countries are linked. Hence, it is particularly important to analyze the sensitivity of the empirical calculations to future increases in the labor costs.

Skilled Labor

The cost of skilled labor is more likely to be influenced by government policy than that of unskilled labor because direct government intervention affects a larger proportion of the skilled labor force. Most West African countries have established wage rate schedules according to skill categories that set minimum salaries for hiring by both the formal private sector and the government. The diversity of skill categories and lack of empirical evidence make it virtually impossible, however, to establish a single ratio that could take account of distortions induced by the government and adjust market wages of skilled labor to shadow price equivalents. If any distortion does exist, it seems more likely that the market wage exceeds the shadow wage, especially in lower-skill categories, causing an overestimation of social costs.

Correction for distortions in product markets is made difficult by not knowing from which sectors skilled labor is withdrawn. Some would be withdrawn from public service, the output of which is not subject to taxation. A large part of the rest would probably come from the tertiary sector, including formal and informal commerce, transportation, and construction. These activities are not tradable, but their imported inputs and their outputs are sometimes taxed. Finally, some skilled labor may be withdrawn from the import substitution sector for which the border prices of output generally are less than corresponding domestic prices. On balance, the deviation due to product price distortions of the shadow price of skilled labor from the market wage depends on the relative magnitude of offsetting terms in Equation (1). In any case, the effect on the empirical results is minimal because skilled labor is a very minor component of total costs.

Capital

Capital inputs consist of investments with economic lives exceeding a single accounting period and working capital to finance current operations. Determining the social cost of both involves selecting a social rate of discount at which to value the services of capital goods or the opportunity cost of the working capital. In addition, the market prices of the capital goods must be corrected for distortions in product markets.

In principle, the shadow rate of interest equals the rate of return on marginal public sector investments. In a perfectly functioning capital market, this marginal rate of transformation of present into future income equals the marginal rate of substitution of present for future consumption in both the public and private sectors. In such situations, the market rate of interest is an adequate guide to the social opportunity cost of capital services.

In contrast to the well-functioning rural labor market, the market for capital in agriculture appears to be highly segmented. Such segmentation, moreover, represents a fundamental imperfection in the capital market which is likely to persist. It is not sufficient, therefore, to calculate a single shadow rate of interest that would be expected to prevail in a perfect and undistorted capital market. A more complex approach is required, based on the reasons for this segmentation.

The market for agricultural credit is segmented for three main reasons. First, major agricultural investments are generally undertaken by the public sector, usually with a large share of foreign funds obtained on concessional terms. These funds are almost always tied to specific projects, which, because the aid is concessional, may not be required to have the highest marginal rates of return. In addition, it is unlikely that the supply of foreign funds, especially concessional aid, is infinitely elastic. Furthermore, foreign aid is limited by the capacity of these countries to absorb public investment projects. Because concessional aid is limited and rationed to certain projects, there is a divergence in the cost of capital between projects financed with foreign aid and those funded out of government revenues or foreign borrowing at commercial rates. Since the sources and terms of foreign aid vary greatly, effective rates of interest vary as well.

Second, there is a divergence in interest rates between formal and informal sector activities. Interest rates for private borrowers in public development projects, for example, are lower than for those obtaining funds from the informal credit market. This divergence results primarily from reductions in the transaction costs of the capital market. Such reductions occur because the administrative structure of the projects makes it easier to monitor loans, to enforce repayment, and otherwise to reduce the risks and costs of lending. In addition, project credit is usually restricted to certain types of purchases, such as animal traction equipment and modern intermediate inputs. There may also be some wealthier farmers who have access to commercial credit through the modern banking sector on terms that are similar to those offered by the public projects for much the same reasons.

Finally, there may be a divergence in interest rates in the informal credit market because the buying price of capital exceeds its selling price. This divergence should be especially marked between implicit interest rates on self-financed investments and the higher rates paid on short-term loans borrowed from local moneylenders. The divergence probably arises less because of monopolistic lending than because of high transactions costs, including defaults, in the traditional market where loans are small and numerous, information is costly, and sanctions are difficult. The situation is analogous to the difference in wage rates for family and hired labor.[21]

Segmentation of capital markets creates an array of nominal market rates of interest shown in Table B.3. At the top end of the scale, credit is available to some governments from the Eurodollar market on commercial terms. Somewhat below this is credit received from the international development banks. Most public investment, however, has been financed with concessionary foreign aid. In the Ivory Coast rice sector, the average rate of interest paid on past loans has been 3.7 percent, which is substantially below the average rate now generally available to that country. Senegal, on the other hand, as one of the poor Sahelian countries, has paid a relatively low average interest rate of about 3.2 percent and can probably expect to continue to do so. The data thus indicate a plausible range for nominal interest rates on projects financed from these sources of 3 to 13 percent.

The second market rate of interest applies to farmers borrowing from the banking system or participating in public development projects and equals the rate charged on loans by commercial or public development banks. Actual rates paid by farmers may differ from this, however, because most public projects have subsidized purchase prices of capital goods, low rates of interest, and special repayment periods.

Other capital costs are for investments made in traditional agriculture or for the kinds of inputs not furnished by projects. Even though participating fully in a development project, farmers usually must finance all the working capital required for family and hired labor. The only sources are family savings, borrowings from friends and relatives, and loans from the traditional credit market. Interest rates observed in the traditional credit market are very high, sometimes in excess of 100 percent per annum.[22] Part of this high rate is a premium to compensate for default and delayed payment on some of the loans. Part of it may also be due to monopoly elements, but the scanty evidence that exists indicates that these are not

[21] Among the numerous discussions of this problem for various parts of the world, see *6, 11,* and *17.*

[22] Borrowing on the traditional money market is expensive but does not carry any implicit costs associated with increased social obligations, as might be the case if money is borrowed from friends.

TABLE B.3. *Nominal Rates of Interest for Public and Private Borrowers of Various Sources of Capital*
(Percent)

Source	Interest rate Public	Private	Source	Interest rate Public	Private
Ivory Coast:			(Mali:)		
Development Bank:			Commercial banks	—	13.0
Financing for crop			Private foreign[b]	—	13.0
purchases	7.0%	—	Traditional	—	25.0
Other uses	10.0	13.0%	Senegal:		
Commercial banks	—	13.0	Development bank	7.5	13.0
Concessional foreign			Public marketing		
aid	3.7[a]	—	agency	—	13.0
Private foreign[b]	13.0	13.0	Commercial banks	—	13.0
Traditional			Concessional foreign		
borrowing	—	30.0	aid	3.2	—
Traditional lending	—	20.0	Private foreign[b]	13.0	13.0
Liberia:			Traditional	—	25.0
Development bank	—	12.0	Sierra Leone:		
Commercial banks	—	12.0	Trading company		
Private foreign[b]	13.0	13.0	(medium-term)	—	43.3
Mali:			Commercial banks	—	12.0
Development bank	7.5	13.0	Private foreign[b]	—	13.0

[a]This rate is based on a weighted average of concessional lending to the Ivory Coast for rice projects during 1963–76.

[b]The interest rate on foreign lending is for funds denominated in U.S. dollars.

very important in West Africa.[23] Perhaps the most important element in these high interest rates, however, is the high unit costs associated with gathering information and transacting loans, each of which is small in magnitude. These transaction costs have the effect of raising the real cost of capital in the traditional sector well above its cost in formal capital markets. The empirical evidence on traditional interest rates in West Africa is very slim, but estimates of effective market rates have been made by adjusting observed rates for default and delayed payment.[24]

The fundamental nature of this segmentation implies that social discount rates will also vary among markets. The shadow rates of interest used in the empirical calculations are given in Table B.4. These rates have been adjusted for expectations concerning inflation. The adjustment does not equal the current rate of inflation, which is quite high, but reflects the longer-term historical rate up until 1972 plus an additional adjustment of 2.5 percentage points to account for the general upward shift in interest rates all over the world after that date.

[23] For a review of the empirical evidence, see 23.

[24] In Sierra Leone, for example, the cost of credit used to finance the purchase of fishing equipment has been estimated at 43 percent after some incomplete adjustment for default (15). Without any adjustment, nominal interest rates for seasonal credit in the Gagnoa region of the Ivory Coast are 60 to 80 percent.

TABLE B.4. *Real Shadow Rates of Interest for Different Types of Capital*
(Percent)

Country	Shadow interest rates			
	General	Publicly financed	Traditional	Commercial
Ivory Coast	—	5	15–25[a]	8
Liberia	15	—	—	—
Mali	—	2.5	20	8
Senegal	—	2.5	20	8
Sierra Leone	—	3.5	20–40[b]	8

[a] The first rate is the selling price of capital and the second is the buying price.
[b] The first rate applies to three-to-five-year investments in rice mills financed by traders in larger towns; the second pertains to investment in farm tools and working capital.

In estimating the shadow price of public capital, the relative importance of future sources of credit is crucial. For the Ivory Coast it seems likely that relatively little foreign aid will be available on terms as concessionary as in the past. On the other hand, the country will probably not have to depend solely on commercial loans for its credit. A moderate discount on the Eurodollar rate might yield 10 percent as a central nominal value. This is consistent with the interest rate on credit now available from the agricultural development bank. The Sahelian countries, on the other hand, can probably anticipate concessional financing for many years. As more investment goes into larger infrastructure projects, however, terms will probably harden somewhat. A shadow price of public capital for Mali and Senegal of 7.5 percent as a central nominal value therefore seems reasonable.[25] Estimates for the other countries should probably fall within the 7.5–10 percent range.

The shadow rate of interest for formal sector credit to private farmers depends on its source. Most farmers who participate in public projects buy their inputs on credit, and the government supplies the capital used to finance the purchase of these inputs. Since most of these credit programs are financed from the same sources as are the public investments that accompany them, the shadow price of capital should be the same. If loans are obtained from commercial banks, on the other hand, commercial shadow rates of interest should apply.

Shadow rates of interest for traditional credit are based on market rates adjusted for expectations concerning inflation. In most countries these market rates have rather conservatively been estimated at 25 percent, with a spread of 10 percentage points between implicit borrowing and lending rates when this appears to be relevant. In the Ivory Coast, where a substantial amount of capital in the traditional sector is invested in cof-

[25] The approach taken here is national rather than global in perspective, since the cost of capital to the borrower is used rather than the social value to the world of the output or consumption forgone from wherever that capital is removed.

fee and cocoa production, interest rates could be adjusted upward because of the export tax on these crops. Given the great uncertainty of the original estimates and the relatively small magnitude of the adjustment (2–3 percentage points), however, such fine-tuning appears inappropriate. Since most marginal additions of capital outside the traditional sector come ultimately from foreign sources, no other adjustments have been made for distortions in the product markets.

There is some arbitrariness in these estimates, but they at least approximate the reality of a highly segmented capital market. In Liberia, however, the capital market is very poorly developed, and in the absence of better data, a real rate of interest of 15 percent was chosen as the shadow price of capital. Although this is substantially greater than the average cost of capital in most of the other countries, it is not inconsistent with Liberia's heavy dependence upon the traditional credit market.

Land

The shadow price of land used for rice production is defined as its return, expressed in equivalent world prices, from its best alternative use outside of rice cultivation. This definition allows different techniques of producing rice to be compared without the cost of one technique being dependent on the cost of others.[26] In contrast to the other primary factors, the shadow price of land is not estimated using its market price as a reference because the market for land is very poorly developed in most areas. Instead, the return to land is calculated as a residual rent after deducting the costs of intermediate inputs and all other primary factors.

The value of this rent depends on three things. The first is whether land of the type required for a given technique of rice cultivation is scarce.[27] In general, much of West Africa is sparsely populated and land appears to be in abundant supply. Of the countries studied, only Sierra Leone has a population density above 25 persons per square kilometer. The traditional, extensive techniques of crop production employed in the region, however, require large areas of fallow to reconstitute soil fertility. Although the shortage of cultivable land is not yet pressing in most of the areas studied, population is growing rapidly in some of the countries such as the Ivory Coast. Thus even if a shadow price of land is estimated at zero under existing conditions, relative profitability may change in the future with increasing land pressure.

Even if land is not used for crop cultivation, it may still have economic

[26] Such a problem could arise, for example, if there were two techniques and the shadow price of land used for one was calculated by estimating its return in the other, and vice versa.

[27] Frequently, land suitable for one technique or crop is not desirable for others. Techniques involving pumping, for example, can usually be undertaken over a wider area than those using natural flooding.

value. Forests can be cut for timber, and wooded land of any sort may be useful for firewood and charcoal. In addition, palms growing in swampland produce palm wine and building materials, and savannah land produces forage for livestock. Calculations for the Ivory Coast, however, indicate that virtually all the revenue earned by collecting, processing, and distributing these commodities, with the exception of commercial forestry, accrues to labor and intermediate inputs, leaving nothing left over for land rent.[28]

Land scarcity could also arise in the Office du Niger in Mali. There it is not the land per se that is scarce but rather the land located near the major irrigation canals built 40 years ago which are treated as sunk investments. Much of this land is not currently in use, primarily because the additional irrigation infrastructure necessary for its cultivation has not yet been constructed. The land that can be irrigated, however, probably does have a substantial scarcity value and not necessarily just in rice production. It may well be that the land could be used most profitably for the cultivation of long-staple cotton.[29] Sugarcane might also be a possibility. Unfortunately, data are not available with which to calculate the residual return to land in these alternative uses.

In addition to physical scarcity, land may also acquire an economic rent because of site value. Even though land might generally be available throughout a region, it may be relatively scarce in the vicinity of large towns and major transportation routes. In this study, site value within a given producing region is ignored, though the location of each region in relation to consumption centers and the major ports is an important determinant of the cost of transporting rice and the inputs used in its production.

Finally, the shadow price of land may be affected by the way in which it is treated as a natural resource. If it is considered to be a renewable resource, extensive methods of production that lead to permanent degradation, acidification, and erosion of the soil have a greater social cost than those that do not. The bias that results from not taking these externalities into account may not be great, however, if land is currently abundant, since the discounted social cost of this resource depletion depends on the social rate of discount and on the length of time before scarcity occurs.

Water

Water is clearly an important factor of agricultural production, but its shadow price depends on both the ability to allocate or control it and its

---

[28] Similar calculations have not been performed for the other countries. Certainly in the Sahelian countries the shortage of firewood is becoming a problem. The available information, however, does not permit dealing with this problem except through sensitivity analysis.

[29] The best indication of this is the importance of cotton cultivation in the Gezira Scheme in the Sudan, where ecological conditions are similar to those of the Office du Niger.

scarcity. Without some form of control, water has no shadow price regardless of its scarcity because it cannot be varied in response to its marginal social value. The land that receives the water may have an accounting price, but the water does not.[30]

Therefore, most of water's economic value in West African rice production is created through capital investments in control systems for allocating available water supplies. Pumping, for example, brings otherwise valueless river water to the fields, where it has value. If this water is being optimally allocated, that value equals the cost of constructing and operating the pumps and therefore is a return to these inputs rather than to the water itself. Hence, even when water control is provided by new irrigation investments, water still has no shadow price. Only in the case of existing investments, the capital costs of which may be considered sunk, might a shadow price be assigned to water alone if the amount that can be delivered to the fields is limited in relation to cultivation needs.[31]

Water should also be shadow-priced if the growth of irrigation demands on a river or other natural water system reaches the point at which water taken from that system becomes scarce. At present, this situation does not appear to exist, except possibly during the dry season in the Senegal River Valley in some years when water being taken from the river upstream causes greater salt incursion downstream. The data are not yet available to estimate this cost, but sensitivity analysis can be used to show the economic effects of increased water scarcity.

Foreign Exchange

The approach used in this study to shadow price commodities and factors of production differs in one important respect from that encountered in most empirical studies of investments in less developed countries, including those of the Asian Rice Project.[32] These other studies most often begin by classifying goods and services into two categories: those whose shadow prices in terms of foreign currency are equal to their domestic currency border prices at the official exchange rate and those whose shadow prices are appreciably different from this. The former category consists of tradable goods and services and is made commensurate with

[30] Water everywhere, of course, has a very high value measured in terms of total consumer surplus. Rainwater used for upland cultivation, for example, may be very scarce in the sense that farmers would be willing to pay a good deal for it. But it cannot be allocated, except insofar as land is distributed. Its economic value in this case is coincident with that of land.

[31] This could be the case in the Office du Niger, where substantial capital investments can be considered sunk, but water is not scarce in relation to current levels of cultivation. If cultivation were to be substantially expanded, however, water could become scarce and should then be priced or rationed accordingly.

[32] See, for example, Pearson, Akrasanee, and Nelson (18).

the latter group of nontradables, which includes primary factors, by converting foreign currency prices to domestic currency equivalents using the shadow rate of exchange. The shadow exchange rate thus emerges as a crucial relative price between tradables and nontradables.

A shadow exchange rate is not required, however, to calculate social profitability with the present methodology. When used to estimate the shadow price of an output, intermediate input, or primary factor, Equation (1) takes into account the effects of distortions in product prices caused by government taxes and subsidies including those on trade. The resulting shadow price, expressed in terms of foreign currency, equals the opportunity cost of the resource on the world market adjusted for changes in consumption and production of other goods and services whose prices are distorted by taxes and subsidies. Rather than a single shadow rate of exchange, there is a multiplicity of shadow exchange rates, each defined as the ratio of the shadow price of the good or service to its domestic market price.[33]

The foreign exchange shadow prices estimated from Equation (1) differ from other formulations in several respects.[34] First, the effects of consumption taxes and subsidies, as well as of distortions affecting international trade, are included. Thus there are separate terms for consumption and production of tradables, as well as a term that shows the effects of distortions in the markets for nontradables. Second, intermediate inputs are explicitly introduced into the analysis, thus relating the result to the concept of effective rather than nominal rate of protection. Third, the formula is expressed in terms of consumption and production rather than of imports and exports.

Adjusting the market prices of resources in this way requires that the main impact of resource reallocation should be felt primarily in a few closely related markets. This is assured by the assumption that the reallocation is sufficiently small that factor prices remain unchanged. Since prices of tradables are also constant, the only prices that change are those of nontradables other than primary factors. Nontradable intermediate inputs pose no problem, since they are assumed to be produced at constant costs and can be broken down into tradables and primary factors. Prices of nontradable consumer goods, on the other hand, can change in response to alterations in demand conditions.[35] It is reasonable to assume, how-

[33] The correspondence of this approach to the well-known Little-Mirrlees criterion is apparent (16). The approach in this study differs, however, in that the shadow exchange rates thus derived avoid considerations involving distortions or imperfections in the factor markets. On this point see Scott (20).

[34] See, for example, Bacha and Taylor (1).

[35] Price changes can occur here even though technical coefficients and tradables and primary factor prices remain constant because of the possibility of shifting techniques for producing the same good. This is most likely to occur under changing demand conditions when there are constraints on altering the level of output using the same techniques.

ever, that under West African conditions, the impact of most of these demand changes is limited to a few related markets.

There is one way, however, in which changing demand conditions could be transmitted to a broad segment of the economy. This would occur if there were changes in foreign exchange holdings induced by the resource reallocation that resulted in widespread changes in consumption. If this were true to an important extent, these changes would have to be estimated and incorporated into Equation (1). This would require determining how any net foreign exchange earned or saved might be allocated and what the response of consumers to this allocation would be.

An alternative approach would be to value primary factors in terms of foreign currency at the shadow exchange rate which incorporates these changes in consumption, in order to make them commensurable with the border prices of traded goods and services. The elasticity approach to estimating the shadow price of foreign exchange in this way is used fairly extensively.[36] It suffers, however, from a number of important defects. First, it ignores the effects of consumption taxes on both tradables and nontradables. Second, it assumes that there is initial balance-of-trade equilibrium (though this assumption is not necessary if the formula is complicated somewhat). Third, and more important, the relevant elasticities should be general, not partial, equilibrium in that they should show the impact of changes in income and prices of nontradables in addition to that resulting from changes in the availability of foreign exchange. In practice, however, only estimates for partial equilibrium are available. Fourth, an even more severe problem exists if effects on consumption and production of a change in foreign exchange availability occur for reasons other than movements in prices. Foreign exchange accruing to the government, for example, may be allocated independently of market considerations. The use of price elasticities would not be appropriate in this case.

Empirical estimates of the shadow price of foreign exchange using the elasticities approach indicate that overvaluation of official exchange rates in these countries is probably not more than 25 percent. To the extent that foreign exchange accrues to the government and is spent on public inputs, the degree of overvaluation is even less because of the absence of taxes on most of these inputs. Therefore, the method outlined in this paper and used in the empirical calculations should yield more reliable results than the elasticity approach.

---

[36] The equation for estimating the shadow exchange rate using the elasticity approach is derived as a special case of Equation (1) in the Annex to this Appendix.

Summary

The estimation of shadow prices used in this study is based on the concept of maximizing economic efficiency. Shadow prices of tradable inputs and outputs equal their prices on the world market. For domestic factors of production, shadow prices are measured as the value in world prices of national output forgone by shifting these factors out of their alternative uses.

Where factor markets function well, the accounting prices for domestic resources have been estimated by adjusting market prices for government-induced distortions in related product markets. These distortions include those affecting the final consumption of tradables and nontradables that are close substitutes for or complements to rice and those affecting the producer prices for tradable outputs and intermediate inputs affected by resource shifts into rice.

The shadow price of rice output has been estimated at $300–$400 (in 1975 U.S. dollars) per metric ton depending on quality and point of delivery. Although fluctuations in world prices are substantial, this range is consistent with long-run projections and should serve as the basis for decisions regarding rice development efforts.

The shadow price of labor, in most instances, is well approximated by market wage rates varying from U.S. $.60 to U.S. $1.80. There appear to be few imperfections in the labor market, and most labor is withdrawn from subsistence production that is neither taxed nor subsidized. Secondary consumption influences appear minimal because of offsetting price and income effects. In the traditional sector, however, the use of the market wage probably overestimates somewhat the shadow price of labor because it includes transfer costs not actually incurred.

The shadow price of capital varies because of highly segmented capital markets caused by foreign concessional aid, the rationing of public and modern sector credit, and high transaction costs in the traditional credit market. In general, real rates of interest vary from 2.5–5 percent for public funds to 15–40 percent in the traditional sector. No adjustments are made for distortions in product markets because marginal additions to public capital are obtained primarily from abroad, and the corrections that should be made to traditional interest rates are small relative to the magnitude of error in these estimates.

Neither land nor irrigation water is scarce in most circumstances, so that shadow prices for these inputs are assumed to equal zero. Sensitivity analysis is important here, however, because scarcity will increase in the future.

The methodology employed in this study does not require estimation of a separate shadow price of foreign exchange. Such a rate could be estimated using the methodology, however, but is unlikely in any country to be more than 25 percent in excess of the official exchange rate. In most instances it would be much less.

Annex

The elasticity approach to estimating the shadow price of foreign exchange is outlined below. If, as a special case, $t_i^c = t_i^d$ and $t_k = 0$, the shadow price of a unit increment of foreign exchange can, from Equation (1), be expressed as

$$r^* = r\left[1 + \sum_{i=1}^{m} p_i^w t_i(dx_i + d\hat{x}_i - d\bar{x}_i)\right] \tag{A-1}$$

$$= r\left[\frac{\sum_{i=1}^{m} p_i^w(1 + t_i)(dx_i + d\hat{x}_i - d\bar{x}_i)}{\sum_{i=1}^{m} p_i^w(dx_i + d\hat{x}_i - d\bar{x}_i)}\right]$$

$$= r\left[\frac{\sum_{i=1}^{m} p_i^w(1 + t_i)dM_i - \sum_{i=1}^{m} p_i^w(1 - t_i^*)dX_i}{\sum_{i=1}^{m} p_i^w dM_i - \sum_{i=1}^{m} p_i^w dX_i}\right],$$

where M is imports for intermediate and final use, X is exports, and t_i^* is the tax rate on exports written with a negative sign, since it results in a domestic price that is lower than the border price. The shadow price of foreign exchange thus equals its market price times a weighted average of market distortions, where the weights are the changes in imports and exports induced by the changing availability of foreign currency. Multiplying both numerator and denominator by $r/\sum p_i^w M_i dr = r/\sum p_i^w X_i\, dr$, from the balance-of-payments constraint, Equation (A.1) can be expressed in elasticity form as

$$r^* = r\left[\frac{\sum_{i=1}^{m}(1 + t_i)u_i\eta_i - \sum_{i=1}^{m}(1 - t_i^*)v_i\epsilon_i}{\sum_{i=1}^{m} u_i\eta_i - \sum_{i=1}^{m} v_i\epsilon_i}\right], \tag{A-2}$$

where $u_i(v_i)$ is the share of the ith import (export) in total imports (exports) and $\eta_i(\epsilon_i)$ is the elasticity of imports (exports) of good i with respect to the exchange rate.

This is the Harbinger-Schydlowsky-Fontaine (HSF) shadow price discussed in Bacha and Taylor (1). It assumes that world prices are fixed so that a country cannot use tax policy to alter its terms of trade. To the extent that world prices are not fixed, the elasticities of demand for imports and supply of exports need to be replaced with the elasticities of demand for and supply of foreign exchange. If the country is able to influence the prices of one of its exports, for example, the elasticity of supply of exports ϵ_x would be replaced with the elasticity of supply of foreign exchange ϵ_f given by

$$\epsilon_f = \frac{\epsilon_x(\eta_x - 1)}{\epsilon_x + \eta_x}, \qquad (A\text{-}3)$$

where η_x is the price elasticity of foreign demand for the country's exports.

Citations

1 Edmar Bacha and Lance Taylor, "Foreign Exchange Shadow Prices: A Critical Review of Current Theories," *Quarterly Journal of Economics*, 85, No. 2 (May 1971).

2 Trent J. Bertrand, "On the Theoretical Foundations of Shadow Pricing in Distorted Economies." Working Paper in Economics No. 29, The Johns Hopkins University, Baltimore, May 1977.

3 Jagdish N. Bhagwati and T. N. Srinivasan, "Optimal Intervention to Achieve Non-Economic Objectives," *Review of Economic Studies*, 36, No. 1 (January 1969).

4 ———, "The Evaluation of Projects at World Prices Under Trade Distortions: QRs, Monopoly Power in Trade, and Nontraded Goods," M.I.T., Cambridge, June 1978, mimeo.

5 Jagdish N. Bhagwati and Henry Wan, Jr., "The 'Stationarity' of Shadow Prices of Factors in Project Evaluation, with and without Distortion," *American Economic Review*, 69, No. 3 (June 1979).

6 Anthony Bottomley, "Interest Rate Determination in Underdeveloped Rural Areas," *American Journal of Agricultural Economics*, 57, No. 2 (May 1975).

7 Robin Broadway, "Benefit-Cost Shadow Pricing in Open Economies: An Alternative Approach," *Journal of Political Economy*, 83, No. 2 (April 1975).

8 Michael Bruno, "Development Policy and Dynamic Comparative Advantage," in Raymond Vernon, ed., *The Technology Factor in International Trade*. Columbia University Press, New York, 1970.

9 Partha Dasgupta and Joseph E. Stiglitz, "Benefit Cost Analysis and Trade Policy," *Journal of Political Economy*, 82, No. 1 (January/February 1974).

10 Partha Dasgupta, Stephen A. Marglin, and A. K. Sen, *Guidelines for Project Evaluation*. United Nations, New York, 1972.

11 C. D. Datey, "The Financial Cost of Agricultural Credit: A Case Study of Indian Experience." Staff Working Paper No. 296, World Bank, Washington, D.C., October 1978.

12 Walter P. Falcon and Eric A. Monke, "The Political Economy of International Trade in Rice." Stanford/WARDA Study of the Political Economy of Rice in West Africa, Food Research Institute, Stanford University, Stanford, July 1979.

13 Ronald Findlay and Stanislaw Wellisz, "Project Evaluation, Shadow Prices, and Trade Policy," *Journal of Political Economy*, 84, No. 3 (June 1976).

14 Food and Agriculture Organization of the United Nations, Commodities and Trade Division, *Rice Trade Intelligence*, various issues.

15 Dean A. Linsenmeyer, "Economic Analysis of Alternative Strategies for the Development of Sierra Leone Marine Fisheries." Working Paper No. 18, African Rural Economy Program, Departments of Agricultural Economics, Njala University College and Michigan State University, Njala, Sierra Leone, and East Lansing, December 1976.

16 I. M. D. Little and J. A. Mirrlees, *Project Appraisal and Planning for Developing Countries*. Heineman, London, and Basic Books, New York, 1974.

17 Millard F. Long, "Interest Rates and the Structure of Agricultural Credit Markets," *Oxford Economic Papers*, 20, No. 2 (July 1968).

18 Scott R. Pearson, Narongchai Akrasanee, and Gerald C. Nelson, "Comparative Advantage in Rice Production: A Methodological Introduction," *Food Research Institute Studies*, 15, No. 2 (1976).

19 Patricia L. Rader, "The Effects of Ethnicity and Policy on Choice of Farming Techniques in Gagnoa, Ivory Coast." Stanford/WARDA Study of the Political Economy of Rice in West Africa, Food Research Institute, Stanford University, Stanford, July 1979.

20 M. FG. Scott, "How to Use and Estimate Shadow Exchange Rates," *Oxford Economic Papers*, 26, No. 2 (July 1974).

21 Lyn Squire and Herman G. van der Tak, *Economic Analysis of Projects*. The Johns Hopkins University Press, Baltimore, 1975.

22 T. N. Srinivasan and Jagdish N. Bhagwati, "Shadow Prices for Project Selection in the Presence of Distortions: Effective Rates of Protection and Domestic Resource Costs," *Journal of Political Economy*, 86, No. 1 (February 1978).

23 U.S. Agency for International Development, *Spring Review of Small Farmer Credit*. Washington, D.C., June 1973.

24 Peter G. Warr, "Shadow Pricing with Policy Constraints," *The Economic Record*, 53, No. 142 (June 1977).